Biomes and Ecosystems

Biomes and Ecosystems

EDITOR
Robert Warren Howarth
Cornell University

ASSOCIATE EDITOR
Jacqueline E. Mohan
University of Georgia, Athens

Volume 3
Articles

Hauraki Plains Wetlands – Pyramid Lake

SALEM PRESS
A Division of EBSCO Publishing
Ipswich, Massachusetts

GREY HOUSE PUBLISHING

Biomes and Ecosystems, 2013, published by Grey House Publishing, Inc., Amenia, NY, under exclusive license from EBSCO Publishing, Inc.

The paper used in these volumes conforms to the American National Standard for Permanence of Paper for Printed Library Materials, X39.48-1992 (R1997).

LIBRARY OF CONGRESS CATALOGING-IN-PUBLICATION DATA

Biomes and ecosystems / Robert Warren Howarth, general editor ; Jacqueline E. Mohan, associate editor.
   volumes cm.
 Includes bibliographical references and index.
 ISBN 978-1-4298-3813-9 (set) -- ISBN 978-1-4298-3814-6 (volume 1) -- ISBN 978-1-4298-3815-3 (volume 2) --
ISBN 978-1-4298-3816-0 (volume 3) -- ISBN 978-1-4298-3817-7 (volume 4) 1. Biotic communities. 2. Ecology.
3. Ecosystem health. I. Howarth, Robert Warren. II. Mohan, Jacqueline Eugenia
 QH541.15.B56B64 2013
 577.8'2--dc23

                                    2013002800

ebook ISBN: 978-1-4298-3818-4

First Printing

PRINTED IN THE UNITED STATES OF AMERICA

Produced by Golson Media

Contents

Volume 2

Part 2: Articles

A

B

Hauraki Plains Wetlands

Category: Inland Aquatic Biomes.
Geographic Location: New Zealand.
Summary: The Hauraki Plains wetlands have been significantly reduced through transformation to agriculture.

Wetlands were once widespread features of the Hauraki Plains landscape on New Zealand's North Island. Less than 20 percent of the original freshwater wetlands remain today because of intensive drainage for agriculture and development over the past century. National efforts are aimed at restoring and protecting wetlands in the Hauraki Plains, and two locations here are recognized internationally for their global conservation value.

River movements and volcanic activity over thousands of years created the current Hauraki Plains landscape. About 20,000 years ago, volcanic material washed into the Waikato River and caused the floodplain to build up; the river to meander; and ash, pumice, and other sediments to be deposited across the plains. After the Waikato River shifted toward the west coast, the Waihou (formerly Thames) and Piako rivers were formed, flowing north to the sea at the Firth of Thames,

a large bay. Bounded by these rivers, the Hauraki Plains filled in with estuarine deposits, river mud, and peat. Key features of the wetland area today are the Kopuatai peat dome and the tidal flats at the Firth of Thames.

At close to 24,711 acres (10,000 hectares), the Kopuatai peat dome is New Zealand's largest unaltered raised bog, as well as the largest freshwater wetland on the North Island. There are two main wetland types included in the site: peatland and mineralized wetland. The peatland is acidic, low in nutrients, and dominated by rainfall. By contrast, the mineralized areas surrounding the bog are nutrient-rich and sometimes flooded by the Piako and Waitoa rivers. The physical features of the peat dome and mineralized swamps play an important role in flood control and protection; they provide storage for floodwater from the river catchments. Managed by the Department of Conservation, Kopuatai is also a lifeline for many endangered and protected plant, bird, and fish species.

Biota

Important plant species found in the Hauraki Plains wetlands include the kahikatea tree (*Dacrycarpus dacrydiodes*). These endemic (found nowhere else on Earth) conifers can grow to a

637

height of 180 feet (55 meters); they were used for building boats, thanks to the long, straight boards produced from the trunks. Other notable plants in the Hauraki Plains wetlands include the manuka or manuka myrtle (*Leptospermum scoparium*), a shrub-like tree that often returns first in cleared areas. Flax and bulrush are also common plants found in the wetlands.

The wetlands are a popular feeding and nesting site for approximately 54 bird species. Ducks and white-faced herons are two characteristic wetlands birds found here. Other birds include the beautifully plumed pukeko bird or purple swamphen (*Porphyrio porphyrio*). Easily recognized by its indigo plumage and red beak and legs, the pukeko is noted for walking or running, rather than flying, from danger. Various eels are also plentiful in the wetlands area.

The shallow tidal flats of the Firth of Thames cover approximately 21,004 acres (8,500 hectares) and lie within the Crown-administered Coastal Marine Area. This zone includes shallow estuarine water and mudflats, shell banks, grass flats, mangrove forest, salt marsh, and limited freshwater swamp margins. One of New Zealand's three most

A pukeko bird, or purple swamphen (Porphyrio porphyrio), in New Zealand in 2011. Pukekos are highly visible with their bright blue chests and red beaks. (Flickr/Sid Mosdell)

important coastal stretches for shorebirds, the area is used by more than 25,000 waterfowl and wader species, including many migratory birds. Aside from offering food and habitat for waterfowl, the Firth of Thames provides an important fishery of local significance.

Threats and Protection

The biggest threats facing the Haruaki Plains wetlands are drainage for agriculture and housing, logging and its associated erosion, pollution from farm runoff, grazing, burning, and decreasing water levels due to the accumulation of sediment. Dairy farming is the main local industry, using two-thirds of the total area of the plains, although cattle and sheep farming also contribute to the regional economy, along with tourism.

Displacement of native plants by invasive species like willow, gorse, and blackberry; and the spread of exotic fish such as mosquito fish and koi carp can further disrupt the ecology of waterways and make wetland restoration more difficult. Potential climate change threats for this area may come from rising sea levels, which can change the nature of the wetlands, producing salt incursion and coastal area erosion. Extreme weather events produced by climactic changes may also result in flooding beyond capacity of the local hydrological system to contain, again harming habitat niches.

In 1976, New Zealand became a party to the Convention on Wetlands of International Importance, the Ramsar Convention. This intergovernmental treaty provides the framework for national action and international cooperation for the conservation and wise use of wetlands. Two of the six Ramsar sites in New Zealand—the Kopuatai peat dome and the Firth of Thames—are in the Hauraki Plains region and were listed in 1990 for their extraordinary ecological significance.

The Hauraki Gulf Marine Park Act of 2000 further protects the waters of the gulf and the Firth of Thames; it established the Hauraki Gulf Forum to manage use, development, and conservation within the park. Soon after, the Waipa Peat Lakes and Wetlands Accord was signed in 2002. This ratified a commitment by multiple government agencies and conservation organizations to work together to promote the sustainable use and conservation of lake and wetland resources in the Hauraki Plains.

Additional strategies to address wetland conservation in the public and private sector come from the National Wetland Trust, established to increase the appreciation of wetlands and their values within the country; the Environmental Initiatives Fund of the Waikato Regional Council, which provides funding for restoration projects; and open-space covenants among landowners to protect their wetlands in perpetuity. Through efforts that recognize wetlands as a vital cultural and ecological resource, the Hauraki Plains remain a stronghold for wetlands in New Zealand.

NICOLE MENARD

Further Reading

Cromarty, Pam and Derek A. Scott. *A Directory of Wetlands in New Zealand.* Wellington: New Zealand Department of Conservation, 1995.

New Zealand Government. "Welcome to Hauraki District." 2012. http://www.hauraki-dc.govt.nz/Overview/welcome_to_hauraki.htm.

Peters, Monica and Beverley Clarkson, eds. *Wetland Restoration: A Handbook for New Zealand Freshwater Systems.* Canterbury, New Zealand: Manaaki Whenua Press, 2010.

Hawaiian Intertidal Zones

Category: Marine and Oceanic Biomes.
Geographic Location: Pacific Ocean.
Summary: Hawaii's multivarious intertidal zones are strongly influenced by their geographic position and local topography.

Located in the center of the Pacific Ocean in the extreme northern corner of the Indo-Pacific Triangle and influenced by currents generally unfavorable for larval settlement, the Hawaiian Intertidal Zones' species richness is low compared with that of other Pacific Islands. Steep coastlines and a small tidal range of about 3 feet (1 meter) discourage the development of intertidal zonation and, thus, habitat and species diversity. Nevertheless, Hawaii's intertidal regions support a unique complex of native and endemic (found nowhere else on Earth) species.

Life in Hawaii's intertidal is harsh. Organisms are exposed to intense sunlight and high air temperatures. Waters are often nutrient-poor, and waves can be punishing. Many organisms are small and camouflaged, hiding under rocks or crevices during daytime low tides. Only urchins and limpets can withstand the wave-beaten areas.

In more protected areas, intertidal zonation can occur. The lowest zones support algae; mussel beds and then barnacles occur higher up. Limpets and nerite snails scavenge in and above the barnacle band. Above this band, the splash zone and littoral fringe support still other, littorine snail species and an isopod crustacean, as well as maritime vegetation well adapted to saline soils and sea spray. Grapsid crabs scuttle and range throughout.

Intertidal Habitats and Biota

While intertidal zonation is generally reduced compared to the Pacific region as a whole, Hawaii's varied topography creates some variety in intertidal habitats. Calcareous (carbonate-based) and basalt shorelines feature conspicuously-zoned horizontal benches that can extend up to 98 feet (30 meters) seaward. Calcareous shorelines also contain pitted limestone pools that are inhabited by a small littorine snail and a blenny fish, as well as thick turfs of algae that support a complex food web of suspension-feeding, herbivorous, and carnivorous mollusks. On basalt benches, alternating growths of algae create colorful mosaics, while out in the surf zone, algae- and coral-encrusted basalt boulders host a variety of wave-tolerant worms, mollusks, echinoderms, and fishes.

Tide pools occur on these calcareous and basalt shorelines. More exposed pools have a sandy bottom bound with blue-green algae; they typically contain two or three species of small mollusks, grapsid crabs, and two fishes: a blenny and a goby. Seaward, pools become progressively more densely turfed with algae, hosting a variety of worms, mollusks, crustaceans, and echinoderms. Especially common are the spaghetti worm and the bandana shrimp. Seaward pools also serve as incubators for juvenile fishes.

Sandy shorelines provide a different range of habitats, with beach slope and sand-grain size largely determining community composition. The upper beach, including the vegetation line, provides habitat for amphipods, isopods, and ghost-crab males, while female ghost crabs and males of a different species occur lower down. The lower beach, continually awash, supports scavenging mole crabs and four species of gastropods that prey on *spionid polychaetes*—a family of small, thin, filter-feeding worms that build tubes of sand and other particles.

Estuarine ecosystems occur where rivers and streams meet the sea. In Hawaii, these brackish ecosystems support many endemic species, especially fishes and mollusks, and are the primary habitat of the edible Samoan crab. They are also important nursery areas for several species of in-shore marine fishes. Native oysters occur in estuaries that receive significant freshwater infusions. Many Hawaiian estuaries, unfortunately, are threatened by invasive exotic plants and animals.

Mangrove habitats are important and imperiled worldwide, but in Hawaii mangroves are introduced, lacking natural predators to control their growth. Thus, they often invade coastal regions, out-competing native plants, trapping sediment, and destroying habitat for native plants and animals. Mangrove-eradication efforts are under way statewide.

Though reefs are generally subtidal, outer edges and reef flats may be exposed at low spring tides, creating intertidal conditions. Hawaii's reefs are composed primarily of crustose coralline algae, tolerant of the relatively lower temperatures found at these latitudes. Outer reef edges are densely fringed with frondose algae, while reef flats are patchy and diverse, containing calcareous algae and cnidarian corals, stands of frondose algae, rubble, and sand patches. Sand patches host burrowing mollusks and spionid polychaetes. Other types of mollusks, echinoderms, and fishes are common among the rubble and living coral, and a large fan worm occurs in pockets and crevices.

Environmental Threats

Threats to Hawaii's intertidal include pollution, invasive species, and climate change. Changes in sea levels brought about by climate change may threaten the delicate ecosystems of the intertidal. All of Hawaii's species, including its intertidal organisms, are especially sensitive to introductions of nonnative species, because they have evolved in relative isolation from these invaders. Yet compared with coral reefs and with intertidal regions elsewhere in the world, Hawaii's intertidal zones have not been well studied. Some are included in marine protected areas, but further protection and management are needed.

MELANIE L. TRUAN

Anchialine Pools

Anchialine, or near-the-sea, pools are unique brackish-water ecosystems that lie above the marine zone but still experience tidal flux through subsurface connections to the ocean. Generally isolated, they feature many rare and endangered species. An entire species complex of tiny endemic shrimp inhabits different pools, traveling to and fro via underground cracks and fissures in the lava substrate. Anchialine shrimp have few natural enemies, and many are threatened by introduced predators. Anchialine pools themselves are threatened by development, wave and tidal action, and other dangers including tsunamis, lava flows, invasive vegetation, senescence, and global warming.

Further Reading

Hawaiian Government. "Wai'anae Ecological Characterization—Intertidal Communities." 2012. http://hawaii.gov/dbedt/czm/initiative/wec/html/sea/marine/inter.htm.

Kay, E. A. "Marine Ecosystems in the Hawaiian Islands." In *A Natural History of the Hawaiian Islands, Selected Readings II*, E. Alison Kay, ed. Honolulu: University of Hawaii Press, 1994.

Hawaiian Tropical Moist Forests

Category: Forest Biomes.
Geographic Location: Pacific Ocean.
Summary: Hawaiian tropical moist forests are known for their volcanic soils, biodiversity, rich vegetation, and large numbers of endemic species.

Hawaii's tropical moist forests are found within of the United States state of Hawaii in the North Pacific Ocean, or Oceania zone, covering close to 3,000 square miles (7,770 square kilometers) of the Hawaiian archipelago. They are within the ecoregion type of tropical and subtropical moist broadleaf forests. The relatively fertile volcanic soil and abundant rainfall together support biodiverse tropical moist forests featuring large numbers of endemic species (those found nowhere else on the planet).

Hawaii contains a variety of forest types at all elevations, with forest cover representing approximately 1.7 million acres (687,966 hectares), or 41 percent of the state's land. Tropical moist forests are located on the windward or northeastern sides of the islands, where abundant rainfall results as the trade winds cross the mountains. Trade winds also serve as a moderating influence on the region's warm temperatures. Although these forests have escaped the more extensive deforestation found at lower elevations within the Hawaiian Islands, conservationists still consider this forest ecosystem to be endangered.

Hawaii is comprised of 137 islands of volcanic origin. The main Hawaiian Islands include Kauai, Oahu, Molokai, Lanai, Maui, and Hawaii. The state contains numerous climate zones; it is one of the very few U.S. states with a climate suitable for supporting rainforests. The climate zones here support numerous habitats that are renowned for their biodiversity. Hawaii tropical moist forests are among the three major habitat areas comprising the islands, as designated by the World Wildlife Fund, alongside Hawaii tropical dry forests and Hawaii tropical grasslands, savannas, and shrublands. Most of the tropical moist forests are located in the upper elevations within the mountains. Tropical moist forests are further divided into five ecological components: Lowland Wet, Lowland Mesic, Montane Wet, Montane Mesic, and Wet Cliffs.

Upland tropical moist forests are vital components of the state's watersheds, helping provide freshwater to the islands by absorbing moisture and feeding into underground and groundwater sources. The forests also help keep the water clean and are an important defense against soil erosion. Containment of development to lower levels and overall watershed protection has resulted in greater maintenance of biodiversity and native species at higher elevations. Protected areas within the ecosystem include the Alakai Swamp on the island of Kauai and the 'Ola'a Rainforest within Hawaii Volcanoes National Park and Akaka Falls State Park, both on the main island of Hawaii. The ecosystem also houses what is generally regarded as the wettest place on Earth: Mount Waialeale on the island of Kauai.

Biodiversity

Hawaii's isolated existence within the vast Pacific Ocean throughout many centuries has resulted in the development of diverse tropical habitats with a variety of endemic flora and fauna. While exotic, or invasive, species have come to dominate the lower elevations, higher-elevation tropical moist forests still feature more than 90 percent unique endemic species. There is also biodiversity among the individual Hawaiian islands, with each island housing unique species. Common trees include o'hia (*Metrosideros polymorpha*) and koa (*Acacia koa*). Plants include tropical fruits and vegetables, shrubs, fungi, mosses, vines, orchids, and ferns. Trees and vines dominate the forest canopy, while the forest floor consists of a spongy layer of decaying ferns, mosses, and other organic matter. The tropical moist forests also contain bogs featuring sedges, grasses, ferns, shrubs, and small trees.

Animals include a variety of snail, invertebrate, bird, and insect species, including more than one-third of the world's known species of fruit flies. Known species unique to the islands include Hawaiian land snails, Hawaiian lobelias, Eupithecia moths, and Hawaiian honeycreepers, the latter of which are among its unique native bird species. The ecosystem is also home to one of Hawaii's native mammals: the Hawaiian hoary bat or 'ope'ape'a (*Lasiurus cinereus semotus*). Exotic animal species present in the forests include the destructive feral pig. Many of Hawaii's endangered native species of flora and fauna live within tropical moist forest areas such as the Alakai Swamp.

Human Settlement

Polynesians were among the first humans to settle Hawaii, arriving on the islands more than 1,500 years ago. The indigenous Hawaiian culture that developed placed strong emphasis on living in harmony with the natural world. Captain James Cook was among the first Europeans to arrive when he landed in 1778, beginning a period of heavy immigration to the islands and the concomitant introduction of non-native species.

Exotic species introduced into the islands have included pigs, dogs, cattle, sheep, goats, cats, rats, mongoose, and a variety of birds and plants. In the 20th and 21st centuries, population growth and widespread development and military training activities in the lowlands have resulted in deforestation and accelerated the loss of native habitats and species. Other modern concerns include the state's reliance on the tourism industry and its environmental effects.

Environmental Threats

Significant threats to Hawaii's forests have included widespread deforestation at lower elevations; land development for residential, agricultural, and military uses; habitat loss; disease; the introduction of invasive nonnative species such

as feral pigs; and tourism and recreational activities. Approximately two-thirds of the original forest cover has been lost. Much of the highland tropical moist forests remain, while many of the lowland and foothill moist forests have been lost to deforestation.

Some of the lowland forests regenerated, but these new forests are largely populated with nonnative species. Hawaii is home to the highest number of endangered species of any U.S. state. Extinct or endangered species include land snails, native forest birds, and the hoary bat, as well as more than 220 types of plants and approximately 250 types of insects.

The potential impacts of climate change are still being studied. Carbon storage in the soil and vegetation remains in delicate balance, and it is likely that rising carbon dioxide levels in the atmosphere will have some type of impact, but effects on the Hawaiian Tropical Moist Forests biome are not well understood.

Conservation Efforts

Protection and restoration efforts within Hawaii's tropical moist forests have included the removal of exotic species such as feral pigs, and land-use zoning. State efforts include the creation of watershed partnerships, an agreement between state, federal, and private landowners to manage forest resources in a sustainable manner designed to protect the vital forest watershed. Watershed protection has been a key avenue of forest preservation in the highland tropical moist forests. The state Forest Stewardship Program is among several measures designed to encourage responsible forest management. Organizations and agencies involved in conservation and restoration efforts include the World Wildlife Fund, various U.S. federal government agencies, and the Hawaii Department of Land and Natural Resources.

Many of Hawaii's highlands are zoned for conservation; some feature national parks and other protected wilderness areas. Protected areas within Hawaii's tropical moist forests include Hawaii Volcanoes National Park, the Hawaiian Islands Biosphere Reserve, the Kalaupapa National Historic Site, Kaloko–Honokohau National Historical Park, and Pu'uohonua O Honaunau National Historical Park. The state's Natural Area Reserves System includes tropical moist forest areas. A conservation district encompasses many of the remaining natural areas dominated by native species, but others are left unprotected.

MARCELLA BUSH TREVINO

Further Reading

Culliney, John L. *Islands in a Far Sea: The Fate of Nature in Hawaii.* Honolulu: University of Hawaii Press, 2006.

Department of Natural Resources and Environmental Management. "Impact of Rising Temperature on Terrestrial Carbon Flux, Partitioning, and Storage in Tropical Wet Forests." College of Tropical Agriculture and Human Resources, University of Hawaii at Manoa. 2012. http://www.ctahr.hawaii.edu/littonc/research.html.

Ricketts, T. H., E. Dinerstein, D. M. Olson, C. J. Loucks, et al. *Terrestrial Ecoregions of North America: A Conservation Assessment.* Washington, DC: Island Press, 1999.

Ziegler, Alan C. *Hawaiian Natural History, Ecology, and Evolution.* Honolulu: University of Hawaii Press, 2002.

Hawaiian Tropical Shrublands

Category: Grassland, Tundra, and Human Biomes.
Geographic Location: Pacific Ocean.
Summary: This ecoregion hosts a series of habitat niches defined by elevation and moisture regimes, resulting in unique assemblages of flora and fauna.

A shrubland is a plant community composed of grasses, non-woody vegetation, shrubs, and small trees, formed as a result of either natural processes or habitat alteration by humans. Shrublands can exist as a stable community type or as a step along a successional trajectory following disturbance. Because the Hawaiian Islands are the most isolated island archipelago on Earth, with the nearest continental land mass approximately 2,360 miles (3,800 kilometers) away, many of the island's plant species evolved after long-distance dispersal events and thus are endemic, or found nowhere else on Earth.

In Hawaii, shrublands are considered to be an important habitat type because they were historically widespread. Today, these shrublands exist as remnant fragments that are home to many rare species as a result of human impact on the landscape. For this reason, many shrubland communities are critically threatened.

The volcanic origin of the Hawaiian Islands has produced a mountainous environment with orographic weather patterns. As a result, elevation, humidity, precipitation, and temperature differ dramatically within and across the islands. This variation has allowed for more native terrestrial ecosystem types than generally would be found in other geographic areas of similar size. In addition, the rain shadow effect, a consequence of trade winds from the northeast, contributes to the range of habitat type. This causes drier conditions on the leeward sides of islands and wetter conditions on windward sides.

Shrubland Ecosystems

Shrubland ecosystem types are found on all eight main islands—Hawaii, Kauai, Kahoolawe, Lanai, Maui, Molokai, Niihau, and Oahu—as well as on several of the smaller northwestern Hawaiian Islands, and in greatest abundance on leeward slopes. Shrublands in Hawaii are often categorized by location or elevation (coastal, low, or high) and precipitation (dry, mesic, or wet). In the coastal zone, the two types of shrubland are dry and wet.

The Coastal Dry Shrubland communities occur on the northwestern Hawaiian islands of Lisianski, Laysan, Necker, and Nihoa, as well as all the main islands below an elevation of 984 feet (300 meters), where annual rainfall is less than 20 inches (500 millimeters). Some noteworthy genera found in these shrublands include *Chenopodium*,

Euphorbia, Eragrostis, Gossypium, Heliotropium, Scaevola, Schiedea, Sida, and the invasive koa haole (*Leucaena*). The two types of Coastal Wet Shrubland habitats occur near freshwater sources on most of the main islands, as well as on Laysan; they are dominated by native *Hibiscus* or non-native *Pluchea.*

Lowland shrublands occur on most of the main islands, and are categorized as dry, mesic, or wet. Lowland Dry Shrublands occupy exposed sites from 328 to 1,970 feet (100 to 600 meters) elevation, receiving 20 to 67 inches (500 to 1,700 millimeters) of rain annually. Species composition is variable, but tends to be dominated by *Bidens, Dodonaea, Sesbania,* or *Wikstroemia.* Lowland Mesic Shrublands are found at 98 to 2,790 feet (30 to 850 meters) elevation, enduring dry summers and winter rainfall of 39 to 79 inches (1,000 to 2,000 millimeters). Dominant genera here include *Dodonaea, Lipochaeta, Metrosideros, Osteomeles, Styphelia,* and *Wilkesia.*

The Lowland Wet Shrublands are present on some exposed ridge sites from 656 to 2,950 feet (200 to 900 meters), where annual rainfall ranges from 150 to 236 inches (3,800 to 6,000 millimeters), with *Metrosideros* and *Pipturus* vegetation dominating. One significant threat to lowland shrublands is the invasive strawberry guava (*Psidium cattleianum*), which forms monoculture stands, preventing native plants from regenerating.

The high shrubland types include montane, subalpine, and alpine. Montane Dry Shrubland ecosystems are dominated by *Dodonaea* or *Metrosideros,* and are restricted to Maui and Hawaii from 2,953 to 8,858 feet (900 to 2,700 meters), where annual rainfall ranges from 16 to 39 inches (400 to 1,000 millimeters). Montane Wet Shrublands are often dominated by a mix of ferns and shrubs; they occur on Hawaii, Kauai, Molokai, Maui, and Oahu around 3,937 feet (1,200 meters) elevation in areas receiving annual rainfall of 98 to 197 inches (2,500 to 5,000 millimeters). Some of the important genera occupying these shrublands include *Dicranopteris, Metrosideros, Rubus, Sadleria,* and sometimes *Gunnera* and *Pritchardia.*

The two subalpine shrubland types are dry and mesic, both of which are restricted to Maui and

Hawaii, within the range of 5,906 to 9,843 feet (1,800 to 3,000 meters). Subalpine Dry Shrublands receive 20 to 59 inches (500 to 1,500 millimeters) of rainfall annually and contain *Chenopodium, Dodonaea, Styphelia,* and *Vaccinium* species. The Subalpine Mesic Shrubland community type is known from only one location on East Maui, where annual rainfall ranges from 51 to 75 inches (1,300 to 1,900 millimeters); it is vegetated predominantly by *Sadleria* and *Vaccinium.*

Finally, Alpine Dry Shrublands receive 30 to 49 inches (750 to 1,250 millimeters) of rainfall annually, and are restricted to elevations of 9,843 to 11,155 feet (3,000 to 3,400 meters) on East Maui and Hawaii. These extremely exposed habitats are home to species of silverswords (*Argyroxiphium*) and *Dubautia.* Plant species growing in these high-elevation sites have adapted to survive harsh and variable conditions. The silverswords exhibit a classic plant adaptation to such environments, having densely pubescent leaves that give the plant a silvery appearance and reflective property, protecting it from intense irradiance. Notable fauna include birds such as the Nene goose (*Branta sandvicensis*) and the Hawaiian dark-rumped petrel, an endangered bird that nests in these high elevations.

Fine, short hairs are visible on the leaves of this *Argyroxiphium* and *Dubautia* hybrid, an adaptation to the environment that helps protect the plant from intense sunlight. (Wikimedia/Forest & Kim Starr)

Biodiversity and Threats

Hawaii is famous for encompassing a range of biotic communities, each with unique characteristics that have allowed for spectacular biotic diversification over time. Hawaii has some of the highest rates of endemism in the world, and also the greatest number of threatened and endangered species in the United States.

The increasing prevalence of invasive species in these shrubland communities, however, is shifting ecological processes and threatening native species. Many native shrubland habitats have already lost a large number of indigenous and endemic species due to human activity. Some of the most critical threats to these habitats are introduced mammals, invertebrates, plants (especially several fire- and drought-adapted species), and human land use.

Additional threats to the Hawaii Tropical Shrublands biome include overgrazing by livestock, and even seemingly benign human activity such as hiking, which can damage delicate plants and lead to erosion. Climate change impacts are still being studied. Rising sea levels are already threatening the long-term existence of some of the outlying, smaller islands here, which could spell trouble for several of the bird species that circulate among the shrublands habitats across the archipelago.

Studies have shown decreasing rainfall and increased average air temperature. However, projections of these trends into the future is an unknown. As a rule, endemic species in such variegated habitat niches are more vulnerable to climate pattern alteration than species that thrive broadly across less unique habitats. Therefore, the species adapted to Hawaii Tropical Shrublands biome niches will tend to be more at risk as the climate continues its warmer, drier trends.

Margaret J. Sporck
Christopher A. Lepczyk

Further Reading

The Nature Conservancy. "Hawaiian High Islands Ecosystem." 2007. http://www.hawaiiecoregionplan.info/ecoregion.html.

Wagner, Warren L., et al. *Manual of the Flowering Plants of Hawaii, Revised Edition*. Honolulu: University of Hawaii Press, 1999.

Ziegler, Alan C. *Hawaiian Natural History, Ecology, and Evolution*. Honolulu: University of Hawaii Press, 2002.

Highveld Grasslands

Category: Grassland, Tundra, and Human Biomes.
Geographic Location: Africa.
Summary: This diverse, high-altitude grassland is becoming increasingly degraded by urbanization, agriculture, and mining activities.

Grasslands once covered the majority of the modern-day African continent, but today a comparatively small remnant of this formerly expansive biome remains. As the African continent shifted over geologic time, grasslands were gradually converted to modern savannas that feature trees and a more tropical climate. In southern Africa, however, a small portion of the grassland biome remains. The Highveld (from Afrikaans, meaning *high field*) grassland is a remarkable ecosystem within the grassland biome. Highveld grassland is defined by high elevation; dry, cold winters; occasional snow; and the summer rainy season, when thunderstorms bring almost daily showers to the area. These features create a unique habitat for many rare species.

The Highveld grassland performs vital ecosystem functions for the human population, including storing and purifying water, a scarce resource. Despite these benefits, the Highveld Grasslands biome is being fragmented and degraded by urban expansion, agriculture, industry, and fuel demands.

The Highveld grassland is set on a landlocked plateau in the heart of South Africa, about 3,940–5,910 feet (1,200–1,800 meters) in elevation. It is bordered by the arid Kalahari and Great Karoo Deserts to the west and southwest; the woody, low-altitude bushveld to the north; and the Drakensberg Mountains to the east and southeast. Compared with the surrounding ecosystems, the Highveld grasslands receive greater annual rainfall, averaging 20–35 inches (500–900 millimeters), but occasionally exceeding 79 inches (2,000 millimeters) in the eastern Highveld. Some of this rain seeps into the ground and is gradually released during the dry winter.

Fire plays a natural role in maintaining the Highveld grasslands, as species that are not fire-adapted, such as many trees and shrubs, have difficulty establishing here. Frost similarly restricts the distribution of many trees and shrubs.

Biota

Grasses dominate this ecosystem, although other small herbaceous plants are common, and shrubs may be found in sheltered areas. The predominant grass species is redgrass (*Themeda triandra*). Other grasses include broom needlegrass (*Triraphis andropogonoides*), sawtooth lovegrass (*Eragrostis superba*), and many others. Hallmark shrubs include bitterkaroo (*Pentzia globosa*), as well as various woody shrubs and dwarf shrubs such as dwarf buffalothorn (*Ziziphus zeyheriana*).

Predictable patchiness of rainfall, fire regime, and frost patterns overlay a variety of soil types on the Highveld grassland. Dividing the resulting mosaic of sub-habitats into two broad categories—sweet and sour grasslands—is helpful when considering their functions and human-use histories. Due to lower rainfall and fewer frost events, sweet grasslands are palatable to grazing animals throughout the year, as they are nutritious and low in fiber.

Sour grasslands, on the other hand, are adapted to environmental conditions that favor underground nutrient storage, making the leaves of these grasses less nutritious outside the growing season. Highveld grasslands characterized by sweet grasses are threatened by overgrazing and trampling by livestock, whereas the sour grasslands tend often to be burned to encourage younger, more nutritious plant growth.

The Highveld is home to a diverse fauna community, many of which are endangered. These include straw-colored fruit bats, the African rock python (*Python sebae*), and the blue crane (*Anthropoides paradiseus*), South Africa's national bird. Reptiles such as pythons, monitors, and Nile crocodiles are common within the biome. An endemic (exists nowhere else on Earth) reptile found in the Highveld is the giant girdled lizard (*Cordylus giganteus*). Major mammals found in the area include the brown hyena, leopard, African civet, mountain zebra, and honey badger.

Effects of Human Activity

Agriculture on the Highveld grasslands began in the 13th century, and remains a dominant activity. Maize is the most important commercial crop, but wheat, sorghum, and sunflowers are also cultivated to a lesser extent. As a result of its intensive use, the Highveld Grasslands biome suffers broadly from erosion and soil degradation.

To salvage exhausted soil, farmed land is often permanently converted to pasture. However, this land-use strategy is less effective in these grasslands than in some other parts of the world, as the native soil here is rarely resilient enough for this transition. After intensive agriculture, the Highveld soil structure is physically altered; microbial communities are irreversibly changed, which make restoration efforts in this ecosystem more difficult.

Urbanization and an increasing need for energy are threatening the remaining Highveld grasslands. The Gauteng province (a part of the former Transvaal), the most densely populated region of South Africa, is situated within the Highveld grasslands. Almost three-quarters of the South African population receives electricity from the Highveld's many power-generation plants, and demand continues to increase. The Highveld plateau possesses gold and shallow coal deposits, both of which are extensively mined. The coal-mining industry practices ecosystem rehabilitation, although success is limited because revegetation is especially slow as organic matter is easily depleted here.

Human activity has also reduced the effectiveness of the Highveld grasslands as a source of water and natural water purification. Though the Highveld receives and stores a significant amount of rain, plantations of exotic trees draw large amounts of water from the ground. Because these trees are ultimately harvested and exported, valuable water is being permanently removed from the region. Peat harvesting also negatively affects ecosystem function by impairing the biome's capacity to purify water. Peat forms slowly over time and acts as a sponge to remove pollutants from runoff—but it is rapidly being removed from this ecosystem.

Ongoing climate change has the potential to remove even more water from the Highveld grasslands, as average annual rainfall will decrease, according to some predictive models. Grasslands would retreat across some parts of the plateau in such a scenario, as arid conditions would overtake groundwater recharge rates. Such a development would combine with the human impacts of continued mining and more intensive agricultural activity to raise the stress level of habitat and species here. The increasing human population is ultimately a vector that could lead to widespread negative tipping points when combined with the effects of global warming here.

Conservation Efforts

Currently, less than 2 percent of all South Africa's grasslands are officially conserved. The Highveld grasslands are home to several threatened endemic species. The sungazer lizard (*Cordylus giganteus*) and the robust golden mole (*Amblysomus robustus*), for example, are both jeopardized by habitat degradation. The Highveld Grasslands biome has also been identified as an ecosystem at high risk of biological invasion.

Plateau-wide conservation organizations like the South African Grasslands Programme, and community-driven initiatives like the recently proposed Highveld National Park, are steps toward conserving this unique ecosystem and its long-term benefits to human residents. New models put forth by such groups identify several key areas within the Highveld grasslands that, if protected, would be highly effective in preserving the ecosystem's functions and ecological diversity.

LINDSEY NOELE SWIERK

Further Reading

Dovers, Stephen, et al., eds. *South Africa's Environmental History: Cases and Comparisons.* Athens: Ohio University Press, 2003.

Egoh, Benis N., et al. "Identifying Priority Areas for Ecosystem Service Management in South African Grasslands." *Journal of Environmental Management* 92, no. 6 (2011).

Low, Barrie and Tony Rebelo, eds. *Vegetation of South Africa, Lesotho and Swaziland.* Pretoria, South Africa: Department of Environmental Affairs and Tourism, 1996.

Himalayan Alpine Shrub and Meadows, Eastern

Category: Grassland, Tundra, and Human Biomes.
Geographic Location: Asia.
Summary: The Eastern Himalayan Alpine Shrub and Meadows biome features spectacular flora that developed between the snowfields and tree lines; it is threatened by changes in land use and climate.

The Eastern Himalayan Alpine Shrub and Meadows biome lies between the tree line and snowline, spreading across parts of India, Nepal, Bhutan, and Myanmar. As one proceeds to higher elevations, tree life is abruptly replaced by a magnificent combination of shrubs and meadows, before reaching the permanent snow-covered areas.

Separated from the western part of the Himalaya Mountains range by the Kali Gandaki River, the eastern altitudinal belt holds an area of outstanding biodiversity and high endemism (home to species found nowhere else on Earth). In fact, the Eastern Himalayas Alpine Shrub and Meadows are home to more than 7,000 species of plants, three times more than the rest of the Himalayas, making it perhaps the most important alpine area in the world, and an obvious conservation priority.

Rhododendrons in the Annapura Conservation Area, with a view of Annapura South in the background. The conservation area was created in order to preserve the biodiversity of the area's rhododendrons. (Thinkstock)

Biota

Since the mid-1800s, expeditions have been made to this area and other parts of the Himalayas in search of new species and medicinal plants. As a result of such voyages, many plant species were introduced to the gardens of the Western world. It is estimated that approximately one-quarter of the popular garden plants found throughout the world originated in this region.

During spring and summer, this habitat explodes with purple, blue, and pink flowers, as several types of rhododendrons and alpine herbs bloom on the steep hills of these massive mountain ranges. Representative plants include *Alchemilla, Anenome, Gentiana, Impatiens, Viola,* and many other herbaceous plants, as well as a range of full-size and dwarf rhododendrons.

Altitude, topography, and aspect encourage a high rate of endemism, and create a broad range of unique local environmental conditions to which different plant forms have evolved and adapted. Among the most significant adaptations is the so-called vegetation dwarfing, in which as altitude increases, shrub species of rhododendron reduce their height and radically change their appearance.

Vegetation dwarfing is associated with maintaining environmental conditions on a range that allows metabolic efficiency at lower temperature. Even though this adaptation allows many species to be distributed on a wider range along the altitudinal gradient, most of the Himalayan alpine shrubs and herbs are restricted to a narrow altitudinal belt of about 656 feet (200 meters) in total.

This restriction hinders their ability to adapt to long-term environmental changes, making the Eastern Himalayan Shrubs and Meadows biome particularly vulnerable to climate change. Small shifts in temperature and precipitation patterns could have dramatic effects on plants and animals adapted to live at a certain altitude. Ongoing changes in climatic conditions may result in massive species loss in mountain ecosystems.

In addition to altitudinal effects, species distribution is determined by particular microclimatic conditions. The eastern Himalayas receive the influence of the southwestern monsoon from September to May, but this general condition varies significantly across the mountain range. The intricate topography that characterizes this region creates local weather conditions with great variations in rainfall patterns, for example. While some areas receive as much as 118 inches (3,000 millimeters) of rainfall per year, others average a mere 12 inches (300 millimeters). This topographic effect is further marked by slope aspect: North-facing slopes receive considerably less sunlight than their southern counterparts, developing far more humid conditions and longer periods of snow cover. This results in starkly different vegetation communities, and also affects the distribution of fauna.

Besides being significant for their plant biodiversity, these shrubs and meadows are important parts of the altitudinal migration routes of many large carnivores. Throughout the year, alpine shrubs and meadows serve as bridges that connect lower areas of evergreen broadleaf or subalpine coniferous forest with permanent snow-cover zones. This connection allows the movement of wildlife between ecosystems, helping maintain their integrity. These mountain grasslands are crossed by big mammals such as snow leopards, wolves, and even Asiatic black bears, all of which chase takins, goats, and blue sheep along the abrupt hills.

Human Impacts

The eastern Himalayas are rich not only in terms of plant diversity, but also in habitation by diverse ethnic groups. The region combines Buddhists, Hindus, Animists, and Christians—members of groups that have always had a spiritual connection with nature. These ethnic groups value many areas of the Himalayas as sacred sites. The cultural complexity of the eastern Himalayas represents a huge challenge to establishing successful conservation initiatives, however.

At first glance, the Eastern Himalayas Alpine Shrubs and Meadows biome may seem well-preserved, given that its extent has been relatively stable during the past few decades. In reality, it is undergoing extremely destructive degradation processes that seriously threaten this distinctive area. Livestock grazing and trampling by yaks have deleterious effects on shrub growth and regeneration. Collection of fuelwood and plant extraction for medicinal purposes affect the survival of many shrubs and alpine plant populations. On top of these more localized environmental problems are the effects of climate change on mountain ecosystems, which are not yet fully understood.

Several protected areas have been established to conserve the habitats of the tallest—and youngest—mountain range in the world. Of particular importance are Sagarmatha National Park, home of Mount Everest; and the Annapurna Conservation Area, created in the 1990s with the objective of conserving rhododendron biodiversity. Even though 20 percent of the overall area is under protection, these measures have had limited success in stopping the depletion of most Himalayan ecosystems. One of the biggest problems is the lack of local biodiversity across the Himalayas, because most of the protected areas are concentrated on the Nepal side, restricting conservation to a particular zone. Recent efforts have been directed toward creating landscape connectivity through biological corridors and transboundary management initiatives.

Large-scale planning, involving mainly the governments of Nepal, India, and Bhutan, is crucial for preserving the area, as well as guaranteeing resources for the rural and urban people who depend on it. These efforts have proved to be effective only in combination with local stewardship components. Since 2000, key programs such as the Sacred Himalayan Landscape and the Transboundary Biodiversity Conservation Initiative, led by the International Center for Integrated Mountain Development, have promoted regional development and biodiversity conservation through integrating local spiritual beliefs and practices into sustainable management of the natural resources of this breathtaking place and fragile ecosystem.

Lucía Morales-Barquero

Further Reading

Schickhoff, U. "The Upper Timberline in the Himalayas, Hindu Kush and Karakorum: A Review of Geographical and Ecological Aspects." In G. Broll and B. Keplin, eds., *Mountain Ecosystems: Studies in Treeline Ecology.* Berlin: Springer, 2005.

Shrestha, T. B. "Rhododendrons." In T. C. Majupuria and R. Kumar, eds., *Nepal: Nature's Paradise (Insights into Diverse Facets of Topography, Flora and Ecology).* New Delhi, India: Vedams eBooks, 1999.

Wikramanayake, E., E. Dinerstein, and E. Colby Loucks. *Terrestrial Ecoregions of the Indo-Pacific: A Conservation Assessment.* Washington, DC: Island Press, 2001.

Himalayan Alpine Shrub and Meadows, Western

Category: Grassland, Tundra, and Human Biomes.
Geographic Location: Asia.

Summary: This mountain ecosystem is threatened by both climate change and grazing practices.

The Himalayan Alpine Shrub and Meadows biome, western section, is a Palearctic bioregion in the high Himalayas of Nepal, Tibet, and India. The biome lies between the tree line and the snowline, and roughly corresponds to 9,843–16,404 feet (3,000–5,000 meters) in elevation; the location of the snowline depends on moisture availability and may change from year to year. It occupies an area of more than 27,027 square miles (70,000 square kilometers). The Western Himalayan Alpine Shrub and Meadows is a montane ecozone with conifer forests immediately below it and snow or permanent ice above.

There is a strong moisture gradient from east to west in the biome, and precipitation generally is derived from the summer Asian monsoon season. Overall, the bioregion is moister than central Tibet, but less moist than the adjacent Eastern Himalayan Alpine Shrub and Meadows biome. The Kali Gandeki river valley is a major break in the Himalayan range; it acts as the eastern boundary between these two Himalayan biomes. Moving west, the biome includes western Nepal, and the Uttarakhand and Himachal Pradesh states in India, ending at the Himalayan divide created by the Sutlej River.

The Himalayan region has a variety of ecosystems and habitats that vary from tropical rainforest to tundra. Uplift of the Himalayas and the Tibetan Plateau not only generates the Asian monsoon, but also creates refuges and barriers for various species. The high altitude of ridges, combined with long deep valleys, has acted as a filtering point for species. More mobile species such as birds are able to penetrate farther from their source populations in India or China, so distribution is often a function of dispersal ability. Grasslands cover more than one-third of the Himalayan region; this particular biome is the midpoint between forest and ice.

The climate is mild, with cool summers and cold winters characterized by numerous microclimates. High summer solar radiation and relatively high precipitation from the Asian monsoon mean that these areas are fertile even with a relatively short growing season. Snow covers the ground in winter, and melts in April and May. This spring snowmelt swells the rivers downstream and provides additional moisture for plants. Unfortunately, the spring melt has begun occurring earlier each year—an effect of global warming—which disturbs many species that depend on the moist conditions that now occur before the day length has increased sufficiently for growth.

Many days, especially during the rainy summer, are cloudy and have thick, long-lasting fogs. As is common in the Northern Hemisphere, the northern slopes are moister and cooler in temperature, and thus more closely resemble the neighboring biome to the east. Strong incoming solar radiation means that this area can have extreme diurnal temperature fluctuations, depending on cloudiness.

Vegetation

The lower elevation vegetation of this biome is dominated by low, shrubby rhododendrons (*Rhododendron* spp.) and birch (*Betula utilis*), with willow (*Salix* spp.) common along the shallow waterways. Many of the small rhododendrons are extremely colorful during their flowering season. Less-common species associated with the dwarf rhododendrons include *Hippophae rhamnoides* and *Cotoneaster microphyllus*, with *Ephedra gerardiana* and juniper (*Juniperus* spp.) occurring at higher elevations. These species often have medicinal or other value.

Scattered shrubs give way to rich alpine meadows at the higher elevations. During the spring and summer, these shrubs bloom with colorful flowers. The alpine pastures are known locally as *bugyal*. Shrubland areas typically host mixed communities with the genera *Anemone, Aster, Delphinium, Doronicum, Gentiana, Meconopsis, Mertensia, Pedicularis, Polygonum,* and *Primula* represented, among others. A large proportion of species in the meadows are endemic, or found in no other place on Earth; many are endangered.

Fauna

The best known of the fauna here is the snow leopard (*Panthera uncia*). This iconic feline species feeds on the numerous large ungulates that are supported by the alpine meadows, such as serow (*Capricornis sumatraensis*), Himalayan tahr (*Hemitragus jemlahicus*), Himalayan musk deer (*Moschus chrysogaster*), and Himalayan blue sheep (*Pseudois nayur*). The biome potentially is also the edge of the range for brown bear (*Ursus arctos*) and Tibetan wolf (*Canis lupus*). Wild yak (*Bos grunniens*) were once plentiful, but are now thought extinct. The predominant fauna consists of small mammals that live in the underbrush, such as the Himalayan and pale weasels, Himalayan palm civet, pikes, and voles.

Avian fauna consists of more than 130 species, including the golden eagle (*Aquila chrysaetos*) and Himalayan griffon (*Gyps himalayensis*). The bird fauna are much richer than the mammal fauna, because the species are able to move in and out of the biome more easily and thus avoid deleterious effects of the harsh winters.

Environmental Threats

A major threat to wildlife here is climate change, with its attendant shifting of temperature zones generally in an upslope direction and shifting of spring snowmelt runoff earlier in the season when downstream habitats cannot fully absorb the benefits. Other threats include fragmentation of habitat by overgrazing of sheep, yaks, and cattle. Nomadic herders use these meadows from May to October, and spend the winter at lower elevations. Poaching and hunting are additional threats.

Secondary negative human effects include firewood gathering in the shrublands by herders, as well as traders who move through the mountains from one country to another on long, centuries-old trade routes; and harvesting of plants for traditional medicines in all three countries. Numerous surveys have shown that the majority of people living in this biome extract at least some medicinal plants for personal use or for sale.

Studies have shown that in the future, changes in soil moisture from climate change will have a bigger effect on vegetation than current grazing patterns and human stressors. Getting a clear picture regarding climate change is difficult, however, as the extremely complex terrain has not been well modeled.

Grassland degradation results in changes in species composition; loss of species diversity; and loss of the dense sod layer, which provides nutrients and protects the vegetation growing in this area. When the sod layer is lost, only bare rock and mineral soils remain, and both these substances are unfertile and unable to store moisture effectively for vegetative growth. These systems cannot recover during human time scales, because of the severe environmental conditions, even if all disturbances were removed. As anthropogenic disturbances continue to grow, these problems will only grow worse.

Threats also include increases in tourism in these fragile areas. Equilibrium in grazing can be reached by local users. When visiting yaks carrying supplies for tourists enter the equation, however, there is little incentive for these interlopers to conserve resources, and increasing degradation can occur.

Conservation Efforts

There is a long history of biological protection in the area, from local resource preservation over the centuries to new national protected areas (PAs). The social customs and mythology in the Himalayan region are based on plants, animals, and the environment, which are often worshiped as minor deities. This has led to a cultural reverence for the natural world and restrained resource use in past centuries.

Grazing has long been the major use of this area, but customary relationships limited its effects. Growing populations and new market opportunities have led to a major increase in grazing, however. Overgrazing leads to a decline in palatable species for livestock and in loss of total biomass. Also, forbs are more prevalent than grasses in heavily grazed areas. Plant litter, which protects the soil from erosion and protects plant roots, is significantly lower in areas of high grazing. Generally, the drier the land, the more vulnerable it is to damage from overgrazing.

Significant PAs here include the Nanda Devi Biosphere Reserve in India and the Annapurna Conservation Area in Nepal. Although this biome has numerous PAs, the majority of them are mainly rock and ice, so there is less actual protected biome than the figures indicate.

JOHN ALL

Further Reading

Cui, Xuefeng and Hans F. Graf. "Recent Land Cover Changes on the Tibetan Plateau: A Review." *Climatic Change* 94, no. 1 (2009).

Gaston, A., P. Garson, and M. Hunter. "The Status and Conservation of Forest Wildlife in Himachal Pradesh, Western Himalayas." *Biological Conservation* 27, no. 1 (1983).

Luo, T., W. Li, and H. Zhu. "Estimated Biomass and Productivity of Natural Vegetation on the Tibetan Plateau." *Ecological Applications* 12, no. 4 (2002).

Rawat, G. S. "Temperate and Alpine Grasslands of the Himalaya: Ecology and Conservation." *Parks* 8, no. 3 (1998).

Sarkar, S., M. Mayfield, S. Cameron, T. Fuller, and J. Garson. "Conservation Area Networks for the India Region: Systematic Methods and Future Prospects." *Himalayan Journal of Sciences* 4, no. 6 (2007).

Shekhar, C. and R. Badola. "Medicinal Plant Cultivation and Sustainable Development." *Mountain Research and Development* 20, no. 3 (2000).

Himalayan Broadleaf Forests

Category: Forest Biomes.
Geographic Location: Asia.
Summary: This is a highly exploited forest biome, dominated by *Shorea robusta* and *Quercus leucotrichophora* forest occurring in the lower mountain slope areas.

The Himalayan Mountains extend into eight different countries in Asia: Afghanistan, Bangladesh, Bhutan, China, India, Myanmar, Nepal, and Pakistan. From the eastern border of western Pakistan to the frontiers of Myanmar, the Himalayas run a distance of 1,491 miles (2,400 kilometers)

in the shape of an arc. The region is divided into three major subregions: the Sub-Himalayas, with an average altitude of 2,952 to 3,937 feet (900 to 1,200 meters); Lower Himalayas, with an average altitude of 12,139 feet (3,700 meters); and Greater Himalayas, with a single range rising above 19,685 feet (6,000 meters) and nine of the 14 highest peaks in the world, including Mount Everest.

The Tibetan Himalayas and the Tibet Plateau achieves the highest altitudes of 13,123 to 19,357 feet (4,000 to 5,900 meters). The annual rainfall across the ranges varies from 59 to 98 inches (1,500 to 2,500 millimeters). Rainfall patterns in the Himalayas are heavily influenced by the summer monsoon season; most rain here falls from mid-June through September.

Forest Vegetation

Broadleaf forests extend along the east-west arc of the region. The forest vegetation varies considerably, depending upon the elevation. At elevations lower than 3,280 feet (1,000 meters), where temperatures are tropical to subtropical, the forests are dominated by sal trees (*Shorea robusta*). Sal forests penetrate through mid-mountain ranges into the far north along river slopes and valleys, and extend across Nepal, Bangladesh, and India. While some researchers consider sal deciduous, others consider it evergreen or a borderline between deciduous and evergreen tree. In contrast to most of the deciduous tree species that are generally summer-flushing, sal is a spring-flushing species; vegetative budding breaks around the spring equinox in the region.

Oaks constitute the second-most dominant broadleaf forests in the region. Banj oak or Himalayan white oak (*Quercus leucotrichophora*) and other *Fagaceae* dominate at approximately 1,640 feet (2,000 meters). Tilonj oak or green oak (*Quercus floribunda*) and Rianj oak or wooly-leaved oak (*Quercus lanata*) are common at 6,230 to 7,870 feet (1,900 to 2,400 meters) in the Himalayas. Karshu oak (*Quercus semecarpifolia*) dominates at 7,870 to 9,840 feet (2,400 to 3,000 meters) in subalpine conditions. Both the *Shorea* and *Quercus* reach a height of about 82 to 130 feet (25 to 40 meters).

As elevation increases, sal is replaced by chir pine (*Pinus roxburghii*), which colonizes bare sites with a wide range of nutrient availability. Among deciduous trees, the subalpine *Betula utilis* dominates extensive areas. Other temperate trees occur in small patches through the forest, including maple, ash, walnut, horse chestnut, and dogwood. Oak trees are generally associated with shrubs of *Rhododendron arboretum* and *Lyonia ovalifolia*. Other major shrubs in these oak-dominated forests include *Arundinaria falcate*, *Myrsine* spp., *Berberis asiatica*, *Randia tetrasperma*, etc.

In the Nepal Himalayas, the subtropical broadleaf forests are dominated by deciduous species of *Terminalia*; the subtropical semi-evergreen species of the *Schima-Castanopsis* community; mixed *Alnus-Dalgergia-Acacia* forest stands; and mixed *Shorea-Terminalia-Anogeissus-Lagerstroemia*.

Fauna

The subtropical broadleaf environment forms an important habitat for mammals such as the Himalayan langur (*Presbytis entellus*), tiger (*Panthera tigris*), Asian elephant (*Elephas maximus*), gaur (*Bos gaurus*) and sloth bear (*Melursus ursinus*). Rich avian biodiversity has also been reported in the Himalayan subtropical forest biome. In the oak forests of Kumaun Himalaya, more than 300 species of birds have been recorded. These include the golden eagle (*Aquila chrysaetos*), collared falconet (*Microhierax caerulescens*), velvet-fronted nuthatch (*Sitta frontalis*), long-billed vulture (*Gyps indicus*), grey-headed woodpecker (*Picus canus*), four leaf warbler (*Phylloscopus* spp.), giant hornbill (*Buceros bicornis*), oriental pied hornbill, (*Anthracoceros albirostris*), Indian grey hornbill (*Ocyceros birostris*), pea-fowl (*Pavo cristatus*), hill myna (*Gracula religiosa*), Asia paradise flycatcher (*Terpsiphone paradisi*), and many others. In addition to resident birds, winter and summer migrant birds are also common. Summer migrants prefer the subtropical broadleaf environment, as compared to winter migrant birds, which show preference for aquatic and scrub habitat in the region.

Biome Threats

To a marked degree, these forests are over-exploited by humans, with mounting evidence of

Opinions vary on whether sal is deciduous or evergreen. These sal leaves are used as plates by local people. (Wikimedia/ Biswarup Ganguly)

habitat fragmentation. With respect to its economic importance, sal forests yield valuable timber. Sal meets the subsistence needs of the local people, including fuelwood, leaves for plates, seed for oil, litter for animal bedding and composting, feed for cattle, resin or latex from heartwood, tannin and gum from bark. Oak is mostly used as firewood, as it burns hot and the coals last long. Locally, oak is also used to make charcoal. Its leaves and wood are known to burn even when green. The leaves and tender branches of oak serve as cattle fodder., Timber is also used for making agricultural implements. The seeds of banj oak are used for treating scabies and urinary disorders; roasted banj oak seed is also used as a coffee substitute.

Himalayan biodiversity is severely threatened due to natural and anthropogenic disturbances. With forest stands fragmented, some wildlife is disappearing. In particular, the subtropical forests that dominate in low elevation areas tend to be more highly populated. Large-scale felling of trees for timber and other industrial raw materials since the colonial period and forest conversion to agricultural lands are the major causes of deforestation.

Oak forests here generally are moist, fire-free and close-canopied. The oak-dominated soils have higher water-holding capacity than the chir pine forests that have been overtaking them in many localities as an immediate-successor colonizing species. Chir pine is both fire-resistant and unpalatable to animals. Chir pine forests in the region are known as xerophytic edaphic climax forests, in other words, ideally adapted for moisture-lacking, nutrition-depleted soils.

Replacement of oak by pine is a major concern and is not preferred by villagers, as once pine is established, it is highly susceptible to fire—also representing increased habitat-destruction risk to animals here. Low regeneration of oak in the region is attributed to excessive browsing and pruning, reduced seed production, and increased seed predation.

Much of the broadleaf forest in the Himalayans has in recent times been converting to scrub vegetation, which dramatically impacts habitat structure. Sensitive ungulates, such as Himalayan musk deer, are under severe pressure, for example, as their food sources and shelter choices here are disrupted. Other factors contributing to the poor wildlife outlook in this region are poaching and trade of animal parts and faulty land-use practices. In addition, recurrent fire is a major disturbance factor impacting the forest structure and function. Intensive conservation efforts are needed to save the remaining forests and wildlife of this biome.

Climate change is projected to impact snow and ice patterns here, with higher average springtime temperatures resulting in faster snowmelt and more water eroding mountainous slopes. Climate change may also affect the rainfall patterns in the Himalayas, causing changes in the duration or quantity of moisture released during the monsoon season. Such alterations to precipitation, humidity, and storm cycles will impose significant consequences on forest make-up and on the capacity of its inhabitants to keep up the pace of adaptation or migration.

Krishna Prasad Vadrevu

Further Reading

Puri, Gopal S. *Indian Forest Ecology, Vols. 1 and 2.* New Delhi, India: Oxford Book and Stationery Co., 1960.

Sharma, Eklabya, et al. *Climate Change Impacts and Vulnerability in the Eastern Himalayas.* Kathmandu, Nepal: International Center for Mountain Development, 2009.

Singh, J. S. and S. P. Singh. "Forest Vegetation of the Himalaya." *Botanical Review* 53, no. 1 (1987).

Sultana, Aisha and Jamal A. Khan. "Birds of Oak Forests in the Kumaon Himalaya, Uttar Pradesh, India." *Forktail* 16, no. 1 (2000).

Troup, Robert S. *The Silviculture of Indian Trees, Vol. I: Dilleniaccae to Leguminosae (Papillonaceae).* New York: Oxford University Press, 1921.

Zobel, D. B. and S. P. Singh. "Himalayan Forests and Ecological Generalizations." *BioScience* 47, no. 11 (1997).

Himalayan Subalpine Conifer Forests

Category: Forest Biomes.
Geographic Location: Asia.
Summary: The varied evergreen forests of the Himalaya Mountains contain distinct ecoregions according to elevation and precipitation patterns.

The Himalayan subalpine conifer forests are an ecoregion about the size of New England, consisting of the temperate coniferous forests of the middle and upper elevations of Nepal, India, and Pakistan. This biome is part of the larger, extremely varied Himalayan ecosystem that extends from high alpine meadows to alluvial grasslands in the foothills. The forests are important both to Himalayan biodiversity and to the seasonal migrations of several bird and mammal species. This region is a transition between the treeless alpine meadows and rocky alpine screes, as well as the Terai and Duar grasslands.

The Himalaya mountain range includes the tallest peaks on Earth; it is also the youngest mountain range. Much of the rainfall in the region comes from the southwestern monsoon that arrives from the Bay of Bengal. The rains pass through the eastern region first; the west receives considerably less precipitation. This difference in moisture has a stark effect on vegetative growth patterns. The tree line ends at 10,830 feet (3,300 meters) in the west, but continues on to 13,120 feet (4,000 meters) in the wetter east.

Flora

The eastern forests are dominated by fir (*Abies spectabilis*), larch (*Larix griffithii*), hemlock (*Tsuga dumosa*), and juniper (*Juniperus recurva* and *J. indica*), with varied rhododendron species in the understory (*Rhododendron hodgsonii, R. barbatum, R. campylocarpum, R. campanulatum,* and *R. fulgens*). The understory may also include maple (*Acer* spp.), rose (*Sorbus* ssp.), and juniper. At elevations of 8,200 to 9,840 feet (2,500 to 3,000 meters), Himalayan hemlock (*Tsuga dumosa*) dominates the forests in mixed stands with fir, and again, the understory is rich with rhododendron species.

In the drier west, the conifer forests are the most extensive, consisting of blue pine (*Pinus wallichiana*), chilgoza pine (*P. gerardiana*), East Himalayan fir (*Abies spectabilis*), rhododendron (*Rhododendron campanulatum*), birch (*Betula utilis*), cypress (*Cupressus torulosa*), deodar (*Cedrus deodara*), silver fir (*A. pindrow*), and West Himalayan spruce (*Picea smithiana*). Chilgoza pine, valued for its edible seeds, is considered to be near-threatened, due to intensive grazing; attempts at artificial regeneration have been largely unsuccessful.

The forests tend to fall into specific types: pure fir forests, mixed oak-and-fir forests, mixed rhododendron-fir-birch forests, and mixed coniferous forests. Cypress and deodar are found mainly at higher elevations, above 7,870 feet (2,400 meters), while the fir forests form a continuous belt at 9,840–11,480 feet (3,000–3,500 meters) on the southern side of central Nepal's main ranges, mixed with a rhododendron understory.

These subalpine areas are rich with economically and medicinally important plants, including a type of bamboo (*Arundinaria*) and *Daphne bholua*, a plant used to make paper and rope.

Wildlife

The eastern forests are moderately more diverse than the west. Three mammals are endemic (found nowhere else on Earth) to the forests: the Himalayan field mouse (*Apodemus gurkha*), which has a range limited to these forests and is a strict endemic; the Bhutan giant flying squirrel (*Petaurista nobilis*); and Hodgson's giant flying squirrel (*Petaurista magnificus*). The Himalayan musk deer (*Moschus chrysogaster*) is also found here; it is hunted for its musk glands. The endangered red panda (*Ailurus fulgens*) can be found in the wet bamboo forests from 9,840 to 13,120 feet (3,000 to 4,000 meters).

Other endangered mammals include the particolored squirrel (*Hylopetes alboniger*) and the tiger (*Panthera tigris*). Tiger populations are more common in the broadleaf forests at lower elevations than in this ecoregion, as the conifer forests do not contain sufficient prey, but tigers frequent these woods in the course of their hunt.

The fauna are not especially diverse in the dry western forests, but the habitat is important for the Himalayan brown bear (*Ursus arctos isabellinus*), which is believed to be the basis for the Yeti myth. Brown bears hibernate from October to April or May. The only mammal endemic to the western forests is the Muree vole (*Hyperacrius wynnei*), found in Kaghan Valley, Swat Valley, and the Murree Hills.

The western region contains habitats for several threatened species, including the wild goat known as the markhor (*Capra falconeri*), the national animal of Pakistan; the goat-antelope; the southern serow (*Naemorhedus sumatraensis*); and the Himalayan tahr (*Hemitragus jemlahicus*), an ungulate related to the wild goat.

None of the 285 bird species of the region is strictly endemic to it, but nine species are rare enough to be found almost nowhere else: Himalayan quail (*Ophyrsia superciliosa*), western tragopan (*Tragopan melaocephalus*), hoary-throated barwing (*Actinodura nipalensis*), white-cheeked tit (*Aegithalos leucogenus*), white-throated tit (*A. niveogularis*), spectacled finch (*Callacanthis burtoni*), immaculate wren-babbler (*Pnoepyga immaculata*), orange bullfinch (*Pyrrhula aurantiaca*), and Kashmir nuthatch (*Sitta cashmirensis*).

The Koklass pheasant (*Pucrasia macrolopha*) and Himalayan monal (*Lophophorus impejanus*) are found throughout the forests. The Himalayan griffon (*Gyps himalayensis*) is an Old World vulture that breeds in the rocky crags and peaks of the Himalayas, laying a single egg, and scavenges carcasses in the surrounding mountains and open areas.

Habitat Threats

Any loss of habitat in this region could be catastrophic for several bird and mammal species that depend upon the area for seasonal migration. Human activity, including cutting trees for firewood, hiking, and tourism, are the major threats to the area. Some areas are too steep for commercial logging, but since the 1980s, road construction in southeastern Tibet has opened up new areas for commercial forestry operations. Studies are needed on the long-term impact of climate change in this biome. Changes in snow accumulation, melt timing and volume, and rainfall patterns are likely to impact the area's soil structure, vegetation communities, and wildlife populations.

BILL KTE'PI

Further Reading

Chang, D. H. S. "The Vegetation Zonation of the Tibetan Plateau." *Mountain Research and Development* 1, no. 1 (1981).

Gairola, Sanjay, R. S. Rawal, and N. P. Todaria. "Forest Vegetation Patterns Along an Altitudinal Gradient in Sub-Alpine Zones of West Himalaya, India." *African Journal of Plant Science* 2, no. 6 (2008).

Sakai, Akira and S. B. Malla. "Winter Hardiness of Tree Species at High Altitudes in the East Himalaya." *Ecology* 62, no. 5 (1981).

Schaller, G. B. *Wildlife of the Tibetan Steppe.* Chicago: University of Chicago Press, 1998.

Singh, J. S. and S. P. Singh. "Forest Vegetation of the Himalaya." *Botanical Review* 53, no. 1 (1987).

Stattersfield, A. J., M. J. Crosby, A. J. Long, and David C. Wege. *Endemic Bird Areas of the World: Priorities for Biodiversity Conservation.* Cambridge, UK: BirdLife International, 1998.

Himalayan Subtropical Pine Forests

Category: Forest Biomes.
Geographic Location: Asia.

Summary: This unique subtropical forest biome is dominated by *Pinus roxburghii*, the chir pine, and has low levels of floristic and faunal biodiversity.

The Himalayan region extends across south-central Asia from Afghanistan to Burma, or Myanmar. The Himalaya Mountain range arcs east and west for 1,491 miles (2,400 kilometers). Altitude zonation breaks out into three major zones: the Sub Himalayas, with elevations of 2,952 to 3,937 feet (900 to 1,200 meters); Lower Himalayas, averaging 12,139 feet (3,700 meters); and Greater Himalayas, rising above 19,685 feet (6,000 meters), including the world's tallest peak, Mount Everest. Annual rainfall here is in the 59–98-inch (1,500–2,500-millimeter) range, and is affected by the June-to-September summer monsoon.

Vegetation

In the Himalayan region, the vegetation transitions from a semideciduous, willow tree species of sal (*Shorea robusta*) in the foothills, through evergreen tree stands based around chir pine (*Pinus roxburghii*), and to oak trees (*Quercus* spp). As elevation increases, evergreen vegetation dominates up to 11,480 feet (3,500 meters), after which alpine meadows occur above the timberline. The pine forests here stretch for almost 1,240 miles (2,000 kilometers) across the lower elevations of almost the entire length of the Himalaya Mountain range.

The predominant pine species in the region is chir pine *(Pinus roxburghii)*. Chir pine dominates in the lower elevations at less than 7,550 feet (2,300 meters) along the central and western Himalayas. At higher elevations, above 7,550 feet (2,300 meters) and up to 9,840 feet (3,000 meters), the blue pine (*Pinus wallichiana*) proliferates. Other, less dominant pine species include chilgoza pine (*Pinus gerardiana*), which occurs in the dry valleys of the Himalayas, eastern Afghanistan, and northern Pakistan; khasi pine (*Pinus kesiya*); and mercus pine (*Pinus merkusii*), which occurs mostly in the eastern Himalayas of Assam state and Burma, or Myanmar.

The dominant chir pine subspecies here, *Pinus roxburghii Sargent,* is a wind-pollinated, subtropical evergreen species. It is also an early colonizer and a fast-growing tree, replacing the oak forests that are affected by negative habitat pressures across the Himalayas. The density of pine trees in a typical forest can vary from 80 to 630 trees per 2.5 acres (one hectare). The biomass density varies widely, depending on the succession stage and site-disturbance regimes.

Chir pine is found interspersed with other trees here, such as sal; the oaks *Quercus leucotrichophora, Q. floribunda,* and *Q. semecarpifolia;* west Himalaya fir (*Abies pindrow*); and Himalayan birch (*Betula utilis*). The chir pine is highly drought-resistant; it is widely planted by locals because of its high survival rate and ease of plantation establishment and management.

Chir pine trees (Pinus roxburghii) dominate this landscape of low hills in south Asia. Pine forests cover close to 1,240 miles (2,000 kilometers) of land at the lower elevations of the Himalaya Mountains. (Wikimedia/Udayanarya)

Although chir pine trees themselves are fire-resistant, the dried pine needles that constitute forest litter are quite fire-prone, due to their high resin content. Chir pine forests are characterized by frequent fires; this often results in low plant biodiversity. To control fires and facilitate growth of grass for cattle feeding, local people use fire as a management tool. The resultant grass layer in the pine forest is dominated by species such as *Arundinella nepalensis, Themeda anathera,* and *Imperata cylindrica,* which all have fodder value.

Human Use of Resources

Pines serve as a valuable timber resource for the human population. Chir pine timber is particularly useful for making items such as railway sleeper cars, wooden poles for telecommunications lines, home furniture and cabinets, and packing cases and crates. Chir pine also is a source of good-quality oleo resin, which after distillation generates turpentine oil and rosin. This turpentine is then used in the preparation of paints, varnishes, polishes, chemicals, and pharmaceuticals; whereas rosin is used extensively in many indus-

tries to make soaps, paper, sealing waxes, oilcloth, inks, and disinfectants.

The resinous wood and pinecones of the chir pine are burned for both fuel and lighting. The seeds are edible. The resin from the leaves has medicinal properties and is used by locals to heal skin lesions. The leaves also are used for purposes such as thatching, packing, cattle bedding, and fertilizer.

The oil of chir pine needles is used as a diuretic by the local people. The bark decoction made from blue pine yields a yellow dye, which is used for dyeing woolen clothes and home textiles. For all of these reasons, pines in the region are widely planted by locals as well as forest officials for economic use as well as ecosystem functions, especially for preventing erosion on the hillsides.

Wildlife

The wildlife supported by the Himalayan subtropical pine forests is relatively poor compared with that of many tropical evergreen forests. The notable wildlife in these forests includes the Asian elephant (*Elephas maximus*), clouded leopard (*Pardofelis nebulosa*), gaur (*Bos gaurus*), golden langur (*Trachypithecus geei*), barking deer (*Muntiacus muntjak*), and goral (*Nemorhaedus goral*).

The forests also support a rich variety of bird species. Among those species unique to this biome and found nowhere else, that is, endemic species here, are the cheer pheasant (*Catreus wallichi*), the chestnut-breasted partridge (*Arborophila mandellii*) and nine other endemic and near-endemic bird species.

Environmental Threats

The Himalayan subtropical pine forests are under intense anthropogenic pressure, due to deforestation, grazing, lopping of tree branches, and removal of litter for fodder and fuelwood, in addition to frequent fires. Most of the pine trees in the region are stunted; the forests are of secondary nature, having been logged or otherwise cleared for agriculture. Prime forests are now restricted to less than five percent of the region.

Some non-anthropogenic causes of forest degradation include geological disturbances such as landslides, soil erosion, and earthquakes. These phenomena affect both ecosystem stability and function. Conservation efforts to mitigate these effects are needed to protect the remaining forests in the region.

Potential climate change threats to this area were once thought to be mainly from melting glaciers, according to a 2007 United Nations report that predicted Himalayan glaciers would disappear by about 2035, cutting into the groundwater supply to this biome. However, a more recent study from the universities of California and Potsdam report that the Himalayan glaciers are advancing, rather than retreating. Other factors beyond average air temperature—including the fields of rocks on the glacial fields—may have greater impact on preserving glaciers than originally thought.

KRISHNA PRASAD VADREVU

Further Reading

Nelson, Dean and Richard Alleyne. "Some Himalayan Glaciers Are Advancing Rather than Melting, Study Finds." *The Telegraph*, January 2011. http://www.telegraph.co.uk/earth/environment/climatechange/8284223/Some-Himalayan-glaciers-are-advancing-rather-than-melting-study-finds.html.

Puri, G. S. *Indian Forest Ecology, Vols. 1 and 2.* New Delhi, India: Oxford Book and Stationery Co., 1960.

Singh, J. S. and S. P. Singh. "Forest Vegetation of the Himalaya." *Botanical Review* 53, no. 1 (1987).

Tucker, Richard P. "The Forests of the Western Himalayas: The Legacy of British Colonial Administration." *Journal of Forest History,* 26, no. 3 (1982).

Hokkaido Montane Conifer Forests

Category: Forest Biomes.
Geographic Location: Japanese archipelago.
Summary: The major Japanese island of Hokkaido features mountain peaks and valleys that are home to a vibrant evergreen ecosystem with high species diversity.

Hokkaido, the northernmost of Japan's major islands, has a rugged landscape of snow-capped volcanoes. Natural high-elevation cold temperature regimes here are mitigated somewhat by the moderating atmospheric influence of the surrounding ocean. Together, elevation zones and strong seasonal changes have provided for rich species diversity. Human impact on the environment here, however, has become profound.

Japan's high levels of species richness and endemism (species found nowhere else) are by-products of island geography, climatic influences, and ice-age refugia. Nearly 2 million years ago, colliding Pacific and Eurasian plates gave rise to snow-capped mountains and volcanoes, many of which are active today. Hokkaido, the northern-most of Japan's 6,852 islands, is bound by the Sea of Japan (west), Sea of Okhotsk (north), and Pacific Ocean (east). Soya Strait, to its south, separates the island from the largest Japanese island, Honshu.

Hokkaido is home to the Hokkaido Montane Forest biome, one of nine such forest ecoregions in Japan. The World Wildlife Fund considers this biome vulnerable, due to high rates of habitat conversion, few intact areas, wildlife poaching, and development pressures.

Geography and Climate

Japan's island ecosystems are distributed along a north-south climate gradient, according to regional differences in temperature and precipitation. An oceanic mild climate is responsible for warm temperate rainforests—also known as evergreen broadleaf or Laurisilva—in the south. To the north lie the cool temperate montane forests, including those on Hokkaido, where the average annual temperature hovers at 46 degrees F (8 degrees C) and average annual precipitation is 45 inches (115 centimeters).

Most of the island experiences cool summers and icy winters. The conifer forests on Hokkaido thrive from sea level up to about 4,900 feet (1,500 meters). In the central highlands is the mostly volcanic Daisetsu mountain range, also known as the Roof of Hokkaido. The elevation of the highest mountain within the range is approximately 7,500 feet (2,290 meters).

Biota

Hokkaido coniferous forests share several species with the Siberian taiga, including snow forest affinities along upper montane and subalpine zones. Dominant conifers include Asian or Jezo spruce (*Picea jezoensis*) and Sachalin fir (*Abies sachalinensis*), mixing with birches (*Betula ermanii* and *B. maximowicziana*) that are sometimes found in pure, post-fire stands. Upper elevations, in the alpine zone, are dominated by Siberian dwarf pine (*Pinus pumila*) interspersed with birch-alder (*B. ermanii-Alnus maximowiczii*) and mountain ash (*Sorbus sambucifolia*). These create thicket and scrub communities with poorly developed understories, but thick ground mosses.

Notably, Asian spruce is widely distributed in East Asia, including the Kamchatka Peninsula in Russia, presumably through an ancient connection to the Russian Far East established during repeated ice ages. Recurring ice ages also resulted in cir-

cumpolar species distributed on Hokkaido, such as bunchberry dogwood (*Cornus canadensis*) and wood sorrel (*Oxalis acetosella*), among others.

In the central part of this ecoregion can be found many rare and endemic (found only here) plant species, including the willow (*Salix paludicola*), the wolf's bane (*Acontium yamazaki*), and the whitlow grass (*Draba nakaiana*). There are also more than 400 species of ferns and 200 species of mushrooms on the island.

Hokkaido's vertebrate species total more than 200 birds and 40 mammals. Species of conservation concern include Stellar's sea-eagle (*Haliaeetus pelagicus*); white-tailed sea-eagle (*H. albicilla*); Blakiston's fish owl (*Keputa blakistoni*), a rare and endangered owl; the endangered Japanese crane (*Grus japonensis*); and an endemic subspecies of hazel grouse (*Tetraste bonasia vicinitas*).

Endemic mammal subspecies include pika (*Ochotona hyperborean yesoensis*), a remnant of past glacial epochs, a threatened sable (*Martes zibellina brachyuran*), and the Sika deer. Hokkaido is also the only place in Japan that has brown bears (*Ursos arctos yesoensis*): the Yezo bear. Most of the other mammals here are small: rodents, rabbits, bats, red fox, and weasels. Hokkaido forests also provide habitat for a variety of frogs, salamanders, lizards, and snakes.

Examples of relatively intact montane forests occur in two main areas on the island: Daisetsuzan National Park, the largest national park in Japan; and the nature reserve Horoka Tomamu Montane Forest, located at the headwaters of the Mu River in the center of Hokkaido. This reserve includes species-rich hemiboreal (containing elements of both temperate and boreal zones) forests that provide good examples of native plant communities.

Human Impact

Present threats to the preservation of wildlife on Hokkaido include poaching, especially for sable fur; construction of roads; tourism; agricultural activity including over-grazing; and industrial development. Another concern is climate change. It is getting warmer in Japan—an increase of about 1.8 degrees F (1 degree C) during the last century—especially in urban areas, where the heat-island effect is combining with global warming to generate unusually hot temperatures. Japan's network of islands is particularly vulnerable to sea-level rise, which could be 1.5 to 3 feet (0.5 to 1 meter) or more by century's end, depending on how global emissions play out, and effects worse when rising sea levels combine with storm surges.

Japan also will likely experience more frequent heat waves, droughts in places, intense rain and stronger typhoons (already on the rise), and a decline in marine productivity because of the degradation of coral reefs.

Upper elevation communities, especially unique subalpine ones, may recede on all but the tallest and most northern-latitude mountains here, due to the rising temperatures. Snowfall may increase on Hokkaido, however, given the island's relatively high latitude and high-elevation areas. Such climatic extremes will affect human health (e.g., rise in tiger mosquito populations, heat exhaustion); agricultural production (e.g., projected decline in rice yields of 12 to 13 percent by mid-century throughout most of Japan); and wildlife vigor (e.g., seasonal timing, growth, survival, and migration routes of Pacific salmon—the Okhotsk Sea is an important life line for regional salmon).

How climate change effects the composition and processes that maintain forests depends on many factors, including site-specific ones, such as microclimate and soils; topographic position, such as slope, aspect, and elevation; and the climate-niche preference of local forest species. It is likely that as conditions warm, species will move up in elevation and/or latitude, where they can.

It would appear that because of Hokkaido's high latitude, it may not experience as dramatic a change in its plant communities as will the more southerly areas. However, regional climate change projections forecast a decline in the climate niche for conifers—especially high-elevation types—and an increase in the climate niche of temperate-favoring broadleaf deciduous trees. Efforts are underway throughout Japan, given its vulnerable status, to address climate-change impacts to agriculture, water storage, fisheries, forestry, and urban areas.

DOMINICK A. DELLASALA

Further Reading

Government of Japan: Environment Agency. "Technology Transfer Manual on Nature Conservation." Overseas Cooperation Panel, Japan, 1999. http://www.env.go.jp/earth/coop/coop/document/08-ttmnce/08-ttmnce.pdf.

Niiyama, Kaoru. "The Role of Seed Dispersal and Seedling Traits in Colonization and Coexistence of *Salix* Species in a Seasonally Flooded Habitat." *Ecological Research* 5, no. 3 (1990).

Uesaka, Shohei and Shiro Tsuyuzaki. "Differential Establishment and Survival of Species in Deciduous and Evergreen Scrub Patches and on Bare Ground, Mt. Koma, Hokkaido, Japan." *Plant Ecology* 175, no. 2 (2005).

Honey Island Swamp

Category: Inland Aquatic Biomes.
Geographic Location: North America.
Summary: Once heavily logged for its cypress trees, today Honey Island Swamp is a healthy second-growth swamp supporting a variety of wildlife.

Honey Island Swamp is a cypress-tupelo gum swamp, located approximately 50 miles from New Orleans. Honey Island Swamp gets its name from swarms of honey bees that were once seen on a nearby island. This subtropical, low-elevation ecoregion is a well recovered wetland more than 27 miles (43 kilometers) long and approximately 7 miles (11 kilometers) across, situated between the Pearl and West Pearl Rivers in the southern coastal plain of Louisiana.

Considered one of the least altered river swamps in the country today, Honey Island Swamp was logged for its cypress when timber industries moved in starting in the early 1700s. The swamp now exhibits the characteristics of a healthy, mature second-growth forest.

Cypress-tupelo gum swamps such as Honey Island Swamp are flooded year round. These wetlands are comprised of muddy-bottom channels with silty, tidal-driven waters. They are characterized by low-gradient elevations that, along with tidal currents, slowly move the water so that aquatic habitats are intermediate between streams and lakes in their morphology and biological characteristics. Regular flooding prevents the forest from being overtaken by upland tree species.

Spanish moss can be seen hanging from the cypress trees that emerge from shallow water that is often covered with a layer of green algae. This blanket of algae on the surface is able to grow because of the slow or stagnant movement of the waters. The water here is often tea-colored from tannic acid produced during the slow decomposition of the large amounts of detritus that remain in the swamp as the tree limbs and twigs age and fall away.

Biota

Honey Island Swamp encompasses dense loblolly and slash pine forests and mixed hardwood forests located in the uplands; these include a robust array of tree species. Among the oaks are water, overcup, nuttall, cow, obtusa, and live oak; other hardwood varieties are bitter pecan, hickory, American elm, and swamp red maple; additional types featured are sweetgum, magnolia, and beech. Mid-story in the mixed hardwoods stands are found ironwood and arrowwood, as well as Virginia willow, swamp privet, buttonbush, and water elm. In the lowest elevations of swampy floodplain grow the bald cypress and tupelo gum, along with swamp blackgum.

The bald cypress is Louisiana's official tree, named in 1963. This deciduous tree's falling foliage adds to the decaying material that adds nutrients to the swamp waters. Adaptations of the cypress tree to the swamp are their flared or buttressed lower trunks. This wide base helps anchor the tree and prevent damage from the wind. Also, as a result of the constant submersion in water, the cypress tree develops "knees" that allow the roots to spread over a larger area. Cypress's ability to resist decay is what make it such a valuable resource as lumber.

Honey Island Swamp supports everything from alligators to bald eagles. This ecoregion sustains a variety of habitats for animals that move through, live on top of, or along its edge. These are low-oxygen waters with little current or mixing. Mosquitoes lay their eggs in the stagnant water, providing part of the base of the food web here. Honey Island Swamp also provides habitat for deer, squirrel, rabbit, fox, beaver, nutria, mink, opossum, raccoon, and bobcat—as well as a population of feral pigs. Cougar and wolf are seen occasionally.

Bald eagle and golden eagle are seen occasionally; swallowtail kite and osprey are more common, as are the red-shouldered hawk, barred owl, and eastern screech-owl. Other year-round resident avians include wood duck, wild turkey, yellow-billed cuckoo, carolina chickadee, tufted titmouse, carolina wren, gray catbird, and brown thrasher. Waders include the yellow-crowned night heron and great egret.

Adding to the variety are northern cardinal and eastern towhee; and a range of woodpeckers such as red-headed, red-bellied, downy, hairy and pileated. In 1999 there was thought to be a sighting of the thought-to-be-extinct ivory-billed woodpecker, but there was little proof of any sightings. The news, however, brought much attention to the area.

A variety of reptiles—including alligators—and amphibians are also found within the swamp's boundaries. The ringed map turtle, designated as a threatened species, is only found in this region. Reptiles and amphibians are especially suited to

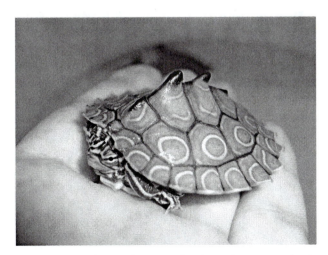

The threatened ringed map turtle, shown here, is endemic to the Honey Island Swamp region in the southern coastal plain of Louisiana. (Wikimedia/ Eekhoorntje)

swamp life, because of their ability to adapt to the changing environment, and to take advantage of a range of water levels. Fish here include large-mouth bass, bluegill, warmouth, red-ear sunfish, alligator gar, freshwater drum, flathead catfish, and buffalo fish. Enormous quantities of crawfish thrive in Honey Island Swamp.

Human Impact

In 1980, President Jimmy Carter authorized the 37,000-acre (15,000-hectare) Bogue Chitto National Wildlife Refuge (NWR), located north of and adjacent to Honey Island Swamp. The swamp is now a permanently protected wildlife area under the control of this refuge, and managed by the Louisiana Department of Wildlife and Fisheries. A larger part of the swamp is claimed as the Pearl River Wildlife Management Area (WMA). The Pearl River WMA, at 35,000 acres (14,000 hectares), is comparable to the Bogue Chitto NWR in size.

Honey Island Swamp remains intact except for its waterways. The swamp is accessible only by boat or on foot; few roads exist in the area, and none within the borders of the swamp. As designated by the Nature Conservancy, Honey Island Swamp is Louisiana's first Nature Preserve (with the pine forests of the northern portion of the area controlled by the Bogue Chitto NWR). This ecosystem provides habitat and food sources for the surrounding plant and animal life, water preserves, water filtering, spawning and fishing areas, as well as flood control from major storm surges.

Global warming poses a real and growing threat to the Honey Island Swamp biome, in particular from the projected impact of sea-level rise. At the most pessimistic estimates, the swamp could remain substantially inundated, such that areas today fed and refreshed by tidal flows would become entirely different, permanently submerged habitats. Saltwater intrusion, and its habitat-changing dynamic, is an associated impact risk that seems imminent.

SANDY COSTANZA

Further Reading

Keddy, Paul A. *Water, Earth, Fire: Louisiana's Natural Heritage.* Bloomington, IN: Xlibris, 2008.

NatureWorks. "Swamps." New Hampshire Public Television, 2012. http://www.nhptv.org/nature works/nwep7i.htm.

Nickell, Joe. "Tracking the Swamp Monsters." *Committee for Skeptical Inquiry* 25, no. 4 (2001).

Hudson Bay

Category: Marine and Oceanic Biomes.
Geographic Location: North America.
Summary: The second-largest bay in the world, Hudson Bay supports diverse cold-water coastal and marine ecosystems.

Named for the Dutch explorer Henry Hudson, Hudson Bay is the second-largest bay in the world after the Bay of Bengal. It is a marine bay and marginal sea in northern Canada that drains most of the central area of the country as well as parts of the upper Midwest of the United States, about 1.5 million square miles (3.9 million square kilometers) in all. Hudson Bay connects with the North Atlantic Ocean via the Hudson Strait, and to the Arctic Ocean via the Northwest Passages.

Hydrology and Geography

Many streams and rivers contribute to its waters, but not nearly enough to offset its brackish salinity. Hudson Bay is surrounded by vast expanses of wetlands, shrubland steppes, taiga or boreal forest, and tundra ecosystems.

Hudson Bay is a relatively shallow saltwater body, with an average depth of 330 feet (100 meters). It is about 900 miles (1,450 kilometers) long north to south—including its southernmost lobe, James Bay—and about 650 miles (1,050 kilometers) across at its widest. Total surface area is approximately 450,000 square miles (1.16 million square kilometers). Hudson Bay is ice-covered for half the year; ice begins to form in early November, and the whole bay has very low year-round average temperatures. The ice cover begins to melt in mid-June, usually clearing from the eastern shores first. Water temperature rises to 46 to 48 degrees

F (8 to 9 degrees C) on the western side of the bay by late summer.

While it is a saltwater or brackish body, Hudson Bay receives freshwater inflow at an annual volume of some 170 cubic miles (700 cubic kilometers); even more comes from precipitation and ice melt. Consequently, and also due to its limited hydraulic exchange with the Atlantic, the bay is less salty than the ocean.

Rich Base of Life

Because of its vast dimensions, the Hudson Bay marine ecosystem has many aquatic and coastal ecozones with varied habitats that are used year-round by Arctic and sub-Arctic species, as well as seasonally by migratory fish, marine mammals, and birds. The sea ice, for instance, supports the seals upon which polar bears depend. Millions of geese and shorebirds feed and breed in vast salt marshes. Productive submarine eelgrass beds provide food for waterfowl migrating to and from breeding habitat in the Arctic Islands, and shelter for crustacean and fish populations. Numerous estuaries provide habitat corridors for anadromous fish like salmon and Arctic char, and support beluga whales.

The Hudson Bay supports great numbers of large mammals, both aquatic and terrestrial. Migratory marine mammal species such as belugas, narwhals, and bowhead whales frequent the region as permitted by ice conditions across the northern channels. Other cetaceans, such as orca and minke whales, are rarer visitors. Large concentrations of up to 20,000 beluga whales are found in the estuaries of the Nelson and Churchill Rivers in July and August.

Ringed seals, bearded seals, and harbor seals, as well as walruses, reside in the bay year-round, while harp and hooded seals visit seasonally. Subsistence harvesting of these marine mammals is important to the aboriginal peoples and the economy of the bay. Arctic foxes and polar bears reside on coasts in the summer, and on sea ice during the rest of the year; they, too, are hunters of the seals.

At least 130 species of birds migrate to the Hudson's waters each summer, and depart in late fall. Waders, water fowl, and raptors share pelagic, intertidal, and wetland habitats here. Most such species—at least 100—find breeding zones on shore. Cranes, plovers, herons, bitterns, loons, pelicans, ducks, geese, swans, osprey, falcons, eagles, and owls share the Hudson Bay.

Less is known about the prevalence and range of marine fish species here, due to the near-absence of commercial fisheries; even traditional subsistence fishing has been somewhat limited. Some 60 or so fish species are known in these waters, however, and many of these are adapted to the relatively shallow, brackish environment here. Of the recognized species, about two dozen stay in the marine environment all their lives; half that many species are marine but frequent the estuaries as seasonal nursery grounds; another nine feed in the brackish coastal realm, but spawn in freshwater zones; as many as 16 freshwater species have evolved salinity tolerances sufficient to spend some time here in brackish coastal reaches or estuaries; and one—the Atlantic salmon—winters in saltwater in between the freshwater spawn and final act of its anadromous life cycle.

Conservation and Climate Change

The governance of the Hudson Bay area is largely a series of compromise decisions and shared responsibilities between federal, provincial, local, and First Nations bodies. As global warming proceeds—and faster here in its effects than in many if not most other parts of the world—such organizations will have to work more closely in concert to help determine the growing number of issues. These will range from plans for hydroelectric dams and other water diversion schemes; new deepwater port facilities and related marine transport strategies; mining, agriculture, grazing, and aquaculture rights; tourism pursuits such as hunting; real estate development; and encompassing all of these areas: conservation.

There has already been considerable focus on the impact of climate change and transboundary contaminants on Hudson Bay. Historic commercial whaling, particularly for bowheads and belugas, depleted these populations, some of which have not entirely recovered. The steady increase in regional temperatures over the last 100 years has caused a lengthening of the ice-free period, and continues to jeopardize the sustainability of ice-dependent species such as polar bears and seals. These ice-dependent marine mammals are vulnerable to both airborne contaminants and global warming. Because these carnivorous animals are high in the food chain, they accumulate contaminants such as heavy metals, carcinogenic hydrocarbons, and radioactive materials in their fatty and other tissues.

Diminished sea-ice cover, decreasing flow from its tributary rivers, and increasing precipitation are some of the long-term trends already detected in the Hudson Bay. Each of these vectors is projected to intensify, along with other effects such as sea-level rise, which will have as yet undetermined but certainly serious impacts on the habitats here.

MAGDALENA ARIADNE KIM MUIR

Further Reading

Ferguson, Steve H., Lisa L. Loseto, and Mark L. Mallory, eds. *A Little Less Arctic: Top Predators in the World's Largest Northern Inland Sea, Hudson Bay.* New York: Springer, 2010.

Ritchie, J. C. *Postglacial Vegetation of Canada.* Cambridge, MA: Cambridge University Press, 1987.

Stewart, D. B. and W. L. Lockhart. "An Overview of the Hudson Bay Marine Ecosystem." Fisheries and Oceans Canada, 2005. http://www.dfo-mpo.gc.ca/libraries-bibliotheques/toc-tdm/314704-eng.htm.

Hudson Bay Lowlands

Category: Forest Biomes.
Geographic Location: Canada.
Summary: The Hudson Bay Lowlands biome is the largest wetlands in North America, but the region faces significant changes as a result of isostatic rebound and climate change.

The Hudson Bay lowlands are the largest wetland area in North America, covering a vast expanse of Canada between the Canadian Shield and the southern shores of Hudson and James Bays. Encompassing parts of the Canadian provinces of Manitoba, Ontario, and Quebec, the lowlands are along the ecotone (overlap) between the boreal forest to the south and Arctic tundra to the north. The area faces significant change, however, from the effects of climate change as well as isostatic rebound (rising ground elevation) of a landscape once burdened by the fantastic weight of continental, ice-age glaciers.

The location of Hudson Bay and the lowlands to its southwest and south correspond with the approximate center of the Laurentide Ice Sheet—the massive ice load that covered North America during the Pleistocene period, before it retreated and melted some 20,000 years ago. The massive weight of that ice, which was as much as two miles thick, compressed the ground below, allowing incursion of salt water in areas far beyond the current shores of Hudson Bay. In addition, the ice scoured the ground beneath it, effectively bulldozing or filling in the original drainage network and ensuring, because of that disrupted drainage, that large swaths of the region would remain wet and bogged down after the glaciers retreated.

The landforms left behind include gravel ridges and hills—usually derived from glacial deposits—raised beaches, occasional rocky outcrops, permafrost hummocks, and extensive lowlands filled with glacial or marine sediments. Upland areas are often dominated by boreal forest vegetation, in particular white spruce—one of the dominant species of the North American boreal forest—with balsam fir, quaking aspen, balsam poplar, and paper birch. Important plant associates include dogwood, willow, lingonberry, bearberry, twinflower, miterwort, false toadflax, wintergreen, or mosses such as *Hylocomium* or *Pleurozium*. Stands of black spruce and jack pine may also be important components of the forests of upland sites.

Lowland and Wetland Types

The type of lowland sites are influenced in large part by site conditions. Tundra, dominated by shrubs such as blueberry, lingonberry, bearberry, and lichens such as *Cladonia*, occurs in relatively well-drained sites in a strip along the Hudson and

James Bay coasts. On poorly drained sites, bogs develop under acidic conditions, and fens develop under neutral or alkaline conditions. Both develop in poorly drained areas, where organic matter may accumulate over time. The Hudson Bay Lowlands biome features an abundance of both.

A common type of wetland in the Hudson Bay Lowlands is the muskeg. The nutrient-deficient substrate often supports an open stand of black spruce, but the muskegs are dominated by mosses such as *Sphagnum*, with a handful of shrubs such as Labrador tea, dwarf willow, and dwarf birch. Another type of muskeg is dominated by *Cladonia* lichens, with shrubs such as Labrador tea and blueberry.

While raised bogs appear to be higher in elevation than the surrounding tundra, their elevation results from accumulation of peat in areas that were once lower in elevation than the surrounding landscape. The sites are poorly drained, much like a raised wetland, and as such are dominated by black spruce, a common resident of wetland areas throughout the North American boreal forest. Another type of raised bog, the palsa, features a permafrost core, with the raised elevation the result of action by the ice. Black spruce is again the more noticeable dominant, but palsas have significant cover by Labrador tea, blueberries, and sedges.

Fens are often dominated by tamarack; with dwarf birch and bog willow; mosses such as *Tomenthypynum*, *Depanocladus*, and *Campylium*; sedges; and cottongrass.

In tidal flats and estuaries along the shores of Hudson and James Bays, a type of sedge-dominated marsh may develop. In addition to sedges, grasses such as tundragrass, alkaligrass, pendant-grass, and arrowgrass; and herbs such as primrose, bog star, and felwort are common.

Other vegetation types in the region include the dwarf birch-willow communities that flank marsh flats and river shorelines, and balsam-poplar communities found on alluvial and beach deposits.

Fauna and Change

Several large rivers flow through the Hudson Bay Lowlands into the Hudson or James Bays. Rivers that flow into Hudson Bay include the Churchill,

The Polar Bear Capital

Polar bears come off the ice and into the Hudson Bay lowlands in the late summer and fall, when ice melts on the adjacent Hudson and James Bays. Male bears tend to stay near the coast, on areas such as beach ridges, while females with cubs move inland to riparian, lakeside, and tundra habitats. The high concentration of polar bears in Churchill, Manitoba, has earned this town the title of Polar Bear Capital of the World.

A polar bear hunts for seals beneath the water. Because the Hudson Bay Lowlands biome is warming, the icepack on the Hudson and James Bays now melts earlier and freezes later. (Thinkstock)

Nelson, Hayes, Severn, and Winisk Rivers. Those that flow into James Bay include Attawapiskat, Albany, Moose, Harricana, and Nottaway Rivers. Whales such as beluga may be spotted in the larger river estuaries, such as that of the Churchill River near Churchill, Manitoba.

The Hudson Bay Lowlands biome is undergoing great change. One cause is geological: The Hudson Bay region is undergoing isostatic rebound, in which the landscape, formerly compressed under the great weight of the Laurentide Ice Sheet, is expanding vertically, in effect, getting higher. Vegetation belts shift northward to colonize land exposed by the retreating waters.

The other cause is climatic. Climate change is causing the landscape to warm. Some of the more striking effects are out over the Hudson and James Bays, where the icepack melts earlier and freezes later. But warmer temperatures disrupt the landscape as well, melting permafrost, increasing loss of water from the land surface via evaporation and transpiration, and increasing the risk of wildfire in the region. While wetlands may be the most heavily affected by the change, upland ecosystems are likely also to be adversely affected.

DAVID M. LAWRENCE

Further Reading

Abell, Robin, et al. "Freshwater Ecoregions of the World: A New Map of Biogeographic Units for Freshwater Biodiversity Conservation." *BioScience* 58, no. 5 (2008).

Barbour, Michael G. and William Dwight Billings. *North American Terrestrial Vegetation, 2nd Ed.* Cambridge, UK: Cambridge University Press, 1999.

Larsen, James A. *The Boreal Ecosystem.* New York: Academic Press, 1980.

Ritchie, J. C. "The Vegetation of Northern Manitoba: II. A Prisere on the Hudson Bay Lowlands." *Ecology* 38, no. 3 (1957).

Sjörs, Hugo. "Bogs and Fens in the Hudson Bay Lowlands." *Arctic* 12, no. 1 (1959).

Hudson River

Category: Inland Aquatic Biomes.
Geographic Location: North America.

Summary: A unique river that flows both north and south simultaneously because of the incoming tide from the Atlantic Ocean.

The Hudson River is a very unusual river in that instead of flowing in typical high- to low-gradient fashion, it flows both ways (northward and southward, in this case) simultaneously, earning its name *Muhheakantuck,* the Lenape-Delaware tribal word for *river that flows two ways.* While freshwater flows down the river from the north, saltwater flows up the river from its mouth at the Atlantic Ocean, creating a layering effect of higher-salinity water on top of lower-salinity water or freshwater.

The river provides a unique habitat for its numerous plants and animals, housing both freshwater and saltwater communities. The diverse flora and fauna population also includes more than 100 nonnative species. The Hudson supports a large human community as well, with people using the river for recreation, drinking water, commercial shipping and fishing. Unfortunately, the Hudson is also heavily polluted from human use, but local organizations as well as federal and local governments are working together to decrease the pollution levels and to preserve the river's ecosystem.

Hydrology and Climate

The Hudson River is 315 miles (507 kilometers) in length, beginning in the Adirondack Mountains from Tear of Clouds Lake. The Hudson River Valley lies almost entirely within the state of New York, except for its last 22 miles (35 kilometers), where it serves as the boundary between New York and New Jersey, before it empties through New York Bay and into the Atlantic Ocean. Over 90 percent of the river's basin drains from New York State, but it also drains from Vermont, Massachusetts, Connecticut, and New Jersey. The basin is connected to the Great Lakes and Lake Champlain through many natural and manmade waterways, which allow for many aquatic species to move back and forth between the Hudson and these lakes.

The Hudson River basin experiences long, cold winters and short, warm summers, and has generally warmer temperatures at the southern end.

The average water temperature during the winter months is about 25 degrees F (minus 4 degrees C); the summer range is 70–75 degrees F (21–24 degrees C). Annual precipitation in rainfall for this region is approximately 39 inches (100 centimeters). The mean annual snowfall for the entire Hudson River Basin varies from about 98 inches (250 centimeters) in the northern Hudson Valley, to about 20 inches (50 centimeters) near Manhattan.

The river is divided into three sections by basin drainage: the Upper Hudson, the Mohawk, and the Lower Hudson/Hudson River Estuary. The Upper Hudson contains freshwater and flows south from Mt. Marcy in the Adirondacks to Troy, New York. The Lower Hudson, which is often referred to as the Hudson River Valley, is an estuary and so contains both fresh- and saltwater mixed together, making for a brackish flow.

The 155 miles (250 kilometers) of the river from below the dam at Troy to the Atlantic Ocean flow both south and north simultaneously because of the incoming tide from the ocean and the river's low gradient in this stretch. While the river begins at a high elevation level in the Adirondacks, by the time the Hudson reaches Albany, New York, it has mostly flattened out and only slopes about 5 feet (1.5 meters) during the 155-mile (250-kilometer) stretch to the ocean. The movement of the water in the lower half of the river assists in recycling nutrients throughout the river, creating a rich environment for its inhabitants.

Biota

The Hudson is home to both freshwater and saltwater communities. The upper half of the Hudson is where freshwater communities are located. Diatoms and cyanobacteria are the foundation for the food web in the upper Hudson; they feed the beetle, mayfly, and caddisfly communities. Diatoms and cyanobacteria are also food for the numerous protozoans, rotifers, and copepods, as well as benthic (bottom-dwelling) organisms such as bivalves and amphipods. The copepods and rotifers in turn provide food for larval and juvenile fish.

While the entire Hudson contains more than 200 species of fish, only about 70 of these species live in the Upper Hudson, as the water is colder here. Fish species in the freshwater Hudson include brook trout, dace, shiner, chub, and catfish.

The lower half of the Hudson has a much wider variety of plant and animal life, due to the mixed salinity of the water, as it hosts both freshwater and saltwater creatures. The phytoplankton community consists of diatoms and cyanobacteria, as well as green algae and dinoflagellates. Zooplankton common here include: *Eurytemora affinia*, *Arcartia hudsonica*, and *Temora longicornis*. Other benthic invertebrates in the Lower Hudson are bivalves, amphipods, barnacles, gastropods (snails), crabs, and shrimp. These creatures living in the bottom of the river burrow into the sediment, which accelerates the breakdown of organic matter and assists in recycling nutrients back into the water.

The saline water here is also home to a large population of cordgrass and other submerged aquatic vegetation (SAV) such as green fleece (*Codium fragile*), sea lettuce, and chenille weed. All of this SAV provide food and shelter to many of the creatures in the river, including the large fish population. The warmer waters of the Lower Hudson are home to many different species of fish, including yellow perch, sunfish, bass, and the anadromous types: sturgeon, shad, and eel. Marine fish are also found in the most southern part of the river; these species include the bay anchovy, cunner bluefish, and winter flounder. Occasionally, marine mammals such as the grey seal, harbor seal, dolphin, and whale can be found in the Lower Hudson, as well.

Human Impact

The human population of the river, especially over the last century, has had a profound effect on the river's ecosystem. Since Henry Hudson discovered the river in 1609, it has been a very important transportation route, and is still used for commercial shipping today. Humans began to physically alter the river to ease navigation beginning in the early 19th century, and deforested the basin to make it more hospitable for human use.

During the 20th century, many companies built factories along the river's banks, and would dump raw sewage as well as other pollutants into

Zebra Mussels

The Hudson River is home to about 100 non-native species, with the zebra mussel having the largest effect on the river's ecosystem. Zebra mussels first appeared here in 1991, and were most likely introduced into the river by the human population. The zebra mussels greatly affected the levels of oxygen in the water, as well as the clarity of the water, which had the greatest impact on the phytoplankton and zoo-benthos, killing off large portions of the communities. In turn, the fish in the river had less available food sources.

Currently, the zebra mussel population is at a much lower level, allowing other communities to return close to pre-invasion levels. The lower level of zebra mussels is attributed to predators of this exotic species, as well as the possibility of a disease or parasite among the mussels.

This 2008 photograph shows many small zebra mussels (Dreissena polymorpha) attached to a larger native mussel. (U.S. Fish and Wildlife Service)

soned, and fishing was banned because of the fish being too toxic to ingest. This dumping went unchecked until the 1970s, when the federal Clean Water Act was created. In 1977, PCBs were banned in the United States, and in 1983, the U.S. Environmental Protection Agency (EPA) declared a 200-mile (322-kilometer) stretch of the river, from Hudson Falls to New York City, to be a Superfund site requiring cleanup.

Other ongoing pollution issues that currently affect the river include: urban and agricultural runoff; and heavy metals, dioxin, furans, pesticides and other pollutants. Government agencies, as well as organizations of concerned citizens, are working together to reduce pollution and try to restore the Hudson to its natural state.

Several programs that have been created recently to address the issues of climate change, which are projected to include sea-level rise, more extreme storms, riverbank erosion, earlier spring snowmelt runoff, and higher average air and water temperatures. These effects will put pressure on a range of habitats and species here. The New York Department of Environmental Conservation works with the Hudson River Estuary Program to help communities understand projected impacts of climate change and to coordinate regional responses.

Elizabeth Stellrecht

Further Reading

Benke, Arthur C. and Colbert E. Cushing, eds. *Rivers of North America.* Burlington, MA: Elsevier Academic Press, 2005.

Henshaw, Robert E, ed. *Environmental History of the Hudson River: Human Uses That Changed the Ecology, Ecology that Changed Human Uses.* Albany: State University of New York Press, 2011.

Levinton, Jeffrey S. and John R. Waldman, eds. *The Hudson River Estuary.* New York: Cambridge University Press, 2006.

Yozzo, David J., Jessica L. Andersen, Marco M. Cianciola, William C. Nieder, Daniel E. Miller, Serena Ciparis, and Jean McAvoy. *Ecological Profile of the Hudson River National Estuarine Research Reserve.* Washington, DC: National Oceanic and Atmospheric Administration, 2005.

the river. General Electric manufacturing facilities at Hudson Falls and Fort Edward, New York, discharged between 105 and 650 tons (95 and 590 metric tons) of extremely toxic polychlorinated biphenyls (PCBs) into the river from 1947 to 1977. Many of the river's inhabitants were severely poi-

Huetar Norte Plantation Forests

Category: Forest Biomes.
Geographic Location: Central America.
Summary: These vast plantation forests in Costa Rica occupy a matrix of rainforest remnants and agricultural land.

In extremely humid conditions toward the north end of Costa Rica and continuing to the Nicaraguan border, very distinct forest ecosystems develop. In many ways, the Huetar Norte region is a typical example of tropical rainforest being replaced by other land covers because of human activities. In the past few decades, cattle ranching converted large expanses of rainforest to pasture lands, although recently this activity has become less profitable. Thus, landowners have favored reforestation to supply the wood market, creating vast extensions of forest plantations.

Today, the landscape is a combination of rainforest fragments, reforestation areas, pasture, abandoned pasture, and agricultural land. Uniformly distributed trees of the same species certainly contrast with the chaotic rainforest that used to dominate the area, creating the attractive and unique picture that defines the Huetar Norte region.

Compromising approximately 29,344 square miles (76,000 square kilometers), or about 14 percent of Costa Rica's surface, the Huetar Norte region includes six ecozones and is an economically important area for this small country. The name *Huetar* is derived from the languages of the original indigenous groups that used to inhabit the central part of the country and a small part of the area known today as the Huetar Norte region. Though the Huetares ethnic group was among the strongest and most organized indigenous group when the Spanish came, their culture practically disappeared, and the adoption of the name for the region is mainly to commemorate the indigenous past.

The economic value of the region comes mostly from livestock grazing and wood extraction. During the late 1990s, it was estimated that the region provided more than 40 percent of the wood used within Costa Rica. The wood supply came partly from logging of natural forest, but most of it was provided by the vast forest plantations, an industry that exploded in the area after 1988—the year in which a tax reduction was approved for land that was undergoing reforestation processes.

Biodiversity

Besides its importance in wood production, in the past two decades the region has been increasingly recognized as an area of high conservation value. In fact, the area is so diverse that the highest tree-biodiversity index for Costa Rica—110 species per 2.5 acres (1 hectare)—has been reported for the Huetar Norte region. From inundated flat lowland up to high mountain areas, unique plant and animal species here have developed within a matrix of native forest, plantation forest, pasture, abandoned pasture, and agricultural lands. This spread of land uses was originally covered mainly by lowland tropical rainforest and very humid tropical premontane forest, extending in elevation from 131 feet (40 meters) in the lowlands to 1,640 feet (500 meters) in the premontane areas.

Joining the main plantation forest species (see section below) are such naturally occurring trees, shrubs, flowering plants, and understory types as palm, magnolia, orchids, begonias, epiphytes, ferns, and such iconic standards as the passion flower (*Passiflora tica*) and ramonean heliconia (*Heliconia ramonensis*).

Among the endangered animal species of the Huetar Norte, predators include puma, jaguar, and cougar. Larger mammals include peccary, Baird's tapir (*Tapirus bairdii*), and three-toed sloth (*Bradypus variegatus*). The area also offers habitat for various monkey species, such as the white-headed capuchin (*Cebus capucinus*) and mantled howler (*Alouatta palliata*). Reptiles are abundant here, in the form of iguana, basilisk, coati, and snake.

Estimates of the number of bird species in the biome range as high as 200 or more; among the more recognizable are the clay-colored thrush (*Turdus grayi*), Montezuma's oropendola (*Psarocolius montezuma*), and black guan (*Chamaepetus unicolor*). These are joined by many parakeets, woodpeckers, toucans, tanagers, and raptors.

Conservation Efforts

This biome has undergone significant depletion of its forest cover and now is a highly fragmented landscape. Significant efforts have been invested to recover connectivity between forest areas in the Huetar Norte region, to promote sustainable development for its inhabitants and to protect animal species such as the great green macaw (*Ara ambiguus*). The best-known initiative is the San Juan-La Selva Biological Corridor. Among the conservation achievements of this corridor is the establishment of the Refugio Nacional de Vida Silvestre Mixto Maquenque, created to halt the disappearance of the very few great green macaws left, and to protect their feeding and nesting sites: the almedro de montana trees.

Another 14 protected areas exist within the Huetar Norte and seek to safeguard the natural resources of the region. Plantation forests have been agents for reestablishing connectivity between these protected areas and other forest remnants, as well as for recovering degraded soils. Plantations of teak (*Tectona grandis*), gmelina (*Gmelina arborea*), chancho (*Vochysia guatemalensis*), laurel (*Cordia alliodora*), almendro (*Terminalia amazonia*), pines (*Pinus* spp.), cipres (*Cupressus lusitanica*), eucalyptus (*Eucalyptus* spp.), and others have been established on previous pasture lands. Even though their original purpose was commercial, these plantations have played an important role in helping regenerate previous pasture lands and creating hospitable environment for wildlife.

Though the ecological services of plantations—including wood production, carbon sequestration, soil protection and conservation, and support for natural regeneration processes—are widely recognized, their use is controversial. In Huetar Norte, as well as in many other areas in the world, there is concern about the establishment of pure plantations of exotic species such as teak and gmelina. The use of mixed plantations of native species is broadly recommended in areas of biological importance such as the Huetar Norte, but pure plantations are preferred over mixed ones by industrialists.

Forest cover and connectivity are also being attempted through Payment for Environmental Services (PES) by the Huetar Norte Forestry Program, which was established more than a decade ago. During 2000–04, around 185,330 acres (75,000 hectares) came under a PES contract. Through the Huetar Norte Forestry Program, thousands of acres (hectares) are being protected from deforestation or are being reforested through plantations. In addition, more than 700 families in the region have benefited from these payments, making plantation forests an economic tool for supporting stability in the area. Hence, it is expected that plantation forests will keep increasing somewhat, or at least be sustained at current levels in the near-future, making these extensive green carpets distinctive elements of the landscape of northern Costa Rica.

However, like other regions of Costa Rica, the Huetar Norte is likely to be impacted by global warming in the form of trends toward warmer, drier weather patterns. Such trends are anticipated to be even more pronounced at the higher, premontane elevations within this biome. At the same time, decreased precipitation in the lowlands is projected to cause managers of some plantation forests to replant further upslope. Such a movement will impact the many integrated flora and fauna within the habitat to migrate to higher elevations. Something of a squeeze will be exerted upon these species, caught between drier lowlands and warmer uplands. The capacity for some species to adapt to quickly changing climate here will remain in question.

LUCIA MORALES-BARQUERO

Further Reading

Gardner, Toby A., et al. "The Value of Primary, Secondary, and Plantation Forests for a Neotropical Herpetofauna." *Conservation Biology* 21, no. 3 (2007).

Montagnini, F. and C. F. Jordan. *Tropical Forest Ecology: The Basis for Conservation and Management.* New York: Springer, 2005.

Villate, Rodrigo, Lindsay Canet-Desanti, Oliver Chassot, and Guisselle Monge-Arias. *El Corredor Biológico San Juan-La Selva: Una Estrategia Exitosa de Conservación (The San Juan–La Selva Biological Corridor: A Successful Conservation Strategy).* San José, Costa Rica: Comité Ejecutivo del Corredor Biológico San Juan-La Selva, 2008.

Humber Estuary

Category: Marine and Oceanic Biomes.
Geographic Location: Europe.
Summary: The Humber is the largest coastal-plain estuary on the east coast of Britain; it supports significant wetland and coastal habitats.

The Humber Estuary provides the outlet for the Ouse and Trent rivers, which together drain one-fifth of the land surface of England. This catchment of more than 9,266 square miles (24,000 square kilometers) provides the largest single supply of freshwater from Britain into the North Sea. The high levels of suspended sediment within its waters feed a highly dynamic geomorphological system with accretion and erosion of intertidal and subtidal areas.

As the largest coastal-plain estuary on England's east coast, the Humber is also one of the longest in Britain, stretching 50 miles (80 kilometers) from Goole to Spurn Head at its mouth. It is 9 miles (14 kilometers) across at its widest point. Due to its position within the North Sea basin, it is macrotidal, with the second-highest tidal range in Britain, after the Severn Estuary. It has a mean spring tidal range of 18.7 feet (5.7 meters) at Spurn Head, which increases up the estuary to more than 23 feet (7 meters) at locations around 28 miles (45 kilometers) inland.

The important habitats within the estuary—such as mudflats, sand flats, and salt marsh—are maintained by the copious amounts of sediment brought in with the tide and carried down the Ouse and Trent rivers. An estimated 1,654 tons (1,500 metric tons) of clay and silt are carried in with every tide from the eroding shores outside the estuary, and more than 1.1 million tons (1 million metric tons) of material is estimated to be present within the water column at any time. This sediment is constantly deposited, worn away, and redistributed within the estuary as part of every tidal cycle. The deposited sediments provide material that is essential to maintain the estuary's important intertidal and subtidal habitats, which are highly dynamic in response to physical and biological variables.

One of the key geographical features of the estuary is the long, curling spit of Spurn Head, which extends from the northern bank of the estuary where it meets the North Sea. This headland is nearly 3 miles (5 kilometers) long but less than 164 feet (50 meters) wide in places. It is made up of sand and shingle eroded from the coastline farther north, which is washed southward by longshore drift to accumulate as a long embankment inside the mouth of the estuary. The spit is occasionally breached by the sea, but then reforms through further deposition. In its lee forms a sheltered area of mudflats, which are used by large numbers of wading birds. The headland itself also forms a major navigational landmark for thousands of migrating birds.

Biodiversity

Within the estuary are significant wetland and coastal habitats. Nearly one-third of the estuary is exposed at low tide, with sand flats in the outer estuary and mudflats further inland. The inner estuary supports substantial areas of reed bed, but areas of salt marsh are limited in extent due to land-use changes over the past 300 years. Where salt marsh is present, it is backed by grazing marsh in the middle and outer estuary; on the south coast, it is backed by low sand dunes with marshy slacks and brackish pools. Beyond these transitional zones, the estuary is bordered mostly by grazing and arable land.

Breeding, wintering, and migrating birds make extensive use of the Humber Estuary, which is recognized as one of the most important such areas in Europe, with nine species of international importance present. There are breeding populations of bittern, marsh harrier, avocet, and little tern. Important numbers of waterbirds, especially geese, ducks, and waders, are supported during the migration periods and in winter. These include species such as the bar-tailed godwit, golden plover, dunlin, knot, bittern, and shelduck. The area regularly supports more than 187,000 individual waterfowl.

Effects of Human Activity

The estuary has two large ports at Hull and Grimsby, and has an average of 40,000 ship movements per year. Its ports make up the United Kingdom's largest harbor complex, transporting 14 percent of the country's overseas trade. Other industries along the estuary include chemical works, oil refineries, and power stations.

This human use of the estuary has involved processes such as dredging, land drainage, and flood management, which have affected the estuary's ecology over the centuries. Approximately 222,400 acres (90,000 hectares) of land around the estuary exists due to historical land reclamation, and, as such, it is below the spring high-tide level, currently protected by 146 miles (235 kilometers) of coastal defenses. In the future, these areas will be at risk from the changing conditions caused by sea-level rise, coastal erosion, and climate change.

Conservation Efforts

Proactive management will be required to protect the estuary and the surrounding area, including the creation of space to enable managed retreat in response to sea-level rise; maintenance of the sediment budget within the estuary; acceptance of dynamic change within the estuary system; maintenance of a range of habitat types, together with healthy populations of the associated species; and enhancement of water quality.

Recent projects on the estuary, at Paull Holme Strays and Alkborough Flats, have indicated how the estuary might be managed in the future. At these two sites, seawalls have been moved back, and more than 1,236 acres (500 hectares) of land have been allowed to flood to create new intertidal areas. This managed realignment will lessen the chance of flooding elsewhere along the estuary shore, preventing damage to homes and businesses, and creating valuable habitat for birds and other wildlife.

CARLOS ABRAHAMS

Further Reading

Allen, J., et al. *The Humber Estuary: A Comprehensive Review of its Nature Conservation Interest.* Hull, UK: Institute of Estuarine and Coastal Studies, 2003.

Fujii, T. "Spatial Patterns of Benthic Macrofauna in Relation to Environmental Variables in an Intertidal Habitat in the Humber Estuary, UK: Developing a Tool for Estuarine Shoreline Management." *Estuarine, Coastal and Shelf Science* 75, nos. 1–2 (2007).

Humber Management Scheme. "Humber Estuary European Marine Site." 2012. http://www.humber ems.co.uk.

Natural England. *National Character Area Profile: Humber Estuary.* Sheffield, UK: Natural England, Government of the United Kingdom, 2012.

Pollard, Michael. *Great Rivers of Britain.* London: Evans Brothers, 2002.

Huntsman Valley (Tasmania) Plantation Forests

Category: Forest Biomes.
Geographic Location: Tasmania.
Summary: The Huntsman Valley plantation forests are a human-made monoculture with a formerly biodiverse ecosystem being replaced by one designed to produce pine trees as the sole forest crop.

European settlement on the Australian island state of Tasmania has had disastrous effects on its forests. While the Tasmanian Aborigines who settled here had long modified the forests through their use of fire-stick farming to maximize their grazing and hunting outputs, European colonization resulted in widespread extinctions; deforestation; transformation of the coastal ecoregions by greatly increased grazing; and the loss of about a third of the forest to agriculture, urban settlements, and intensive forestry. Today in the Huntsman Valley here, the older forests have been thoroughly replaced by pine plantations.

Though it currently is isolated from mainland Australia, Tasmania was connected to the mainland by the Bass Strait land bridge off and on during the Pleistocene, as sea levels fluctuated following glaciations. Although the land bridge was arid during most of that time, and would have been inhospitable to many species, nevertheless, Tasmania has been fully isolated for only the past 13,000 years, and there are strong connectivities and congruences between the ecosystems of Tasmanian forests and those of the mainland. These connections are strongest in the parts of Tasmania nearest the Bass Strait, such as the Huntsman Valley.

The Tasmanian temperate forest ecoregion is a transition between the dry mainland and cooler, more humid western Tasmania. The greatest endemism (species found exclusively in this biome) is found in the forest understory, though much of this endemism was jeopardized by the conversion of the Huntsman Valley to plantation forests. These plantations are monocultures—forests managed to produce a single crop, in this case pine trees. The king billy pines, white gums, and myrtles that once grew within the Huntsman Valley plantation forests have all been replaced.

Monoculture Controversy

Monoculture forestry provides more effective growth and greater yields than more-diverse forests do; human-made monocultures stress uniformity, and lack the diversity of tree sizes that occurs in natural monocultures. They also deprive the greater ecosystem of the niches provided by dead trees and meadow-like openings. Also, because the trees are all the same size, they are readily harvested by clear-cutting, which dramatically affects the habitat.

Mechanical harvesting compacts soils, destroying much of the remaining understory ecosystem. Opponents of monocultures claim that they are also ideal breeding grounds for pests and disease, because the ecosystem lacks many pests' natural predators or the appropriate, multifaceted defenses against disease. To encourage replanting, each new crop of trees in a monoculture plantation is government-subsidized, attracting more criticism.

The photograph shows a wide swath of deforested land, likely caused by illegal clear-cutting, along the banks of a winding river in Tasmania. About a third of Tasmania's forest has been destroyed in the aftermath of European colonization, mostly through agriculture, human settlement, and intensive forestry. (Thinkstock)

On the other hand, the use of plantation forests can in theory benefit overall biodiversity, by limiting logging activities to those plantations and leaving other forests alone—much like farming in specific controlled areas, rather than destroying numerous ecosystems by attempting to produce the same yields through foraging.

Wildlife

The Tasmanian devil (*Sarcophilis harrisii*) and the Tasmanian thylacine (*Thylacinus cynocephalus*) were once widespread in mainland Australia, but became extinct there long before the arrival of Europeans, most likely through competition with the dingo. The dingo is not found in Tasmania, where both of these species survived. The Tasmanian thylacine, the largest marsupial carnivore, can be found throughout Tasmania's temperate forests.

Other mammals of note are the duck-billed platypus (*Ornithorhynchus anatimus*), echidna (*Tachyglossus aculeatus*), long-nosed potoroo (*Potorous tridactylus*), spotted-tail quoll (*Dasyurus maculatus*), eastern quoll (*D. viverinus*), red-necked wallaby (*Macropus ruogriseus*), and wombat (*Vombatus ursinus tasmaniensis*), all of which are widespread throughout Tasmania.

Bird populations here include two vulnerable species: the swift parrot (*Lathamus discolor*) and the forty-spotted pardalote (*Pardalotus quadragintus*). The Tasmanian native hen (*Gallinula mortierii*), black-headed honeyeater (*Melithreptus affinis*), and yellow wattlebird (*Anthochaera paradoxa*) are found in few other places.

Several lizard species live on the forest floor, including the mountain dragon (*Tympanocryptis diemensis*). Rawlinson's window-eyed skink (*Pseudemoia rawlinsoni*) is endemic to Tasmania's temperate forests. The Tasmanian tree frog (*Litoria burrowsi*) and the Tasmanian froglet (*Crinia tasmaniensis*) are both endemic.

BILL KTE'PI

Further Reading

Chen, Henry C. L. and Pete Hay. "Defending Island Ecologies: Environmental Campaigns in Tasmania and Taiwan." *Journal of Developing Societies*, 22, no. 3 (2006).

Reid, J. B., R. S. Hill, M. J. Brown, and M. J. Hovenden. *Vegetation of Tasmania*. New York: CSIRO Publishing, 1998.

Williams, W. D., ed. *Biogeography and Ecology in Tasmania*. New York: Springer, 1974.

Huron, Lake

Category: Inland Aquatic Biomes.
Geographic Location: North America.
Summary: Lake Huron, the second-largest of the Great Lakes, has seen dramatic changes in its fish species mix, and climate change could accelerate the process.

Lake Huron, in terms of surface area, is the second-largest of the Great Lakes of North America, and the fourth-largest freshwater lake in the world. Lake Huron has a surface area of approximately 23,000 square miles (60,000 square kilometers). The lake contains a volume of 850 cubic miles (3,540 cubic kilometers). Lake Huron has the longest shoreline of any of the North American Great Lakes, at 3,827 miles (6,157 kilometers), including the shorelines of its islands, which number more than 30,000. The surface of Lake Huron is situated 577 feet (176 meters) above sea level.

The average depth of the lake is 195 feet (59 meters); its greatest depth is 750 feet (229 meters). Lake Huron is 206 miles (332 kilometers) in length, and it has a breadth of 183 miles (295 kilometers) at its greatest width. The total area of the Lake Huron drainage basin is 51,700 square miles (133,900 square kilometers).

Regional Geography

Along the northeast end of Lake Huron is Georgian Bay, separated from the main part of Lake Huron by Bruce Peninsula and Manitoulin Island. Georgian Bay is so large that it is sometimes referred to as the sixth Great Lake; it is bordered by the Canadian province of Ontario. The shoreline of

Georgian Bay is strikingly similar to the shore-line of Lake Superior, possessing rugged, rocky cliffs above relatively unspoiled waters. Manitoulin Island is the largest freshwater lake island in the world, and in turn contains its own lakes. A smaller bay, Saginaw Bay, extends southwest from Lake Huron into the state of Michigan.

At the northwest corner of Lake Huron are the Straits of Mackinac, a deep trench of water that connects Lake Michigan with Lake Huron. The Straits of Mackinac equalize the water levels of the two Great Lakes, making them essentially two parts of the same lake, in a geological sense as well as hydrologically. Because of these connections between Lakes Michigan and Lake Huron, the two are sometimes collectively referred to as Lake Michigan-Huron. Taken together in this way, Lake Michigan-Huron, with a surface area of 45,300 square miles (117,000 square kilometers), qualifies as the world's largest freshwater lake by area.

Lake Superior is at a slightly greater elevation than both Lake Michigan and Lake Huron. Lake Superior drains into the St. Marys River at Sault Ste. Marie; the St. Marys then empties into Lake Huron. Lake Huron itself empties into the St. Clair River at its southernmost point; the St. Clair flows into Lake St. Clair, a very shallow lake with a surface area of 430 square miles (1,114 square kilometers), which in turn sends water south into Lake Erie.

Lake Huron, as were the other Great Lakes, was formed by the melting and retreating of continental glaciers at the end of the last ice age. Before the last glaciation event, Lake Huron emptied into what is currently the Ottawa River Valley and then into the St. Lawrence River, close to present-day Montreal. Lake Huron's drainage shifted to the south when the land surface rebounded with the retreat of the glaciers. In some areas around Lake Huron, the land surface is still rising by approximately 13 inches (350 millimeters) per century.

When the first French explorers saw Lake Huron, they named it *La Mer Douce,* which means *sweetwater sea.* A map published in 1656 called the lake Karegnondi, an indigenous Wendat word that is translated generally as *Freshwater Sea,* or *Lake of the Hurons,* referring to the indigenous Huron people who lived along the shores. The Huron had established trading networks across the region, and the French fur trade accelerated the founding of settlements at many points around the lake.

Ecology and Threats

The ecology of Lake Huron has been subjected to numerous major changes and disturbances during the last 100 years. The lake formerly was home to a robust deepwater fish community, with the lake trout being the dominant top predator. The lake trout consumed cisco, sculpin, and other native prey. However, invasive fish species, such as alewife, rainbow smelt, and the parasitic sea lamprey, became quite numerous in the lake during the early 20th century, arriving via shipping lanes from the Atlantic Ocean via the St. Lawrence River. Already weakened by overfishing, then devastated by the sea lamprey, the lake trout population disappeared in the lake by 1950. Except for the bloater, all native species of deepwater cisco also were extirpated here.

Since the 1960s, the Pacific salmon, which is exotic to the Great Lakes region, has been stocked in Lake Huron. Lake trout have also been the focus of restocking programs, in an attempt to restore this species and rebuild its population in the lake, but a significant amount of natural reproduction of lake trout in Lake Huron has not been observed since the restocking and release programs have been initiated.

More recently, other nonnative species have been introduced into Lake Huron and the other Great Lakes; these new invaders are having devastating impacts on the regional ecology. The invasive species include the zebra and quagga mussels, the round goby, and the spiny water flea. The deepwater fish community of Lake Huron was on the verge of collapse by 2006; there have also been a number of documented changes to the zooplankton community here. Catches of chinook salmon have declined precipitously in recent years, and lake whitefish are less abundant and generally in poor condition. Most or all of these negative, recent changes in the ecology of Lake Huron may be due in large part to these newer invasive species disrupting the structure and function of the native aquatic communities.

Lake Huron has many wetlands that supply nesting and gathering areas for 30 species of wading and shoreline birds, and 27 species of ducks, geese, and swans. Enormous numbers of birds visit these habitats during migrations through the region. Large numbers of fens and bogs, together with other wetland types, sustain diverse plant and animal communities throughout the shoreline areas.

Most of the more than 30,000 islands in the lake are relatively undisturbed habitats, some hosting rare species of plants and insects. Saginaw Bay is the largest freshwater coastal wetland in the United States, at 1,143 square miles (2,961 square kilometers). A large proportion of Lake Huron's fish species use this wetland during their growth and development periods.

Two of the unique ecosystems of Lake Huron are the alvars and the Pinery, an Ontario Provincial Park. Alvars are extremely rare hostile environments, consisting of exposed limestone bedrock and very thin, poorly drained soils. Because temperatures fluctuate dramatically in alvars, these ecosystems are population by species of very rare, specifically adapted plants and lichens. A few species of conifers live on alvars, and some of these individual trees are among the oldest living trees in the Great Lakes region. Alvars are fragile and easily disturbed by even minimal human activities.

The Pinery Provincial Park, on the shoreline in southern Ontario, includes a black oak savanna, a type of ecosystem that combines sand dunes and oak meadows. Oak savanna ecosystems are extremely rare; one other such habitat in the region can be found at the western end of Lake Erie, in the Oak Openings Toledo Metropark.

The Pinery offers protection to the sand dunes, meadows, and oak trees; it is critical habitat for the five-lined skink, the only known lizard species in Ontario, and the bluehearts flower, a plant that is extinct everywhere with the exception of the Pinery. In total, 300 species of birds and 700 species of plants have been identified and recorded in this unusual, endangered ecosystem.

Climate change effects are in evidence in Lake Huron and the other Great Lakes. Higher average water temperatures, already sustained for years, are projected to lead to mostly ice-free winters in the future, an extreme departure from seasonal patterns that have built and integrated many habitat niches here. Another key natural cycle, that of stratification of water levels during summer, has been occurring two weeks earlier in the year than during the mid-20th century average time of onset. This, too, is pressuring the nutritive load and food web vigor in the lakes—and could lead to a spread of oxygen-depleted zones.

The overall water volume seems to be declining as well, likely due to faster rates of evaporation caused by higher air and water surface temperatures. Some scientists predict an increase in such fish species as smallmouth bass and yellow perch here—at the expense of lake trout and other species that favor colder water.

Daniel M. Pavuk

Further Reading

Beckett, Harry. *Lake Huron.* Vero Beach, FL: Rourke Corporation, 1999.

Grady, Wayne, Bruce M. Littlejohn, and Emily S. Damstra. *The Great Lakes: The Natural History of a Changing Region.* Vancouver, Canada: Greystone Books, 2011.

Ylvisaker, Anne. *Lake Huron.* Mankato, MN: Capstone Press, 2004.

Iberian Conifer Forests

Category: Forest Biomes.
Geographic Location: Europe.
Summary: This outstanding, diverse set of forest communities is distributed across the whole Iberian Peninsula, from the Mediterranean coastline to the high mountain areas.

Because of its location in southwestern Europe and its mountainous topography, the Iberian Peninsula of Portugal and Spain has a diverse gradient of climatic conditions that becomes evident in three main biogeographical domains. First, the Mediterranean bioregion is the largest in the Iberian Peninsula, and covers about three-quarters of its territory from the northeast (Catalonia) to southwest (Andalusia and south-central Portugal). Second, the Euro-siberian or Atlantic bioregion extends from northwestern Galicia and northern Portugal to the north-central Iberian Peninsula (Cantabrian Mountains and Euskadi). Finally, the Boreo-alpine bioregion occupies a reduced territory in the northeastern Iberian Peninsula, mostly in the area of the Pyrenees, and in a scattered manner in the summit areas of the Iberian and Central mountain systems and the Sierra Nevada.

Conifer forests occur in all three of these bioregions, which are also conditioned by human intervention, especially land-use changes linked to forestry. Examples of this influence are reforestation plans, promoted in the Iberian Peninsula since the 1950s to control soil erosion; water-flow management; and timber production in which conifers, especially pines, have been widely used.

In the Mediterranean bioregion, general bioclimatic conditions are characterized by mild temperatures, which average 61 degrees F (16 degrees C) annually, and seasonal rainfall that averages less than 24 inches (600 millimeters) annually. Rainfall occurs mostly during the spring and autumn with a typical summer drought period.

Flora

The Iberian conifer forests are dominated by thermophilous species that have adapted to low water and poor soils. In addition, they show outstanding resprouting capacity as an adaptation mechanism to wildfires.

The most extended conifer is Aleppo pine (*Pinus halepensis*), which covers large areas of the eastern Iberian coastline and the Ebro River valley. It is the most extended conifer in the whole Mediterranean basin, due mainly to its high colonization capacity.

677

Two other important pines are the maritime pine (*Pinus pinaster*) and the stone pine (*Pinus pinea*); the first appears on the Mediterranean coastline, while the second occupies the northwestern and central Iberian Peninsula.

Junipers are present from the coast to about 3,281 feet (1,000 meters). Cade juniper (*Juniperus oxycedrus*) and Phoenician juniper (*Juniperus phoenicea*) appear in basal areas of the central and eastern Iberian Peninsula, while Spanish juniper (*Juniperus thurifera*) occupies higher areas in central Spain. Finally, Mediterranean cypress (*Cupressus sempervirens*) has relict populations; most of the time, it has been planted as an ornamental tree. In certain coastal environments of the Murcia region, characterized by water scarcity, sparse relict stands of barbary thuja (*Tretraclinis articulata*) remain.

In mountain areas from approximately 3,281–5,906 feet (1,000–1,800 meters), Mediterranean bioclimatic conditions are replaced by Eurosiberian conditions, similar to those present in central Europe. This area is characterized by more abundant rainfall with an annual average of 39 inches (1,000 millimeters), and cooler temperatures averaging 50–53 degrees F (10–12 degrees C).

Two main conifer species form broad forest stands: Scotch pine (*Pinus sylvestris*) and European black pine (*Pinus nigra*). Scotch pine occupies large areas of the northeastern Iberian Peninsula, especially in the Pyrenees, Iberian, and central mountain systems; it has also been widely planted in the Cantabric Mountains and in the Sierra Nevada. Nevertheless, due to its high tolerance to a wide range of environmental conditions, and to recent land-use changes such as the abandonment of traditional grazing, this conifer is increasing its extension. European black pine appears in the Pyrenees south of the Iberian mountain system, and in the highest Mediterranean coastal mountains; it has also been planted in northern and central Spain. At this altitude, common juniper (*Juniperus communis*) occurs in the understory of northeastern evergreen holm oak (*Quercus ilex*) forests. Yew (*Taxus baccata*) forms sparse stands predominantly mixed with beech (*Fagus sylvatica*) forests on shady and humid slopes, mainly in the Cantabric Mountains.

High-mountain conifer forests include those geographically limited to the highest ranges from about 5,906 feet (1,800 meters) to the timberline at about 7,546 feet (2,300 meters), just below the summit areas occupied by alpine grasslands where dwarf juniper (*Juniperus nana*) sometimes appears. At that altitude, bioclimatic conditions are Boreoalpine, characterized by cold temperatures year-round below 50 degrees F (10 degrees C), high rainfall of up to 59 inches (1,500 millimeters) per year, and abundant snowfall during the winter months.

Three conifer forest communities are represented at this stage. First, European silver fir (*Abies alba*) forests occur on cold and shady slopes of the central and western Pyrenees, and also in two little relict areas in the Guara and Montseny massifs, which represent the southern stands of this conifer in its whole European distribution range. Another singular forest, dominated by Spanish fir (*Abies pinsapo*), occurs in a small area of southern Spain. This species represents a flora relict from cold periods in three mountain systems: Sierra de las Nieves, Grazalema, and Sierra Bermeja. Finally, mountain pine (*Pinus mugo* ssp. *uncinata*) forests reach the highest altitudes in the Iberian forests. Its geographical distribution is limited to areas up to 5,577 feet (1,700 meters), predominantly in the Pyrenees and to a lesser extent in the Iberian mountain system.

Exotic conifer plantations have been increasing in mountain landscapes, mainly in the northern Iberian Peninsula, because of their high timber production capacity. The most commonly used species are Monterrey pine (*Pinus radiata*) and Douglas fir (*Pseudotsuga menziessi*).

Fauna

The Iberian Conifer Forests biome offers a wealth of diversity, especially among bird species, of which at least 150 are present. These include the endangered griffon vulture (*Gyps fulvus*), the Spanish imperial eagle (*Aquila heliaca adalberti*), and many others. Mammals common to the area include several endangered species, such as the Gredos ibex (*Capra pyrenaica victoriae*) and the wolf. Red and roe deer are quite common in the forest areas.

Threats

Modern threats to this biome are mainly from overuse, such as clear-cutting trees for the lumber industry; road construction; and, more recently, tourism. Climate change effects predicted for the Iberian conifer forests include changes in the distribution of various plant and animal types; certain species may be forced to migrate to find suitable habitat at different altitudes, and latitudes, than they previously occupied, due to altered temperature, moisture, and fire regimes. Changes in the onset timing of seasons is also a concern. Species adapted to warm, Mediterranean climate areas, for example, may some day populate areas that are currently too cool to accommodate them—which will apply stress to today's inhabitants of those zones.

F. JAVIER GÓMEZ

Further Reading

Andersson, F. A., ed. *Coniferous Forests. Ecosystems of the World,* Vol. 6. Amsterdam: Elsevier Science, 2005.

Benito Garzon, Marta, Rut Sanchez de Dios, and Helios Sainz Ollero. "Effects of Climate Change on the Distribution of Iberian Tree Species." *Applied Vegetation Science,* February 6, 2008. http://www.uam.es/proyectosinv/Mclim/pdf/MBenito_AVS.pdf.

Eckenwalder, James E. *Conifers of the World: The Complete Reference.* Portland, OR: Timber Press, 2009.

Farjon, Aljos. *A Handbook of the World's Conifers, Vols. 1 and 2.* Boston: Brill Academic Publishers, 2010.

Iceland Boreal Birch Forests and Alpine Tundra

Category: Forest Biomes; Grassland, Tundra, and Human Biomes.
Geographic Location: North Atlantic Ocean.

Summary: The sparse vegetative cover across most of Iceland, along with increasing concerns about soil erosion, put sustainable human practices under increased scrutiny here.

The Republic of Iceland is an island nation located in the mid-subarctic region of the North Atlantic Ocean, just south of the Arctic Circle. With a total land area of 64,089 square miles (103,125 square kilometers), it is the second-largest European island, and the 18th-largest island in the world.

Iceland's predominant biome is tundra; the country lies within a subarctic ecoregion of boreal mountain birch forest and alpine tundra that is unique to the countries of northwestern Europe, also including Scandinavia and Greenland. The climate here is temperate and is considered marginal, as differences of only a degree can trigger irreversible losses in vegetation and topsoil. The island phytogeography is sparse, reflecting both its isolation from the larger land masses of North America and Europe, and the ravages of past periods of intense glaciation. Summers and winters are chilly and damp, with extremes of sunlight and darkness.

Icelandic horses grazing in a meadow in Olafsvik on the Snaefellsnes peninsula in Iceland. The nearly 700,000 ewes and 53,000 horses that graze on the land threaten Iceland's sensitive soils and vegetation. (Thinkstock)

Flora

Only about 25 percent of the land on Iceland is covered with vegetation; several hundred species are found, half of which can be dated to the last ice age. There is evidence to suggest that Iceland once supported broadleaf and conifer forests; however, with successive periods of glaciation beginning with the Pleistocene era, primitive forests passed to extinction and were replaced by expanses of downy birches (*Betula pubescens*). Pockets of mixed forests survived in nunatak mountain areas, the ice-free rocky locations that jutted out from early glacial surroundings to the present day. Iceland is noted for sparse land cover that includes ferns, club mosses, willows, herbs with free petals, gentians, daisies, monocotyledons, and water plants.

In the 9th century, a widespread volcanic eruption in the southern region of Iceland left a thick blanket of ash that covered much of the country. This deposit, called *tephra*, created a baseline from which the ecosystem impacts can be measured by archaeologists. Before European colonization, nearly 60 percent of the island was covered in forests of dwarf birch (*Betula pubescens*), mosses, bogs, heaths, and wetlands of varied grasses and sedges. Willow (*Salix* spp.), poplar (*Populus* spp.), and ash (*Sorbus acuparia*) were also native at that time. Today, just 28 percent of the total land area of Iceland is covered with vegetation, with only approximately 1 percent covered with woody species.

Fauna

The arctic fox (*Alopex lagopus*) is the island nation's only indigenous land mammal. Thousands of marine mammals, however, regularly find refuge along the coastal fjords and cliffs. Atlantic salmon and other pelagic fish are a fundamental component of the food web here.

Iceland's location in the mid-subarctic region of the North Atlantic Ocean provides an ideal location for migrating birds. Its stark vegetation and chilly summer climate is not conducive to extensive speciation, but the southern and western regions are known for remarkably mild winters, and Iceland's extensive cliffs and coastlines are important rookeries for European and North American birds.

Today, the European grey heron migration from Norway is an attraction for tourists. Cliff birds include the auk, puffins, guillemots, kittiwakes, fulmars, gannets, and sea eagles; further inland, marshes and heaths provide sufficient habitats for various species of gulls and loons. Inland freshwater lakes are populated by a variety of dabbling and diving duck species including eiders, longtails, harlequins, and the common scoter. Stilts and waders can be found in the marshes and lowland regions.

Threats

The ecological disruption caused by wide-scale clearing of the land is of great concern to Icelanders. Early settlers used fire as a primary agent that made the transformation to agriculture and pastureland possible in less than 50 years. Trees were also burned for charcoal, a substance used extensively in iron smithing and the whetting of scythes for the harvesting of hay.

Attempts to reforest the island have included the introduction of the nootka lupine and lyme grasses—with only marginal success. Continuing erosion of degraded landscapes is directly correlated to the impacts of Viking settlement, with particular periods of degradation noted during the 15th and 17th centuries. Contemporary models of climate change suggest that at the time of the Viking migration, the island was undergoing a climate-induced process of erosion and land degradation that was accelerated by grazing and farming. Today, nearly 700,000 ewes and 53,000 horses graze on sensitive soils and vegetation.

As a member of the European Economic Area (EEA), Iceland has in recent years strengthened its management of land resources and environmental planning, in accord with EEA standards. Key industries include fisheries, hydropower, geothermal energy, and ecotourism; each is being examined with an eye toward ecological sustainability. The Ministry for the Environment published a national environmental program in 1993 titled "Towards Sustainable Development," and a National Sustainable Development Action Plan in 1997.

Many of Iceland's wetlands are protected under the Ramsar Convention. However, several of these wetlands are at risk from excessive drainage for farming; this tends to reduce wetland area available for birds. Plans to build a hydroelectric plant are also under scrutiny, as this would divert waterways, stress anadromous fish species, and potentially damage some floristic habitats. At the same time, climate change may bring to Iceland a similar dramatic glacial melt as is occurring in Greenland; this would only add to erosion and concerns over soil stability and habitat vigor.

VICTORIA M. BRETING-GARCIA

Further Reading

Amorosi, Thomas, Paul Buckland, Andrew Dugmore, Jon H. Ingimundarson, and Thomas H. McGovern. "Raiding the Landscape: Human Impact in the Scandinavian North Atlantic." *Human Ecology* 25, no. 3 (1997).

McGovern, Thomas H., Gerald Bigelow, Thomas Amorosi, and Daniel Russell. "Northern Islands, Human Error, and Environmental Degradation: A View of Social and Ecological Change in the Medieval North Atlantic." *Human Ecology*, 16, no. 3 (1988).

Ólafsdóttir, Rannveig, Peter Schlyter, and Hördur V. Haraldsson. "Simulating Icelandic Vegetation Cover During the Holocene—Implications for Long-Term Land Degradation." *Geografiska Annaler: Series A, Physical Geography* 83, no. 4 (2001).

Smith, Kevin P. "Landnam: The Settlement of Iceland in Archaeological and Historical Perspective." *World Archaeology* 26, no. 3 (1995).

Indian Ocean

Category: Marine and Oceanic Biomes.
Geographic Location: East Africa, South Asia, and Australia.
Summary: The Indian Ocean connects one-third of the world's human population and offers vast richness of habitat, yet it is under threat from unregulated runoff and pollutants as well as climate change.

The Indian Ocean is the third-largest of the world's five main oceans, and comprises approximately one-fifth of the total ocean area of the world. The Indian Ocean rim region is home to around one-third of the world's population. It is bordered by the Indian subcontinent (from which it derives its name), and the southeast Asian peninsula; Indonesia; Australia; East Africa; and the Arabian Peninsula of western Asia. The southern portion is bordered by the Antarctic or Southern Ocean.

Major marginal seas and lobes, or extensions, of the Indian Ocean include: the Arabian Sea, Red Sea, Persian Gulf, Bay of Bengal, Andaman Sea, and the Great Australian Bight. The Indian Ocean has four access waterways of great significance: the Suez Canal (Egypt), Bab el Mandeb (Djibouti-Yemen), the Strait of Hormuz (Iran-Oman), and the Strait of Malacca (Indonesia-Malaysia).

Factors such as area's ecological and topographical complexity have led scientists to conclude that the Indian Ocean is generally characterized by unique species and high levels of biodiversity. This is supported by the high level of endemism—species found nowhere else on Earth—that has been recorded in adjacent, shallow habitats.

Climate and Resources

The majority of the continental shelves of the Indian Ocean are somewhat narrow, with an average width of only around 124 miles (200 kilometers), except off the western coast of Australia, where the continental shelf extends to more than 620 miles (1,000 kilometers). Given its vastness, the area's climate is also diverse. The area north of the Equator has a monsoon climate with semiannual reversing winds, which can bring violent rains to India from June to September. Just south of the Equator is the trade-winds zone, where weather conditions are more consistent year round.

Climate zones here include tropical, temperate, sub-Antarctic, and Antarctic. A large part of the Indian Ocean lies within the tropical and temperate climatic zones, where numerous corals and other organisms that depend upon coral reefs are found. The tropical shore areas are home to ecologically important mangrove thickets that provide breeding and nursing grounds for many

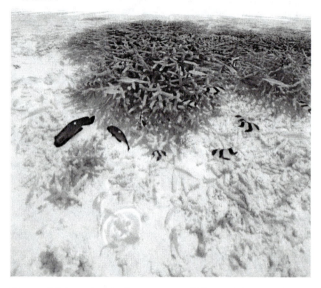

Tropical fish in the shallow waters off the Maldives in the Indian Ocean. The Maldives are one of several Indian Ocean island groups with shallow fringing reefs that provide habitat for commercially valuable fish. (Thinkstock)

7,725 meters) for the Sunda Trench. Ridges that rise up from the seafloor to 13,123 feet (4,000 meters) in depth also provide unique and varied deep-water habitats across much of the southern part of the region.

The trench areas found in the Indian Ocean are of great ecological and economical importance, with their cold, nutrient-rich upwellings supporting a range of commercial fisheries focused on albacore and bluefin tuna, sharks, whiting, snapper, salmon, rock lobster, deep sea crab, abalone, and other marine types. Tuna coral reefs provide another rich area for large-scale fishing. Madagascar, for example, has extensive reefs of all types off of its coasts, as does India in the Gulf of Mannar reef constellation; other areas blessed with important reef zones include the Andaman Sea; the Arabian Sea and Gulf of Oman; and Maldives, Mauritius, Reunion, Seychelles, and Comoros Islands with shallow fringing reefs.

Marine reptiles of the Indian Ocean include the green, olive ridley, hawksbill, leatherback, loggerhead, and flatback turtles. Marine mammals here feature the endangered dugong, spinner dolphin, humpback dolphin, killer whale, sperm whale, blue whale, and the southern elephant seal.

Birds occur in large numbers in the many island, coastal, and estuarine environments around the Indian Ocean. These range from resident species such as the Mauritius parakeet and kestrel; to shorebirds and waders like heron and ibis; to the far-traveling albatross, lesser frigate, and tern.

marine species. During the winter in the near-Antarctic areas furthest to the south, surface ice formation occurs. However, overall the Indian Ocean is the warmest of the Earth's oceans.

Another important feature of this ocean is the depth of many of its trenches, including the Diamantina Fracture Zone, also known as the Diamantina Trench or Diamantina Deep, and the Sunda Trench, which was formerly known as the Java Trench. These are deep areas of complex topography, featuring troughs to depths that scientists argue range from 24,245 to 26,401 feet (7,390 to 8,047 meters) for the Diamantina Trench and from 23,813 to 25,345 feet (7,258 to

Threats

Worldwide, coral reefs are under the threat of depletion due to global climate change; direct human pressures, both commercial and recreational; and inadequate governance, awareness, and political will, something that is especially true in the Indian Ocean. Overexploitation of fishery resources has been reported off the shores of Tanzania and Mauritius, while the coral reefs near Kenya, Tanzania, and Mauritius also suffer from destructive fishing practices. Trampling of coral by fishermen has degraded many reefs in the region.

Coral mining and coral sand mining in Mafia (Tanzania), Comoros, Mauritius, and Madagascar have caused damage to the reefs here. More than 551,156 tons (500,000 metric tons) of coral sand is excavated annually from Mauritius; most beaches in Comoros also have been scarred by this destructive activity. More than 276 tons (250 metric tons) of shells and corals are exported from Tanzania in a typical year. Exploitive collection and coral mining has moved from the depleted areas off of East Africa to the Maldives Islands, where it is used for road construction, and to Sri Lanka and India, where tens of thousands of tons are removed annually for a variety of purposes.

Sedimentation from agricultural practices is yet another ecological concern throughout the region, one that affects the reefs off the coasts of the Seychelles, Dar es Salaam, and Zanzibar. Eutrophication, which is the overabundance of nutrients leading to unsupportable blooms of aquatic plant matter, derives from the runoff of fertilizer and sewage; this has resulted in the disruption of habitats in East African waters, the Arabian Sea, Bay of Bengal, and elsewhere around the Indian Ocean.

Seagrass beds that are found throughout the region—another vital habitat type—are also under pressure, from intensive use of bottom traps and beach seine nets, fishing practices using explosives, sand mining, and dredging off coastlines and in shipping channels. The high degree of water turbidity has driven down the rate of growth of seagrasses off the western coast of India, while recent rapid fluctuations in salinity levels from river runoff have had deleterious effects on the beds near Bangladesh.

Other activities that exact a high environmental cost in the Indian Ocean include petroleum exploration from large reserves of hydrocarbons, especially in the Arabian Sea; deforestation, leading to higher runoff levels; coastal development, which undermines mangrove and seagrass bed habitat; and excessive levels of motorized tourism, which leads to persistently more pollution. Since 1963, approximately 50 percent of India's mangroves have been destroyed. The government of India in 1984 established Sundarbans National Park, looking to protect one of the world's biggest and richest mangrove areas.

The Indian Ocean provides major sea routes connecting the Middle East, Africa, and east Asia with Europe and the Americas. Given the significant amount of international trade that takes place throughout the region's waters, the Indian Ocean is vulnerable to high levels of pollution caused by deliberate waste disposal at sea and by accidental oil spills. This poses a difficult-to-regulate threat to the direct survival of marine organisms and on the ongoing buildup of toxins in the marine ecosystem.

Because the Indian Ocean is bordered by the fast-growing economies of such nations as India, Thailand, and Indonesia, there is a high risk that more industrial, agricultural, and municipal waste and runoff will be released into the Indian Ocean in the future because of a lack of regionally coordinated enforcement of environmental protection standards. Each of these threats is likely to be multiplied or complicated by climate change. The direct impacts of global warming upon the Indian Ocean will likely include rising sea levels; rising air and water temperatures; more and heavier tropical cyclones; altered timing of monsoon season onset; and increasing acidity of seawater and the negative effects of this on coral, mollusk, and lower-food-web species reproduction.

ALEXANDRA M. AVILA

Further Reading

Campbell, M. L. "An Overview of Risk Assessment in a Marine Biosecurity Context." In Gil Rilov and Jeffrey A. Crooks, eds., *Biological Invasions in Marine Ecosystems.* New York: Springer, 2009.

Ginsburg, R. N. *Global Aspects of Coral Reefs: Health, Hazards, and History.* Miami, FL: University of Miami, 1994.

Jameson, Stephen C., John W. McManus, and Mark D. Spalding. *State of the Reefs—Regional and Global Perspectives.* Townsville, Australia: International Coral Reef Initiative, 2005.

Laipson, E. "The Indian Ocean: A Critical Arena for 21st Century Threats and Challenges." In E. Laipson and A. Pandya, eds., *The Indian Ocean—Resource and Governance Challenges.* Washington, DC: Henry L. Stimson Center, 2009.

Indian Ocean Islands (Southern) Tundra

Category: Grassland, Tundra, and Human Biomes.
Geographic Location: Indian Ocean.
Summary: These islands are home to flourishing Antarctic and sub-Antarctic populations of seabirds and seals, but are vulnerable to introduced species and climatic changes.

The Southern Indian Ocean Tundra biome occurs on islands in the sub-Antarctic region of the Indian Ocean. These islands belong to different countries, but have similar climates and species. They include Prince Edward Islands (PEI), a territory of South Africa; the Crozet and Kerguelen archipelagos, which belong to France; and the Heard and McDonald islands, property of Australia. Few people live here; those who do are mostly scientists or monitors. The islands have unique environments and some have conservation designations or environmental restrictions for visitors. The climate here is extreme, with the strong, cold winds of Antarctica bringing rain, snow, and rough seas.

PEI is the westernmost island group in the Southern Indian Ocean Tundra biome. Its two components, Marion Island and Prince Edward Island, are 112 square miles (290 square kilometers) and 17 square miles (45 square kilometers), respectively. These islands are the remains of a shield volcano from the center of the West Indian Ocean Ridge, and are the youngest of all these islands.

The PEI and Crozet groups are relatively close to each other. Crozet archipelago comprises 193 square miles (500 square kilometers) on five main volcanic islands: the western group of Île aux Cochons, Îles des Pingouins, and Îlots des Apôtres; and the eastern group of Île de la Possession and Île de l'Est. The two eastern islands are the oldest of the Crozets, at 8 million years. Among the western islands, Île aux Cochons is 400,000 years old, Îles des Pingouins is 1.1 million years old, and Îles des Apôtres is 5.5 million years old. Île de la Possession is the largest island in this archipelago, at 50 square miles (130 square kilometers) and with a maximum height of 3,064 feet (934 meters). Île de l'Est is the second-largest island, at 46 square miles (120 square kilometers).

The Kerguelen Islands are located 932 miles (1,500 kilometers) east of the Crozet Islands; they range as high as 6,070 feet (1,850 meters). The entire land area is 2,703 square miles (7,000 square kilometers). The group is basically one large island—Grand Terre, of 2,548 square miles (6,600 square kilometers)—and about 300 other islets. There is a permanent ice cap, and the archipelago is approximately 39 million years old.

Heard Island is 147 square miles (380 square kilometers) in area and includes two active volcanoes. The McDonald Islands are 25 miles (40 kilometers) west of the Heard Islands; they are composed of McDonald Island, Flat Island, and Meyer Rock. McDonald is the body here, at about 1 square mile (2.6 square kilometers), formed of two parts joined by a narrow isthmus.

Biodiversity

The Southern Indian Ocean Tundra islands are characterized by low biodiversity. All ecological niches may not be filled, due to the geographical isolation here, or in some cases due to their geologically recent formation. Some species may not be fully adapted in the best ways to their ecosystems, leaving them vulnerable to competition from aggressive introduced species. Additionally, there may be short food webs with fewer species at each level, and individual species can be absolutely essential. A marine example is caridean shrimp off the coast of PEI, which are consumed by the top predators on the food chain: penguins. Similarly, the Marion flightless moth aids in the release of nutrients from plants. Removal or depletion of either one of these species could lead to rapid devastation of the entire biota.

All these islands are rocky but support plants such as grasses, lichens, liverworts, and mosses. Tussock grasses, in particular, are well-adapted to snow and cold. In past times, sailors visiting these islands consumed Kerguelen cabbage, an important source of vitamin C.

Marine vegetation flourishes in this area, and it has become a vital aspect of supporting the

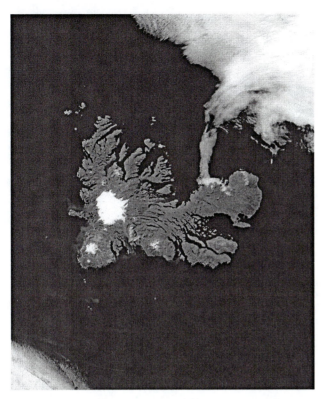

This satellite photo shows Grand Terre and its permanent ice cap in the Kerguelen Islands. This archipelago is thought to be about 39 million years old. (NASA)

terrestrial ecosystems. Giant brown kelp forms underwater forests at up to 164 feet (50 meters) in hard-bottom subtidal areas, and hosts many colorful invertebrates, while the cochayuyo bull kelp attaches to rocky coasts. Storms tear off large quantities of the giant algae, which then rots on the beaches and stones; it provides vital nutrients to plants, animals, birds, and insects.

The islands attract wildlife such as penguins, seals, and seabirds. The largest bird colonies of the biome are on PEI, followed by the Crozet Islands, which have breeding colonies of all six species of albatross and a very large portion of the world's population of king penguins.

All six Antarctic seal types can be found here on the islands. The fur seals and southern elephant seals breed in the region, and they are increasing in numbers now that seal hunting has mainly stopped. Large numbers of leopard seals winter on the rocks of Heard Island. The colony of southern elephant seals on Heard Island and the Kerguelen Islands is one of the three largest such colonies in the world.

Related Antarctic and sub-Antarctic species and subspecies coexist, living alongside one another but rarely breeding with one another. The sooty albatrosses and light-mantled albatross, the northern giant petrel and southern giant petrel, and the Antarctic fur seal and sub-Antarctic fur seal live together on the islands.

Environmental Threats

These Southern Indian Islands are nearly pristine, but their species are not varied or perfectly adapted, so they are quite vulnerable to human activity and to introduced plant and animal species. Invasive grasses have become dominant in parts of the islands, while introduced mice and cats prey on breeding seabird colonies. Climate change, particularly warmer temperatures and seasons, could encourage these nonnative species and adversely affect seabirds and their reproduction. Diving seabirds are already under stress, often caught by long line fishing for the Patagonian toothfish, also known as Chilean sea bass, that occurs around the islands. The gray-headed albatross and white-chinned petrel may be especially damaged as they habitually follow fishing boats. Sea-level rise and greater storm surges, too, hold the potential to disrupt the intertidal and shoreline habitats here.

Magdalena Ariadne Kim Muir

Further Reading

Bester, M. N., et al. "A Review of the Successful Eradication of Feral Cats From Sub-Antarctic Marion Island, Southern Indian Ocean." *South African Journal of Wildlife Research* 32, no. 1 (2002).

Chandra, Satish, V. Suryanarayan, and B. Arunachalam, eds. *The Indian Ocean and Its Islands: Strategic, Scientific and Historical Perspectives.* Thousand Oaks, CA: Sage, 1993.

Sinclair, Ian and Olivier Landgrand. *Birds of the Indian Ocean Islands.* London: New Holland Publishing, 2008.

Indochina Dry Forests, Central

Category: Forest Biomes.
Geographic Location: Southeast Asia.
Summary: Indochina's central dry forests are known globally for their large vertebrate fauna, reminiscent of those found in African savannas. Land use practices are making this biome vulnerable to degradation.

Indochina boasts a wide variety of forest types. Highly diverse, subtropical moist forests are found in the north, while dry deciduous forests and dry evergreen forests are found in southeastern Indochina. The dry deciduous forests of central Indochina are the most dominant forest type of them all, covering an area of 123,600 square miles (320,123 square kilometers).

Historically, these forests formed a contiguous ecosystem extending into India—until habitat conversion to agriculture in Thailand began to isolate various segments from one another. Large blocks of this forest type remain in north, northeastern, and south-central Cambodia, whereas most of the similar natural habitat is cleared away in Thailand, Laos, and Vietnam. Despite long-term human pressures that have degraded much of Indochina's natural ecosystems, the fragments of dry forest that remain contain an extraordinary diversity of flora and fauna.

These deciduous forests are adapted to a dry season that lasts several months, followed by several months of monsoon rains. The climate is warm; temperatures rarely fall below 68 degrees F (20 degrees C). Annual rainfall is less than 60 inches (1,524 millimeters). Unlike nearby semievergreen forests, these forests withstand both a drier and a longer dry season of approximately four to five months.

Fires occur frequently from December to early March, when conditions are driest. The forests grow on thin, sandy, or rocky soil in a landscape that is generally flat, with a few scattered hills and seasonally inundated grasslands crossed by temporary streams.

Forest Composition

Another name for the central dry forest is *deciduous dipterocarp forest*, so called because it is dominated by deciduous dipterocarp tree species. *Deciduous* refers to tree species that shed their leaves for part of the year; this adaptation helps the trees reduce water loss when water availability is low. This behavior is common among deciduous trees in North America during the fall. Unlike North American deciduous trees, however, dry-forest deciduous trees do not shed all of their leaves at the same time.

Dipterocarp refers to a family of tropical tree species that are highly diverse and widespread throughout southern and southeast Asia, with two species that extend into central Africa. Of the nearly 550 identified dipterocarp species, 48 are found in Indochina. Only six species of dipterocarps are deciduous; all six occur in Indochina. They are the most ecologically as well as economically important tree group in the Southeast Asian lowland forests; they are the most dominant species, and they provide valuable tropical hardwoods. The wood is known for strong, light, straight, and knot-free timber. Oily aromatic resins are harvested from all the Asian species. Cambodians collect the resins by cutting holes at a tree's base and periodically burning the area to stimulate resin flow.

A combination of drought stress and frequent fires have created low, open forests with understories dominated by grasses, cycads, shrubs, or bamboo beneath a closed or semi-closed canopy. The deciduous dipterocarp tree trunks are smooth and unbranching, with cauliflower-like crowns that often emerge above the canopy. Large buttresses at the base of some trees allow them to grow tall despite the thin soil.

Compared with their evergreen relatives, deciduous dipterocarps are shorter, about 30 to 120 feet (9 to 37 meters), with broader, thicker leaves. They are adapted to frequent fires, with thick, corklike bark, and the ability to resprout from roots. With increased stress from human pressure and repeated fires, these fire-adapted deciduous dipterocarp forests become degraded and transformed into savanna woodlands. Such a community retains the most fire-resistant species, interspersed with grasses and shrubs.

Biodiversity and Threats

Despite decreases in populations and species, this region is still considered to be one of the most zoologically diverse regions of the world. Many bird species are found here, including the giant ibis, white-rumped ibis, white-shouldered ibis, lesser adjutant, Sarak's crane, green peafowl, and long-billed vulture.

In the early 1900s, scientists and big-game hunters described the game abundance in northern and eastern Cambodia's dry deciduous forests as second only to the game abundance in the Serengeti. Large populations of mega-herbivores such as the Asian elephant, wild water buffalo, gaur, baten, kouprey, Eld's deer, and both the Javan and Sumatran rhinoceros grazed these forests. They were hunted by large carnivores, including tigers, clouded leopards, leopards, and wild dogs. By 1936, the American Museum of Natural History's Fleischmann-Clark Indochina Expedition reported declines in the numbers of wild water buffaloes and widespread, indiscriminate shooting of Eld's deer.

Habitat loss from agriculture and human settlement, as well as hunting for local and international trade, has all taken a large toll on animal populations here. Both rhinoceros species have become extinct in the area, and the kouprey is considered to be extinct globally, with an occasional sighting in remote areas of Cambodia. All the other mega-herbivores, as well as the tiger, are endangered.

The greatest threats to the region are from human activities; the area is densely populated, and the impact of human habitation is widespread. For example, fire is used to clear large areas for building and other purposes, which decreases the natural habitat. Climate change models offer an uncertain glimpse into the future for this region, particularly as they may impact rainfall, which in turn impacts the delicate balance of life in this biome.

FERN RAFFELA LEHMAN
JACQUELINE E. MOHAN

Further Reading

Sivakamur, M. V. K., H. P. Das, and O. Brunini. *Impacts of Present and Future Climate Variability and Change on Agriculture and Forestry in the Arid and Sub-Arid Tropics*. Dordrecht, Netherlands: Springer, 2005.

Sterling, Jane Eleanor, Martha Maud Hurley, and Le Duc Minh. *Vietnam: A Natural History*. New Haven, CT: Yale University Press, 2006.

World Wild Fund for Nature. "Southeastern Asia: Thailand, Cambodia, Laos, and Vietnam." 2012. http://worldwildlife.org/ecoregions/im0202.

Indochina Subtropical Forests

Category: Forest Biomes.
Geographic Location: Asia.
Summary: This biome has diverse fauna, with several endemic species. Because of land conversion from forest to agriculture and hunting, many plants and animals here face an uncertain future.

The Indochina Subtropical Forests biome consists of forests across the highlands of northern Myanmar, Laos, Vietnam, and the southern Yunnan province of China. The landscape is mountainous, with hills and river valleys extending from Yunnan into the northern Indochina region. These river valleys contain the middle catchments of the Red, Mekong, and Salween Rivers. Intrusive igneous rocks and Paleozoic limestone are dominant parent material in these areas. Mountain ranges here generally peak around 6,600 feet (2,000 meters) in elevation, with the river valleys at elevations of 656–1,312 feet (200–400 meters). The Yunnan Plateau can reach as high as 9,840 feet (3,000 meters), however.

The climate in northern Indochina is dominated by summer monsoons that bring 47–98 inches (1,200–2,500 millimeters) of rain annually. The monsoon brings much of the annual precipitation from the Bay of Bengal and South China Sea from April to October. The entire region, including the Yunnan Plateau, has an extended cool and dry season from November to April. The spring pre-monsoon (April to May) is the warmest time of the year; January is the coldest. Frost forms infrequently and only at high elevations.

Vegetation

The unique climate and geology of this region have given rise to lush, diverse forests. This biome has tropical, subtropical, and temperate groups of plants with distinct origins; these floristic groups are: Indian, Malesian (or Sundaic), Sino-Himalayan, and Indochinese. At lower elevations of 1,970–2,625 feet (600–800 meters), subtropical broadleaf evergreen forests contain trees of the dipterocarp, birch, oak, and zapote families.

At higher elevations of 2,625–6,600 feet (800–2,000 meters), a distinctive moist montane broadleaf forest exists, with communities dominated by species of the oak and laurel families, over an understory of small-statured bamboo and ferns. Much higher in elevation on the Yunnan Plateau, Yunnan pine (*Pinus yunnanensis*) is dominant in open-canopy forests. In northern Vietnam, elevations greater than 6, 600 feet

Large fauna of the Indochina subtropical forests, like this Malayan sun bear, depend on extensive forests to survive, but these habitats are becoming increasingly fragmented. (Wikimedia/BirdPhotos.com)

(2,000 meters) have a unique fir-and-hemlock community that contains species from the maple, oak, magnolia, and laurel families.

The vegetation in this ecoregion is generally shaped by the monsoonal precipitation and the dry season; some evidence suggests that lower-elevation seasonal forests are adapted to frequent low fires. *Phonix* palms, for example, are well adapted to survive ground fires. The level of natural frequency of fire is difficult to predict currently, as most present-day fires are human-caused (anthropogenic). At higher elevations, open-canopy pine forests on dry slopes are also thought to burn periodically. There is speculation that the forest species composition would be quite different without the anthropogenic fires.

Wildlife

This biome boasts diverse fauna, with several endemic (found nowhere else) mammals and birds. Notable endemic and near-endemic mammals are a species of slow loris, the white-cheeked gibbon, Ouston's civet, Roosevelt's muntjac, Tonkin snub-nosed monkey, and Anderson's squirrel. Once thought to be extinct, the Tonkin snub-nosed monkey is critically endangered, with fewer than 250 individuals remaining.

The region boasts several iconic species such as the Asian elephant; the Indochinese tiger; and the smallest of all bears, the Malayan sun bear. Once found across the ecoregion, Indochinese tigers are extinct in the Yunnan province and number fewer than 350 individuals elsewhere. Other notable animals are two wild cow species (gaur and banteng), southern serow, pygmy loris, Asian black bear, clouded leopard, Asiatic wild dog, black striped weasel, smooth-coated otter, and three macaque species.

The region has diverse avifauna, including the near-endemic short-tailed scimitar babbler. Several pheasant and hornbill species need intact mature forests. Hornbills, which eat large fruits, play a critical role in dispersing the seeds of large-seeded mature forest trees and liana species. This region has healthy populations of white-rumped and slender-billed vultures, which have declined dramatically elsewhere in southern Asia.

Environmental Threats

The Indochina Subtropical Forests ecoregion has several conservation challenges. Shifting cultivation agriculture has been practiced here for millennia. With the human population rising, the proportion of cleared land has increased, leaving less land for the forest to regrow. In recent years, the forests have become fragmented, especially at lower elevations, as elsewhere in the tropics. Consequently, habitat is declining for species such as tigers, elephants, gaurs, and bears that need large intact forests or forests connected with landscape corridors.

Hunting is a critical component of local cultures, and game provides important protein in human diets. International demand for animal parts needed in Chinese traditional medicine has also contributed to poaching. Tigers are especially vulnerable, as tiger body parts are highly prized and sought after for Chinese traditional medicine. As a result, conservationists have identified "empty forest syndrome," wherein the current forest structure, largely dependent on adult trees, remains visibly unaffected—but many of the animals that exert critical herbivory and predatory roles for future forest structure are gone from the area.

Despite this region's biodiversity and the important ecosystem services it provides, such as watershed protection for the Mekong River and its delta area, the effects of climate change on the forests of this region are poorly understood. More research is needed to help predict looming changes in temperature, moisture, storm patterns, fire regime, and seasonal cycle timing.

Shafkat Khan
Jacqueline E. Mohan

Further Reading

Pimm, S. L., et al. *Terrestrial Ecoregions of the Indo-Pacific: A Conservation Assessment.* Washington, DC: Island Press, 2001.

Sharma, Eklabya, et al. *Climate Change Impacts and Vulnerability in the Eastern Himalayas.* Kathmandu, Nepal: International Center for Mountain Development, 2009 .

Sterling, J. E., et al. *Vietnam: A Natural History.* New Haven, CT: Yale University Press, 2006.

Indus River

Category: Inland Aquatic Biomes.
Geographic Location: Asia.
Summary: The backbone of agriculture and food production in Pakistan, the Indus River biome has been damaged by human activity; its future is jeopardized by global warming.

The Indus River is one of the major rivers of south Asia. Its name derives from the Sanskrit word *Sindhu,* which means simply river or ocean. The Indus basin largely depends on the snows and glaciers of several key mountain ranges: the Himalayan, Karakoram, and Hindu Kush ranges of Tibet, northern Pakistan, and Afghanistan. The river originates near Mount Kailash on the Tibetan Plateau and travels southwestward approximately 1,900 miles (3,050 kilometers) before draining into the Arabian Sea, a northern part of the Indian Ocean.

Geography and Hydrology

The annual flow of the Indus is 7.3 trillion cubic feet (207 billion cubic meters)—twice the volume of the Nile River. The Tibetan Plateau ice field contains the largest area under perennial snow outside of the polar regions. Here, the Indus River headwaters are found, at an elevation of about 18,000 feet (5,500 meters). The glacial melt flows into the territory of Kashmir in India before entering Pakistan. Then it runs along the slopes of the Karakoram Range and the Nanga Pabat massif, forming gorges that reach depths of 15,000–17,000 feet (4,600–5,200 meters).

The river continues as a mountain river in Pakistan's North West Frontier Province, before reaching the Punjab Plain. (The word *Punjab* means water of five rivers.) There, it receives five important tributaries—Jhelum, Chenab, Ravi, Beas, and Sutlej—that make it much larger, often resulting in flooding during the monsoon season. Another tributary of significance is the Kabul River, which rises in Afghanistan, crosses the Afghan-Pakistan border near the Khyber Pass, and flows into the main branch of the Indus at Attack, above the confluences of the five tributaries.

The Indus River basin covers about 200,000 square miles (518,000 square kilometers), extending from the foothills of the Himalayas to the Arabian Sea. Flooding during the rainy season is very common in the region; much of the ecosystem richness along the river is sustained by these regular flooding episodes. However, in 2010, extreme flooding in the basin and adjacent areas affected more than 20 million people and deluged nearly one-fifth of Pakistan. The 2010 flood was determined to be the most destructive in the history of Pakistan.

Biota

The Indus River basin covers a wide range of ecosystems, from the Tibetan Plateau alpine steppe through scrubland, the Indus Valley Desert, xeric woodlands, and broadleaf forests, to subtropical pine forests and the Indus Valley mangroves. Near the foothills, the dominant tree species is the deodar cedar (*Cedrus deodara*). In the plains, the most common tree species are acacia (*Acacia arabica*), shisham (*Dalbergia sissoo*), thistles, and tamarisk or salt cedar. Several palm species are found in the delta region.

The Indus River, including the delta region, is rich in fish species. There are 22 endemic (found nowhere else on Earth) fish species among the approximately 150 fish species that have been identified here. The endemics include Indus baril (*Barilius modestus*), Indus garua (*Clupisoma naziri*), and Rita catfish (*Rita rita*). The most common, and one of the commercially valuable, species found in the river is the Hilsa (*Tenualosa ilisha*), a local favorite edible species. Other common fish here are the palla, sukkur, thatta, and kotri, and several varieties of both catfish and snakehead.

An endangered dolphin species, the Indus River dolphin (*Platanista minor*), is endemic to the Indus River. In recent years, this cetacean has been threatened, due to its small population size, fragmentation of the river by dams and barrages, catching and poaching, pollution, and habitat degradation.

The delta region is rich in marine life, and is considered to be one of the most important ecological regions in the world. The Indus River Delta–Arabian Sea mangroves support a high-salinity estuary, due to the salts brought to the coast by the Indus from the desert, a process that has intensified as the river is increasingly used for irrigation and water supply. Mangroves are an important habitat that support a range of plants specialized to survive in this salty environment. The mangroves here also support large numbers of fish, crustaceans such as freshwater shrimp, and invertebrates that find food, shelter and oxygen in the waters beneath the tree roots. Mangroves are also home to about 25 amphibian species and large numbers of migratory seabirds.

Human Impact

The Indus River has influenced the history, culture, geography, and economy of the region for thousands of years. Historically, the Indus Valley is known as one of the earliest hubs of human civilizations. The ancient cities of Harappa and Mohenjo-Daro, dating back to at least 3300 B.C.E., were discovered here.

There is evidence that ancient civilizations constructed an extensive network of irrigation canals in the region. The British East India Company initiated modern irrigation in 1851, with two major systems, the Guddu Barrage and Sukkur Barrage. The partition of India and Pakistan in 1949 led to the signing of the Indus Waters Treaty in 1960, which guarantees Pakistan water from the Indus, Jhelum, and Chenab Rivers, and gives India the rights to the water of the Ravi, Beas, and Sutlej Rivers. The treaty also permitted the construction of two huge dams in Pakistan. Overall, there are three major dams, 23 barrages, 12 inter-river canals, 48 canals, and 106,000 watercourses in the Indus River basin to either generate power or to irrigate. Unfortunately, many of these constructions have degraded the immediate and extended habitats around the Indus.

Pakistan depends on the Indus, especially because the lower plains and arid lands receive little rainfall. In recent years, the Indus River and the surrounding area have been affected by extensive deforestation, industrialization, urbanization, and climate change. Because of deforestation, the river is even shifting its course westward. Industrialization and urbanization have polluted the waters, affecting aquatic life.

Extensive construction of dams and barrages has altered the volume and flow of the river in the delta. This alteration of water flow is having a direct effect on the natural ecosystem and the lives of people in the region. The effect of climate change on the glacial and snow mountains of the Himalayas will further affect the flow of water in the river, and will have a huge effect on the people living in the lowlands.

More research on the impacts of climate change is required to predict effects in the region. However, there will be additional pressure, as the population of Pakistan has significantly expanded in recent decades, to nearly 190 million, while that of India has reached 1.2 billion people. Feeding this number of people, many of whom live in semiarid environments, will pose a great challenge and will further stress their main water sources.

Adding to the tension, parts of the Indus River and several of its major tributaries flow through India, with whom Pakistan has had a long-standing border dispute in the Kashmir region. The two countries have fought two wars over the issue; despite the 1960 treaty, Pakistan fears that India might divert water from the upper reaches of the Indus.

Pakistan requires an integrated water-management policy to manage the river, both to meet its people's demands and to support plant and animal life in the plains. Some options include increasing water storage by improving infrastructure, improving agricultural production by growing more food with less water, preventing increased salinity in the basin and delta, and developing better conservation standards with farmers, municipal managers, and ecologists.

KRISHNA ROKA

Further Reading

Albinia, Alicia. *Empires of the Indus: The Story of a River.* New York: W. W. Norton and Co., 2008.

Archer, D. R., N. Forsythe, H. J. Fowler, and S. M. Shah. "Sustainability of Water Resources Management in the Indus Basin Under Changing Climatic and Socio-Economic Conditions." *Hydrology and Earth System Sciences Discussions* 7, no. 1 (2010).

Michel, Aloys Arthur. *The Indus Rivers.* New Haven, CT: Yale University Press, 1967.

Qureshhi, Asad Sarwar. "Water Management in the Indus Basin in Pakistan: Challenges and Opportunities." *Mountain Research and Development* 31, no. 3 (2011).

Revees, Randall R. and Abdul Aleem Chaudhry. "Status of the Indus River Dolphin *Platanista minor.*" *Oryx* 32, no. 1 (1998).

Indus Valley Desert

Category: Desert Biomes.
Geographic Location: Asia.
Summary: One of the least hospitable ecoregions, the Indus Valley Desert is home to only a few plants and animals.

The Indus Valley Desert is almost completely uninhabited; it is considered to be one of the most inhospitable ecoregions of Asia. The Indus Valley itself is named for the Indus River, which flows through Pakistan and part of northern India, originating in western China's Tibetan Plateau. The ancient Indus Valley civilization developed along this river, which is a vein of moisture through an otherwise arid region, and subsequent civilizations have followed that pattern.

The Indus Valley Desert covers about 7,500 square miles (19,425 square kilometers) between the Indus and Chenab Rivers. It receives about 25 inches (64 centimeters) of rain a year—a fairly large amount by desert standards—but the level of evaporation still exceeds precipitation due to the extremely hot summers, which reach 113 degrees F (45 degrees C).

Vegetation

The dominant vegetation here is the *Prosopis* genus of flowering shrubs, making the Indus Valley Desert plants cousins of the mesquites of the American deserts. Most of the *Prosopis* individuals in Pakistan are *Prosopis cineraria,* also known as ghaf or kandi; the short, flowering tree is the provincial tree of Pakistan's Sindh province. The white mulberry (*Morus alba*), native to northern China, grows here also, and is noted for the manner in which its stamens act as catapults, launching pollen at 350 miles per hour (563 kilometers per hour).

Other plant life includes sareenh (*Alibizia lebbeck*); saltcedar (*Tamarix*), a shrub that thrives in the saline soil of the desert and grows leaves encrusted with salt; numerous *Acacia* shrubs and trees; bushy evergreen trees called jall (*Salvadora oleiodes*), which fruit in March; and caperbush (*Capparis* spp).

Wildlife

There are 32 mammal species in the desert. Though there are red fox, wild boar, and several species of rodents in the Indus Valley Desert, there are only five large mammals: the Punjab urial (*Ovis orientalis punjabensis*); Indian wolf (*Canis lupus pallipes*); hyena (*Hyaena hyaena*); caracal (*Caracal caracal*); and Indian leopard (*Panthera pardus fusca*), the only one of the subcontinent's five big cats to be found in the desert.

The Punjab urial is a wild sheep, sometimes called the arkars, related to the mouflon. Males have large horns turning almost a full revolution. Each September, they mate with four to five ewes, instinctively spreading their genetic material as much as possible. Leopards prey on the urial, as do Indian wolves.

The Indian wolf is little understood. Exterminated ruthlessly as a pest in some past eras, it has been revered and protected in others. Smaller than European wolves, but larger than Arabian wolves, about 3 feet (1 meter) long and 2 feet (0.6 meter) high at the shoulder, they have generally reddish fur. Indian wolves are sometimes hunted, because although they are protected as an endangered species (there are 3,000 to 4,000 wolves in the country), they have a history of attacking livestock and children. Because they have little commercial value, the population in the desert tends to be safe.

Camel riders in the Indus Valley Desert, which extends about 7,500 square miles (19,425 square kilometers) and is largely uninhabited. In spite of that, the biome is still threatened by human activity, especially by poaching and by agricultural practices that exacerbate desertification, such as livestock overgrazing and logging. (Thinkstock)

The leopard, on the other hand, is nearly threatened due to habitat loss, conflict situations, and rampant poaching. India and Pakistan lead the world in leopard poaching by a wide margin, with poachers selling leopard skins and body parts on the black market. Poaching is one of the only direct human threats to this ecosystem, as the region is inappropriate for settlement, industry, mining, grazing, or agriculture. In the Indus Valley Desert, leopards compete principally with hyenas and wolves.

There are a significant number of reptile species here in the desert, including Sindh krait, Indian star tortoise, Indian cobra, black pond turtle, yellow monitor lizard, and gharial. Nearly 200 bird species are found in the Indus Valley Desert, including the red-necked falcon (*Falco chicquera*) or turumti, a bird of prey found throughout the Indian subcontinent and sub-Saharan Africa. Medium-size, with a broad wingspan, the falcon often hunts in pairs and reuses corvid tree nests. Other, more common avian species include the black kite, barn owl, myna, Alexandrine parakeet, red-vented bulbul, shelduck, rock pigeon, hoopoe, shikra, and Indian peafowl.

Human Impact

Although humans have occupied this region for thousands of years, it has only been during the last two centuries that long-lasting damage has been inflicted upon this desert environment. Desertification, a process that is already occurring in parts of semiarid and drought-prone Pakistan, can lead to natural resources depletion, agricultural losses, food shortages, and, ultimately, hardships for affected populations.

There are both natural and manmade causes of desertification: reduction in rainfall; reduction in river and tributary outflow; inappropriate agricultural practices, such as improper use of fertilizers and pesticides, and livestock overgrazing; overuse of resources; and reduction in vegetative cover from various causes, such as logging. These problems, along with a rapid increase in human population and the accompanied demand for food in Pakistan, are placing great strains on the desert ecosystems.

The effects of global warming are not entirely clear in this region—while air temperatures may rise and monsoon events gain in severity, the previously recorded glacial retreat in Tibet has recently been seen to have reversed. Therefore, predictive models of future precipitation in the Indus Valley Desert remain uncertain. Nevertheless, there is a serious possibility that the Indus Valley Desert will spread and the Indus River floodplain will desiccate, if glacial retreat does resume, and if increasing evaporation rates reduce the volume of the river by half, as some past projections have indicated.

BILL KTE'PI

Further Reading

Ezcurra, Exequiel, ed. *Global Deserts Outlook.* Nairobi, Kenya: United Nations Environment Programme, 2006.

Grewal, B., ed. *Insight Guides: Indian Wildlife.* Singapore: APA, 1992.

MacKinnon, J. *Protected Areas Systems Review of the Indo-Malayan Realm.* Canterbury, UK: World Bank Publications, 1997.

McIntosh, Jane. *The Ancient Indus Valley.* Santa Barbara, CA: ABC-CLIO, 2008.

Panuganti, Sreya. "Desert Solitaire: Why India And Pakistan Should Collaborate To Combat Desertification." Stimson Center, April 10, 2012. http://www.stimson.org/spotlight/desert-solitaire -why-india-and-pakistan-should-collaborate-to -combat-desertification-/.

Irrawaddy Moist Deciduous Forests

Category: Forest Biomes.
Geographic Location: Southeast Asia.
Summary: Although this region is fairly rich in biodiversity and has a moderate number of endemic species, it suffers from inadequate management of resources, which poses a serious threat to its future.

Found in the central areas of the Southeast Asian country of Burma (Myanmar), the Irrawaddy Moist Deciduous Forests biome covers an area of 53,400 square miles (138,300 square kilometers). Due to its remote geographical location, limited access, rugged landscape, and political instability, this area remains one of the least explored and scientifically studied ecoregions of the world. This ecosystem can be classified under both tropical and subtropical moist broadleaf forests zones that are found on well-drained hilly slopes or undulating lands located above 3,000 feet (1,000 meters) in elevation.

The Irrawaddy is the 10th-largest river in the world in terms of water discharge, and the most important commercial waterway of Burma. Geographically, the ecoregion is located within the Irrawaddy river basin and its major tributaries: the Chindwin, Salween, and Sittang rivers. It extends over the catchments of Bago Yoma and the foothills of Rakhine Yoma. Notably, all these rivers originate in the north, run more or less southward almost parallel to one another, before emptying into the Gulf of Martaban and the Andaman Sea.

Nearly the whole of Burma lies within the tropics; therefore, the country enjoys a tropical climate characterized by rain-bearing southwest monsoon winds. However, the north-to-south orientation of its hills and mountains brings about some minor, isolated climatic differences. Burma has three seasons. The cold season runs October through February, with average temperatures of 68–75 degrees F (20–24 degrees C). The hot season, March through May, features average temperatures of 86–95 degrees F (30–35 degrees C). The wet season, running June through September, sees average temperatures of 77–86 degrees F (25–30 degrees C). The ecoregion lies within a dry belt where the average annual rainfall is approximately 59 inches (150 centimeters), interrupted by frequent dry spells.

Biota

The climatic conditions and the soil composition of this area support the growth of closed high forest. Canopies often reach a height of up to 120 feet (37 meters). Teak (*Tectonia grandis*) and pyinkado or ironwood (*Xylia kerri*) are the most important

canopy species of the biome, and are leading items of export, promoting Burma to considerable economic importance. In addition, the Irrawaddy Moist Deciduous Forests region is home to bamboo groves and some key evergreen species: *Berrya ammonilla, Ginelina arborea, Homalium tomentosum, Lannea grandis, Millettia pendula, Mitravgyna rotundifolia, Odina wodia, Pterocarpus macrocarpus, Salmalia insigni, Terminalia belerica, T. pyrifolia* and *T. tormentosa.*

In the lower Burma region, the common bamboo species are *Bambusa polymorpha* and *Cephalostachyum pergracile*; in the north, *Dendrocalamus hamiltonii, D. membranaceus,* and *Cephalostachyum pergracile* are common species of bamboo. The dominant undergrowth of these forests includes *Leea* spp., *Barleria strigosa,* and the family *Acanthaceae*, a taxon of flowering plants. Importantly, when extraction of timber leaves gaps in the forest canopy, the cleared areas often become hubs of the harmful invasive weed *Eupatorium odoratum.*

The ecoregion harbors a number of large animals. Asian elephants (*Elephas maximus*) are geographically widespread here—but their numbers have declined in recent years due to large-scale poaching and smuggling of ivory and skin, especially into Thailand. Tigers (*Panthera tigris*) and Eld's deer (*Cervus eldi*), quite common species until the turn of the century, are now critically endangered.

Other characteristic fauna—now found most readily in the national parks—are the gaur (*Bos gaurus*), sambar (*Cervus unicolor*), serow (*Capricornis sumatrensis*), golden cat (*Felix temmincki*), masked palm civet (*Paguma larvata*), marbled cat (*Felis marmorata*), leopard cat (*Felis bengalensis*), spotted linsang (*Prionodon pardicolor*), Himalayan black bear (*Selenarctos thibetanus*), Himalayan sun bear (*Helarctos malayanus*), binturong (*Arctictis binturong*), dhole (*Cuon alpinus*), and capped langur (*Presbytis pileatus*).

The biome remains home to 350 species of forest and waterbirds. Recent research unfolds that the white-throated babbler (*Turdoides gularis*), belonging to the family *Timaliidae*, is the single near-endemic bird species of the region; it is found in few if any other places. Other species of forest birds common to the region are woodpecker, laughing thrush, babbler, oriole, drongo, parakeet, barbet, pigeon, dove, and magpie. Waterbirds found in the Irrawaddy include red-wattled lapwing (*Vanellus indicus*), wagtail, sandpiper, forktail, and river chat (*Thamnolacea leucocephala*).

Five protected areas, with a total area of 1,700 square miles (4,420 square kilometers), have been delineated within this ecoregion: Pego Yoma, Alaungdaw Kathapa, Shwesettaw, Kyatthin, and Minwun Taung.

The region is moderately rich in mineral resources; precious stones such as ruby, sapphire, and amber are mined in Chindwin and the Irrawaddy valley.

Human Impact

Despite increasing concerns about overarching threats such as political instability, poaching, illegal logging, cutting trees for fuelwood and fodder, unsustainable harvesting, and unsound mining practices—and despite local and international efforts to find solutions to these problems—the region continues to suffer from environmental threats. As a consequence, local authorities have turned to the idea of establishing more protected areas and new methods of forest management, in the hope of making the area economically sustainable while maintaining the biodiversity. However, public indifference remains a problem. Insufficient financial resources, lack of community education, and scant enforcement continue to contribute to inadequate environmental management of the region.

On a global level, climate change is already effecting the Irrawaddy drainage basin, as much of the water source here is from the melting Himalayan glaciers. Due to this accelerated melt, increases in water discharge are expected initially—along with the risks of flooding and landslides—to be followed later by seasonal water shortages and limited water supplies for downstream communities. Habitat fragmentation from these forces looms as a major threat to the Irrawaddy Moist Deciduous Forests biome.

Rituparna Bhattacharyya

Further Reading

Adams, William M. *Green Development: Environment and Sustainability in the Third World, 2nd Ed.* London: Routledge, 2001.

Clarke, J. E. "Biodiversity and Protected Areas—Myanmar." Mekong River Commission, 2011. http://www.mekonginfo.org/assets/midocs/0002035-environment-biodiversity-and-protected-areas-myanmar.pdf.

Water Resources eAtlas. "Watersheds of Asia and Oceania." http://pdf.wri.org/watersheds_2003/as13.pdf.

Xu, Jianchu, et al. "The Melting Himalayas: Cascading Effects of Climate Change on Water, Biodiversity, and Livelihoods." *Conservation Biology* 23, no. 3 (2009).

Irrawaddy River

Category: Inland Aquatic Biomes.
Geographic Location: Southeast Asia.
Summary: The main river running entirely through the country of Burma, the Irrawaddy is relatively undeveloped, but human activity is approaching a tipping point that could devastate species such as the Irrawaddy river dolphin.

The Irrawaddy River (*Ayeyarwaddy* River in Burmese) is the principle river of Burma (or Myanmar), and is the most important commercial shipping route in the country. The name *Irrawaddy* is believed to come from the Sanskrit word meaning elephant river. The Irrawaddy River drains most of Burma, through a region known as the Chindwin Valley, and also drains parts of India and Thailand. It originates at the confluence of the N'mai and Mali Rivers, which are fed by glacial runoff from remote mountains in the northern regions of the country and in Tibet, China. The river flows generally from north to south, emptying into the Andaman Sea area of the Indian Ocean.

The Irrawaddy River is approximately 1,350 miles (2,170 kilometers) long. Between the Burmese cities of Myitkyină and Mandalay, the

Rowboats along the Irrawaddy River in Burma. The stalled Myitsone dam project and six additional dam projects proposed for the river pose a significant threat to villages on the river and marine life. (Thinkstock)

Irrawaddy flows through three narrow gorges, or defiles. Its total drainage area is 158,000 square miles (411,000 square kilometers). At its mouth, the river discharges between 82,000 and 1,152,000 cubic feet (2,300 to 32,600 cubic meters) of water per second into the Irrawaddy Delta. The delta begins around 58 miles (93 kilometers) north of Hinthada; the edges of the delta are formed by the southern tip of the Arakan Mountains to the west and the Pegu Mountains to the east. The delta's nine arms cover 26,870 square miles (69,600 square kilometers). The Irrawaddy is the fifth-most heavily silted river in the world, and this makes the delta incredibly fertile.

The Irrawaddy River embodies two climate zones: the northern sections of the river run through a humid subtropical climate, while the south passes through a humid tropical climate. Both zones are affected by the South Asia summer monsoons, which bring heavy rain from May to October. During this six-month span, the region receives between 60 and 100 inches (150 and 250 centimeters) of precipitation.

At the Burmese capital Rangoon (Yangon), in the lower half of the river, temperatures fluctuate between an average of 77 degrees F (25 degrees C) in January to 86 degrees F (30 degrees C) in June. At Myitkyina, in the upper basin of the river, the temperatures averages 64 degrees F (18 degrees C) in January, and 82 degrees F (28 degrees C) in June. Temperatures decrease further north, as altitude increases toward the glaciers that feed the rivers, averaging below 68 degrees F (20 degrees C) annually.

The Irrawaddy River passes through several cities between Myitkyinā and Rangoon. The river is Burma's principal shipping route, as the Chindwin Valley has no railroad. In fact, the Irrawaddy has been used for transport and trade since the 6th century. Only a handful of the cities have large port facilities; the other towns can only accommodate one or two small barges at a time. It is likely the government will push to upgrade all these facilities, which will bring pressure on river habitats, but also present the opportunity to "get it right" ecologically.

Because the delta is a major rice growing area, rice is an important commodity in trade along the river. One of the Irrawaddy's downstream tributaries, the Mu river, has been used for agriculture since the 9th century. The Mu Valley irrigation project is the largest such activity in the country, allowing for dry season growing of corn, peanuts, wheat, sesame, millet, cotton, and other dry crops. Native flora along the river ranges from hardwoods like teak, to conifer stands, swaths of bamboo, and in the delta area, extensive mangrove swamps.

Wildlife

The Irrawaddy River is home to a wide range of wildlife, including 79 known species of fish and four known endemic bird populations. New species of fish are still being discovered, including a new species of hill stream catfish. The biodiversity of the river is at risk due to human expansion, dam building, and pollution. The two species most at risk are the Irrawaddy dolphin and the Irrawaddy river shark.

The Irrawaddy dolphin (*Orcaella brevirostris*) is most closely related to the orca whale (*Orcinus orca*). It has a sparse distribution in the shallow, coastal waters of the Indian Ocean from the Philippines to northeastern India. Freshwater populations are isolated to the Irrawaddy River; Mahakam River in Indonesia; Malampaya Sound in the Philippines; the Mekong River in Laos, Cambodia, and Vietnam; and Songkhla Lake in India. Marine populations stretch from India all the way to Australia.

The Irrawaddy shark (*Glyphis siamensis*) is a river shark only found in the Irrawaddy River near Rangoon. There is little information on the shark; the only studied specimen was found in the late 19th century at the mouth of the Irrawaddy River. It exists only as a museum specimen. Though no other specimens have been found to confirm that it is not extinct, it is classified as Critically Endangered on the ICUN Red List.

Human Impact

The Irrawaddy River has its share of problems and environmental concerns. In this energy-driven age, power sources are a major commodity in the region. This river is a potential power source in which China continues to show great interest. The government in Burma put the $3.6-billion Myitsone dam project on hold in September 2011. As of March 2012, the Chinese government has been increasing pressure on Burma to restart construction. Thousands of individuals have already been displaced from various small villages and farmland surrounding the river. The implications of blocking the river are far-reaching for local populations and the environment. Besides the displacement of thousands of people, harm to local fish and dolphin populations, damming would have a large impact on water flow through the delta, and therefore to delta agriculture.

Another current environmental concern comes from overzealous gold mining along the banks of the river. Mercury and cyanide are both harmful byproducts of blasting for gold. This, and oil leaking from the boats that run along the river, badly

The Irrawaddy Dolphin

The Irrawaddy dolphin bears a strong resemblance to the beluga whale but with a darker coloration, with a pale to dark gray back and a white underbelly. Laotian cultures believe them to be the reincarnated spirits of their departed. The Irrawaddy dolphin is less active than most other dolphins like the bottlenose, only making occasional low leaps out of the water. They feed together in small groups, between six and 15 individuals. In the 1970s, there were reports of fishermen working in tandem with the local dolphin populations. Fishermen would tap the sides of their boats to signal the dolphins, who would chase fish toward the nets. In exchange, the fishermen would share their catch with their aquatic assistants.

The Irrawaddy dolphin is classified vulnerable (VU) on the IUCN Red List. The main threats facing the Irrawaddy dolphin come from fishermen—as bycatch in fishing nets—and from global warming, which plays havoc with estuarine habitats through saltwater intrusion, bank erosion, increased turbidity, and related factors. Irrawaddy dolphin populations in the Philippines are down to fewer than 70 individuals; the total population of the species is estimated to be below 1,000 individuals. The Convention on International Trade in Endangered Species (CITES) has placed a ban on live capture of the Irrawaddy dolphin for display in aquariums, due to the dwindling populations in the wild.

disrupt the ecosystem, killing off both fish and dolphins and causing serious health problems for people living in nearby villages.

Finally, local fisherman use methods that can be harmful to wildlife. Electrode-fishing is a method utilizing an electrode to run a high voltage current through the water, paralyzing or killing fish. The fish then float to the surface, where they are netted. Dolphins also fall prey to the current. Another method involves the use of gill nets; some dolphins, usually young calves, become trapped in the nets, where they soon drown. Overfishing also can deplete fish populations, causing drastic and negative effects on the food chain.

Local governments, environmental and human rights groups are doing what they can to stem environmental damage and perhaps to block the building of the Myitsone dam project. There are six additional dam projects proposed for the Irrawaddy River. Already, unregulated mining, deforestation, and erosion have taken a heavy toll here. Burmese activists fear that if the dam project moves forward, its impact will extend to communities downstream that rely on the river flow and adequate nutrients to sustain the rice production on which the country depends. Meanwhile, the Chinese and the China Power Investment Corporation are lobbying persistently to restart construction.

WILLIAM FORBES
ANDREW OSBORN

Further Reading

Burma Rivers Network. "Irrawaddy River." 2012. http://www.burmariversnetwork.org/burmas -rivers/irrawaddy.html.

Smith, Brian D. and Randall R. Reeves. "River Cetaceans and Habitat Change: Generalist Resilience or Specialist Vulnerability?" *Journal of Marine Biology* 2012, no. 1 (2012).

Smith, Jeff. "Two Rivers: The Chance to Export Power Divides Southeast Asia." National Geographic Daily News, October 25, 2011. http://news.nationalgeographic.com/news/ energy/2011/10/111026-mekong-irrawaddy -hydropower-dams.

Thant, Mynt-U. *Where China Meets India—Burma and the New Crossroads of Asia.* New York: Farrar, Straus & Giroux, 2012.

iSimangaliso Wetlands (St. Lucia Estuary)

Category: Marine and Oceanic Biomes.
Geographic Location: Africa.

Summary: The iSimangaliso Wetlands is a tidal estuary of unparalleled beauty and diversity, whose Zulu name aptly means "miracle and wonder."

The iSimangaliso Wetlands biome in South Africa is a tidal estuary ecosystem on the Indian Ocean, situated along the northeast coast of the KwaZulu-Natal province. It stretches from Kozi Bay in the north to St. Lucia in the south. Virtually the whole of the iSimangaliso Wetlands biome has been incorporated into iSimangaliso Wetland Park, and has been recognized as a World Heritage site by the United Nations Educational, Scientific and Cultural Organization (UNESCO).

iSimangaliso is a Zulu word meaning "miracle and wonder," which appropriately describes this beautiful and diverse tidal estuary. One of South Africa's largest protected areas, it encompasses 1,266 square miles (3,280 square kilometers), and extends 174 miles (280 kilometers) along the coast. The estuary area includes some of the world's largest forested coastal dunes; they reach up to 656 feet (200 meters) high. The park also includes all of Lake St. Lucia, the St. Lucia and Maputaland Marine Reserves, the Coastal Forest Reserve, and the Kosi Bay Natural Reserve.

As a tidal estuary, iSimangaliso has interconnected habitats and wetlands that vary seasonally, with fresh- and marine-water inputs and varied salinity levels. There are swamps along Lake St. Lucia; these and other areas are fed by water seeping through the coastal dunes, that provide freshwater habitats year-round, irrespective of the changing salinity of the lake and estuary. There are also swaths of grassland habitat here.

Biodiversity

The iSimangaliso Wetlands biome is important as a breeding ground for giant leatherback and loggerhead sea turtles. Other species include 100 corals, 1,200 fish, 36 snakes, 80 dragonflies, 110 butterflies, and approximately 520 bird types. Important birds include the sea-eagle, pink-backed and white-backed pelican, greater and lesser flamingo, and the fish-eagle. There are 11 animal species endemic (found nowhere else) to the biome and at least 100 species that are near-endemic; these are also found in other places around South Africa.

The large submarine reefs harbor many fish and corals, with particularly strong coral diversity in Sodwana Bay. Whales and dolphins migrate through off-shore waters. Lake St. Lucia, 50 miles (80 kilometers) long and up to 14 miles (23 kilometers) wide, is home to hundreds of hippopotamuses and crocodiles, as well as pelicans, flamingos, ducks, waders, and fish.

In addition to modern species, the iSimangaliso Wetlands biome contains a living fossil: the coelacanth. This is a fish species virtually unchanged from millions of years ago; it was known only from the fossil record and presumed to be extinct—until a live specimen was found off the southeast African coast in 1938. It is still a rare protected fish, but several living specimens have been found and photographed in a submarine canyon off the coast near Sodwana Bay.

Threats and Conservation

The iSimangaliso Wetlands is vulnerable to anthropogenic changes. Extensive use and alteration of the rivers flowing into Lake St. Lucia for agriculture and other purposes have reduced freshwater flows. At the same time, the wetlands and particularly the lake were subject to drought for more than 10 years, ending in 2011. For much of this time, the estuary was closed off from the Indian Ocean, but due to low freshwater inflow and high evaporation, salinity levels rose throughout the ecosystem here. This led, in June 2009, to a red-and-orange algal bloom in parts of Lake St. Lucia that lasted for at least 18 months, ending only with heavy summer rains in 2011. The extent and persistence of the algal bloom illustrates a likely effect of global warming, and the fragile balance of the wetlands ecosystems.

Other threats include damage by excessive tourism and overfishing. As a result of heavy traffic, the wetlands have been zoned into three ecotourism use zones: a zone of low-intensity use in the wilderness, where all visitor access is by foot; a moderate-use zone, where visitors can view wildlife from vehicles, camps, and shelters; and a high-intensity use zone, where (in seven clusters) there are roads, interpretative and educational displays, guided walks, accommodations, and other tourist facilities. Overall human activity is confined to one-third of the wetlands here.

The iSimangaliso Wetlands biome is relatively well managed under national and provincial regulations. South Africa protects wetlands under the World Heritage Convention Act of 1999. The wetlands also contain four Ramsar sites that recognize the ecological functions of wetlands and their importance as economic, cultural, scientific, and recreational resources. There is intensive management, research, and monitoring of human activity in the wetlands. Sufficient funds

A sign warning swimmers not to enter a crocodile habitat in the iSimangaliso Wetland Park. About one-third of the park is open to visitors. (Wikimedia/Christian Wörtz)

are in place nationally and through internationally-funded projects for the ongoing operation of the wetlands park.

MAGDALENA ARIADNE KIM MUIR

Further Reading

Saarinen, Jarkko, Fritz Becker, and Haretsebe Manwa, eds. *Sustainable Tourism in Southern Africa: Local Communities and Natural Resources in Transition.* Bristol, UK: Channel View Publications, 2009.

Somers, Michael J. and Matthew Hayward, eds. *Fencing for Conservation: Restriction of Evolutionary Potential or a Riposte to Threatening Processes?* New York: Springer, 2011.

Surhone, Lambert M., Mariam T. Tennoe, and Susan F. Henssonow, eds. *iSimangaliso Wetland Park.* Beau Bassin, Mauritius: Betascript Publishing, 2001.

World Heritage Centre. "iSimangaliso Wetland Park." http://whc.unesco.org/en/list/914.

Italian Sclerophyllous and Semi-Deciduous Forests

Category: Forest Biomes.
Geographic Location: Europe.
Summary: This broadly distributed and diverse mixed forest biome of Mediterranean climates occurs throughout the coast of the Italian peninsula.

Although the term *Mediterranean* refers to a specific region that includes parts of Africa, Asia, and Europe, the climate of this type—cool or mild, wet winters and hot, dry summers—is actually common to temperate coastal systems throughout the world. As climate is an important determining factor in vegetation, some ecosystems in coastal California and Australia, for example, are actually quite similar structurally and functionally, though quite different compositionally, to the coastal forests typical of the Italian and French coasts and valleys.

At low elevations, these coastal forests typically are characterized by a mixed evergreen or semi-deciduous canopy with a diverse understory and herbaceous layer. While superficially similar to other coastal forests, certain local and regional factors, particularly the presence of mountains, make the sclerophyllous (thick-leaved scrub vegetation) and semi-deciduous forest types specific to the Italian peninsula and coastal France a unique biome.

This biome covers much of the Italian peninsula (as well as nearby islands, such as Ventotene) and parts of the southern coast of France, a total of 39,500 square miles (102,200 square kilometers). Most regional climatic variation here is the result of mountain ranges near the coast. With increasing altitude, local zones tend to sustain more precipitation increases and lower temperature. Variation in edaphic (soil) conditions is more related to volcanic activity and the underlying bedrock types, which include dolomite, limestone, marl, sandstone, and schist-marl.

Thus, it is elevation that primarily dictates forest zonation and volcanic activity, while bedrock type dictates forest composition. Land cover here is a mix of broadleaf deciduous woodland (30 percent), evergreen woodland (30 percent), cropland (30 percent), dry grassland (five percent), and inland water bodies (five percent).

Vegetation

Forests at the lowest elevations, generally closest to the coasts, are dominated by a mix of evergreen sclerophyllous (hard, waxy) broadleaf species, such as the commercially important cork oak, and deciduous broadleaf species, such as downy oak and Manna ash. At higher elevations further inland, the cooler temperatures and greater precipitation favor deciduous species, and forest compositions shift away from sclerophyllous vegetation.

As elevations increase still more, the cold-tolerant European beech becomes a dominant species, but the diverse woody flora include the endangered Lobel's maple and coniferous species such as the locally abundant yew and relict populations of silver fir in the Apennine Mountain.

Forests at highest elevations are no longer considered to be part of the Italian Sclerophyllous and Semi-Deciduous Forests biome; instead, they are included in the Apennine Deciduous Montane Forests biome. Many of these montane forests are also dominated by beech, and the distinction between the two biomes can be unclear. As the timberline—the elevation at which tree species drop out altogether in favor of subalpine and alpine vegetation—is approached, forest composition shifts from beech to pine and juniper, clarifying differences between these two biomes. Only near the summits of the highest mountains do elevations surpass the timberline.

As expected of a heterogenous forest type that covers such a broad range of climatic conditions, the understory and herbaceous flora of this biome are extremely diverse. Variation in aspect, as well as elevation, allow the massifs to support many species not found elsewhere, in other words, endemic. Many of the 3,300 plant species, perhaps 10 to 20 percent, found in this region are endemic here, with ranges restricted to small pockets of suitable habitat at various elevations in the mountains. The region's abundance of orchid species is particularly noteworthy.

Wildlife

Italian sclerophyllous and semi-deciduous forests also contain a charismatic and ecologically important faunal component. In addition to an abundance of moth and butterfly species, more than 40 native mammal species occur in these forests, including the only populations of the Apennine brown bear; the storied Italian wolf; two marten species; and the Italian chamois, a species similar in form and function to the mountain goat.

Endemic frog and salamander species occur at elevations throughout the mountains, and many large birds of prey, such as the golden eagle, use rocky outcroppings as nesting habitat.

Human Activity Impact

Although many of the high-elevation forests remain intact, much of the forest at lower elevations has been structurally altered, and old

A juvenile Italian chamois (Rupicapra pyrenaica ornata). Chamois are one of 40 types of native mammals found in the Italian sclerophyllous and semi-deciduous forests. (Thinkstock)

growth is rare. Thus, although most of the region is forested, a plethora of even-aged stands, the result of clear-cutting during the late 19th and early 20th centuries, are in evidence. A large portion of the region is currently designated for conservation and is sparsely populated by humans, however, so given time and appropriate ecosystem management, it is possible that many of these even-aged stands can attain old growth.

Nonetheless, these forests are not safe from anthropogenic effects. Although human habitation of the mountainous areas is low, human densities in the coastal regions are growing, and forests in the mountains represent lucrative recreational development opportunities, such as ski resorts. Extraction of antimony and the planned construction of an underground nuclear power plant are typical of industrial activity that not only reduces water supply and carries the risk of soil and water pollution, but also necessitates the building of access roads and tunnels. Landscape fragmentation, a result of road building and ski-resort development, potentially threatens functionally important animal species that require broad home ranges, such as the Apennine brown bear and Italian wolf.

Finally, there are some threats to this system that humans can no longer control. Hot, dry summers and previous livestock grazing put these ecosystems at risk of both natural and human-ignited forest fires. Forest composition in some even-aged stands has shifted toward rockrose, a shrub species. Fires in rockrose stands can be particularly severe, because the species has ladder fuels, allowing fires to reach its canopy and causing fire to spread rapidly across the landscape. During the 21st century, regional temperatures are projected to increase due to global climate change. Severe climatic events, such as droughts, are projected to become more frequent, further increasing the likelihood of severe forest fires, and thus drastic habitat alteration, fragmentation, erosion, and species stress.

Conservation Efforts

In recognition of these threats, large parts of this biome have been classified for preservation, conservation, and protection. Examples of protected areas within the biome include the national parks of Abruzzo, Majella, Sibillini, Gran Sasso, and Laga Mountains, Foreste Casentinese, and Gargano. An astounding 28 percent of the region of Abruzzo alone is protected as parkland. Although much of this parkland is actually classified within the Apennine deciduous montane forest, this demonstrates that ecosystem conservation has been recognized as an important goal, and projected rates of loss and conversion of Italian sclerophyllous and semi-deciduous forest are lower.

JESSE FRUCHTER

Further Reading

Funk, S. M. and J. E. Fa. "Ecoregion Prioritization Suggests an Armoury Not a Silver Bullet for Conservation Planning." *PLoS ONE* 5, no. 1 (2010).

Lieberman, M. R. *Walking the Alpine Parks of France and Northwest Italy*. Seattle, WA: The Mountaineers, 1994.

Loy, A., P. Genov, M. Galfo, M. G. Jacobone, and A. Vigna Taglianti. "Cranial Morphometrics of the Apennine Brown Bear (*Ursus arctos marsicanus*) and Preliminary Notes on the Relationships with Other Southern European Populations." *Italian Journal of Zoology* 75, no. 1 (2008).

Merlo, M. and L. Croitoru. *Valuing Mediterranean Forests: Towards Total Economic Value*. Wallingford, UK: CABI Publishing, 2005.

World Wildlife Fund. "Mediterranean Forests, Woodlands and Scrubs." 2012. http://worldwildlife.org/biomes/mediterranean-forests-woodlands-and-scrubs.

Itigi-Sumbu Thicket

Category: Forest Biomes.
Geographic Location: Africa.
Summary: This poorly known, highly threatened thicket ecosystem, with a high level of endemism, occurs in small blocks in Zambia and Tanzania.

The Itigi-Sumbu thicket is a poorly known, distinctive ecosystem characterized by dense vegetation and high levels of endemism (species found nowhere else). A wide variety of mostly drought-deciduous, spineless shrubs form a low, intertwined canopy. This ecosystem type occurs mainly in two small isolated patches in Zambia and Tanzania, often in mosaics with other woodlands and savannas. The thicket differs markedly in physical structure and species composition from these adjacent ecosystems. These distinct attributes are thought to be the product of the underlying duricrust or hardpan soil (hardened, mineral products of weathering).

The small extent of Itigi-Sumbu thicket has been further reduced and degraded by ongoing threats, especially clearing for agriculture. Unlike in Tanzania, blocks of the ecoregion in Zambia are in protected status, although that has not always prevented encroachment.

The thicket covers approximately 3,000 square miles (7,800 square kilometers) in two distinct locations. One part is located near the town of Itigi, Singida Region, in central Tanzania. The second, and larger, occurrence is in northeastern Zambia, in the Kaputa District of Northern Province, between the southern tip of Lake Tanganyika and Lake Mweru Wantipa, approximately 370 miles (600 kilometers) southwest of the Tanzanian location. The Zambian stand extends slightly northward into the Democratic Republic of the Congo.

Elevations in the ecoregion range from 3,100 to 4,300 feet (950 to 1,300 meters) above sea level. The climate is characterized by a seasonal wet-dry pattern typical of this part of tropical Africa. Alternating wet and dry seasons result from the presence of the rain-bearing Intertropical Convergence Zone from November to April, and the dry-season subtropical high-pressure cell from May through October. Annual precipitation in the Zambian portion averages about 55 inches (140 centimeters), with considerable year-to-year variability; the Tanzanian section is drier, with average annual rainfall of approximately 20–28 inches (50–70 centimeters).

A distinctive characteristic of the thicket is its duricrust or hardpan soils. These are well aerated, sandy-textured, acidic soils at depths of 2 to 10 feet (0.6 to 3 meters) that desiccate in the dry season. A close spatial correlation exists between thicket vegetation and duricrust soils, suggesting that this factor may explain the distinct geographical pattern of the ecoregion as distinguished from surrounding woodlands.

Flora

The Itigi-Sumbu Thicket biome is characterized by dense, almost impenetrable, vegetation, mainly of spineless, multibranched shrubs that form a main canopy 10–16 feet (3 to 5 meters) high, with some woody plants reaching as high as 20–40 feet (6 to 12 meters). The canopy is dominated by deciduous species that shed their leaves during the dry season. At least 100 woody plant species occur, including *Albizia petersiana, Baphia burttii, B. massaiensis, Bussea massaiensis, Combretum celastroides orientale, Grewia burttii,* and *Pseudoprosopsis fischeri.*

The thicket is often spatially interspersed with xeric (dry-adapted) woodlands, especially of the miombo type. Although these mosaics are all part of the broader Zambezi River regional ecosystem that drives endemism, the Itigi-Sumbu thicket is distinct in terms of species composition, its dense physical structure, and the minor role of fire. The thicket also tends to occur in discrete patches with abrupt boundaries, rather than grading into surrounding ecosystems.

Fauna

Compared with the flora, the vertebrate fauna are relatively indistinct, and large mammals in the country, such as elephants, are uncommon, owing in part to the difficulty of movement in the dense thicket. However, these thickets were once habitat for the black rhinoceros—though poachers have eradicated the rhinoceros in this ecoregion. On the other hand, in this relatively small area, 365 bird species have been recorded; termites are abundant; and endemism has been documented for some reptiles and butterflies.

Environmental Threats

The Itigi-Sumbu Thicket biome is considered to be a zone of regionally important ecological distinctiveness. Because relatively little research has been carried out here, some uncertainty remains concerning its physical integrity and conservation status. One evaluation describes the thicket as "relatively stable," with potential for successful conservation intervention; another concludes that it is under significant stress and severely threatened.

A remote sensing analysis conducted in the Zambian portion indicates that its surface area decreased by 71 percent over the last quarter of the 20th century, and suggests that the remainder of this portion could be gone by the end of the first quarter of the 21st century—if observed degradation continues apace. It is presumed that the Tanzanian portion has experienced similar deterioration; one study suggests that half of its original area has been cleared.

Major threats to the thicket stem from increased demand for land and resources associated with agricultural exploitation, charcoal production, and human settlement. Current demographic statistics report annual population growth in Zambia and Tanzania at approximately 3 percent—more than twice the world average. Despite the fact that the Zambian portion of the thicket is located in officially protected areas—including Mweru Wantipa National Park, Tabwa Reserve, and Nsumbu National Park—protected status apparently has done little to control encroachment into the area. In Tanzania, no part of the thicket falls under protected status. While portions of the biome are somewhat well-adapted to semiarid conditions, its habitats will certainly be under greater stress as global warming imparts higher annual temperature averages and, potentially, lower overall precipitation here.

ANDREW M. BARTON
E. MARK PIRES

Further Reading

Almond, S. *Itigi Thicket Monitoring Using Landsat TM Imagery.* London: University College, 2000.

Chidumayo, Emmanuel and Davison Gumbo, eds. *The Dry Forests and Woodlands of Africa: Managing for Products and Services.* London: Routledge, 2010.

Kideghesho, J. R. "The Status of Wildlife Habitats in Tanzania and its Implications to Biodiversity." *Tanzania Wildlife* 21, no. 1 (2001).

Japan, Sea of

Category: Marine and Oceanic Biomes.
Geographic Location: Pacific Ocean.
Summary: The sea of Japan's complex but isolated ecosystem supports a rich variety of marine and avian life.

A marginal sea in the Pacific Ocean between the archipelago of Japan, the Russian and Korean mainland, and the Russian island of Sakhalin, the Sea of Japan has several unique characteristics due to its near-complete enclosure by land and resulting isolation from the rest of the Pacific. Just as islands are home to species found nowhere else, the Sea of Japan has its own unique ecosystem. It is significantly less saline than the surrounding ocean, and, like the Mediterranean Sea, its tidal activity is very low. The Sea of Japan lacks any major bays, capes, or large islands.

Hydrology and Climate

The contribution of river discharge to the water exchange is low—about 1 percent. Most of the rivers flowing into the sea are from the Japanese archipelago. Japan's four largest rivers—the Shinano, Ishikari, Agano, and Mogami—all empty into the Sea of Japan. The largest river from mainland Asia flowing into the sea is the Tumen River, which begins in Mount Baekdu and forms part of the boundary separating China, North Korea, and Russia.

The water balance of the Sea of Japan is maintained by the inflow and outflow through its five straits: the Strait of Tartary, between the Asian mainland and Sakhalin; La Perouse Strait, between Sakhalin and Hokkaido; the Tsugaru Strait, between Hokkaido and Honshu; the Kanmon Straits (also known as the Straits of Shimonoseki), between Honshu and Kyushu; and the Korea Strait, between South Korea and Kyushu.

Once a landlocked sea, the Sea of Japan became joined to the Pacific Ocean over the Miocene period, as the land fragmented sufficiently for Japan to detach from mainland Asia. The Tsugaru and Tsushima straits formed first, about 2.6 million years ago; La Perouse is the youngest, forming as recently as 11,000 years ago and disrupting the migration of mammoths to and from Hokkaido. Even today, all the straits are shallow, limiting the exchange of water with the Pacific and the migration of marine life to and from the sea.

The isolation of the sea leads to clearly separated layers in the water, except during winter in the northern reaches. Winter temperatures in the central and southern sea hover close to 32 degrees F (0 degrees C) at the bottom of the sea; 34–35 degrees F (1.1–1.6 degrees C) at depths of 1,310–1,640 feet (400–500 meters); 36–40 degrees F (2.2–4.4 degrees C) at 656–1,310 feet (200–400 meters); and as warm as 50 degrees F (10 degrees C) 328 feet (100 meters) below the surface. Because of the counterclockwise circulation of the sea currents, the eastern waters are consistently warmer than the western waters. The difference is greatest—about 10 degrees—in the winter.

The waters of the Sea of Japan thus are predominantly warm, and when the northwestern monsoon wind brings cold, dry air over the sea from the Asian mainland (from October to March), powerful storms and typhoons can result. The western coasts of Japan are regularly pummeled with waves in excess of 25 feet (8 meters). The monsoon also carries seawater evaporation south to the Japanese mountains, where it precipitates as snow and adds to the convection of surface waters.

In the northern area of the sea, particularly in the Strait of Tartary, where the water is especially shallow, the surface freezes for months, beginning as early as October and lasting as late as June. The domination of evaporation over precipitation also leads to higher levels of salinity in the southern sea in the winter; this more-saline water is gradually carried to the north, balancing out the introduction of less-saline water as winter ice thaws in the spring. Summer winds blow in the opposite direction, as warm, humid air from the Pacific is carried over the Sea of Japan to the Asian mainland.

Though tidal activity is limited, it is also complex. The tidal flow of the Pacific pushes through the Korea Strait and the northern portion of the Strait of Tartary twice a day, and pushes through the Tsugaru Strait once a day, causing a daily tide on the Koreas' eastern shores, the Far East of Russia, and Honshu and Hokkaido in the Japanese archipelago. The amplitude of these tides is low but varies considerably across the Sea of Japan. Near the Korea Strait, for example, it is 10 feet (3 meters); at the southern tip of the Korean peninsula, it is only half that.

Biodiversity

Along with the nutrient richness that is aided by its unique combination of temperature and salinity gradients and tidal cycles, the sea's waters are extremely rich in dissolved oxygen—as high as 95 percent of the saturation point just below the surface and still as high as 70 percent at depths of 9,840 feet (3,000 meters). This leads to a thriving environment for some 3,500 species of fauna and 800 of marine flora, including nearly 1,000 species of crustaceans.

While lacking the deepwater fauna richness of the greater Pacific Ocean, the Sea of Japan biome does support cod, Atka mackerel, sardines in great abundance, and the anadromous fish such as salmon and trout. Sea lions and cetaceans are here in great populations. Squid spawn in the East China Sea before migrating north into the Sea of Japan via the Korea Strait.

Coastal areas of the northern reaches of the Sea of Japan provide a wintering haven for Steller's sea-eagle (*Haliaeetus pelagicus*), and vital nesting areas for the endangered spotted greenshank (*Tringa guttifer*). Cranes and shorebirds use Furugelm Island, off the Russian coast, as a stopover in migrations.

Along the southern waters of the Sea of Japan, the wetlands and rice paddies of the Lower Maruyama River (in Japan's Hyogo prefecture) were designated a Ramsar Wetland of International Importance in 2012. Authorities say several thousand species live in or depend on the 1,380-acre (560-hectare) area. It is vital, for instance, to such threatened fish species as chum salmon (*Oncorhynchus keta*), three-spined stickleback (*Gsterosteus aculeatus*), and Japanese rice fish (*Oryzias latipes latipes*). It is here that the previously extirpated oriental white stork (*Ciconia boyciana*) was successfully reintroduced in the 1950s.

Common fish include saury, mackerel, sardines, anchovies, herrings, sea bream, salmon, trout, squid, cod, pollock, and Atka mackerel. The sardine is the most plentiful pelagic fish, constituting more than 70 percent of the total catch. Squid spawn in the East China Sea before migrating to the Sea of Japan. Seals and whales are common mammals. Because the straits are so shallow, there is no characteristic oceanic deep-water fauna.

At the surface of the Sea of Japan, levels of dissolved oxygen can reach 95 percent. The high levels of oxygen, which can remain as high as 70 percent even at great depths, have led to waters rich in marine plants and animals. About 3,500 different species of fauna and 800 species of flora have been counted in the Sea of Japan. (Thinkstock)

Threats and Controversy

A significant ecological problem in the Sea of Japan is oil pollution, particularly along major shipping routes, where the number of accidents of seagoing vessels has increased in the 21st century. There are inconclusive reports of radioactive pollution from Soviet-era reactor equipment and materials disposal at sea. Mercury contamination is a recurring problem, particularly in the Agano River, which feeds into the Sea of Japan near Niigata, site of a mass poisoning event in 1965.

Climate change is projected to warm the average seawater temperatures here, a process that seems to be already underway. Major outbreaks of giant jellyfish (*Nemopilema nomurai*) have become far more common than they were historically in the Sea of Japan, to the point where commercial fisheries are registering their repeated swarms as a serious threat.

Coral reefs in the Sea of Japan are extending their range further to the north, but this buildup is counterbalanced by the fact that reef-dwelling fish are migrating here from more tropical waters, likely at the expense of at least some of the cold-water species native to the Sea of Japan. Additionally, even the new reefs are threatened by the crown-of-thorns starfish (*Acanthaster planci*), the classic foe of coral reef-building organisms. This starfish only thrives in warm-water conditions.

While the name *Sea of Japan* is the most commonly used, other nations have pushed for different nomenclature. South Korea prefers the name *East Sea*, while North Korea refers to the sea as *East Sea of Korea*. The South Korean government specifically argues that the name *Sea of Japan* serves as a reminder of the period when Korea was occupied by imperial Japan. Although *Sea of Japan* has been the most common name in international use for more than two centuries (in large part because of Japan's prominence in international trade), the term *East Sea* was found on Western maps as early as the 1700s.

Sea of Korea was actually the most common Western use in that century, likely because of Japan's strict isolationist policies from 1633 to 1853, which made Western interaction with the country almost negligible. *Oriental Sea* was about as common in Western usage as *Sea of Japan* until the opening of Japan to international trade in the mid-19th century.

BILL KTE'PI

Further Reading

Fukuoka, N. "On the Distribution Patterns of the So-Called Japan Sea Elements Confined to the Sea of Japan Region." *Journal of Geobotany,* 15, no. 1 (1966).

International Hydrographic Organization. *Limits of Oceans and Seas, Special Publication No. 28.* Monte Carlo, Monoco: International Hydrographic Organization, 1953.

Tang, Q. and K. Sherman, eds. *The Large Marine Ecosystems of the Pacific Rim: Assessment, Sustainability, and Management.* London: Blackwell Science, 1999.

Java-Bali Rainforests

Category: Forest Biomes.
Geographic Location: Southeast Asia.
Summary: A highly fragmented southeast Asian rainforest, home to several endemic species, this biome faces significant threat from rapid encroachment of high-density human populations.

Set between the eastern Indian Ocean and the Java Sea is the Indonesian island of Java, an elongated and rugged strip of land with one of the highest-density human populations in the world. Off its eastern coast, across the 1.5-mile-wide (2.4-kilometer-wide) Bali Strait, is another Indonesia island, the much smaller Bali. Eastern Java and Bali are home to the biome known as the Java-Bali rainforest, which takes the form of scattered patches of deciduous and semi-evergreen forests. These support many endemic (found nowhere else on Earth) and near-endemic species; the lush tropical ecosystem is under pressure from agriculture, poaching, logging, and related human land-use activity.

The climate of the Java-Bali region, classified as tropical wet and dry, and affected by seasonal monsoons, receives less rainfall than adjacent rainforest biomes in western Java. The region is densely mountainous and actively volcanic, with the highest elevations being 12,060 feet (3,840 meters) at Mt. Semeru in Eastern Java, and 10,300 feet (3,140 meters) at Mt. Agung on Bali. Both volcanoes have histories of recorded activity over the last 100 years.

Vegetation

Variations in elevation and its subsequent effects on rainfall result in microclimates across the region. Mountainsides that face toward the south receive higher rainfall and are able to better support forest growth. Such localized variations result in tiny, isolated pockets of dry deciduous, moist deciduous, and semi-evergreen forests scattered in parts of the lowlands and on the mountains.

Most of the original plant growth in these rainforests has been cleared through logging or burning for agricultural purposes. Although volcanic soil is highly fertile, repeated burning and shifting cultivation and grazing under the control of the increasing human population have contributed to prevalent loss of habitat and soil erosion.

The diversity of the flora and fauna in the Java-Bali rainforests is considered low to moderate, especially compared to other rainforest biomes in southeast Asia. In Bali, the small size of the island, dwindling natural growth, and human disturbance of what remains of the forests put extreme pressure on the survival of the few endemic species.

The benchmark strata and characteristics of a healthy rainforest ecosystem, such as a distinct emergent layer, canopy, understory, and forest floor, do not exist or are difficult to identify in the Java-Bali Rainforest biome. Although a few semi-evergreen forests survive at elevations higher than 1,600 feet (490 meters), the remaining deciduous forests in the lower elevations are composed mostly of light stands of trees, only a few of which are taller than 80 feet (24 meters). Such stands have open canopies, allowing sunlight to reach the forest floor and support herbaceous undergrowth. There is no dominant tree type in this biome, unlike in many equatorial or tropical rainforests. Some of the common genera here include *Homalium, Acacia, Ailanthus,* and *Dipterocarpus.*

Characteristic Animals

There are at least 100 mammal species here, with two that are strict endemics: the critically endangered bawean or Kuhl's deer (*Axis kuhlii*), and the endangered Javan warty pig (*Sus verrucosus*). The wild banteng (*Bos javanicus*), though not endemic, is also endangered. The now-extinct Balinese tiger (*Panthera tigris balica)* was the smallest and rarest tiger subspecies in the world, recorded now only through a few bones and skins kept in museums. Once strictly endemic to the island, it is believed to have succumbed to hunting and human encroachment on its already-limited habitat.

The Javan tiger (*Panthera tigris sondaicus),* which was larger in size and had a more extensive habitat than its Balinese cousin, is also extinct, while a Javan leopard subspecies is considered endangered. Today, no large predators exist on Bali. The largest undomesticated mammals that remain on the island are the Javan rusa deer (*Rusa timorensis*) and the wild boar (*Sus scrofa*), both of which thrive in stable populations here and elsewhere.

Fifteen bat species have been identified in the Java-Bali region. These include fruit bats and flying foxes, which residents hunt both for their meat and for medicinal use. The cave fruit bat is economically important as a pollinator of durian trees, which contribute to the region's agriculture and economy. In spite of this, the fruit bats are exterminated by orchard owners, who believe that the animals are harmful to their crops.

Approximately 350 bird species are present in the biome, 10 of which are endemic or near-endemic. Although some of the near-endemic species are also found in the montane rainforests of western Java, they continue to face significant conservation pressure there as well, with threat levels ranging from threatened to critically endangered. Among the bird species here that are under threat are the Bali starling, the Bali myna, Javanese lapwing, and Javan hawk eagle. The critically endangered Bali starling, a well-known songbird which is endemic to the island, has been successfully bred in captivity, and subsequently released into the forest. However, illegal poaching continues to take its toll, and it is unknown how well the released birds have been able to breed in the wild.

Conservation and Threats

Eighteen protected zones, the largest of which is Meru Betiri National Park, cover a total of 900 square miles (2,330 square kilometers), or 4 percent of the land area of the biome. Several of these zones are rather small, 12 square miles (31 square kilometers) or less, and are scattered throughout the region. Despite the designation of these protected sections, poaching and settlement continue apace.

The Java-Bali Rainforest biome faces significant ecological threats. The biome's conservation outlook is negatively affected by the absence of much of its original forest cover, and the extreme and continuing fragmentation of remaining stands. Ironically, lost rainforest means lost carbon sink, thus less ability to "apply the brakes" on the increasing global greenhouse gas concentrations in the Earth's atmosphere.

Climate change presents uncertain challenges that complicate strategies for how to mitigate the largely human-caused stresses this biome already endures. It is projected that East Java and Bali will experience rising average air temperatures, later seasonal onset of rainfall and monsoons, decreased rainfall in East Java but increased rainfall in Bali, a longer dry season in both areas, and more extreme weather events overall. Each of these phenomena has been recorded as already underway.

Impacts will include expanding hillside erosion, greater insect and disease stress upon some forest flora, shift in habitat elevations, and opportunities for invasive species at the expense of some native ones. For example, the Javan hawk-eagle is known to breed slowly, and to prefer living in a single, familiar area throughout its adult life. These tendencies make the species vulnerable to each of the habitat impacts listed above.

MARICAR MACALINCAG

Further Reading

Newman, Arnold. *Tropical Rainforest*. New York: Checkmark Books, 2002.

Van Balen, S., Vincent Nijman, and Resit Sozer. "Conservation of the Endemic Javan Hawk-Eagle *Spizaetus Bartelsi* Stresemann, 1924 (Aves: Falconiformes): Density, Age-Structure and Population Numbers." *Contributions to Zoology* 70, no. 3 (2001).

Wikramanayake, Eric, Eric Dinerstein, Colby J. Loucks, et al. *Terrestrial Ecoregions of the Indo-Pacific: A Conservation Assessment*. Washington, DC: Island Press, 2002.

Jordan River

Category: Inland Aquatic Biomes.
Geographic Location: Middle East.
Summary: The Jordan River is a unique aquatic ecosystem whose excess water exploitation has resulted in a decrease in its outflow, a reduction in its quality, and a modification in its biota.

The Jordan River is a 156-mile (251-kilometer) river in the Middle East, lying within Lebanon, Syria, Jordan, Israel/occupied Palestine. About 80 percent of the river is situated in Jordan, Israel, and the occupied Palestine territory. Its headwaters are in the area of the Lebanon–Syria–Israel border junction, and as an endorheic river—not draining to an ocean, but to an inland area—the Jordan River ends at the Dead Sea, which lies at 1,388 feet (423 meters) below sea level and is one of the saltiest bodies of water on Earth.

Hydrology and Geology

The Jordan River rises in Mount Hermon, at the juncture of Syria and Lebanon north of the Golan Heights. It is formed by the confluence of three main rivers: Nahr Hasbani, which flows from Lebanon; Nahr Baniyas, which arises from a small spring at Baniyas, Syria; and Nahr al-Liddani, which is the largest tributary of the Jordan River within the Israeli territory. Running north-south, it flows into Hula Emeq (Lake Hula) and into its surrounding swamps within Israel and occupied Palestine.

Before its drainage in the 1950s, Hula Emeq had a length of 3.3 miles (5.3 kilometers) and a width of about 2.5 miles (4 kilometers). It was about 10 to 13 feet (3 to 4 meters) deep, comprising an area of 5 square miles (14 square kilometers), and its surrounding swamps occupied an additional 14 to 19 square miles (37 to 49 square kilometers). Hula Emeq, with its surrounding swamps, was one of the richest ecosystems in terms of biodiversity in the Levant, with more than 500 species of aquatic invertebrates and vertebrates, and more than 90 species of aquatic vegetation. It was also an important stop for many migratory birds and a nesting place for others.

Exiting the former Hula Emeq, the Jordan pushes through the basalt barrier until it reaches Lake Tiberias in occupied Palestine/Israel after 11 miles (17 kilometers). Lake Tiberias, also known as the Sea of Galilee, or Kinneret, is the largest natural lake in the Middle East and the lowest freshwater lake on Earth: 686 feet (209 meters) below sea level, with a length of 14 miles (22 kilometers), a maximum width of 7 miles (12 kilometers), a depth of 141 feet (43 meters), and a surface area of about 66 square miles (170 square kilometers). It is worth noting that Lake Tiberias has a limited biotic diversity (such as reeds; phytoplankton; zooplankton; benthic or bottom-dwelling animals such as snails; and fish species such as *Acanthobrama terraesanctae*, *Tristramella simonis*, and *Tilapia*, known as St. Peter's fish) relative to that of Hula Emeq.

Subsequently, the Jordan River meanders through the plain of al-Ghawr. In this section, many western and eastern tributaries feed the Jordan River, the principal ones being Nahr Yarmuk and Nahr az-Zarqa, which enter the river from the east. The majority of these tributaries are wadis or seasonal streams, some with saline water of about 20 parts per thousand and high temperatures above 68 degrees F (20 degrees C).

The Jordan River then discharges into the Dead Sea, the deepest hypersaline lake in the world—almost 1,312 feet (400 meters) below sea level, with a length of about 47 miles (75 kilometers) and a width of 6 to 9 miles (10 to 15 kilometers).

The Jordan River runs along the Jordan Rift Valley from its headwaters to its termination in the Dead Sea. A remarkable feature of this valley is its exposure to different natural climatic environments. Its northern part, down to Lake Tiberias, is characterized by a mediterranean climate, with hot, dry summers and cool, rainy winters. Farther south, the climate gets drier. Such variations, along with ample, if seasonal, water supply and fertile soils, make this valley—especially the al-Ghawr plain—an extensive agricultural site that has attracted many civilizations for approximately 10,000 years.

Biodiversity and Human Activity

Being located in the Middle East, which is a unique transition zone between three major biogeographic domains—the Palaearctic, Afrotropical, and Oriental realms—the Jordan River has a high number of faunal and floral elements of all three

A butterfly clinging to wildflowers growing beside a shallow restored section of the Hula Emeq in Israel in November 2010. (Wikimedia/Dror Feitelson)

domains, with the additions of some local and euryhaline taxa (organisms that can endure and thrive in a wide range of salinity conditions), such as approximately 100 species of reptiles, six species of amphibians, and 25 species of freshwater fish.

Because of rapid population growth, significant increases in urbanization and industrialization, aggressive expansion of irrigated areas, and prevailing drought in the region, demands on water resources have been great. The Jordan River is an important water source for Lebanon, Syria, Jordan, Israel, and occupied Palestine. Considering its marked importance, the Jordan River remains a major subject in any solution to conflicts across the Middle East.

In an attempt to divert water from the river for irrigation, drinking, and other purposes, Israel, Lebanon, Syria, and Jordan have long and often made plans for water development and drainage, including the construction of dams (such as the King Talal Dam across Nahr az-Zarqa), canals (such as the eastern Ghawr Canal), and pumping and power stations (such as the pumping station on Nahr Hasbani). The largest projects were the drainage operations carried out by the Jewish National Fund and National Water Carrier (NWC) in Israel.

The drainage of Hula Emeq started in 1948 and continued for a decade. The first phase of the latter stages involved deepening, widening, and straightening of the Jordan River south of Hula Emeq; excavation of two main drainage canals, diverting the Jordan north of the lake; and making smaller canals in the swamp and peat areas.

This manmade project led to the disappearance of some aquatic plants (e.g., *Hydrocotyle vulgaris*), extinction of some endemic fauna of the lake (e.g. *Discoglossus nigriventer* and *Acanthobrama hulensis*), and had other unforeseen effects that rendered the reclaimed swampy land unsuitable for agriculture. Many environmentalists had called for the preservation of a part of the lake and swamplands. Following the drainage, a nature reserve and a national park with a large shallow pond and swamps, encompassing an area of about 43,060 square feet (4,000 square meters), was set up to restore a small part of the previous ecosystem.

Israel's NWC in the 1950s also started transporting water for irrigation and drinking from Lake Tiberias to populated and arid areas in Israel. It was the largest water project in the country— a system of huge pipes, canals, reservoirs, and pumping stations. The excess exploitation of water, as well as drought, caused shrinking of the lake and stressed its ecology. This led to the disappearance of some of the biota present there (e.g., *Nemacheilus galilaeus*).

The construction by NWC for the diversion of Jordan River water led to tension with Syria and Jordan. In an attempt to block the water flow into Lake Tiberias, Syria intended construction of a Headwater Diversion Plan, which was met by a physical attack from Israel in 1965. In close sequence, Jordan diverted Nahr Yarmuk at its upper reaches into the eastern Ghawr Canal for irrigation purposes.

The construction of dams, canals, and pumping stations have decreased the flow of the Jordan River by more than 90 percent, and heavily modified its hydrology. Moreover, the dumping of sewage and agricultural runoff has reduced its water quality and affected its biota. Such disastrous effects have resulted in a call for the conflicting countries to strike a better balance between water resources, and to recognize and act to secure the ecological and historical importance of the Jordan River.

NISREEN H. ALWAN

Further Reading

Dimentman, Chanan, et al. *Lake Hula: Reconstruction of the Fauna and Hydrobiology of a Lost Lake.* Jerusalem: Israel Academy of Sciences and Humanities, 1992.

Fischhendler, Itay. "When Ambiguity in Treaty Design Becomes Destructive: A Study of Transboundary Water." *Global Environmental Politics* 8, no. 1 (2008).

Horowitz, Aharon. *The Jordan Rift Valley.* Rotterdam, Netherlands: A. A. Balkema Publishers, 2001.

Inbar, Moshe. "A Geomorphic and Environmental Evaluation of the Hula Drainage Project, Israel." *Australian Geographical Studies* 40, no. 2 (2002).

Krupp, Friedhelm, et al. "The Middle Eastern Biodiversity Network: Generating and Sharing Knowledge for Ecosystem Management and Conservation." *ZooKeys* 31, no. 1 (2009).

Por, Francis Dov. *The Legacy of Tethys: An Aquatic Biogeography of the Levant.* Dordrecht, Netherlands: Kluwer Academic Publishers, 1989.

Stevens, Georgiana. *The Jordan River Valley.* New York: Carnegie Endowment for International Peace, 1956.

Junggar Basin Semi-Desert

Category: Desert Biomes.
Geographic Location: Asia.
Summary: This unique ecosystem on the southern edge of the Eurasian steppe belt represents a transition from the steppe to the deserts of central Asia.

A unique ecosystem that represents a special transition between steppe and desert can be found in the Junggar Basin of northwestern China. This area provides home for some well-adapted animals of central and inner Asia, such as the Bactrian camel, Asian wild ass (kulan), goitered gazelle, saiga antelope, and specialized jerboas.

The Junggar Basin was one of the last habitats of Przewalski's horse, which is now extinct in the wild. Most of the original ecosystem here was destroyed and turned into agricultural fields; intensive grazing and irrigated agriculture still threaten the remaining fragments. Wheat, barley, oat, and sugar beet are grown in the fields, irrigated largely by runoff from melted snow from the permanently white-capped Altai and Tian Shan mountains. Cattle, sheep, and horses are raised in the natural and seminatural grazing territories here. The Junggar Basin is also marked by large oil fields, the extraction of which is one of the biggest threats to the remaining ecosystem.

Geography and Climate

The Junggar Basin is bounded by the Tian Shan Mountains to the south, the Altai Mountains to the northeast, and the Tarbagatai Mountains to the northwest. Its territory also extends into western Mongolia and eastern Kazakhstan. In China, it corresponds to the northern half of Xinjiang province. The whole basin is easily accessible through the valley of the upper Irtysh River from the north, the Junggarian Gate from the west, or from the direction of Gansu and the rest of China from the east.

As the Junggar Basin is open to the northwest through a series of large gaps in the bordering mountainous ranges, moist air masses that originate in Siberia flow in, providing these lands enough water to remain a semidesert rather than becoming a true desert. Thus, the Junggar Basin has colder temperatures, more precipitation, and a more extensive flora than the enclosed basins to the south, such as the Tian Shan Desert in the Tarim Basin.

In the center of the Junggar lies the Gurbantunggut Desert, representing the most arid part of the territory, with 3 to 4 inches (80 to 100 millimeters) of mean annual precipitation. On the fringes, precipitation is in the range of 4 to 10 inches (100 to 250 millimeters), supporting dry-steppe habitats. Around several runoff-supplied lakes, there is also more pronounced variety and vitality in the biome. Meadows, riparian environments, and wetlands long ago were particularly rich in species diversity—but most of these habitats have by now been heavily altered for agricultural use.

Flora

The flora of the Junggar Basin consists mostly of low scrub types. Taller shrublands of saxaul bush and the gymnosperm *Ephedra przewalskii* can be found near the margins of the basin. Streams descending from the Tian Shan and Altai ranges do support stands of poplar trees, together with *Nitraria* bushes, tamarisk, and willow trees.

These same, moisture-craving vegetation communities can be found on the shorelines of the lakes and in the occasional oasis. Even the most arid parts of the basin are moist enough to support some vegetation, except for approximately 5 percent of the biome that is covered by shifting sand dunes: the Gurbantunggut Desert.

Fauna

The fauna of this ecosystem are a unique mixture of species from the Eurasian steppe and from the deserts of Central Asia. In addition to the great number of jerboa and gerbil (jird) species—typical desert inhabitants—species of ground squirrels, hamsters, voles, and field mice are also present here. Further steppe species are the steppe and marbled polecat, marmot, several species of pica, and the Tian Shan birch mouse. Subterranean rodents such as the zokor and mole vole are also characteristic members of steppe ecosystems.

In the Junggar Basin, Bactrian camels, Asian wild asses (kulans), goitered gazelles, and saiga antelopes live together. The typical carnivores are the wildcat and the Pallas's cat, the Eurasian lynx, the red and the corsac fox, the dhole (Asian wild dog), the wolf, and the brown bear.

The greatest zoological values of the Junggar Basin are the world's largest remaining herds of Asian wild ass, as well as herds of goitered gazelle, and some of the last remaining wild Bactrian camels. The Junggar Basin is a logical place where Przewalski's horse could

The Junggar Basin, where some of the last remaining wild Bactrian camels roam, could once again be home to Przewalski's horse, shown here, if it were to be reintroduced in the wild. (Thinkstock)

be reintroduced in the future, as it was one of the last biomes where this horse was known to survive in the wild. Today, several international efforts are under way to reintroduce this species to its historic homeland in central Asia.

The critically endangered Cheng's jird is the only endemic mammal of the Junggar Basin today. The several species of jerboas and jirds that inhabit the Junggar also make outstanding value for nature conservation. Other species of special significance are the critically endangered sunwatcher toadhead agama; the plate-tailed gecko, a lizard endemic to central Asia; and the Tartar sand boa.

Protected Areas

The northeastern part of the Junggar Basin lies in Mongolia; it includes the Junggarian section of Great Gobi National Park, an International Biosphere Reserve. Other important protected areas are the Bogdhad Mountain Biosphere Reserve (China), the Arjin Mountain Nature Reserve (China), and the Markakol'dkiy State Nature Reserve (Kazakhstan).

Although the Junggar Basin has its natural values, it is a very sad fact that nearly all of the original meadow, marsh, and riparian habitats in the basin have been converted to irrigated agricultural fields. A further threat is human immigration and translocation to this region from eastern China. The looming impacts of climate change are also being studied. Any changes to the Siberian region to the north are liable to have profound impacts here, as so much of the Junggar Basin moisture flows in from that source.

ATTILA NÉMETH

Further Reading

Laidler, L. and K. Laidler. *China's Threatened Wildlife.* London: Blandford, 1996.

MacKinnon, J., Meng Sha, C. Cheung, G. Carey, Zhu Xiang, and D. Melville. *A Biodiversity Review of China.* Hong Kong: World Wide Fund for Nature International, 1996.

Mongolia Ministry for Nature and Environment. *Mongolia's Wild Heritage.* Boulder, CO: Avery Press, 1996.

Murzayev, E. M. "The Deserts of Dzungaria and the Tarim Basin." In *World Vegetation Types,* S. R. Eyre, ed. New York: Columbia University Press, 1971.

Kafue Flats

Category: Inland Aquatic Biomes.
Geographic Location: Africa.
Summary: Kafue Flats is one of Africa's largest wetlands and among the richest wildlife areas in the world. Dam construction and operation is causing population declines among some species.

One of the most studied and unique riverine ecosystems, Kafue Flats is an extensive floodplain characterized by cyclically inundated grass-sedge associations, expansive lagoons, reed marshes, and oxbow lakes. The ecosystem ranks among Africa's largest wetlands and the world's richest wildlife areas. It is especially famous for the endemic (found nowhere else) Kafue lechwe (*Kobus lechwe kafuensis*) antelope species and an abundance of waterbirds, including large aggregations of the wattled crane (*Grus carunculatus*). Within the landscape lie two parks that constitute a Ramsar Wetlands of International Importance site, because of the parks' importance for resident and migratory species.

Human impact here is considerable. The river supports a large human population, tourism, and hydroelectric power generation. Water flow has been greatly regulated following the construction of two large dams at opposite ends of the biome in the 1970s. These dams have markedly altered the ecological dynamics, leading to declined populations of at least some ungulates.

Geography and Hydrology

The ecosystem is located midway along the Kafue River, a major tributary of the Zambezi. It occupies a low-lying plain in Zambia, stretching about 158 miles (255 kilometers) long and 25–37 miles (40–60 kilometers) wide, and covering approximately 2,510 square miles (6,500 square kilometers).

With rainfall averaging less than 31 inches (800 millimeters) per year, moisture is sustained mainly by direct rainfall in the upper river catchment, where precipitation is much heavier. Maximum inundation occurs with a time lag of up to several weeks after peak rainfall in the catchment, reaching a peak from April to June, although this pattern varies considerably from year to year. Water may take up to two months to pass through the gentle profile, whose elevation drops by only 20–33 feet (6–10 meters) along its entire length.

Soils are predominantly alluvial clays, with hydrology varying considerably from high moisture content most of the year in the finer clays, to

715

less waterlogged in the more coarse clays. Hard pans up to 12 inches (300 millimeters) or more in depth occur on the outer margins at the end of the dry season.

Flora

The general vegetation is a diverse mosaic of woodlands interspersed with miombo trees and shrubs such as acacia and bushwillows; the legume mopane tree; and grasslands characterized by species such as *Vossia cuspidata* and *Oryza barthii*, as well as representative species of *Hyperrhenia*, *Setaria*, *Cyperus*, and *Typha*.

A variety of aquatic plants predominate in the open water, including the blue water lily (*Nymphaea capensis*), pondweed (*Potamogeton* spp.), and water hyacinth (*Eichhornia crassipes*), the last being a potentially serious invasive. Floating mats often occur as a result of breakoffs from aquatic and semiaquatic riverine vegetation on the banks.

The structure and composition of vegetation within the immediate vicinity of the river depend largely on soil type and topography. Primary productivity here is much higher than in the more nutrient-poor soils further away from the river. The transition to woodland above the permanent floodline, with scattered trees and shrubs dominated by *Termitaria*, is closely associated with conspicuous termite mounds and some common woody shrub weeds such as *Mimosa pigra*.

Fauna

Kafue Flats is a globally unique wildlife paradise, containing numerous mammal and bird species, some of which remain permanently, while others migrate as part of their annual cycles and in response to drought or food scarcity elsewhere. About half of all the remaining lechwe (*Kobus lechwe*) in Africa are found here. However, the population for the endemic Kafue lechwe, one of three key subspecies, is estimated to have declined from about 100,000 in the 1970s to fewer than 40,000 now. This decline is attributed largely to changes in the flooding regime occasioned by construction of the Itezhi-Tezhi dam, which is used for storage of peak-season flows. The storage is undertaken to maximize hydropower produc-

tion downstream at the Kafue Gorge, Zambia's primary power source.

Other large mammals on the floodplain and in the adjacent woodlands include the hippopotamus (*Hippopotamus amphibious*), blue wildebeest (*Connochaetes taurinus*), Burchell's zebra (*Equus burchelli*), African buffalo (*Syncerus caffer*), and greater kudu (*Tragelaphus strepsiceros*). Smaller ungulates include the sitatunga (*Tragelaphus spekei*), southern reedbuck (*Redunca arundinum*), and oribi (*Ourebia ourebi*). Some carnivores reside permanently in the area, including the spotted hyena (*Crocuta crocuta*), lion (*Panthera leo*), serval cat (*Felis serval*), and side-striped jackal (*Canis adustis*), while the wild dog (*Lycaon pictus*) visits occasionally.

The Kafue Flats biome hosts the greatest abundance of waterbirds in Zambia, including large aggregations of the wattled crane, one of the most threatened birds, and the largest and rarest of the six crane species in Africa. Other notable birds include the vulnerable slaty egret (*Egretta vinaceigula*) and long-tailed cormorant (*Phalacrocorax africanus*). With more than 400 migratory bird species estimated to pass through each year, the area has been designated an Important Bird Area.

More than 50 fish species have been recorded, some of which are endemic to southern Africa, including the Kafue killifish (*Nothobranchius kafuensis*). Some migrate locally out onto the floodplains, taking advantage of increased habitat and protective vegetative cover.

Lochinvar and Blue Lagoon National Parks, located on the south and north banks of the river, respectively, have been recognized for their importance as wetlands and habitat for resident and migratory birds. They form part of the Kafue Flats Ramsar site.

Effects of Human Activity

More than 1.3 million people live within the wider watershed, about 25 percent of whom rely directly on the wetlands for their livelihoods in cattle grazing, hunting, fishing, and tourism. Sugar-cane irrigation and processing causes effluent discharges back into the river, causing nutri-

A Kafue lechwe, which is endemic to the Kafue flats area. Its population is thought to have fallen from as many as 100,000 in the 1970s to less than 40,000 today. (Thinkstock)

ent enrichment and invasive plant growth, clogging the waterway, and suffocating fish. Animal poaching and overfishing are problems, as is the elimination of peak floods and some extreme drought conditions, because of uneven water management of the area's two dams.

As effects of global warming—both those already detected, and the potential future ramifications to temperature, precipitation, air currents, and seasonal onsets—become better understood, it is clear that human interaction with the hydrological cycle here will become an even sharper instrument that can enhance or harm these wetlands and their dependent species.

The human interference of the natural wet and dry cycles has already disrupted fish reproductive cycles, and affected the birds and local population that depend on these fish. As a result, species have been lost as the wetland habitat has degraded. Conservation programs have been undertaken by local communities, governmental and non-governmental agencies that are committed to developing coordinated programs to rehabilitate the environment and its resources.

Evans Mwangi

Further Reading

Fishpool, L. D. C. and M. I. Evans, eds. *Important Bird Areas in Africa and Associated Islands: Priority Sites for Conservation.* Ormond Beach, FL: Pisces Publications, 2001.

Mumba, M. and J. R. Thompson. "Hydrological and Ecological Impacts of Dams on the Kafue Flats Floodplain System, Southern Zambia." *Physics and Chemistry of the Earth* 30, nos. 6–7 (2005).

Smardon, R. C. *Sustaining the World's Wetlands: Setting Policy and Resolving Conflicts.* New York: Springer, 2009.

Thieme, M. L. *Freshwater Ecoregions of Africa and Madagascar: A Conservation Assessment.* Washington, DC: Island Press, 2005.

Kakadu Wetlands

Category: Inland Aquatic Biomes.
Geographic Location: Australia.
Summary: Kakadu contains more than 10,000 insect species, in excess of 280 bird species, 117 kinds of reptiles, 60 mammals, 53 freshwater fish, and more than 1,700 plant types.

The Kakadu Wetlands has been called a climate change hot spot. Situated in the Alligator Rivers region of Australia's Northern Territory, the Kakadu Wetlands has a tropical monsoonal climate. Humidity is low and rain rare in the dry season (April to September). Build-up or transition months include high temperature and high humidity, with violent lightning storms. The rainy season is from January to March and sometimes April; it is both warm and wet. Rainfall in Kakadu averages 51–61 inches (1,300–1,565 millimeters) per year.

The Kakadu Wetlands biome is part of an extensive network of habitats here that feature stone

Galah cockatoos in the 1.7-million-acre Kakadu National Park. Over 280 bird species are found within the Kakadu Wetlands biome. (Thinkstock)

plateaus and escarpments, floodplains and billabongs, tidal flats, monsoonal rainforests, and coastal beaches.

Kakadu has nearly 310 square miles (500 square kilometers) of coast and estuary lined mostly with mangrove forest. Creeks and rivers influenced by tides can extend 62 miles (100 kilometers) inland. The estuaries and tidal flats grow in the dry season and erode in flood season, which also moves silt out to sea, making the waters along the Kakadu coast muddy. Estuaries and tidal flats are home to animals and plants that have adapted to oxygen-deficient salt mud.

Biota

Of the known plant types in Kakadu, 67 species are rare or vulnerable. While certainly not unaffected by invasive flora, Kakadu is one of the most weed-free areas of the country, with only 5.7 percent of known plants here classed as weeds. Unlike most of Australia, Kakadu has largely avoided both plant and animal extinctions, with a few exceptions.

Floodplains plant communities here feature spike rush and other sedges, freshwater mangrove, paper bark, pandanus, and water lilies. Mangroves are also found in the estuaries and tidal flats; they stabilize the coastline while offering feeding and breeding areas for many species of fish. Tidal flats behind the mangroves—including the isolated patches of freshwater monsoon forest

of mangrove, fig, and kapok—support grasses, sedges, and hardy succulents.

Mangrove swamps and glasswort flats are most common, but coastal and riverbank freshwater springs allow pockets of coastal monsoon rainforest. Birds are plentiful in both, as are flying foxes—and the crocodiles that feast on fallen flying foxes. Beaches and the coastal area are home to turtles, snakes, and a variety of sea cow.

Other wetlands and floodplains here are noted for their extreme seasonal changes. Floodplains that are wet for two to six months support grasses, sedges, and mangroves, while floodplains under water for six to nine months produce varieties of water lilies. The creek banks and permanent waterholes (billabongs) support paperbark trees as well as mangroves and water pandanus.

Wet season rains create a shallow freshwater sea that extends far over the plains. When the water recedes, life concentrates in the remaining pools and wet areas. Among such species are reptiles, including varieties of turtles, snakes, and both fresh- and saltwater crocodiles. The area also is home to wallabies and a wide variety of birds. Monsoon pockets are home to different avian varieties—figs, rainbow pittas, and the like—than those of the other areas.

The floodwaters provide nutrients to the floodplains, which in combination with abundant sunlight and groundwater make these habitats naturally abundant in animal and plant life. When the dry season reduces water to billabongs, creeks, and rivers, the floodplains remain vital feeding grounds and refuges for waterbirds. Some waterbirds abundant in Kakadu cluster in a narrow band along the northern coast of Australia, among them the magpie goose, green pygmy goose, wandering whistling duck, and Burdekin duck. Commonly seen in the area, too, are the comb-crested jacana (lotus bird), cormorant, egret, heron, and ibis. About 30 species of migratory birds from the Northern Hemisphere visit these wetlands each wet season.

Kakadu boasts more than 280 bird species, some of which range across habitats, while others are unique to only one. The birds spotted in

Kakadu constitute about one-third of Australia's bird species. Over 6,835 square miles (11,000 square kilometers) of the savanna here is home to endangered or restricted range species.

The region's mammal population includes rare, endemic (found only here), vulnerable, and endangered species. Time of day or time of year affects the activity of many Kakadu mammals. Sixty species of placental mammals and marsupials are known to inhabit the park; mostly they are forest and woodland types, either nocturnal or active mainly in the cooler parts of the day. Nearly all mammal species here are considered to be in decline, although in Kakadu they have found an ideal environment to make their stand.

Kakadu also has 25 species of frog, 117 different kinds of reptile, 32 types of fish in a single creek system, and 10,000 species of insect. Because of the intense daytime heat, most snakes are active only at night. The crocodile population of Kakadu is tremendous, and the saltwater varieties are found in most Kakadu waterways. Accordingly, swimming is discouraged but not prohibited in plunge pools and gorge areas.

Pressures and Preservation

Reptile populations have tumbled since the introduction of the cane toad, and once common lizards and snakes are now rare. Invasive species include the water buffalo; wild pig; cane toad; giant sensitive tree (*Mimosa pigra*); and para grass, a weed that is taking over areas once covered with legumes, grasses, and other food for the native birds. Similarly aggressive nonnative vegetable species have corrupted parts of the floodplain and water systems.

Climate change also is impacting the area, with a greater incidence of cyclones that periodically devastate the environment (as well as local businesses, including tourism). More erratic and intense rains are expected to disrupt tourism in Kakadu National Park; the potential impact and mitigating actions are being studied. In addition, the woody component of the savannas has thickened, possibly as a result of elevated atmospheric carbon dioxide, which favors the growth of some species, while others have declined due to increased fire frequency. Climate change may also be blamed for the spread of fire-prone weeds such as gamba grass.

Since 1992, Kakadu National Park has been a World Heritage Area, a United Nations Educational, Scientific and Cultural Organization (UNESCO) site. Listed for both its natural and cultural heritage, the park's nearly 1.7 million acres (683,000 hectares) have been designated a Ramsar-protected Wetlands of International Importance, covered under the 1971 convention that identified 1,950 such sites around the world as needing conservation attention. Although savanna woodlands cover 80 percent of the park, the wetlands are of paramount interest.

The park's South Alligator River is thought to be the only large riverine system in the world to be contained and protected entirely within a national park, and Kakadu may be the only national park in the world containing a full river system catchment area.

JOHN H. BARNHILL

Further Reading

Australian Government. "Floodplains and Billabongs." Department of Sustainability, Environment, Water, Population and Communities, 2012. http://www.environment.gov.au/parks/kakadu/nature-science/habitats-floodplains.html.

Finlayson, C. Max and Isabell von Oertzen, eds. *Landscape and Vegetation Ecology of the Kakadu Region, Northern Australia*. Dordrecht, Netherlands: Kluwer Academic Publishers, 1996.

Steffen, Will. *Australia's Biodiversity and Climate Change*. Colingswood, Australia: Csiro Publishing, 2009.

Kakamega Rainforest

Category: Forest Biomes.
Geographic Location: Africa.
Summary: The Kakamega rainforest is the only remaining fragment of Guinea-Congolian

rainforest in western Kenya. Surrounded by a growing population, it is rich in endemic and rare species.

The Kakamega rainforest is a relict tropical rainforest fragment situated in the Western Province of Kenya, close to the border with Uganda. It is the last large patch remaining in Kenya of the Guinea-Congolian rainforest, which once was connected to the rainforests of the Congo River basin and Central Africa. This forest is protected in a reserve and national park, but as such covers less than 10 percent of its original area, with only about 45,220 acres (18,300 hectares) in the reserve—of which only 29,650 acres (12,000 hectares) are forest.

Large portions of the forest are secondary growth in glades resulting from logging, agricultural clearing and burning, and commercial forestry plantations. Areas of indigenous forest remain in the northern part of the rainforest and along the River Yala Nature Trail in the southern part of the reserve.

The forest lies 5,085–5,410 feet (1,550–1,650 meters) above sea level, and receives high rainfall spread over two distinct rainy seasons each year. The Kakamega Rainforest biome lies within the Lake Victoria catchment area.

Biota

Scientists estimate that about 24,700 acres (10,000 hectares) of closed-canopy indigenous forest are left standing here, comprised of a wide range of both hardwood and softwood tree species. Some 300 arboreal species have been identified. The most common genera of trees include *Antiaris, Ficus, Croton, Celtis, Trema, Harungana,* and *Zanthoxylum*. Exploitation of hardwoods has left species like Elgon teak rare and localized. There are some plantations of hardwoods like *Milicia* and mahogany within the forest reserve.

The forest currently is most famous as a bird sanctuary, with at least 16 species found in Kakamega and nowhere else in Kenya. Rare species include the endangered African gray parrot, with only a few individuals remaining. There also are several bird species, such as the blue-headed bee-eater and Turner's eremomela, that are absent from

neighboring Uganda. This fact further supports the idea that the Kakamega rainforest is a relict patch and an important biodiversity reservoir.

Birdwatching is a growing and important activity in the forest, attracting visitors from around the world. Tourism is supported by local guides and guide associations made up of outstanding local naturalists. The tourist attraction provides a source of income for human communities near the forest, as well as further incentive for conserving the biome. Visitors who are interested in birds are able to spot such varieties as the great blue turaco, Ross's turaco, black-and-white casqued hornbill, double-toothed barbet, blue-shouldered robin-chat, yellow-bellied wattle-eye, Jameson's wattle-eye, Ayres's hawk eagle, African crowned eagle, and numerous other species.

The forest is rich in butterflies, with more than 350 species recorded here, including many with Guinea-Congolian rainforest affinities. Endemic (found nowhere else) butterflies include king forester (*Euphaedra kakamegae*) and a skipper butterfly, *Metisella kakamegae*.

Primates abound in the forest, including black and white colobus monkeys, red-tailed monkeys, blue monkeys, potto, bushbabies, and the rare De Brazza's monkey in the northern part of the forest at Kisere. Reptiles include the Jameson's mamba, Gaboon viper, and rhinoceros viper, as well as the rare Gold's cobra. The forest has been the site of long-term studies of primates, with some unique observations of interactions between different primate species.

Effects of Human Activity

The Kakamega Rainforest biome lies within one of the most densely populated zones of East Africa. It is surrounded by a human population exceeding 175 individuals per 0.4 square mile (one square kilometer). Most of the households surrounding the forest depend on it for a wide range of needs, including firewood, which is taken from fallen branches and small felled trees. Each day, the roads and tracks leading into the forest are filled with women and children collecting wood for use as fuel for cooking. Research indicates that more than $100 million worth of product is derived from the forest by the surrounding communities each year.

Encroachment remains a problem; most of the forest is unmanaged and relatively unprotected. Exploitation of commercial species is widespread, even with the protected reserves. Other issues include burning grassy areas to keep trees from encroaching, as desired by livestock herders; mismanagement of water resources; and snaring and poaching of wildlife for bushmeat. Currently, no large mammals remain in the forest, and species of duiker and monkey continue to be poached.

The Kakamega rainforest remains one of Kenya's most important biodiversity areas. It is also a microcosm of the problems facing forest conservation in many tropical regions: overexploitation, degradation, overharvesting of wild species, and encroachment due to population growth. Climate change poses another challenge to the area, impacting reforestation projects and habitat improvement for affected fauna and flora as local, national, and international agencies work to overcome the human effects on the area.

DINO J. MARTINS

Further Reading

Beentje, H. *Kenya Trees, Shrubs and Lianas.* Nairobi, Kenya: National Museums of Kenya, 1994.

Bennun, L. and P. Njoroge. *Important Bird Areas in Kenya.* Nairobi, Kenya: East Africa Natural History Society, 1999.

Cords, M. "Mixed Species Association of *Cercopithecus* Monkeys in the Kakamega Forest, Kenya." *University of California Publications in Zoology* 117, no. 1 (1987).

Emerton, L. *Summary of the Current Value of Use of Kakamega Forest.* Nairobi: Kenya Indigenous Forest Conservation Programme, 1994.

Kenya Indigenous Forest Conservation Programme (KIFCON). *Kakamega Forest: The Official Guide.* Nairobi, Kenya: KIFCON, 1994.

Shorrocks, B. *The Biology of African Savannahs.* Oxford, UK: Oxford University Press, 2007.

Tsingalia, M. H. *Animals and the Regeneration of an African Rainforest Tree.* Berkeley: University of California, 1988.

Kalaallit Nunaat Arctic Tundra

Category: Grassland, Tundra, and Human Biomes.
Geographic Location: Arctic.
Summary: A demanding ecosystem where only very rugged species can survive; the fragile equilibrium, close to the freezing point, makes it highly vulnerable to climate change.

Kalaallit Nunaat is the indigenous Kalaallistut name for Greenland, the world's largest island. Greenland, politically an autonomous unit of Denmark, has a total area of 836,330 square miles (2.2 million square kilometers) and is bordered by the North Atlantic Ocean to the southeast, the Arctic Ocean to the north, Greenland Sea to the east, and Baffin Bay to the west. Greenland boasts the world's largest national park: Northeast Greenland National Park, with an area of 375,291 square miles (972,000 square kilometers). It also has many hot springs distributed throughout the territory. The area around this habitat is highly vegetated, with willow scrubs and a wide variety of heath species.

Greenland's climate is Arctic, with summer temperatures below 41 degrees F (5 degrees C). Only on the southwestern coast of the island do summer temperatures rise above 50 degrees F (10 degrees C), producing a small sub-Arctic region. The northern part of Greenland is subject to four months of semidarkness during winter, and midnight sun during summer. Close to Melville Bay here, the aurora borealis, often referred to as the northern lights—caused by the collision of energetic charged particles with atoms in the high-altitude atmosphere—can almost constantly be observed in the night sky.

The Kalaallit Nunaat Arctic tundra is a treeless biome with the lower soil layer permanently frozen; this is known as permafrost. Plants are adapted to extreme climate conditions, including dark winters, low precipitation, and low Arctic temperatures. The ice-free areas are restricted to the coastal fringes, while 80 percent of the land is beneath the second-largest ice cap in the world.

Low and High Arctic Tundra

Greenland is divided in two ecoregions: low and high Arctic tundra. The low Arctic region lies south 70 degrees north latitude, found particularly at the heavily-sculpted inlet system known as Scoresby Sund on the east coast, and south of 75 degrees north at Melville Bay on the west coast—while the high Arctic tundra lies north of these latitudes.

The high Arctic tundra itself is divided into three latitudinal regions along the east coast: Peary Land from the northernmost point of Greenland to 79 degrees north, the midcoast at 72–79 degrees north, and Jameson Land at the transition from low Arctic to high Arctic. In Peary Land, as an example, precipitation is always in the form of snow, in a range of 1–8 inches (25–200 millimeters) per year.

Flora

Very little of any zone is covered with vegetation. Typical species that are present in Peary Land include snowbed varieties of mosses, heaths of Arctic bell heather, fellfield, and *Carex* stands. The midcoast region is characterized by halophytic, or salt-tolerant, vegetation; flora communities here include fellfields, wet fens, dwarf shrub heaths, grasslands, snowbeds, and *Dryas* heaths. Jameson Land has 75 percent vegetation cover; it is rich in mosses, fellfield, and dwarf shrubs, which together make it an important foraging area for musk oxen in both winter and summer.

In the low Arctic tundra ecoregion, vegetation is dominated by dwarf shrubs, *Betula* spp., and *Dryas* spp. at the coast, with various willows (*Salix* spp.) and sedges (*Carex* spp.) dominating inland. Along watercourses, thickets of willow grow 13–16 feet (4–5 meters) tall.

Fauna

Despite the challenging climatic conditions, many animal species have successfully adapted to the environment. Greenland is home to the Arctic fox, a predator that has white fur during the winter and brown fur during the summer; caribou; Arctic hare, a rabbit that can run up to 40 miles (64 kilometers) per hour; Arctic wolf; the threatened polar bear, the world's largest land carnivore, which spends most of its time at sea; musk ox; wolverine, a powerful predator able to kill prey many times its size; ermine; and collared lemming. Dwelling near the coast are a few marine mammals, including the harp seal, ringed seal, bearded seal, hooded seal, and Atlantic walrus.

Greenland's avifauna is highly diverse; many of the migratory bird species come during the breeding season in summer. They feed mainly on mosquitoes and other insects. Among the bird species are the barnacle goose, king eider, pink-footed goose, common eider, snowy owl, gyrfalcon, rock ptarmigan, knot, sanderling, and common raven. As winter approaches, the migratory birds on the west coast migrate further southward across North America, while those on the east coast migrate to western Asia.

Musk oxen foraging in Greenland's arctic tundra. While most areas lack significant vegetation, 75 percent of Jameson Land is covered in mosses, fellfield, and dwarf shrubs, which provide food for musk oxen. (Thinkstock)

Conservation and Environmental Threats

Greenland has a human population of about 56,000 inhabitants. It is generally a land of subsistence fishing, herding, and small-scale farming. Northeast Greenland National Park was created in 1974 to preserve Greenland's biodiversity from hunting pressure and to protect the vital habitats on which its fauna depend. In 1977, it was designated an international Biosphere Reserve. About 40 percent of the world's population of musk oxen reside within the limits of the park, and many bird species have found a breeding ground there as well.

Other conservation efforts include the designation of 11 Ramsar Wetlands sites throughout Greenland: Aqajarua and Sullorsuaq, Qinnquata Marra and Kuussuaq, Kuannersuit Kuussuat, Kitsissunnguit, Naternaq, Eqalummiut Nuaat and Nassuttuup Nunaa, Ikkattoq, Kitsissut Avalliit, Heden, Hochstetter Forland, and Kilen. The latter two are located within Northeast Greenland National Park.

One of the major threats to Greenland is climate change. As temperatures increase, there will be a lengthening of the growing season, thawing of the permafrost, enhancement of soil microbial activity, and subsequent expansion of shrubs northward, all of which would alter the landscape. A few high-Arctic plants are expected to become extinct, such as the *Ranunculus sabinei,* which is limited to the narrow coastal zone. Although the recent climatic history of Greenland shows a trend toward cooling, a warming trend is projected through 2100. Snowfall may increase under this warming scenario.

In recent years, the musk ox and caribou populations crashed in southern Greenland as a result of an altered freeze-thaw cycle of the upper permafrost layer, which resulted in ice crusting and decreased forage availability. This issue is of concern due to the high probability that such events will increase in frequency as a result of short-term fluctuations in temperature. Scientific projections based on actual air temperatures show that western Greenland's goose population may decline as a result of initial cooling in their breeding grounds, accompanied by cooler summers, later snow melt, and less snow-free space for feeding and breeding.

ROCIO R. DUCHESNE

Further Reading

Bliss, L. C. and N. V. Matveyeva. "Circumpolar Arctic Vegetation." In *Arctic Ecosystems in a Changing Climate: An Ecophysiological Perspective,* edited by F. S. I. Chapin, et al. San Diego, CA: Academic Press, 1992.

Callaghan, Terry V., et al. "Arctic Tundra and Polar Desert Ecosystems." In *Arctic Climate Impact Assessment,* Cambridge, UK: Cambridge University Press, 2005.

Egevan, Carsten and David Boertmann. "The Greenland Ramsar Sites: A Status Report." Ministry of Environment and Energy, National Environmental Research Institute, Denmark, February 2001. http://www2.dmu.dk/1_viden/2_publikationer/3_fagrapporter/rapporter/fr346.pdf.

Kalahari Desert

Category: Desert Biomes.
Geographic Location: Africa.
Summary: The vast red sand country of the Kalahari Desert supports some of Africa's most iconic wildlife.

The Kalahari Desert is a vast region of porous, sandy soils that covers much of south-central Africa, from the Orange River in South Africa through Botswana, Zimbabwe, Namibia, and Angola, to the Congo. This 965,255-square-mile (2.5 million-square-kilometer) region is 10 times the size of Great Britain. The southern section of the Kalahari supports iconic rolling red dunes that are intersected by ancient dry riverbeds lined with camelthorn acacia trees 49 feet (15 meters) tall.

The Kalahari generally exists in drought, with 10 inches (250 millimeters) of annual rainfall in the southwest region, to approximately 26 inches (650 millimeters) in northeastern Botswana. Occasional thunderstorms during the wet season, October to March, bring life to the dry earth.

The Kalahari was formed during the breakup of the Gondwana supercontinent, and by 65 million years ago, it had begun to assume its modern dry and dusty character. The desert is cut by two dry rivers: the Nossob and the Auob. These rivers flowed until the tectonic movements that created Africa's Great Rift 10–15 million years ago tilted the ground in such a way that the rivers lost momentum, and their water soaked into the Kalahari sands.

Biodiversity

Despite the aridity of the environment, the Kalahari supports diverse fauna that are adapted to desert conditions. Large herbivores like the gemsbok (*Oryx gazella*), a large antelope with 3-foot- (1-meter-) long spear-like horns to provide protection against lions, are adapted to survive long dry periods when temperatures here exceed 113 degrees F (45 degrees C).

Ground squirrels of the Kalahari have an innovative way to keep cool: They protect themselves from the sun by holding their bushy tails over their heads like a parasol on hot days. This behavior allows them to be active in the sun for longer periods. Some arid-adapted mammals, like springbok and gemsbok, possess pelage (hair) patterns that are thought to reflect heat.

The southern parts of the Kalahari were once home to one of the world's greatest animal migrations. Springbok, blue wildebeest, and red hartebeest once moved from the Orange River in the

This African ground squirrel photographed in the Kalahari Gemsbok National Park is using its tail to shade its back from the sun while it feeds. (Thinkstock)

northern part of the Kalahari in numbers that were estimated to be 80 million animals covering 130 miles (210 kilometers), with a 14-mile (22-kilometer) front. Although these estimates may be exaggerated, the migration clearly was an impressive sight, with people sitting on their doorsteps watching the herds travel by for three days. Hunting and human expansion have decimated these herds.

Giraffes have been reintroduced to the Kalahari. Lions, leopards, brown and spotted hyenas, cheetahs, and (on the more mesic, or moist, eastern side), African wild dogs that prey on the still-rich supply of herbivore fauna. Smaller and more unusual predators abound, including Cape foxes, black-backed jackals, caracals, African wildcats, and honey badgers. Other animals of the Kalahari include black-footed cats, yellow mongoose, insect-eating aardwolf, and velvet monkey.

Boreholes sunk along the dry riverbeds of the Auob and Nossob rivers 100 years ago in search of water have proved to be beneficial to wildlife today; for example, they have sustained Burchell's sandgrouse. Normally, these birds get enough water from the seeds they feed on, but during the breeding season, males fly on a 75-mile (120-kilometer) round trip daily from their nests across the desert to collect water around these boreholes and then return to their chicks. The feathers of the sandgrouse have special barbs that act like a sponge to hold up to 1.4 ounces (40 milliliters) of water. Raptors like lanner falcons rocket like fighter planes into the midst of the drinking birds, looking for a meal.

Hundreds of arid-adapted plant species exist in the Kalahari, the best-known of which is the camelthorn (*Acacia erioloba*), which has numerous uses, from candy and a coffee surrogate, to beams for fencing posts and firewood, with the plant's roots used to help alleviate toothaches and as a drink to ward against tuberculosis. The pods are fodder for elephants. Other plants include black thorn (*Acacia mellifera*), shepherd's tree (*Boscia albitrunca*), tsama melon (*Citrullus roastes*), belle mimosa (*Dichrostachys cinerea*), Kalahari currant (*Searsia tenuinervis*), and buffalo thorn (*Ziziphus mucronata*).

Human Settlement

Humans have inhabited the Kalahari at least since the San, or Khoisan, bushmen settled in the Kalahari more than 3,000 years ago. They are renowned for their prowess in hunting large and dangerous prey with the aid of small poison arrows. When an animal such as an eland or gemsbok was hit with the arrow, the San people would track the animal over many miles (kilometers) until it succumbed to the poison.

The Khoisan knew where to obtain water from the environment and from plants such as tsama melon, gemsbok cucumber, and wild cucumber, and they stored and carried the water in blown ostrich eggs. While the people and animals of the Kalahari benefit from the water stored in plants, the plants themselves benefit by having their seeds transported in the dung of the animals that eat them.

The Khoisan people were followed into the Kalahari by Bantu, and then by European pastoralists starting in the 1600s. The Europeans classified the bushmen as "vermin" and killed an estimated 200,000 within the first two centuries of their occupation of the Capeland. A people of hunter-gatherers then, only a small number of the San follow their traditional way of life in the Kalahari today. Modern civilization has invaded the Kalahari and is threatening the natural resources and habitats of the area. Mineral companies have discovered large coal, copper, and nickel deposits in the region; one of the largest diamond mines in the world is located at Orapa in the Makgadikgadi, a depression of the northeastern Kalahari.

Protected Areas

The core Kalahari Desert is largely protected in Kgalagadi Transfrontier Park, which is composed of the Gemsbok National Park in Botswana and the Kalahari Gemsbok National Park in South Africa. The transfrontier park is a large wildlife preserve and conservation center that is jointly managed. The ecosystem extends through Botswana to the Nxai and Makgadikgadi salt pans and the Central Kalahari Game Reserve.

Kalahari Gemsbok National Park receives 30,000 visitors each year. Its aim is to be a self-contained ecosystem, but even at the size of 8.9 million acres (3.6 million hectares), the park is not large enough to support a viable population of lions, because predator population density is related to resources, and deserts are resource-poor areas. This situation is increasingly problematic, as much of the park is fenced to minimize human-animal conflict.

The effect of artificial waterholes is another problem faced in the region. These areas create halos of overgrazing around them. Most mammals can persist only with regular access to water, so they graze all vegetation in the vicinity of the bore. This overgrazing leads to erosion, salinity problems, and altered vegetation communities. Scientists are analyzing the effects of rising temperatures as a result of climate change, and already have noticed how rising atmospheric carbon dioxide levels are impacting the soil, advancing the leaching of nutrients in some cases. This type of adverse condition can upset subsistence pastoral farming and natural grazing patterns, adding to habitat stress on a range of plant and animal species.

MATT W. HAYWARD

Further Reading

Hayward, M. W. and G. I. H. Kerley. "Prey Preferences of the Lion (*Panthera Leo*)." *Journal of Zoology* 267 (2005).

Hayward, M. W., J. O'Brien, and G. I. H. Kerley. "Carrying Capacity of Large African Predators: Predictions and Tests." *Biological Conservation* 139 (2007).

Knight, M., P. Joyce, and N. Dennis. *The Kalahari.* Cape Town, South Africa: Struik, 1997.

Mills, M. G. L. *Kalahari Hyaenas: Comparative Behavioural Ecology of Two Species.* London: Unwin Hyman, 1990.

Owens, M. J. and D. D. Owens. *Cry of the Kalahari.* London: Collins, 1985.

Redfern, J. V., C. G. Grant, A. Gaylard, and W. M. Getz. "Surface Water Availability and the Management of Herbivore Distributions in an African Savanna Ecosystem." *Journal of Arid Environments* 63 (2005).

Kalahari Xeric Savanna

Category: Grassland, Tundra, and Human Biomes.
Geographic Location: Africa.
Summary: The Kalahari Xeric Savanna biome hosts a low variety of plant and animal species with multiple adaptations to harsh environmental conditions; it is increasingly in danger of desertification.

The Kalahari Xeric Savanna ecosystem is one of the least biodiverse regions in southern Africa, as it is one of extremes: highly variable low mean annual rainfall that occurs primarily during winter months, and high temperature differences between day and night. These climatic conditions create a harsh environment, requiring both plant and animal species to develop specific adaptations to survive and reproduce.

This biome extends throughout the Kalahari Desert from the northern Cape region in South Africa, north to the Democratic Republic of Congo, covering large expanses of land in northwestern South Africa, southern Botswana, and southeastern Namibia. The term *xeric* refers to *dry;* these savannas exist around the core of the extremely arid Kalahari desert, and many species here have evolved to survive in such conditions.

The area is characterized by large night-day temperature fluctuations, with daytime highs up to 113 degrees F (45 degrees C), and after-dark lows falling near freezing. Rainfall often comes during short

A springbok (Antidorcas marsupialis) in the Kalahari Gemsbok National Park. Springbok are the primary prey for cheetah in the Kalahari Xeric Savanna. (Thinkstock)

and infrequent, but powerful, thunderstorms. Rainfall is regionally highly variable, ranging from 6 inches (150 millimeters) in the northeast to up to 20 inches (500 millimeters) in the southwestern Kalahari. The Kalahari Desert itself consists primarily of reddish-brown, highly oxidized, and nutrient-poor soils.

Biodiversity

Typical savanna vegetation is represented by open grasses from families *Aristinda, Eragrostis,* and *Stipagrostis,* along with scattered trees and shrubs. Although very low in endemic (found only here) species, the Kalahri is well known for such plants as the camelthorn tree (*Acacia erioloba*)—the largest and best-adapted tree species in this biome—interspersed with silver cluster-leaf, umbrella thorn, shepherd's tree, and many types of shrubs.

Among the animals are gemsbok (*Oryx gazella*), Kalahari lion (*Panthera leo*), cheetah, leopard, spotted and brown hyena, wild dog, and jackal. Birds here include secretary bird, sociable weaver (*Philetairus socius*), and raptors such as eagles, owls, falcons, and kites. Most of these species, even though they are not different from those in adjoining regions, have developed a set of special adaptations. The Kalahari lion, for example, has larger spatial ranges, hunts smaller prey, and lives in smaller groups than its neighbors.

The sociable weaver builds immense communal nests up to 20 feet (6 meters) long and 7 feet (2 meters) high, weighing as much as 1.1 tons (1,000 kilograms). These structures house up to 300 laying birds and their young at a time. These giant nests are well insulated to buffer the extreme temperature swings.

The wild dog, an endangered species, is the Kalahari's most threatened carnivore. Although restrictive measures such as tracking collars are in place to protect their dwindling numbers, local farmers admit to shooting the dogs and destroying their collars. Also affecting the dog population is rising temperatures from climate change, which can reduce the abundance of prey, and sharper competition among higher-up predators such as lions and leopards.

Protected Areas and Human Activity

Approximately 18 percent of the Kalahari Xeric Savanna ecoregion falls within protected areas, the largest of which is Central Kalahari Game Reserves in Botswana. Kgalagadi Transfrontier Park, shared by Botswana and South Africa, is Africa's first Peace Park. This transboundary park was created to enable better animal and human movement across protected areas; it has so far been a successful conservation and rural development strategy in both countries.

Human settlements expanding into these regions, as well as the fences people build (primarily to separate cattle from wildlife for fear of foot-and-mouth disease), have had detrimental effects on a number of the wildlife species that need to migrate long distances to obtain water, such as the wildebeest and hartebeest. The problem is especially serious in Botswana, where the beef industry is a major source of revenue.

Degradation of vegetation, sometimes leading to desertification, is a major concern. According to the United Nations Convention to Combat Desertification, four main processes are primarily responsible for the condition: deforestation; overharvesting of high-value timber products, medicinal plants, and large fauna; soil erosion; and nutrient depletion. The underlying causes of this include rapid population growth, inappropriate government policies, poor agricultural practices in unsuitable regions, overgrazing, woodcutting, environmental change, and mining.

A first diagnostic signal of degradation is an increasing rate of thorny-shrub establishment, a process referred to as bush encroachment. This problem is especially severe throughout southern Africa, primarily because it means that large grazing areas are lost (or reduced in carrying capacity), accompanied by a degradation of habitats and reduction in species diversity.

Conservation Efforts

A fair proportion of this biome is under conservation and protection, but far from all or even most of it. As a result, there are wide areas experiencing increasing land degradation. One of the main approaches to improving the state of degraded shrublands is the mechanical removal of encroaching species and controlled burns. Alternatively, some areas that become bush-encroached in riparian grasslands can more easily be returned to a nonwoody state after large floods drown the woody species; they are then removed mechanically.

A race against time is developing, as global warming adds to the encroachment threats, disrupts growing seasons, and exerts pressure on growers and herders, who in turn are likely to support poaching, water diversion, and less than ideal, short-term land-use practices.

Narcisa G. Pricope

Further Reading

Knight, M. and P. Joyce. *The Kalahari: Survival in a Thirstland Wilderness.* Cape Town, South Africa: Struik Publishers, 1997.

Schulze, R. E. "Climate." *Vegetation of Southern Africa.* Cambridge, UK: Cambridge University Press, 1997.

Van der Walt, P. and E. Le Riche. *The Kalahari and Its Plants.* Pretoria, South Africa: Van der Walt and Le Riche, 1999.

Kalimantan (Indonesia) Rainforests

Category: Forest Biomes.
Geographic Location: Southeast Asia.
Summary: These extremely valuable rainforest habitats are spread across much of the world's third-largest island, Borneo.

Kalimantan is the Indonesian state that takes up the southern two-thirds of the great island of Borneo. Two basic types of rainforest here combine for one of the oldest intact rainforest ecosystems on Earth: the Borneo montane rainforests and the Borneo lowland rainforests. The Kalimantan Rainforests biome spreads across some 155,000 square miles (400,000 square kilometers), with the Equator running directly through the center.

Geography and Climate

Borneo is the third-largest island in the world. It is located in the middle of the great Australasian archipelago, and its territory is split between three nations; Indonesia controls the greatest extent, followed by Malaysia and Brunei. Kalimantan, which serves as the Indonesian name for the entire island of Borneo, is also an administrative name for its four provinces here, which together comprise 210,000 square miles (543,000 square kilometers).

This land includes coastal mangroves, broad alluvial plains dotted with peat swamp forests, a generally hilly interior, and central mountain ranges that in some cases, such as the Muller and Schwaner Mountains, climb as high as the tallest peak of Mount Raya—7,474 feet (2,278 meters)—but generally rise to the level of 3,940–5,580 feet (1,200–1,700 meters). Among these ranges are the Kayan Mentarang, Apokayan, Meratus, and Sambeliung. The land tends to be very well watered, with networks of small streams and extensive wetlands. Major rivers fan out from the interior; these include the Kahayan, Barito, Mahakam, Kapuas, Katingan, and Rajang.

The climate of the biome is humid tropical, or wet equatorial; moist year round but with a somewhat drier season from July to October and the rainy season extending through much of the other eight months. Rainfall occurs in the range of 110–134 inches (280–340 centimeters) annually. Temperatures are consistently warm and vary little throughout the year, generally staying within the range of 73–88 degrees F (23–31 degrees C).

Biodiversity

The Kalimantan lowland rainforests are replete with dipterocarp tree species, those with twin-winged fruiting seeds. An open canopy is one result of the plant diversity here; it is also quite tall. The forest is quite dense and tends to be dark below the canopy, however, with intertwined lianas (vine plants), mosses, ferns, and hardwood trees. Bornean ironwood, renowned for its hardness, is a prized species for the timber industry.

Sadly, they grow quite slowly, meaning that logged stands may not see a re-growth of ironwood for many decades. Nestled into hollows between tree roots and among the twisting lianas are pitcher plants, many species of which feature spiky traps for flies and other insects.

Animals that spend much of their lives high in the canopy here include flying lizards; Kuhl's gliding gecko, a very well-camouflaged species; Prevost's squirrel, a striking, inky black one; Colugo lemur; harlequin flying frog; and Wallace's flying frog, which does spend time on the forest floor, mainly to lay eggs in ponds and pools. Moving across the forest floor, as well as the higher levels, are such endangered fauna as the proboscis monkey and the orangutan—easily the most iconic animal of Kalimantan.

Montane rainforests in Kalimantan, generally those occurring above 3,000 feet (900 meters), are populated by myrtle and laurel species, as well as oaks that would be considered stunted if found at lower elevations. In the understory, ferns and rhododendrons dominate, except at the highest altitudes where heathers are more in evidence. Pitcher plants are also common on the forest floor in these higher-altitude forests.

Kalimantan montane rainforests support an amazing abundance and diversity of species. Of orchids, the biome is estimated to host more than 2,500 species, with perhaps one-third of these endemic (found nowhere else).

Bats and rodents boast the highest number of individual species among mammals here. Larger and medium-sized mammals include the Borneo elephant, sun bear, Sunda clouded leopard, bearded pig, Sambar deer, red muntjac, banteng, banded palm civet, short-tailed mongoose, yellow-throated marten, Malaya badger, and smooth-coated otter. Birds include the mountain blackeye and pygmy white-eye—both endemic genera—mountain serpent eagle, long-tailed parrots, and a wide range of hornbills.

Threats and Conservation

Logging, coal and mineral mining, and petrochemical extraction and processing are among the heavy industries that have increasingly plagued the ecosystems of Kalimantan since the mid-20th century. The spread of human population in from the coasts, and outward from inland industrial centers—combined with road-building, illegal logging, and periodic uncontrolled land-clearance fires—have resulted in increasing deforestation, which threatens both habitats' degradation and fragmentation.

Kalimantan holds the greatest expanse of tropical rainforests in Indonesia, and the thick jungles and myriad insects have tended to discourage large-scale agriculture. However, in 1983, large swaths of Kalimantan's tropical rainforests were destroyed by wildfire, causing heavy damage to the ecosystem. Then, in 1996, the government sponsored a program to convert peat swamp forest to rice paddies, through massive excavations that backfired by draining the areas—and making them far more fire-vulnerable. Fire as a clearing tool in agriculture continues to cut away at rainforest, but logging and mining are far more aggressive than the planting of rice, tobacco, sweet potatoes, and sugarcane.

Climate change effects, combined with more frequent El Nino oceanic and atmospheric events, are projected to heavily impact the Kalimantan Rainforests biome. Higher water temperatures in the Indian Ocean are predicted to combine with El Nino events to reduce rainfall over Borneo—even potentially to drought levels, effectively making annual dry seasons much drier. While wet seasons may see heavier rain as climate change sets in, the soils in most of Kalimantan are not likely to store enough water to overcome the dry-season precipitation deficits, according to some researchers.

Rainforest tree species are not known for their drought-resistance; therefore, the impact could well accelerate itself, as insects, disease, and invasive species work to decimate the native rainforest habitat—and undercut its contribution to slowing global warming by virtue of eroding its large carbon-absorption capacity. The increasing incidence of fire events, coupled with more pronounced habitat fragmentation, will also work against rainforest sustainability here.

Kalimantan administrators have partnered with the World Wildlife Fund (WWF) to help maintain the region's rich biodiversity by implementing new protected areas or adding on to existing ones. Among the WWF's focal animal groups for habitat preservation priority are primates and the bird family *Bucerotidae,* known as hornbills. Management policies aimed at protecting their habitats will reinforce managing the protected areas, as well as surrounding landscapes, as integrated systems. Freshwater swamps are under-represented among the protected areas, as are peat swamp forests.

Indonesia has established a network of parks and reserves that buttress efforts to protect the rainforests and related biomes here. Located in the Central Kalimantan province is the Tanjung Puting National Park, which serves as the oldest conservation site of Kalimantan's flora and fauna. In the Gunung Palung National Park, close to Mount Raya, there are a great number of rare and beautiful flora and fauna, including perhaps more than a dozen species of *Rafflesia,* the celebrated, if carrion-scented, giant parasitic flower that makes its home upon various vines. *Rafflesia* grow to 3 feet (1 meter) across or more, and can weigh 22 pounds (10 kilograms). Other preserves in Kalimantan include the Danau Sentarum; Baka Bukit Raya; Kutai; and Kayan Mentarang National Parks.

WILLIAM FORBES
JOEL COVINGTON

Further Reading

Guhardja, Edi, Mansur Fatawi, Maman Sutisna, and Tokunori Mori, eds. *Rainforest Ecosystems of East Kalimantan: El Nino, Drought, Fire and Human Impacts.* New York: Springer, 2000.

Kumagai, Tomo'omi and Amilcare Porporato. "Drought-Induced Mortality of a Bornean Tropical Rainforest Amplified by Climate Change." *Journal of Geophysical Research* 117, no. 1 (2012).

Laurance, William F. and Carlos A. Peres, eds. *Emerging Threats to Tropical Forests.* Chicago: University of Chicago Press, 2006.

World Wide Fund for Nature (WWF). "Conservation in Borneo." 2012. http://wwf.panda.org/what _we_do/endangered_species/great_apes/ orangutans/?uProjectID=ID0169.

Kara Sea

Category: Marine and Oceanic Biomes.
Geographic Location: Arctic Ocean.
Summary: A low-productivity ecosystem covered by seasonal ice for much of the year, the Kara Sea is characterized by a variety of habitats.

A marginal sea in the Arctic Ocean, the Kara Sea is north of the Siberian coast, with the Barents Sea to its west and the Laptev Sea to its east. The Kara Strait and the Novaya Zemlya archipelago separate it from the Barents Sea, while the Severnaya Zemlya archipelago divides the sea from the Laptev. Though the Barents is warmed by Atlantic currents, the Kara Sea has more in common with the Laptev. Very cold year round, it is frozen about nine months of the year, and has low salinity that varies seasonally with the freshwater inputs from the Ob, Yenisei, Pyasina, and Taimyra rivers. The water exchange with the Arctic Ocean and periodic river inputs contribute to providing a range of marine habitats, despite the ecosystem's low productivity.

Much of the Kara Sea lies on the Arctic shelf, and while it is deeper than the East Siberian Sea or much of the Laptev, the average depth is still only 328 feet (100 meters). This has contributed to the prevalence of ice cover, and until modern icebreaking technology, the sea was considered to be non-navigable except in summer.

While the other marginal seas of the Arctic have few islands except along their coasts, the Kara is an exception. Its islands include Uedineniya, or Lonely Island; Vize; and Voronina, which is located in the open sea. The Nordenskjold Archipelago consists of more than 90 islands, including five subgroups.

Biodiversity

Unlike the Laptev, the Kara is an important fishing ground despite the sea's mostly ice-bound condition, but it is not as commercially developed as the warmer Barents. Common fish include flatfish, smelt, scorpionfish, and Arctic cod.

The inland tundra serves as breeding grounds and nesting colonies for as many as 16 million migrating seabirds. Local bird species include the little auk (*Alle alle*), barnacle goose (*Branta leucopsis*), pink-footed goose (*Answer brachyrhynchus*), Sabine's gull (*Xema sabini*), and white-billed diver (*Gavia adamsii*). Also, 16 whale and seven seal species, respectively, have been recorded in the Kara. Polar bear, walrus, and narwhal also are common in great numbers; they use the coastal areas as breeding grounds. Polar bears roam the frozen edges of the sea, foraging, fishing, and hunting seals.

Ice algal activity accounts for 50 percent of total primary productivity in the Kara Sea, as in the Laptev and East Siberian, and is depended on by other Arctic biomes through sedimentation and life cycles. Especially abundant are protozoan and metazoan ice meiofauna, such as nematodes, crustaceans, rotifers (wheel animals), and turbellarians (flatworms).

The larvae and juveniles of benthic, or bottom-dwelling, animals such as mollusks and polychaetes (a class of marine worms) migrate into the ice matrix to feed on ice algae in Kara's shallow waters, forming a sophisticated and diverse ice invertebrate community.

Some species of shrimplike crustaceans here are nearly endemic (found nowhere else) to the Arctic marginal seas; they live on the underside of ice floes. They occur seasonally and transport particulate organic matter from sea ice to the water column through their feces. Pelagic fish such as the Arctic cod feed under and near sea ice, depending on ice crustaceans and other small fauna. The decreased sea ice due to warming waters threatens these communities and the other elements of the food chain that depend upon it. This is an aspect of climate change that may also open the door to the large-scale encroachment by exotic species here.

Environmental Threats

Whitefish have declined significantly, especially in the west, where numbers dropped by half from the 1980s to the 1990s. Overfishing is a growing concern, and coastal dumping of radioactive material from nuclear power plants and submarines has put millions of fish, shellfish, star-

A close view of a barnacle goose, a species common to the Kara Sea. About 16 million migrating seabirds use the nearby inland tundra for nesting. (Wikimedia/Ltshears)

fish, and marine mammals in jeopardy. Unlike the Laptev and East Siberian Seas, the Kara Sea is fairly close to the Russian industrial centers, which are the most polluting in the Arctic region. From the 1960s to the 1990s, the Kara Sea was a dumping ground for high-level nuclear wastes from nuclear-powered submarines as a routine part of their refueling operations. Nuclear reactors suffering malfunctions also were discarded in the sea. At least 16 reactors (10 power-generation and six submarine reactors) were dumped on the sea bottom, and there is a good chance that more have gone unreported.

The petroleum and natural gas reserves detected and surveyed here present an extension of the West Siberian oil basin, but they have not been developed. The reserves are believed to be vast enough that development is inevitable as oil demand increases. If peak-oil models prove valid, the development will have a major impact on terrestrial and marine ecosystems here and in adjoining seas.

As one of the seas on the Arctic shelf, the Kara is expected to be among the Earth's first areas to feel the effects of climate change. Warming surface temperatures, changes in the mixed layer, and reduction of sea ice will greatly impact the distribution and timing of ice-related pelagic production and the deposition of carbon. In time, these changes will affect the deeper waters of the Arctic Ocean since they receive their carbon from the shelf.

Warmer waters and a reduction of ice will affect the life cycle of plankton, microflora, and microfauna, which will have a cascading effect on higher trophic levels such as seabirds and marine mammals. Eventually, climate change will shake up the whole ecosystem. Migration will be affected, too, as longer summers or more plentiful polynyas (expanses of free water surrounded by ice) will encourage more wildlife to remain in place to feed, rather than traveling south.

Protected Areas

The region's Franz-Josef-Land nature reserve is one of the largest marine protected areas in the Northern Hemisphere, and home to diverse habitats. The former Soviet Union had restricted the rights of indigenous peoples to use much of their traditional reindeer-pasturing land around the coasts here, while allowing nonindigenous arrivals to exploit the area for oil and mineral reserves. The collapse of the Soviet Union, however, has made it more difficult for the indigenous people to collectivize and strengthen their political autonomy. Without controls implemented by the former Soviets, the chief factor preventing environmental deterioration of the Kara Sea region is its remoteness.

BILL KTE'PI

Further Reading

Akulichev, Viktor Anatol'evich. *Far Eastern Seas of Russia*. Moscow: Nauka Press, 2007.

Butler, William. *Northeast Arctic Passage*. New York: Sijthoff & Noordhoff, 1978.

Heileman, S. and I. Belkin. *Large Marine Ecosystems of the World: Kara Sea*. Washington, DC: National Oceanic and Atmospheric Administration, 2009.

Kazakh Steppe

Category: Grassland, Tundra, and Human Biomes.
Geographic Location: Asia.
Summary: This transition area between deserts and forests contains many closed basins used by migratory waterfowl. The biome has been damaged by ill-conceived irrigation projects and nuclear bomb testing, leaving a harsh legacy for plants and animals.

Set squarely in the center of Asia, the Kazakh Steppes biome extends from the Ural Mountains in the west to the Altai Mountains in the east, and includes areas of both Russia and Kazakhstan along the border between the two countries. This area constitutes the world's largest continuous grassland. Much of the region has been plowed. Thousands of lakes and wetlands dot the terrain, however, providing vital habitat for waterfowl in a dry region. In the flat expanse of the southern Kazakh steppe, the Irghyz and Turgay rivers flow for more than 100 miles (161 kilometers) before evaporating.

The steppes are the transition from the central Asian desert to the vast boreal spruce-fir forest that covers northern Europe and Asia. They spread across an area of extreme weather that has hot, windy summers with periodic droughts and very cold winters. The steppes contain patches of forests, called *koloks*, both in the northern reaches and in upland areas, that include aspen, birch, and pine.

The Kazakh steppes are believed to be where horse domestication took place about 6,000 years ago. Archaeological evidence indicates that horses were first bridled here, and that a secondary activity was related to processing mare's milk and carcass products, such as leather.

Biodiversity

Wilder regions of the steppes support diverse populations of rodents, including ground squirrels, hamsters, voles, and lemmings. They also support herds of deerlike saigas, boars, lynx, and badgers. Millions of waterfowl nest and use the wetlands and lakes for migration. Notable waterfowl sites are the Naurzum Nature Reserve in Kostanay Oblast (Province) and the Korgalzhyn Nature Reserve in Akmola Oblast, both of which are part of the Saryarka World Heritage area; the lakes of the lower Turgay and Irgiz Rivers in Aktobe Oblast; the Tobol-Ishim forest steppe of Tyumen Oblast and North Kazakhstan Oblast; and Chany Lake, Shchuchy Lake and the Lower Bagan wetlands in Novisibirsk Oblast. These areas are designated as Wetlands of International Importance and somewhat protected under the Ramsar protocol of 1971.

Wetlands in the steppe region typically are composed of mixtures of saline, brackish, and freshwater lakes covered with reeds. The Tengiz-Korgalzhyn lakes contain the northernmost breeding area for greater flamingoes, along with breeding Dalmatian pelicans. Both species migrate south for the winter.

Scientists estimate that 800 species of plants can be found in the Kazakh Steppe biome, but there has been little concerted research review. Some of its unique species are xerophytes, plants that have adapted to water-deprived environments, and halophytes, plants that adapted to salty environments, such as in saline soils and saltwater lakes. Among the area's grasses are *Stipa zalesskii* and furrowed fescue (*Festuca rupicola*), and sagebrushes such as *Artemisia marschalliana*. The most dominant vegetation is compact turf or cushion-like plants, plus lilies (*Liliaceae*), the bulb-generated flowering herbs of family *Amaryllidaceae*, and the genera *Tulipa*, *Ornithogalum*, *Gagea*, *Ixiolirion*, and *Eremurus*. The area also is home to varieties of Russian thistle or tumbleweed.

Water Use

On the south edge of the steppe is the Aral Sea, which is rapidly shrinking from the diversion of its two water sources: the Syr and Amu rivers. The area of the formerly freshwater-to-brackish inland sea shown on most maps is inaccurate now, because most of the water has been diverted for irrigation. The drainage from agricultural land is collected in a series of canals that ultimately flow to the site of the former Kara Salt Lake or to Saryka-mish Lake in Turkmenistan. The Kara Salt Lake has been renamed Golden Age Lake in anticipation of its successfully collecting this drainage water.

One side effect of the diverted water is that irrigation has saturated the ground and brought salt to the surface throughout the region. Numerous saline lakes have formed from the saturated ground, but it is hoped that a new drainage scheme will cause the water table to drop, allowing for reclamation of saline soils. Meanwhile, the Kok-Aral Dam diverts water from the Syr River into the North Aral Sea in an effort to reestablish a smaller, more stabilized lake and return the fishing industry to the area. In 2011, carp, pike, flounder, and perch were at last caught again after a long hiatus in what is now called the Northern Aral Sea.

Vozrozhdeniya Island in the Aral Sea was used as a bioweapons laboratory. There, the former Soviet Union tested anthrax, plague, and other bacteria for weapons use. Other research was on vaccines and how long micro-organisms would survive in the soil. The anthrax-contaminated area was neutralized in 2002.

Land Use

Kazakhstan's Bayan-Aul National Park in Pavlodar Oblast contains eroded rock formations that resemble toadstools and pillars, as well as freshwater lakes in the grassland-forest patches of the steppe. On the south edge of the steppe, where it grades into the central Asian desert, is the Baykonur Cosmodrome. The former Soviet Union built it as a space center in the 1950s, and Russia now rents it from Kazakhstan under an agreement that will continue through 2050. A new alternative facility in Amur Oblast, Russia, is under construction to replace the current site.

Environmental Challenges

The human impact on the Kazakh Steppe biome continues to pose environmental challenges, ranging from fallout from the nuclear testing programs that cause many areas to deal with significant radioactive pollution to the huge irrigation projects that caused the Aral Sea level to drop so substantially that its diminished size has changed the climate in

"Atomic Lakes"

To the east of Kazakhstan's Bayan-Aul National Park is the former Semipalatinsk nuclear test site, which covers some 7,000 square miles (18,130 square kilometers). There, 456 nuclear tests were conducted between 1949 and 1989 by the former Soviet Union. More than 100 of these tests were of the aboveground type and released radiation directly into the atmosphere.

Two "atomic lakes" were created by nuclear-weapons explosions here, as part of a study to see whether rivers could be diverted and canals dug by using such explosions. Plutonium hot spots are found throughout the Semipalatinsk site, which was closed in 1991. Research reactors here are now owned and operated by the government of Kazakhstan. The nuclear test tunnels have been closed and sealed.

the area and left wide swaths of land now subject to erosion. Acid rain from petrochemical industry sites, too, has damaged the environment within Kazakhstan and affected neighboring countries. Pollution from industrial and agricultural sources has compromised the underground water supply to an unknown extent.

Some types of wildlife here are in danger of extinction due to overall pollution levels, and the shape and dynamics of the ecosystem are expected to change with global warming pressing the drying trend further. These habitat stresses will be a continuing challenge throughout the 21st century.

HAROLD DRAPER

Further Reading

Ellis, William S. "The Aral: A Soviet Sea Lies Dying." *National Geographic* 177 (1990).

Outram, Alan K., et al. "The Earliest Horse Harnessing and Milking." *Science* 323 (2009).

Pala, Christopher. "In Northern Aral Sea, Rebound Comes with a Big Catch." *Science* 334 (2011).

Stone, Richard. "A New Great Lake—Or Dead Sea?" *Science* 320 (2008).

Kennebec Estuary

Category: Marine and Oceanic Biomes.
Geographic Location: North America.
Summary: This unique estuarine resource in northern New England provides critical fish, plant, and waterfowl habitat off the Gulf of Maine.

Maine's Kennebec River Estuary is one of northern New England's most significant natural areas, where the fresh water from several of Maine's largest and most important rivers meets the salty North Atlantic Ocean. This meeting creates a blending that provides a unique estuarine environment; it is critically important for numerous plants, fish, and other wildlife, including imperiled species such as the bald eagle, spotted turtle, harlequin duck, tidewater mucket, ribbon snake, redfin pickerel, and Atlantic salmon.

The Kennebec Estuary—the largest tidal estuary north of the Chesapeake Bay on the east coast of the United States—contributes, on a seasonally adjusted basis, more than 6 billion gallons (22.7 billion liters) of freshwater to the Gulf of Maine daily. It also provides a marine habitat that has a substantial effect on the health of Maine's coastal waters and fisheries, and beyond. The area contains over 20 percent of Maine's tidal marshes, a large percentage of Maine's sandy beaches, dune habitats, and globally rare pitch pine woodland communities.

Overall, this estuary on the midcoast of Maine is comprised of two distinct components: Merrymeeting Bay and the lower Kennebec River.

Merrymeeting Bay and the Kennebec River

Located on the upper section of the river, Merrymeeting Bay is formed by the confluence of six rivers meeting the Atlantic Ocean. These rivers—the Kennebec, Androscoggin, Cathance, Abagadasset, Muddy, and Eastern—drain 40 percent of Maine's land, as well as part of New Hampshire. Overall, Merrymeeting Bay is an inland bay of fresh- and saltwater approximately 15 miles (24 kilometers) from the Atlantic Ocean.

Merrymeeting Bay provides an extensive expanse of shallow and nutrient-rich waters; vital plant cover; and more than 4,500 acres (1,821 hectares) of habitat for waterfowl, supporting critical foraging, nesting, and wintering activities. The benefits accrue to a significant number of waterfowl within the Atlantic Flyway, as well as a host of other flora and fauna.

The section downstream from Merrymeeting Bay consists of the lower Kennebec River as it runs between a series of forested uplands and rocky shorelines for approximately 15 miles (24 kilometers). The deep water and strong currents that typify the lower Kennebec provide additional types of habitat, including critical shelter for fish such as the Atlantic salmon, alewife, herring, and shad. Notably, the high salt content and significant tidal shifts of more than 5 feet (1.5 meters) prevent the Kennebec River from freezing over in this section, which allows many bird species to winter in this area.

Biota

The Kennebec Estuary biome is home to a wide range of plant life offering distinct rare and exemplary natural communities, including alder thicket, pitch pine bog, dune grassland, mixed saltmarsh, salt-hay saltmarsh, pitch pine woodland, both brackish and freshwater tidal marshes, maritime spruce-fir forest, and rose maritime shrubland.

Characteristic plants in the ecosystem include mudwork, yellow pond-lily, pygmy weed, tidal spikerush, water pimpernel, mountain laurel, long-leaved bluet, estuary monkeyflower, smooth sandwort, Long's bitter-cress, sweet pepper-bush, and horned pondweed.

There are significant wildlife habitats here, hosting such hallmark species as the seabird roseate tern (*Sterna dougallii*) seabird, and providing deer wintering area.

More than 50 species of freshwater fish and 10 species of anadromous fish (those that migrate from the ocean to spawn in freshwater rivers) use Merrymeeting Bay, including Atlantic salmon (*Salmo salar*), shortnosed sturgeon (*Acipenser brevirostrum*), Atlantic sturgeon (*A. oxyrinchus*), striped bass (*Morone saxatilis*), American shad (*Alosa sapidissima*), the herring species alewife (*Alosa pseudoharengus*), blueback herring (*Alosa aestivalis*), and rainbow smelt (*Osmerus mordax*).

At least one rare mussel species, the tidewater mucket (*Leptodea ochracea*), lives here. One of the bay's small tributaries is Maine's only known location for the redfin pickerel (*Esox americanus*). American eels, declining in much of their geographic range, are abundant here.

Freshwater marshes ring the bay; these provide wild rice and other plant foods for waterfowl, and support a number of rare plants. Among these are American bulrush (*Schoenoplectus pungens*), wild rice (*Zizania aquatica*), pickerelweed (*Pontederia cordata)*, bullhead lily (*Nuphar variegatum*), Parker's pipewort (*Eriocaulon parkeri*), water pimpernel (*Samolus valerandi)*, pygmy weed (*Crassula aquatica*), estuary bur-marigold (*Bidens hyperborea*), Eaton's bur-marigold (*Bidens eatonii*), mudwort (*Limosella australis*), and spongy arrowhead (*Sagittaria calycina* ssp. *spongiosa*). Submerged aquatic vegetation includes tapegrass (*Valisneria americana*) and pondweed (*Potamogeton perfoliatus*).

Effects of Human Activity

Historically, the Kennebec River estuary was a significant cod fishery, providing both a spawning and nursery habitat for many marine species. Over time, however, human settlements and the corresponding heavy use of the estuary's resources have imposed a significant and lasting toll. By the mid-20th century, the biota of several of the rivers contributing to Merrymeeting Bay had collapsed. Massive summer fish kills resulted from depleted dissolved oxygen levels in the waterway, while pesticide use in the watershed had strong negative reproduction effects on native bald eagle populations, among other harmful effects stemming from industrial activities surrounding the waterway.

This degradation, as well as similar harm to river habitats across the United States, led to the passage of the Clean Water Act in the early 1970s. The Act focused public attention on the negative impacts of overfishing, land clearance, damming, and water pollution, all of which have caused environmental damage to this ecosystem. Fortunately, the attention has led to at least a partial recovery of this resilient biome, although a full recovery may never occur or may require a substantially longer time window. New challenges may later manifest other obstacles, such as the recent appearance of harmful algae blooms (red tides) that have periodically paralyzed the region's biota.

Ongoing, the estuary faces pressure from residential development that is damaging some habitats, as well as inappropriate recreational use, and the introduction of nonnative species of fish and plants. Climate change and the subsequent rise in sea level pose lasting threats to the Kennebec. Tidal marshes and other shoreline habitats may flood from rising seas, while the entire ecosystem could be altered by rising air and water temperatures, shifts in ocean circulation, saltwater intrusion, and increased erosion from more intense storms.

Conservation priorities include preserving habitat for migratory fish; maintaining undeveloped shoreline for bald eagle nesting and roosting; and maintenance of beaches and dunes, freshwater and saltwater tidal marshes, and the upland forests that buffer these shoreline ecosystems.

JESS PHELPS

Further Reading

Burroughs, Franklin. *Confluence: Merrymeeting Bay.* Gardiner, ME: Tilbury House, 2006.

Coffin, Robert P. Tristam. *Kennebec: Cradle of Americans.* Camden, ME: Down East Books, 2002.

Köster, D., J. Lichter, P. D. Lea, and A. Nurse. "Historical Land-Use Effects on a Freshwater Tidal Ecosystem." *Ecological Applications* 17 (2007).

Lichter, J., H. Caron, T. Pasakarnis, S. Rodgers, T. S. Squiers, Jr., and Charles S. Todd. "The Ecological Collapse and Partial Recovery of a Freshwater Tidal Ecosystem." *Northeastern Naturalist* 13 (2006).

McLane, Charles and Carol McLane. *Islands of the Mid-Maine Coast.* Gardiner, ME: Tilbury House, 2003.

Kimberley Tropical Savanna

Category: Grassland, Tundra, and Human Biomes.

Geographic Location: Australia.

Summary: The tropical heat of the region and varying rainfall regimes have resulted in unique flora and fauna, which are threatened by land degradation and introduced species.

The Kimberley region of Australia is located in the northern part of the state of Western Australia; it covers some 163,521 square miles (423,517 square kilometers), an expanse slightly larger than the country of Japan. The area comprises the rugged sandstone and limestone ranges of northwestern Australia, the more arid lowlands of Dampier Land in the southwest, and the Daly Basin to the northeast.

The Kimberley is one of the hottest parts of Australia, with an average annual mean temperature of 81 degrees F (27 degrees C). About 90 percent of the region's rainfall occurs during the wet season of November to April, when cyclones are common (especially around Broome) and the rivers flood.

The annual rainfall is highest in the northwest, where Kalumburu and the Mitchell Plateau average 50 inches (1,270 millimeters) per year. Precipitation is lowest in the southeast, at about 20 inches (520 millimeters), and semiarid conditions are particularly harsh during the May-to-October dry season.

Flora

This region is mainly covered by open savanna woodland with boab trees (*Adansonia gregorii*)—known for their hardiness and longevity—in the drier areas, and Darwin stringybark (*Eucalyptus tetrodonta*) and Darwin woollybutt eucalypts (*E. miniata*) in other parts of the Kimberley.

Canopy heights are usually 16–49 feet (5–15 meters). Understories are dominated by such tall grasses as sorghum, heteropogon, themeda, chrysopogon, aristida, and family *Eriachne*. Northern parts of the Kimberley region are known for their dense subcanopy of sand palms (*Livistona eastonii*).

Various tree species and stands of tropical dry broadleaf forests are located along the banks of the Ord and Fitzroy rivers here. Other patches of tropical dry broadleaf forest, called monsoon forests, along with deciduous vine forest or vine thicket, are located in sheltered gorges up north where there is a lot of rain.

Vegetation in the ecoregion is linked to soil, geological factors, and rainfall. The red sandy soil of the Dampier Peninsula is known for its pindan wooded grassland, while the more fertile areas such as the Ord Valley are grasslands consisting of *Chrysopogon*, *Aristida*, *Dicanthium*, and *Xerochloa* (rice grass) in the wetter valleys. Plant varieties are greatest along the river banks.

Many introduced plant species, including the castor oil plant (*Ricinus communis*), cocklebur (*Xanthium strumarium*), and parkinsonia (*Parkinsonia aculeata*), are most commonly found in areas around the catchments of major rivers.

A boab tree (Adansonia gregorii) in Australia. Much of the Kimberley Tropical Savanna biome is made up of open land like this, dotted with boab trees.(Thinkstock)

Fauna

Typical of Australia, the animal population includes marsupials. The Kimberley is home to bilbies (*Macrotis lagotis*), also known as rabbit-eared bandicoots. The bird life in the Kimberley region includes emu, parrots, cockatoos, various migratory species, and raptors.

Reptiles here include the saltwater crocodile (*Crocodylus poroxus*)—the largest of all living reptiles, growing as long as 18 feet (5.5 meters) and weighing up to 2,200 pounds (1,000 kilograms)—and the freshwater crocodile (*Crocodylus johnstoni*), which reaches a maximum length of 10 feet (3 meters) and weights of up to 200 pounds (90 kilograms).

The area also includes goannas (monitor lizards), which grow to up to 8 feet (2.5 meters) long, as well as blue-tongue lizards and other reptiles that have adapted to survive in the dry environment and drink very little water.

The Kimberley is host to the egg-laying function of a range of marine turtles, particularly the green turtle (*Chelonis mydas*), loggerhead turtle (*Caretta caretta*), flatback turtle (*Chelonia depressa*), and hawksbill turtle (*Eretmochelys imbricata*). The Kimberley Tropical Savanna biome and nearby regions host some 25 species of tree frogs and 51 species of ground frogs—one-third of the known Australian frogs. In addition, the Kimberly has vast numbers of insects, especially ants.

Local fauna includes species introduced to the area over the past two centuries, such as foxes, rabbits, cats, dogs, and bristly-back razorback pigs. These animals are present throughout Australia, much to the detriment of indigenous creatures. Not only do these introduced species compete with native animals for food and water, some have become predators.

Environmental Conservation and Threats

The rich diversity of the Kimberley's flora and fauna has led to the establishment of six national parks—Drysdale River, Geikle Gorge, Hidden Valley, Mitchell River, Purnululu (Bungle Bingle), and Tunnel Creek—as well as the Prince Regent Nature Reserve. Conservation largely resides in the area's remoteness and difficulty accessing the gorges, escarpments, and deeply dissected sandstone plateaus of the northern reaches. Accessibility is increasing, however, due mostly to poorly managed tourism and ongoing proposals for mining developments.

Climate change has resulted in rising water levels in coastal areas. Rainfall has increased by as much as 10 inches (250 millimeters) per year since 1967, but recent studies suggest Asian-sourced particulate air pollution—and not global warming—is the main culprit.

The major threats, however, result from current human activity such as land clearance, mining, and introduced species. The largest area of cleared land occurred in the Ord Valley, but the expansion of sheep grazing in the catchment of the Fitzroy River also has resulted in land degradation, with topsoil loss. By 1976, nearly a third of the catchment area had degraded to poor or very poor condition.

JUSTIN CORFIELD

Further Reading

Beard, J. S., K. A. Clayton, and K. F. Kenneally. "Notes on the Vegetation of the Bougainville Peninsula, Osborn and Institut Islands, North Kimberley District, Australia." *Vegetation* 57, no. 1 (1984).

Burt, Jocelyn. *The Kimberley: Australia's Unique North-West.* Nedlands, Australia: Tuart House, 1999.

Edwards, Hugh. *Kimberley: Dreaming to Diamonds.* Swanbourne, Western Australia: Hugh Edwards, 1991.

Lehmann, Caroline E. R., Lynda D. Prior, and David M. J. S. Bowman. "Fire Controls Population Structure in Four Dominant Tree Species in a Tropical Savanna." *Oecologia* 161, no. 3 (2009).

McGonical, David. *The Kimberley.* Terrey Hills, New South Wales: Australian Geographic, 1990.

Tyler, Ian. *Geology and Landforms of the Kimberley.* Kensington, Western Australia: Department of Conservation of Land Management, 2005.

Winter, W. H. "Australia's Northern Savannas: A Time for Change in Management Philosophy." *Journal of Biogeography* 17, nos. 4–5 (1990).

Kinabalu Montane Alpine Meadows

Category: Grassland, Tundra, and Human Biomes.
Geographic Location: Asia.
Summary: The highest mountain region between the Himalayas and New Guinea is a center of plant diversity and endemism; it is especially known for orchids, giant rafflesia, and pitcher plants.

The Kinabalu Montane Alpine Meadows biome is located on Borneo, the third-largest island in the world. Borneo is filled with mountains in a tropical wet climate zone, with widespread rainforests that provide a wide variety of unique habitats promoting plant and animal diversity. Borneo is home to 15,000 species of flowering plants, 34 percent of which are endemic, or exclusive to the island. It is a hotbed of scientific discovery, with an average of three new species discovered every month.

The highest mountains are in the area of northern Borneo dominated by Mount Kinabalu, the Trus Madi Range, and the Crocker Range, all of which are in the Malaysian state of Sabah. Other high mountains are found in the Tawau Hills and Maliau Basin of Sabah, the extensive plateaus along the border between Indonesia and Malaysia, and in the isolated Hose and Dulit ranges of the Malaysian state of Sarawak.

Vegetation

Mount Kinabalu and the Crocker Range exhibit changes in vegetation with increasing elevation. Below 3,281 feet (1,000 meters) is tropical lowland rainforest. In the range of 3,281–8,530 feet (1,000–2,600 meters), a montane zone exists, with oaks, chestnut, myrtles, eucalyptus, and cloves. Because of the mixture of lowland and mountain species, these areas have the greatest plant diversity.

Meadows here are especially known for endemic figs, orchids, laurels, heathers, nettles, pitcher plants, buckthorns, stone oak (actually a beech family), rhododendron, and magnolia plants. At 8,530–10,499 feet (2,600–3,200 meters)

is a zone of dwarf shrubs, with great abundance of lichen, moss, and ferns. Above 10,499 feet (3,200 meters) is a rocky area, with thin soils supporting mostly rhododendron shrubs and herbs such as Elatostema, a genus of nettle.

South of Mount Kinabalu, montane rainforests occupy other areas in the mountains of Borneo above 3,281 feet (1,000 meters). The same elevation belts as on Mount Kinabalu are found, with a montane zone above 3,281 feet (1,000 meters), a rhododendron belt at higher elevations, and an alpine meadow on the highest peaks.

Biodiversity

The Kinabalu Montane Alpine Meadows ecoregion is a high-elevation biome with 4,500 species of flora—among the richest in the world—along with 114 mammal and 180 bird species. There are 78 species of fig, 750 orchid species, and 30 species of pitcher plant. Kinabalu National Park protects meadow habitats in some of the highest mountains between the Himalayas and New Guinea, topped by Mouth Kinabalu at 13,455 feet (4,101 meters). Half of all families of flowering plants are represented here. Notable plants are orchids; giant rafflesia flowers, which grow up to 3 feet (1 meter) in width; and the world's

A parasitic rafflesia in bloom in a Borneo forest. The flowers can be as large as 3 feet (1 meter) across and smell much like rotten meat. (Thinkstock)

largest moss, which grows up to 3 feet (1 meter) tall. While the parasitic rafflesia may be the largest flower in the world, it is not a pleasant one: To attract its pollinator flies, it emits a smell of rotten meat.

Of the bird species found here, 23 are endemic. Mammals native to the area include squirrels, a shrew, and a ferret-badger. Endemic birds include the mountain serpent eagle, red-breasted partridge, crimson-headed partridge, and Bornean spiderhunter. The world's longest stick insect, Chan's megastick, was discovered in forests near the meadows here.

Protected Areas

Protecting the Kinabalu montane alpine meadows is Crocker Range National Park, the largest protected area in Sabah. Both it and the Kinabalu National Park—a World Heritage site—are Important Bird Areas for endemic birds. Near Kinabalu is the Rafflesia Forest Reserve and Information Centre, which has trails to the world's largest flowers.

The Danum Valley Conservation Area is an important scientific research center. Mammals found include rhinos, elephants, orangutans, mouse deer, pigs, leopards, and leopard cats. There are 275 known bird species, including nine endemics. A new bird species—the spectacled flowerpecker—was recently discovered in Danum Valley.

To the south of Mount Kinabalu is the Maliau Basin Conservation Area, a saucer-shaped basin surrounded by cliffs up to 6,234 feet (1,900 meters) high. Because access to the area is difficult, it is known as Sabah's Lost World. Anyone able to venture to this region will have access to 280 species of birds, with many endemics. Pulong Tao National Park is a project funded by Japan and Switzerland in the Kelabit Highlands of eastern Sarawak, Malaysia. Notable plants are a gymnosperm timber, orchids, pitcher plants, gingers, and rhododendrons.

Kayan Mentarang National Park includes an extensive mountain plateau on the border between East Kalimantan and Sarawak. It is known for sun bears, which are the smallest of the bears; pangolins, mammals that eat ants and termites; and wild pigs. There is one volcano in the Borneo highlands, located in Tawau Hills National Park of Sabah. This park is noted for waterfalls, hot springs, and a volcanic landscape. Bombalai, Borneo's only active volcano, is found here.

Despite the many protected areas, environmental threats persist. Some of the slopes outside the parks are being cleared for agriculture, development, mining, and wildlife trades—often without buffer zones between the various uses. Insufficient staff to enforce the various conservation and protection measures also is a factor. Climate change is beginning to affect species habitat, range, and reproduction. Forests under stress from temperature shifts exhibit increased occurrence of disease and harmful pests. Pollution and changing weather patterns combine to alter both rainy and dry periods, which is stressing plant and animal reproductive cycles.

HAROLD DRAPER

Further Reading

Beaman, Reed S. *Phylogeny and Biogeography of Elatostema (Urticaceae) From Mount Kinabalu, Sabah, Malaysia.* Gainesville: University of Florida, 2000.

Garbutt, Nick. *Wild Borneo.* Cambridge, MA: MIT Press, 2006.

Ildos, Angela S. and Giorgio G. Bardelli. *Great National Parks of the World.* Heathrow, FL: AAA Publishing, 2001.

Normile, Dennis. "Saving Forests to Save Biodiversity." *Science* 329, no. 1 (2010).

Klamath Estuary

Category: Marine and Oceanic Biomes.
Geographic Location: North America.
Summary: The Klamath Estuary is located within the Redwood National Park and the Yurok Reservation in northern California and is an important habitat area for native peoples, salmon, and other wildlife.

The Klamath Estuary connects the Klamath River with the Pacific Ocean in northern California. The area at the mouth of the estuary is shared by the Yurok Indian Reservation and Redwood National Park. The estuary and river are important for fish migration; this was once the third-most-productive salmon river on the Pacific Coast, as more than one million salmon returned here to spawn each year. The estuary is home to coho salmon, a federally protected species.

The estuary area has been home to the Yurok Tribe of native Americans for 7,000 years. The Yurok depend on the estuary fishery for sustenance, and on culturally important plants including some of the willows (*Salix* spp.) and ferns (*Pteridophyta*), which are essential for basket-making and regalia. Many plants here also are important for their medicinal qualities, used for healing and in Yurok ceremonies.

The Klamath Estuary biome provides habitat support for a great many animal and plant species, as well as opportunities for hunting, fishing, hiking, and bird watching. It plays important overall ecological roles as well; the wetlands here control floods by hydrologic absorption and storage of large volumes of water. They collect sediment, adding to vital bank and bottomland structure, while filtering water so that channels within the estuary are clear for species requiring low-turbidity conditions, and water is clean for the many species that live in it.

Fauna

The estuary is an important habitat to resident and migratory birds, particularly waterfowl. This is a vital part of the Pacific flyway between the Klamath and Siskiyou mountains. The endangered willow flycatcher (*Empidonax traillii*) is one species that makes its summer home in these wetlands. The great blue heron and mallard ducks are common to the area, and the Yurok use their feathers in ceremonies.

The Klamath River has higher diversity of anadromous fish—which are born in the river, migrate to the sea, where they live, and then return to their birthplace to lay their eggs—than any other in California. The population of such fish has declined over the years, and the ongoing restoration programs require strong partnerships between government and local people, all dedicated to preserving their numbers. Coho (*Oncorhynchus kisutch),* chum (*Oncorhynchus keta*), and chinook salmon (*Oncorhynchus tshawytscha*), as well as steelhead species, are key anadromous fish found here.

Chinook salmon, the largest species of Pacific salmon, is often referred to as king salmon because it can grow to 100 pounds (45 kilograms) or more, and 22–56 inches (56–142 centimeters) long. Chum salmon—from the chinook word *tzum,* which means *spotted*—are the second-most-abundant salmon species of all the Pacific varieties, and are sometimes called dog salmon.

Salmon populations are at risk when any part of their life cycle is disrupted. All species require adequate spawning habitats, such as a clean river location that is cool and well oxygenated. Without these conditions, salmon roe (eggs) may not develop properly. The salmon species of the Klamath estuary require such food as plantonic diatoms, copepods, kelps, seaweeds, jellyfish, and starfish that are found in these brackish waters. When young, they feed on insects, amphipods, and other crustaceans. They typically feed on other fish when older.

Today, there is great controversy over water use in the Klamath Basin. Dams have damaged the salmon migration, and much of the water from the upper basin is diverted for irrigation. Industrial pollution, mining, roads, and logging also have contributed to poor water quality. Most of the river basin is sparsely populated and rural. This may make it easier for restoration of the salmon populations, compared to rivers in more urban areas.

Flora

Vegetation of the Klamath Estuary biome occurs in five layers: aquatic, short, medium, tall, and extra tall vegetation. The aquatic layer includes macrophytes such as ditchgrass (*Ruppia cirrhosa*), water buttercup (*Ranunculus aquatilis*), and leafy pondweed (*Potamogeton foliosus*), which create floating or buoyant canopies at or near the water surface. This layer also includes non-rooted aquatic plants

Swans on a lake in the Lower Klamath National Wildlife Refuge in the Klamath River basin. Other birds that visit the area include great blue herons, mallard ducks, and the endangered willow flycatcher. The refuge, founded by President Theodore Roosevelt in 1908, was the first national waterfowl refuge in the United States. (U.S. Fish and Wildlife Service)

such as duckweed (*Lemna* spp.) and water hyacinth (*Eichhornia crassipes).*

Short vegetation, plants taller than 20 inches (50 centimeters), includes small emergent vegetation and plants such as watercress (*Rorippa nasturtium aquaticum*), saltgrass (*Distichlis spicata*), jaumea (*Jaumea carnosa*), creeping buttercup (*Ranunculus flamula*), and arrowhead (*Sagitaria* spp.), among others.

Medium vegetation never exceeds 29 inches (75 centimeters) tall; this includes pickleweed (*Salicornia virginica*), saltbrush (*Atriplex* spp.), rushes (*Juncus* spp.), and curly dock (*Rumex crispus*).

Tall vegetation, plants up to 59 inches (1.5 meters) high, includes emergent vegetation and the larger shrubs such as broad-leaved cattail (*Typha latifolia*), bulrush (*Scirpus californicus*), California blackberry (*Rubus ursinus*), and coyote brush (*Baccharis piluaris*).

Very tall vegetation includes plants such as shrubs, vines, and trees that are taller than 59 inches (1.5 meters) high. These include trees like western sycamore (*Plantanus racemosa*), Fremont

cottonwood (*Populus fremontii*), red alder (*Alnus rubra*), blue elderberry (*Sambucus mexicanus*), and hazelnut (*Corylus californicus*).

In addition to the native plants, the Klamath estuary is home to a federally endangered plant, the western lily (*Lilium occidentale*).

Because of habitat degradation and disturbances from development, the region's wetlands can be invaded by aggressive, highly salt-tolerant, nonnative vegetation, such as reed canary grass (*Philaris Urundinacea*), purple loosestrife (*Lythrum salicaria*), water hyacinth (*Eichornia crassipes*), and salvinia (*Salvinia molesta*). The most prevalent invasive species in the area, across several of the height layers, are reed canary grass (*Philaris urundinacea*), Himalayan blackberry (*Rubus procerns*), common reed (*Phragmites australis*), and the yellow pond lily (*Nuphar lutea*).

Environmental Restoration

In 2010, restoration and recovery projects began in the Klamath River basin. In the estuary, efforts were initiated to improve in-stream and streamside river habitats in order to benefit threatened coho salmon, as well as chinook salmon and steelhead trout. These included tree planting and expansion of a local plant nursery by the Yurok Tribe that was started in 2009 with funding supplied by the American Recovery and Reinvestment Act.

The tribe also installed complex in-stream structures and stabilized more than 1,000 feet (304 meters) of eroding stream bank. Ponds were built off these streams to provide winter habitat for the threatened coho salmon. Preliminary results show that 250 juvenile coho and 1,500 young chinook salmon were using this new habitat within the first year.

The Klamath Basin Restoration Agreement also started that year; it included a plan to remove four dams. Scientific studies and reports are underway to determine the impact of dam removal as a way to increase the chances of fish survival as they migrate upstream to spawn.

Climate Change

A recent report by the National Research Council suggested that the sea level offshore in this region will rise by 4 inches (10 centimeters) by 2030, and as much as 2 feet (61 centimeters) higher by 2100, which may cause wetlands within the Klamath Estuary biome to disappear. The plants in these areas now require more sediment to keep up with the rise in sea level. The deposition of sand and sediment is unlikely to change, however. Therefore, eventually these wetlands will be under water. Naturally, the river delivered sand to the estuary, but today much of this sediment is trapped behind upstream dams.

The wetlands also could also migrate inland if there is sufficient room. Removing upstream dams would increase sediment flow into the estuary and could offset sea-level rise in part. Further, sea-level rise in northern California is lessened slightly by upthrust in the land through plate tectonics and the ongoing uplift along the San Andreas Fault.

Climate change already is affecting the fragile estuary; scientists continue to prepare weather and climate models to help officials at all levels forecast and deal with the oncoming changes. Three such models agree that future summers will be drier than historical records, with precipitation declining 11–24 percent. Vegetation is likely to shift to favor grasslands such as sagebrush and juniper in the upper basin; oaks and madrone will be favored over maritime conifer forest of redwood, Douglas fir, and Sitka spruce, which are all projected to decline. Projections also indicate that wildfires will increase 11–22 percent by late century. Heat waves, severe precipitation, and prolonged drought are all expected to increase.

Scientists are working to find ways to help increase the resilience of local plant and animal communities, to blunt the negative impacts of climate change. This sense of environmental stewardship seems to be on the upswing in the Klamath River area.

GILLIAN GALFORD

Further Reading

Barr, Brian R., Marni E. Kooperman, Cindy Deacon Williams, Stacy J. Wynne, Roger Hamilton, and Bob Doppelt. *Preparing for Climate Change in the Klamath Basin.* Ashland, OR: National Center for Conservation Science & Policy and the Climate Leadership Initiative, 2010.

Patterson, William D. *Klamath River Estuary Wetlands Restoration Prioritization Plan v1.0.* Klamath, CA: Yurok Tribe Environmental Program, 2009.

Terence, Erica, et al. "Restoring the Shasta for Coho Salmon." *Klamath River News* 8 (Summer 2012).

Klamath-Siskiyou Forests

Category: Forest Biomes.
Geographic Location: North America.
Summary: Among the most important temperate coniferous forests in the world, the Klamath-Siskiyou faces unprecedented threats from land-use decisions and climate disruptions.

Renowned for its extraordinary biological diversity, the Klamath-Siskiyou is among the top coniferous forests on Earth. Straddling the California-Oregon border near the Pacific Ocean and covering an area of some 9.9 million acres (4 million hectares), the region is at the junction of the uplift Coast Ranges, the volcanic Cascades, and the ancient volcanic roots of the Sierra Nevada.

The Klamath-Siskiyou tops the charts among coniferous forests globally with accolades such as Area of Global Botanical Significance, proposed World Heritage Site and International Biosphere Reserve, Global Centre of Plant Diversity, and Global 200 ecoregion as anointed by the World Conservation Union (or IUCN), scientists, and the World Wildlife Fund, respectively. More conifer species and endemic (found nowhere else) plants occur here than nearly any other temperate conifer forest on Earth. The region's roadless areas have been dubbed the Pacific Coastal Outback, reflecting its status as one of the wildest areas remaining near the Pacific coast of the United States.

Geography, Climate, and Threats

The region's remarkable biological diversity is the result of the interplay of ancient, complex geology; varied but stable (long-term) climate; and natural disturbances like wildland fire. Crisscrossing mountain ranges, sometimes called the Klamath Knot, along with varied topography, allow plants with widely different environmental tolerances to persist in close proximity within this biome. Along the coast, temperate rainforests prevail, while a short way inland reveals Jeffrey pine savannas on dry slopes.

Overlapping mountain ranges allow plants from distant areas to fit together like pieces in a jigsaw puzzle. The serpentine and ultramafic bedrock geology, produced by colliding tectonic plates that uplifted mountain ranges when dinosaurs roamed, provide for unusual soils: deficient in some nutrients, but highly toxic in others. Extreme soils here have served as a barrier against weed invasions and as a maternity ward for endemic plant species specializing on them, although this is changing due to human-caused climate disruptions.

The Klamath-Siskiyou Forests biome has several properties that may aid in weathering the coming climate storm. Its central position—sandwiched between the Coastal Mountains,

Shasta Valley-Sierra, and Cascades—rugged terrain, and complexity of soils and microclimates might provide habitats of refuge for some species. However, climate-related models predict hotter summers—as much as 9 degrees F (5 degrees C) warmer—and drier conditions year round by 2100. Potentially, there could be a diminishment in coastal fog; this could trigger extinction of 10 percent or more of the known moisture-loving plants, invertebrates, and salamanders here, and perhaps even the loss of or heavy stress upon fog-dependent coastal redwoods.

Prior to European occupation, old forests covered up to two-thirds of the Pacific Northwest region. Today, less than 20 percent remains, due to industrial logging. Habitat fragmentation impacts many of the remaining forests, where intact forest tends to be of insufficient size and scope to provide adequately for a range of animal species. These forests generally support over 1,000 species of plants, invertebrates, and wildlife; are among the most carbon-dense ecosystems on Earth, important in ameliorating climate change; purify drinking water; and perform a myriad of life-giving services to people and wildlife.

Biodiversity

The region's varied climate has allowed rainforests to flourish nearly undisturbed for centuries along the coast because large, severe fires occurred infrequently, while inland more frequent and severe fires allow chaparral and savannas to prevail.

The oldest stage in forest succession occurs when forests are dominated by large old trees, snags and fallen logs, and have multi-layered or fairly continuous canopies from the ground up. Precisely when this stage in forest succession occurs varies based on site productivity and disturbance conditions; however, forests show maturation generally at 80 years or so.

Much attention rightly has focused on old-growth forests, but, surprisingly, the plant assemblages immediately following fire also support high levels of biological diversity due to presence of legacy structures like big fire-resistant trees; large, dead standing trees or snags; and fallen logs that persist for decades to centuries after fire.

Such structures aid plant establishment, as the roots of even dead trees continue to anchor soils, while snags provide shade for conifer seedlings. Fallen logs in streams provide hiding cover for salmon (*Oncorhynchus* spp.) and those on land soak up moisture to provide to lichens, salamanders, fungi, and invertebrates, which are the pioneers of post-fire succession. When logs eventually decay, they act as living nurseries ripe with plant propagules and organic matter for new soils. Thriving in these conditions are fire-dependent specialists such as black-backed woodpeckers (*Picoides arcticus*) and mountain bluebirds (*Sialia currucoides*) that feed on destructive insects, and shrubs like various *Ceanothus* spp., and manzanita (*Arctostaphylos* spp.) that return nitrogen to soils and restore mycorrhizae—the symbiotic fungi that attach to plant roots and help them absorb nutrients—to below-ground processes necessary for new plant growth.

Klamath-Siskiyou vegetation is a highly diverse mosaic of plant communities in different stages of succession from prior disturbances. Examples of canopy trees in the mixed evergreen forest type include Douglas-fir (*Pseudotsuga menziesii*), white fir (*Abies concolor*), ponderosa pine (*Pinus ponderosa*), and western red cedar (*Thuja plicata*). These mix with understory trees of canyon live oak (*Quercus chrysolepsis*), Pacific madrone (*Arbutus menziesii*), tanoak (*Lithocarpus densiflorus*) and golden chinkapin (*Chrysolepis chrysophylla*). Endemic conifers include Brewer's spruce (*Picea breweriana*) and Port Orford cedar (*Chamaecyparis lawsoniana*) found on serpentine soils, which typically support high levels of endemic plants. Species-rich fens contain the carnivorous cobra lily (*Darlingtonia california*), along with several rare plant associates.

In sum, the biome features the following:

- approximately 3,500 plant species, including 281 endemics;
- nearly two-thirds of the entire California floristic province on just 10 percent of the land mass of California;
- as many as 40 conifer species, one of three regions globally with such richness;
- up to 114 species of butterfly, one of three such areas in western North America;
- dragonflies, flightless beetles, arthropods, and bees are also thought to be exceptionally rich—but are far less catalogued;
- at least 235 mollusk taxa, including at least 60 percent that are endemic;
- nearly 80 percent of all amphibian and reptile species in the Pacific Northwest;
- one of the highest diversities of dwarf mistletoe (11 taxa) in the United States;
- one of the greatest concentrations of ultramafic bedrock and serpentine geology in western North America.

Clearly, the Klamath-Siskiyou Forests biome is a region where the whole is greater than the sum of its parts.

DOMINICK A. DELLASALA

Further Reading

DellaSala, Dominick, ed. *Temperate and Boreal Rainforests of the World: Ecology and Conservation.* Washington, DC: Island Press (2011).

Olson, David, et. al. "Climate Change Refugia for Biodiversity in the Klamath-Siskiyou Ecoregion." *Natural Areas Journal* 32, no. 1 (2012).

The carnivorous cobra lily (Darlingtonia california), shown here in northern California, can be found in fens in the Klamath-Siskiyou forests. (Wikimedia/Noah Elhardt)

Strittholt, James, et. al. "Status of Mature and Old-Growth Forests in the Pacific Northwest, USA." *Conservation Biology* 20, no. 1 (2006).

Kuitpo Plantation Forest

Category: Forest Biomes.
Geographic Location: Australia.
Summary: This community forest here is designed to be sustainable, balancing production of forest products with conservation of fauna and flora. Human interaction, however, threatens this biome.

The Kuitpo Plantation Forest is a community forest initiated in 1899 in the Mount Lofty range of South Australia, within tourism distance of the state capital of Adelaide. It is intended to be managed as a sustainable plantation that produces timber products while maintaining forest cover and operating in the context of the maintenance of native flora and fauna, increasing demand not just for forest products, but also for consumption of leisure activities in natural environments. Proximity to urban Adelaide and free public entry have contributed to pressure on all South Australian forests, however, and the growing intensity and incidence of wildfires, some deliberately set, also represent significant threats.

European settlement in South Australia from the 1830s was swiftly followed by deforestation, resulting from demand for wood for construction and the introduction of foreign species in the search for rapidly growing replacements. By 1873, a law was passed to encourage tree planting, and further legislation established the forestry service, protection of native species, and the establishment of native reserves.

Since 1882, ForestrySA (as it is now known) has been the government agency responsible for managing forest lands and the flora and fauna contained within them. The Kuitpo Plantation Forest is one of several plantation areas in the Mount Lofty Ranges, which is itself one of several state-level forestry projects.

Kuitpo now covers some 8,896 acres (3,600 hectares) of land, with approximately 60 percent devoted to softwood plantation. The forest management system combines the production of more than 882,867 cubic feet (25,000 cubic meters) of timber annually with the maintenance of native wildlife and traditional community cultural practices and concerns.

Biodiversity

Native forest reservations within the Kuitpo forest include Mount Panorama, Christmas Hill, and Knott Hill, which total around 1,236 acres (500 hectares) of land. To reduce the risk of local extinction from fire or other forms of environmental degradation, nature reserves are linked by artificially created biodiversity corridors, averaging 131–262 feet (40–80 meters) wide, along which animals may move within a familiar environment. Species intended to benefit from this project include sugar (*Petaurus breviceps*), yellow-bellied (*P. australis*), and feathertail gliders (*Acrobatus pygmaeus*), as well as the southern emu wren (*Stipiturus malachurus*) and the brown treecreeper (*Climacteris picumnus*). The Radiata pine (*Pinus radiata*) system being introduced in South Australia has been found to double butterfly populations elsewhere.

Ongoing Challenges

In common with much of Australian forest land, low rainfall represents an important environmental constraint in Kuitpo. The introduction of new trees and more intense exploitation of existing species has also raised awareness of the danger of invasive fungi and pests, such as nematodes, some of which are being observed for the first time. The long-term interaction between trees and soil is also a matter of interest. These issues are taking on increased urgency as global warming brings the threat of a warmer local climate, faster evaporation that will bring stress on plant species here, soil erosion from harsher storm events, and other impacts yet to be determined.

Balancing stakeholder interests in this way is conducted in line with official national and international management standards. Neighboring areas, such as the McLaren Vale Wineries, also represent opportunities for sustainable exploitation of natural resources. Tourism development is intended to be conducted within this framework. Outdoor activities—including camping and trekking, dog walking, and horseback riding—have to be managed with respect to the fragility of the environment.

Habitat fragmentation and predation by feral carnivores have led to local extinctions of some species and threatened others. Vandalism has also become problematic in some local areas. The presence of protected creatures also poses some threat to visitors, who must deal with potential snakebite and other dangers.

JOHN WALSH

Further Reading

McCarthy, Kevin, Lauria Cookson, and Damian Scown. *Natural Durability of Six Eucalypt Species from Low Rainfall Farm Forestry.* Kingston, Australia: Rural Industries Research and Development, 2009.

Paull, D. "The Distribution of the Southern Brown Bandicoot (*Isodon obesulus obesulus*) in South Australia." *Wildlife Research* 22, no. 5 (1995).

Smailes, Peter J. and Derek L. Smith. "The Growing Recreational Use of State Forest Lands in the Adelaide Hills." *Land Use Policy* 18, no. 2 (2001).

Zhao, Zeng Q., Kerrie A. Davies, Ian T. Riley, and Jackie M. Nobbs. "Laimaphelenchus Australis Sp. Nov. (Nematoda: Aphelenchina) from Exotic Pines, Pinus Radiata and P. Pinaster, in Australia." *Zootaxa* 1248, no. 1 (2006).

Ladoga, Lake

Category: Inland Aquatic Biomes.
Geographic Location: Europe.
Summary: This large Russian freshwater lake has high primary productivity and is known for its wide variety of fish and a native inland seal species.

A freshwater lake in northwest Russia, Lake Ladoga is the largest lake in Europe, with a surface water area of 6,908 square miles (17,891 square kilometers) and a volume of 201 cubic miles (837 cubic kilometers). The lake is home to about 660 islands with a combined total land area of 168 square miles (435 square kilometers), including the Valaam Archipelago. Several of the islands, such as Kilpola, are large enough to contain lakes themselves.

Lake Ladoga is part of the Neva River catchment; it is sited in a generally flat, low-lying and marshy region. There are tens of thousands of much smaller lakes and ponds throughout the area, in many cases linked by some of the thousands of rivers and streams. Three such rivers are the main inlets of water to Lake Ladoga. The Vuoksi River flows in from Finland to the west; the Volkhov River flows in from the south; and largest of all, the Svir River, flows in from the east.

The Neva River drains Lake Ladoga, sending waters to the Gulf of Finland and thence to the Baltic Sea. The lake is part of both the Volga-Baltic Waterway and the White Sea-Baltic Canal system; by a series of navigable natural rivers, lakes, and manmade canals, this network links the Baltic, the Arctic Ocean, and the Caspian Sea—a fact that has brought about opportunities for many species to migrate and settle into new habitats, with some impacts both negative and beneficial to various plants and animals.

Lake Ladoga has generally strong layering in its water column, with warm surface temperatures developing seasonally, but a benthic or deepwater layer that stays in the range of 34–39 degrees F (1–4 degrees C) year-round.

Biodiversity

The Ladoga ecosystem is home to a wide variety of fish—some 48 species, including an endemic (found nowhere else) species of smelt, and the endangered European sea sturgeon, usually found in saltwater and brackish coastal regions. The caviar trade was one of the contributing factors to the sturgeon's endangerment, and overfishing in general has been a problem in Lake Ladoga in the past. Today, commercial fishing in the lake has declined considerably because of overfishing early in the 20th century. Fish farms and recreational fishing remain widespread. Roach, carp, bream, European perch, ruffe, and salmon are among the leading types of fish in the lake today. Lake whitefish dwell quite comfortably in the colder waters of the central lake portion—except for the Valaam whitefish, an endemic species that prefers the southern shallows.

The Ladoga seal, a subspecies of the ringed seal, is a hallmark endemic animal in the lake. As one of the extremely few freshwater seals on Earth, the population of this pinniped was decimated in the 20th century; highly protected now, the few thousand members of the Ladoga seal community are still jeopardized by side-effects of industrial fishing and shipping, as well as poaching.

Muskrat, beaver, river otter, and flying squirrel are quite numerous along the shores, on the many islands, and in the reedy wetlands in the fringe areas of the lake. As the core freshwater feature of the region, the Lake Ladoga biome also supports elk, sika deer, wild boar, brown bear, European lynx, pine marten, red fox, and European wolf in the bogs and woodlands around the lake proper.

At least 250 identified bird species dwell or migrate through the Lake Ladoga biome. Swamps along the east and south of the lake attract golden plover, jack snipe, redshank, curlew, black-tailed godwit, and osprey. Shorebirds in particular find nesting zones across the southern portion of the lake, where the three main tributary rivers have their mouths. Rails, terns, gulls, ducks, and grebes congregate here—and must remain vigilant under the hungry eyes of osprey, marsh harrier, and white-tailed sea eagle.

The shallow areas of the lake offer favorable reed and rush stands for crustaceans and mollusks. Through a succession of mud flats, fens, and wet meadows, these habitats transition to the typical shrub meadows and mixed coniferous-deciduous forests.

Threats and Conservation

Ladoga is a eutrophic lake: Excessive nutrients have led to extremely high primary productivity, that is, the production of organic compounds from carbon dioxide, principally through photosynthesis—in this case, by the lake's algae. This productivity hurts water quality and reduces the amount of dissolved oxygen—a key factor in marine health, particularly for marine fauna and microfauna.

A muskrat at the edge of Lake Ladoga. The lake is the largest in Europe, with a surface water area of 6,908 square miles (17,891 square kilometers), and about 660 islands. The Nizhnesvirsky Nature Reserve protects 161 square miles (416 square kilometers) on the eastern shore of the lake, almost the same amount of land as all the islands combined. (Thinkstock)

Climate change is in evidence here, as spring temperatures have been arriving earlier—with potential stress delivered to the region's food chain through altered plant germination schedules; changes in migratory bird arrivals, breeding and hatching patterns; uneven availability of nutrients and natural fertilizers; and related factors. These impacts will require many species, both flora and fauna, to adapt and/or migrate elsewhere.

The Nizhnesvirsky Nature Reserve is a 161-square-mile (416-square-kilometer) reserve established by the former Soviet Union and given its highest level of protection: *zapovednost*, a term that doesn't translate well into English, but amounts to a nature reserve intended to be kept forever wild, with human use restricted to limited amounts of scientific research or educational display. (The term *zapovednik* refers both to the nature reserve itself and to the staff assigned to manage and protect it.)

Nizhnesvirsky occupies the eastern shore of Lake Ladoga, covered principally in the same Scotch pine as many of the islands, and is home to many of the migratory birds that stop to feed and drink from the lake's waters. The presence and size of the untouched reserve—nearly as much land as there is in Lake Ladoga's islands—are important features of the ecosystem.

BILL KTE'PI

Further Reading

Kudersky, Leonid K., Juha Jurvelius, Markku Kaukoranta, Pekka Tuunainen, and Kyosti Makinen. "Fishery of Lake Ladoga: Past, Present, and Future." *Hydrobiologia* 322, nos. 1–3 (1996).

Rukhovets, Leonid and Nikolai Filatov. *Ladoga and Onego: Great European Lakes.* New York: Springer, 2009.

Saarnisto, Matti and Tuulikki Gronlund. "Shoreline Displacement of Lake Ladoga: New Data from Kilpolansaari." *Hydrobiologia* 322, nos. 1–3 (1996).

Sparks, T. H., F. Bairlein, J. G. Bojarinova, O. Huppop, E. A. Lehikoinen, K. Rainio, L. V. Sokolov, and D. Walker. "Examining the Total Arrival Distribution of Migratory Birds." *Global Change Biology* 11 (2005).

Lake Télé-Lake Tumba Swamps

Category: Inland Aquatic Biomes.
Geographic Location: Africa.
Summary: This large freshwater ecosystem is the only landscape in Africa with three great-ape species. The biome is threatened by industrial logging, rampant commercial bushmeat hunting, and lack of law-enforcement activities.

Lake Télé-Lake Tumba is an inland equatorial ecosystem located at the border of the Republic of Congo (ROC) and the Democratic Republic of Congo (DRC), 65 percent being in the DRC. At 48,649 square miles (126,000 square kilometers), it occupies most of the Cuvette Centrale, or Central Basin, and lies in the Guineo-Congolian vegetation zone, a complex and diverse botanical region. The Lake Télé-Lake Tumba biome supports some 1,500 to 2,000 vascular plant species, of which 10 percent are endemic (found nowhere else); the region is a regional center of endemism. It also hosts abundant fauna species—including three great-ape species, making it unique in Africa.

Flora

Some 60–65 percent of the landscape north of 1 degree, 30 minutes south is a vast zone of seasonally inundated and permanent swamp forest. By contrast, in the zone south of that latitude, comprising some 35–40 percent of the Lake Télé-Lake Tumba biome, are found mixed terra firma forest types, including mature and secondary forests along valleys and on small hills. These ombrophile (shade-tolerant) and semi-deciduous forests occur in areas between the major river systems here.

Both types are composed of stands of leguminous trees, characterized by hardwood timber species such as *Staudtia stipitata, Polyalthia suavaeoleus, Scorodophloeus zenkeri, Anonidium mannii,* and *Parinari glaberrimum,* which serve as important fruit and seed sources for great apes and other wildlife.

Between the Ubangi and Ngiri Rivers are spots of forest dominated by the evergreen mbau tree (*Gilbertiodendron dewevrei*), providing an important food base for an array of mammals including humans. Areas of seasonally inundated forest and numerous large clearings occur along the major riverbanks in the north, east, and west of the biome. The southern reaches of the landscape are a forest-savanna ecosystem on the Bateke Plateau, a relatively drier area. Evolutionarily savanna-adapted species share their habitats with forest-adapted species, which explains why this ecosystem is one of the most diverse in central Africa.

Biodiversity

The Lake Télé-Lake Tumba biome is the only place where three endangered species of great ape—bonobo (*Pan paniscus*), central chimpanzee (*Pan troglodytes*), and western lowland gorilla (*Gorilla gorilla gorilla*)—are sympatric in their wild range, though separated by the Congo and Ubangi Rivers. A relatively large population of forest elephants also resides in this biome, along with important populations of forest buffaloes, bongos, sitatunga, bushbucks, leopards, lions, seven species of duikers, and seven species of diurnal monkeys.

Freshwater biodiversity here is composed of more than 150 fish species and includes the hippopotamus, African slender-snout crocodile, dwarf crocodile, Nile crocodile, water chevrotain, giant otter shrew, Congo clawless otter, and Hartlaub's duck.

Much of the biodiversity of Lake Télé-Lake Tumba remains poorly described. Relationships among that biodiversity, and environmental and human conditions that influence the distribution and the abundance of that biodiversity, remain inadequately elucidated. This lack of knowledge can be attributed to the fact that Lake Télé-Lake Tumba became a focus of conservation and research only in the recent past. Fresh studies, however, have found that its forests may have remained wetter year-round over the past 20 centuries.

Various research has described the most important continuous population of bonobos in the wild, the presence of lions in savannas within the forest, species of fish new to science, and one new species of bird that was thought to occur only in east Africa. Additional studies have, for the first time, described the discovery of chimpanzees dwelling in swampy habitats. These findings, taken together, indicate that there may be more species waiting to be uncovered and understood in this fecund region.

Human Cultures

Lake Télé-Lake Tumba supports high cultural diversity. The DRC side comprises 30 Bantu tribes of common historical background, and the autochthonous Batswa and Balumbe. Human population densities vary from a mean of 3.5 people per 0.4 square mile (1 square kilometer) in the north, to a mean of eight people per 0.4 square mile (1 square kilometer) in the south. In the north, people are spread across a vast forested area, while in the south, they are distributed among permanent settlements along the rivers, including the shores of Lake Tumba and Lake Maindombe.

In the north, human livelihoods consist primarily of fishing, cultivation of cassava, and hunting in adjacent forests. The people of the south live in the large and midsize towns (Mbandaka, Inongo, and so on), which are markets for forest products. The Batswa and Balumbe are located mostly in the administrative territories of Bikoro, Inongo, and Kutu. Their livelihood is heavily dependent on exploitation of forest products.

Environmental Threats

Biodiversity is threatened by a cohort of anthropogenic factors, including industrial logging and a high level of poverty among local populations, leading to direct pressure on scant resources. The most endangered wildlife species in the Lake Télé-Lake Tumba area exhibit a natural vulnerability from low reproduction rates due to prolonged gestation and long inter-birth intervals, as observed in all three species of great ape, as well as the forest elephants.

The southern part of the DRC side of this biome is heavily logged. Despite the lack of data on the effects of logging on bonobos, the bonobos and

chimpanzees resident on the DRC side are closely related, and may be highly vulnerable to stress caused by disturbance from logging activities, as was demonstrated in the case of chimpanzees in Gabon. Logging operations threaten prime habitat for bonobos in the southern zone of the biome; most of the suitable habitat for bonobos occurs within logging concessions.

Seeking economic opportunities connected to industrial logging operations, immigrants have been increasingly settling in this area over the past two decades. Increased human populations in the south have wiped out local forest-elephant populations. Human immigrants have occupied zones that were suitable for elephant seasonal migrations, leading to increased human-elephant conflict caused by competition for forest fruits and corridor land. Commercial bushmeat hunting is rampant, and is facilitated by increasing availability of weapons and ammunition, ease of access to the landscape through the river network, and the complicity of the army and local political leadership.

BILA-ISIA INOGWABINI

Further Reading

Inogwabini, Bila-Isia, et al. "The Bonobos of the Lake Tumba–Lake Maindombe Hinterland: Threats and Opportunities for Population Conservation." In Takeshi Furuichi and Jo Myers, eds., *The Bonobos: Behavior, Ecology, and Conservation*. Thompson. New York: Springer, 2007.

Inogwabini, Bila-Isia, et al. "Great Apes in the Lake Tumba Landscape, Democratic Republic of Congo: Newly Described Populations." *Oryx* 41, no. 4 (2007).

Inogwabini, Bila-Isia, et al. "Protected Areas of the Democratic Republic of Congo: A Habitat Gap Analysis to Guide the Extension of the Network." *Endangered Species Update* 22, no. 2 (2005).

Twagirashyaka, Félin and Bila–Isia Inogwabini. "Lake Télé-Lake Tumba Landscape." In Carlos De Wasseige, et al. *The Forests of the Congo Basin—State of the Forest 2008*. Luxembourg: Publications Office of the European Union, 2009.

Lakshadweep Coral Reefs

Category: Marine and Oceanic Biomes.
Geographic Location: Indian Ocean.
Summary: This archipelago of coral islands, atolls, and reefs is home to diverse marine life, but under considerable anthropogenic and climate pressure.

The Lakshadweep Coral Reefs biome is associated with the Lakshadweep Islands, formerly known as the Laccadive Islands. This Indian Ocean archipelago consists of 36 small coral islands along with 12 coral atolls, three coral reefs, and a handful of submerged banks. The entire group lies some 186 miles (300 kilometers) off the west coast of the state of Kerala in southern India, separated from the mainland by the Lakshadweep Sea. They comprise the smallest union territory of the Republic of India.

The Lakshadweep islands and atolls are formed wholly from coral; together with the nearby Maldives and the Chagos Archipelago, the islands form a terrestrial ecoregion. The coral reefs themselves have formed over thousands of years and are regarded as the richest corals in India. There are some 100 species of coral here, with the branched staghorn and the ridged brain coral being among the most remarkable.

The combined lagoon area of Lakshadweep is about 1,620 square miles (4,200 square kilometers); most is relatively shallow and largely separated from the open sea. The immediate territorial extent of the oceanic waters surrounding the archipelago is on the order of 7,700 square miles (20,000 square kilometers).

Biodiversity

There are many marine species around the Lakshadweep Archipelago, including a wide range of fish; approximately 600 species have been identified here. These species include the butterfly fish, clownfish, parrotfish, and lagoon triggerfish, as well as moray eels. In addition, ink-blue starfish, sharks, dolphins, tuna, wahoo, and swordfish live nearby. There are some 90 species of sponge and at least 70 species of echinoderms (star fish, sea urchins, sea cucumbers, and sand dollars).

The Lakshadweep bird sanctuary, Pakshi Pitti, protects 3 acres (1.21 hectares) of sand bank where four species of tern breed: sooty, great crested, bridled, and noddy (Anous stolidus), shown here. (Wikimedia/Phil Guest)

The land is also important for significant numbers of sea turtles, which lay their eggs on the beaches, and for hermit crabs, which come ashore periodically. These beaches are vital in particular for the egg-laying functions of the leatherback, olive ridley, hawksbill, and green sea turtles. Bird life, some of which feeds on these beach-bound species or their eggs, includes pelagic birds such as the brown noddy (*Anous stolidus*), the greater crested tern (*Sterna bergii*), and the lesser crested tern (*Sterna bengelensis*).

Plant life on the combined 12 square miles (32 square kilometers) of land surface is rather sparse. Among the chief intertidal-zone flora is Cymodocea (*Cymodocea rotundata*), a beach grass well adapted to both mud and dry-sand environments. Stands of stunted palms crowd some islets. Undersea vegetation includes seagrass, primarily *Cymodocea rotundata*, as well as five other species. Various forms of algae are also found in the waters near the coral reef, and well over 100 species of seaweed thrive around the reefs and further offshore.

Human Settlement

Androth Island, Kavaratti Island, and Minicoy Island have some 10,000 people living on each of them; together, they make up half the population of the entire archipelago. The other people living in the area hail from seven other islands: Amini, Agatti, Kadmat, Kalpeni, Kiltan, Chetlat, and Bitra. The remaining 26 islands, islets, and atolls are all uninhabited.

Most of the local people on the islands are related to those in the Maldives. They have traditionally lived by subsistence fishing. On larger islands there has also been extensive cultivation of coconuts, as well as bananas, vegetables, and edible root crops. As the population has grown, human activity has started to impinge on wildlife, with commercial tuna fishing beginning to cause problems, as some skipjack tuna are sought for sale to markets in Japan, Hong Kong, and elsewhere. Because the islands lack cold storage facilities, however, the rate of fishing for export is lower than in many other areas of the world.

The reefs and atolls have become a popular tourist destination, attracting both swimmers and divers since the area was opened in 1974. This has both positive aspects—wider publicity for ecological conservation, better income for indigenous people—and negative ones, such as increased litter, fuel spills, and other wastes injected into the environment, and the potential for direct physical damage to the coral.

Environmental Threats

There has been an increase in pollution resulting from the dumping of rubbish in marine areas, and from refuse leaching through from burial pits. This problem is accentuated by the sandy, porous soil, which has led to the contamination of some groundwater and the spread of waterborne diseases.

Problems have also arisen from the increase in tourism, although this has been checked by limiting the number of tourists and restricting diving, motorboating, surfing, and some other water sports. In addition, to help reduce the effect on the environment, a low-temperature desalination plant was opened on the island of Kavaratti—one of the first such plants in the world.

The major threat to the coral reefs comes from global warming and climate change. Weather has always affected the region. Indeed, parts of Kalpeni Island were formed in 1847 after a storm dumped large amounts of coral on the beaches, raising the banks on the eastern and the southern shores of the island. Climate change has led to seawater temperature warming, which set the scene for coral bleaching, including a particularly severe event in 1998 that affected roughly half of all live coral here; especially hard-hit were table coral, genus *Acropora;* and brush coral, genus *Pocillopora.* Thankfully, surveys are showing a doubling of the population of pre-1998 live corals in many areas that suffered bleaching, and projections are for a full recovery by 2018.

The overall rise in water levels will affect all the Lakshadweep Islands, causing problems for turtles, crabs, and other forms of life in the water and on the shrinking land areas. None of the islands rises much higher than 3–6 feet (1–2 meters) above the sea now.

The government of India has, for purposes of conservation administration, classified the Lakshadweep Coral Reefs biome as a wetland; this brings it under jurisdiction of the National Wetland Conservation and Management Programme of its Ministry of Environment & Forests. There is a bird sanctuary, Pakshi Pitti, in the form of a protected sand bank of some 3 acres (1.21 hectares). Four species of tern—sooty, great crested, bridled, and noddy—are known to breed here, a quite unusual occurrence for any Indian territory.

JUSTIN CORFIELD

Further Reading

Desai, Vijay J., Deepali S. Komarpant, and Tanaji G. Jagtap. *Distribution and Diversity of Marine Flora in Coral Reef Systems of Kadmat Island in Lakshadweep Archipelago, Arabian Sea, India.* Washington, DC: National Museum of Natural History, Smithsonian, 2003.

Frazier, J. "Exploitation of Marine Turtles in the Indian Ocean." *Human Ecology* 8, no. 4 (1980).

Ghosh, Shekhar. "Changing Law in a Changing World: Case of Mid-Ocean Archipelagos." *Economic and Political Weekly* 22, no. 3 (1987).

Goreau, Tom, Tim McClanahan, Ray Hayes, and Al Strong. "Conservation of Coral Reefs After the 1998 Global Bleaching Event." *Conservation Biology* 14, no. 1 (2000).

Jeromi, P. D. "Tyranny of Distance: State of Economy of Lakshadweep." *Economic and Political Weekly* 41, no. 30 (2006).

Kurup, K. K. N. "Sequestration of Laccadive Islands." *Journal of Indian History* 50, no. 1 (1972).

Rao, T. A. and J. L. Ellis. "Flora of Lakshadweep Islands Off the Malabar Coast, Peninsular India, with Emphasis on Phytogeographical Distribution of Plants." *Journal of Economic and Taxonomic Botany* 19 (1995).

Laptev Sea

Category: Marine and Oceanic Biomes.
Geographic Location: Arctic Ocean.
Summary: A mostly frozen sea north of Siberia, the Laptev is considered to be a high-productivity ecosystem despite the seasonally limited availability of sunlight and nutrients.

A marginal sea of the Arctic Ocean, the Laptev Sea lies between the Arctic Ocean and the northern coast of Siberia, with the Kara Sea to its west and the East Siberian Sea to its east. Like those seas, it is frozen for most of the year, leading to limited populations of flora, fauna, and humans.

The Lena River, which begins in the Baikal Mountains and is one of the longest rivers in the world, is the largest river flowing into the Laptev. The craggy shores are full of gulfs and bays, and the coast becomes mountainous in places. Half the sea is shallow, resting on the continental shelf, which keeps average depths around 164 feet (50 meters). In the north, the sea bottom drops suddenly, and almost a quarter of the sea has a depth of 0.6 mile (1 kilometer). The large freshwater inputs of the Lena and other rivers have significant effects on the ecosystem: About 60 percent of the diatom flora consists of freshwater species transported here from Siberian lakes.

Because of its proximity to the North Pole, polar night on the Laptev lasts three months in the

Polynyas and Climate Change

Like the rest of the Arctic, the Laptev Sea has been affected by climate change. The Laptev has slowly, steadily warmed since at least 1957, and the 1980s and 1990s were especially warm. Arctic land areas have seen winter warming of as much as 10 degrees, changing the levels of precipitation, sea-ice cover, and sea-level pressure, all of which are critical factors in the balance of the local ecosystem, as well as affecting factors such as the flow of water and ice southward into other ecosystems. Areas of open water surrounded by sea ice, called polynyas, have occurred with increasing frequency in the Laptev. Polynyas have always been a feature of the Laptev Sea, as in other parts of the Arctic. They form when warm southern winds keep stretches of water warm enough to resist freezing. The Great Siberian Polynya recurs every year, and in some years, it can stretch for hundreds of miles (kilometers).

Seasonal polynyas become important habitats for marine mammals, allowing them to remain in the north instead of migrating south. The localized lack of ice also permits greater amounts of sunlight to reach below the surface, allowing microalgae to bloom and creating an environment of intense feeding by herbivorous fish, which are fed on in turn by Arctic cod, seals, walruses, narwhals, and whales. While polar bears are not necessarily thought of as marine animals, they have been known to swim more than 30 miles (48 kilometers) across a polynya in search of food.

This August 2009 photograph from a National Oceanic and Atmospheric Administration expedition to the Arctic shows patches of ice floes and melt pools at left and center, and a polynya, or area of open water, at right. Polynyas are appearing more often in the Laptev Sea because of climate change. (NOAA/Pablo Clemente-Colon)

south and five months in the north, when the sun disappears entirely for the winter; the midnight sun of summer is as lengthy. As elsewhere in the Arctic region, this affects the life cycles and available sunlight for what local life there is. Even in the warmest months, temperatures are only slightly above 32 degrees F (0 degrees C), except on the coasts. Winter's coldest months drop to minus 29 degrees F (minus 34 degrees C).

The Laptev contributes more sea ice than the Barents, Kara, Chukchi, and East Siberian seas combined, with the outflow of ice varying considerably—from 96,912 square miles (251,000 square kilometers) in 1985 to 282,627 square miles

(732,000 square kilometers) in 1989, for example. Ice formation begins in October in the coastal south, and September in the Arctic north, creating a continuous sheet of ice covering about 30 percent of the sea.

Biodiversity

Fish populations in the Laptev are scanty because of the ice and the poor availability of sunlight. The lack of commercial fishing in the area diminishes the body of available data, but most of the pelagic fish in the region are whitefish species of genus *Coregonus*.

Pollution in the sea is slight and results principally from chemical spills in coastal areas. River runoff contributes some industrial pollution because of the lack of local facilities in the coastal settlements for processing waste, but because the human population and industrial activity are so sparse in the first place, this is not a problem of nearly the same magnitude that it would be in a more urban or industrialized area. It could become a serious concern should the Siberian coast become more populous in the future.

Human Impact

The Laptev shores were inhabited for millennia by the indigenous Yukaghirs, Evens, and Evenki, who herded reindeer and fished the frigid waters. It was previously known as the Tatar Sea, the Lena Sea, the Siberian Sea (from whence the East Siberian Sea was disambiguated), the Icy Sea, and the Nordenskjold Sea. Mammoth remains are well preserved on many of the sea's islands, and may have been hunted by the Yukaghirs's prehistoric ancestors. Russian explorations began in the 17th century along rivers like the Lena, Anabar, Olenyok, and Khatanga that empty into the Laptev Sea.

Today, there is little commercial activity in the area other than navigation along the Northern Sea Route, which passes through, and mining on the shore. The lengthy winter prevents the feasibility of commercial fishing. Ice typically melts only in August and September in this remote part of the world.

BILL KTE'PI

Further Reading

Alexandrov, Vitaly Y., Thomas Martin, Josef Kolatschek, Hajo Eiken, Martin Kreyscher and Alexandr P. Makshtas. "Sea Ice Circulation in the Laptev Sea and Ice Export to the Arctic Ocean: Results from Satellite Remote Sensing and Numerical Modeling." *Journal of Geophysical Research* 105, no. C7 (July 2000).

International Hydrographic Organization. *Limits of Oceans and Seas, Special Publication No. 28.* Monte Carlo: International Hydrographic Organization, 1953.

Kassens, H., H. A. Bauch, and Heidemarie Kassens, eds. *Land-Ocean Systems in the Siberian Arctic: Dynamics and History.* New York: Springer-Verlag, 1999.

Leeward Islands Xeric Scrub

Category: Forest Biomes.
Geographic Location: Caribbean Sea.
Summary: These unique and highly threatened dry woodlands historically covered much of the Caribbean's Leeward Islands; their endemic species have been compromised by imported exotic types.

This subtropical dry forest and scrub biome once covered much of the Caribbean's Leeward Islands, but today they occur as isolated forest remnants following a long history of agricultural encroachment, mining, infrastructure, tourism, and invasion by exotic species. Leeward Island xeric scrub typically is found at low elevations, characterized by low rainfall and an arid climate that is strongly influenced by the trade wind belt. These forests typically are found along the coastlines of these islands, but on some islands, they extend well inland.

This biome covers most of Antigua, Anguilla, Guadeloupe, Saba, St. Martin, St. Bart's, and the eastern portions of the British and U.S. Virgin

Islands. Compared with the neighboring Windward Island xeric scrub type, this biome is relatively large, spanning a greater geographical area.

Biota

The dominant vegetation ranges from herbaceous and woody shrublands to savannas and even woodlands, depending on soil and other localized conditions. Climate is arid throughout. Few remnants of native, pre-colonial vegetation remain in this region following a history of exploitative land use in the Caribbean.

The vegetation form that today characterizes the biomes is a combination of hearty successional native plants, mixed with nonindigenous species that were introduced for food, forage, building materials, or other anthropogenic uses. Various species of *Acacia* now dominate the former sugarcane and cotton fields, together with copperwoods and other members of family *Burseraceae, Pisonia fragrans,* Chilean mesquite (*Prosopis chilensis*), and guava (*Psidium guajava*).

Although there are several endemic (not found elsewhere) species in the Leeward Islands Xeric Scrub biome, the local biogeography of a large number of relatively proximate small islands is not nearly as diverse as on the Windward Islands. With their history of monocultural plantations as the dominant type of land use, the Leewards have developed what can be termed an overly simplified ecosystem. Even so, there are exceptions, and the protected areas that do retain native vegetation are globally unique and deserve more attention for conservation.

Fauna

The larger islands in the group, such as Guadaloupe, tend to have relatively higher diversity and a higher degree of endemic flora and fauna than smaller islands. The smallest island in the Leeward chain, for example, is Saba, which is home to one frog (*Eleutherodactylus johnstonei*) and one endemic lizard (*Anolis sabanus*). Larger islands are home to many more.

This biome has no native terrestrial mammals; all native mammals are bats. A growing number of introduced mammals, however, are becoming a threat here. Rats, mongooses, goats, pigs, fallow deer, and agouti all wreak havoc on the increasingly rare native flora and fauna.

Conservation Efforts

The conservation status of the biome and its native species varies greatly from island to island and across landscapes. In places like the U.S. Virgin Islands, a network of protected areas extends across nearly one-fifth of the land area, and has protected a significant number of local species.

There is clearly a need to take preventive measures to save the endangered native flora and fauna from continued habitat loss and introduction of aggressive exotic species. In the wake of its agricultural past, and with the more recent disruption from tourist resort development, biodiversity here is in jeopardy. Even plants and animals adapted to the severe arid climate and less-than-ideal soils find it hard to flourish when habitats are continually disrupted. Climate change impacts will add to these stresses, as heavier hurricane damage, coastal soil erosion, and saltwater intrusion will combine to increase habitat pressure here.

Jan Schipper

Further Reading

Donovan, S. K. and T. A. Jackson. *Caribbean Geology: An Introduction.* Kingston, Jamaica: University of the West Indies Publishers Association, 1994.

Harris, D. R. "The Invasion of Oceanic Islands by Alien Plants: An Example from the Leeward Islands, West Indies." *Transactions and Papers, Institute of British Geographers* 31 (1962).

Johnson, T. H. *Biodiversity and Conservation in the Caribbean: Profiles of Selected Islands.* Cambridge, UK: International Council for Bird Preservation, 1988.

Malhotra, A. and R. S. Thorpe. *Reptiles and Amphibians of the Eastern Caribbean.* London: Macmillan Education, 1999.

Stoffers, A. L. "Dry Coastal Ecosystems of the West Indies." In E. Van der Maarel, ed. *Ecosystems of the World 2B: Dry Coastal Ecosystems Africa, America, Asia and Oceania.* Amsterdam: Elsevier Science Publishers, B.V., 1993.

Lena River

Category: Inland Aquatic Biomes.
Geographic Location: Asia.
Summary: This Siberian river is one of the largest in the world, and its delta supports an important habitat for migratory and nesting bird species.

The Lena River is located in eastern Russia. After the Ob River and the Yenisei River, it is the third-largest catchment area in Asia, at 926,645 square miles (2.4 million square kilometers). It originates in the Baikal Mountains, west of Lake Baikal in the Siberian Plateau, and discharges via a delta approximately 249 miles (400 kilometers) wide into the Laptev Sea and the Arctic Ocean. As a result of mixing river water in the Laptev Sea, there is a brackish surface plume of more than 217 miles (350 kilometers) over the saline seawater. The vast Lena River delta alone includes more than 6,000 watercourses, more than 50,000 lakes, and 1,600 islands.

The Lena River's catchment area is predominantly underlain by 78–93 percent permafrost. More than 70 percent of the catchment area is covered by forests, of which more than 90 percent are needle-leaved deciduous forest comprised mainly of larch. In the delta area that is dominated by tundra, vegetation is sparser, and the regularly flooded areas with waterlogged soil are, if vegetated, covered by grassland or woody vegetation. Most of the overall catchment area is dominated by a continental climate, but a polar climate predominates toward the Arctic Ocean.

After winding some 2,800 miles (4,500 kilometers) from its source, the Lena makes a massive discharge into the Laptev Sea; its volume at the mouth is sufficient to make it one of the 10 largest flows among major rivers of the world. The Lena River accounts for more than 70 percent of the total freshwater input to the Laptev Sea, and for about 19 percent of the input to the Arctic Ocean. The flow shows considerable seasonal differences, however, because the river is frozen for about eight months per year, from October through May.

The Lena River has three main tributaries: the Vilyuy, Aldan, and Vitim Rivers. The main channel of the Lena River is not much affected by flow regulation. Its tributary the Vilyuy has been saddled with several dams, and so has one minor tributary. In an international comparison, the effect of fragmentation and flow regulation in the Lena River is classified as moderate.

The permafrost in the catchment, both active-layer and upper-layer, is sensitive to temperature. Increases in air temperature affect snow depth and rainfall, and result in increased soil moisture. This alters the properties of the active permafrost layer and induces warming of the upper permafrost. In turn, this potentially affects the chemical properties of the river. In comparison with other major rivers of the world, the Lena is characterized by low concentrations of particulate material and by high concentrations of dissolved organic carbon—thus it has ecological service value as a significant carbon sink, one that could help counter global warming.

Biodiversity

The Lena River holds more than 35 fish species, of which several are important for sport and commercial fishing, including grayling (*Thymllus* spp.),

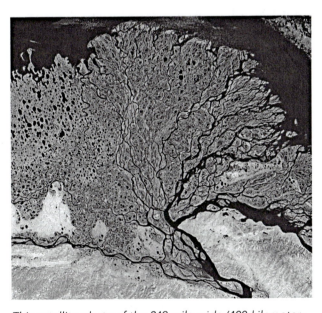

This satellite photo of the 249-mile-wide (400-kilometer-wide) Lena River Delta shows its complex structure, which includes 50,000 lakes and 1,600 islands. (NASA/USGS EROS Data Center Satellite Systems Branch)

sheefish (*Stenodus leucichthys*), burbot, chum salmon, sturgeon, nelma (*Stenodus leucycthis*), and Arctic cisco (*Coregonus autumnalis*). Some marine biologists consider the Lena River Delta to be the original cradle of the Arctic cisco, which is anadromous and is present in all marginal seas around the Arctic. The delta fish populations in general are potentially threatened by overfishing, especially as global warming begins to provide more time annually when waters here will be ice-free.

The Arctic lamprey (*Lethenteron japonicum*) is one of the more unusual species found in the Lena; an eel-like fish that feeds mainly on invertebrates and algae, it features two large teeth on its tongue.

The tundra in the delta area that is flooded each year is an important habitat for many migratory and nesting bird species; it is one of the richest Arctic areas for species diversity and breeding densities. As recently as 1997, a survey observed 16 mammal and 122 bird species in the delta. Of these bird species, including swans, geese, loons, ducks, gulls, terns, black brant, Steller's eiders, plovers, sandpipers, and raptors, 67 bred at least once during the survey period.

The riparian zone and the banks of the Lena River host several endemic (found nowhere else) plant species. Endemic herbs include *Redowskia sophiifolia* of the magnolia clade, and *Ceratoides lenensis*, a tundra fescue. Dandelion, thought of as a weed in many parts of the world, grows abundantly along the Lena's banks and is considered, along with other endemic herbs here, a medicinal plant.

Threats and Conservation

The Lena River catchment is sparsely populated; settlements and villages are mostly located along the riverbanks. Initially, the banks were inhabited by paleo-Asiatic tribes and later by the Sakha or Yakut people, whereas today, Russians are most numerous. Subsistence fishing, farming, and hunting are still among the most significant activities of humans in much of the Lena River catchment.

Threats to this biome include pollution from mining, logging and related industrial activities; agricultural runoff; and climate change. The Lena's waters remain pristine, and indeed are considered some of the cleanest in the world. Keep-ing those waters clean for future generations is of prime consideration. Climate change impacts on the Lena River include changes in the water flow that feeds the Lena due to changes in temperature, evaporation, and precipitation.

In the Lena River Delta area, prevailing pollutants that have been detected include phenols, copper, oil products, and pesticides. The main pollutants are wastewater from industrial companies that are heavy water users, including mining and oil-producing enterprises. The pollution level of the Lena River is generally regarded to be low overall, however.

The Lena River is protected by three federal natural reserves. The Baikalo-Lensky Nature Reserve protects the headwater area; the Olyokminsky Nature Reserve protects the middle stream area; and the Ust-Lensky Nature Reserve protects the delta area. With its area of 23,552 square miles (61,000 square kilometers), the Ust-Lensky Nature Reserve, founded in 1986, is the largest nature reserve in Russia.

FRAUKE ECKE

Further Reading

Dynesius, Mats and Christer Nilsson. "Fragmentation and Flow Regulation of River Systems in the Northern Third of the World." *Science* 266, no. 4 (1994).

Gilg, Olivier, Raphaël Sané, Diana V. Solovieva, Vladimir I. Pozdnyakov, Brigitte Sabard, and Dmitri Tsanos, et al. "Birds and Mammals of the Lena Delta Nature Reserve, Siberia." *Arctic* 53, no. 2 (2000).

Ijima, Yoshihiro, Alexander N. Fedorov, Hotaek Park, Kazuyoshi Suzuki, Hironori Yabuki, and Trofim C. Maximov, et al. "Abrupt Increases in Soil Temperatures Following Increased Precipitation in a Permafrost Region, Central Lena River Basin, Russia." *Permafrost and Periglacial Processes* 21 (2010).

Nikanorov, A., V. Bryzgalo, L. Kosmenko, and O. Reshetnyak. "Anthropogenic Transformation of Aquatic Environment Composition in the Lena River Mouth Area." *Water Resources* 38, no. 2 (2011).

Limpopo River

Category: Inland Aquatic Biomes.
Geographic Location: Africa.
Summary: This large river flows from the heart of southern Africa through four countries and into the Indian Ocean, watering a mosaic of land rich in biodiversity.

The Limpopo River meanders eastward from highlands in southern Africa through savannas, agricultural and rural homestead areas, finally emptying into the Indian Ocean in Mozambique. Along its course, it waters the borderlands of South Africa, Botswana, and Zimbabwe. The area is characterized by relatively high human population density with the major land use consisting of seasonal subsistence farming. The Limpopo River is world-renowned for its abundance of wildlife.

At more than 154,441 square miles (400,000 square kilometers), an area larger than Germany, the river forms one of the largest watersheds in Africa. It is the second-largest river in Africa to empty into the Indian Ocean, flowing for 1,087 miles (1,750 kilometers) from one of its main source catchments: the Waterberg, or Water Mountain, massif, north of Johannesburg, South Africa. The river basin is characterized mainly by flat or undulating plains with African savanna.

Climate is generally warm and tropical, and rainfall varies from 8 to 47 inches (200 to 1,200 millimeters) per annum across the course of the Limpopo. Through most of the year, the water flows sluggishly, containing much silt and many sandbars. During the dry season, from April to October, much of the upper sections of this annual river run dry. The rainy season from November to February and the large catchment area ensure that most of the river flows with a

A one-year-old marabou stork (Leptoptilos crumeniferus), one of about 750 bird species that can be found in the Great Limpopo Transfrontier Park. (Thinkstock)

high water volume during the wetter months. The region is also prone to flooding in wet years, like the catastrophic floods in 2000, in which almost 800 people died, and the 2007 floods, which displaced nearly 90,000 people.

From below the confluence of its major tributary in Mozambique, the Olifants, or Elephant, River, the Limpopo River mainly remains permanently navigable. Climate change has impacted the Limpopo River in several ways, most notably reducing water supplies during the dry months.

Biodiversity

A salient feature of the Limpopo River is the wealth of biodiversity within its waters and also within the surrounding ecosystems. Near the river mouth in Mozambique, two biodiversity hot spots intersect: the East African Coastal Forest biome and the northern tip of the Maputaland-Pondoland-Albany hot spot area.

The river also bisects Great Limpopo Transfrontier Park, a collaborative protected area of nearly 13,514 square miles (35,000 square kilometers), roughly the area of Belgium, covering the international borders of South Africa, Botswana, Zimbabwe, and Mozambique. This region has exceptional diversity and abundance of large mammals and, to a lesser extent, plant species. It is also world-renowned as a birding destination, with roughly 750 species to be seen along the watercourse.

The Limpopo hosts at least 50 species of freshwater fish and at least 18 additional introduced species. Species like bull shark (Carcharhinus leucas) have been sighted in the river almost 93 miles (150 kilometers) inland from the sea.

The Limpopo River region is relatively well conserved via formal protected areas and private reserves. Its natural beauty and wealth of wildlife are stimulating a burgeoning ecotourism industry. Large amounts of water hyacinths (Eichhornia crassipes) grow in the slow-moving waters, creating such a dense cover that it can reduce the oxygen available to fish in the river. Riparian forest cover dots the banks of the Limpopo, creating good shelter and diverse habitats for the creatures that live there.

Human Settlement

In 1932, the remnants of an ancient culture on the shores of the Limpopo River were discovered. Subsequent research has shown that the kingdom of Mapungubwe had extensive building complexes, as well as mastery of stone masonry and gold forging, and far-flung trading partners. In the period around 1000 C.E., people were likely attracted to the area for its vast agricultural opportunities, and because it contained many elephants (*Loxodonta africana*) with the allure of ivory as a major trading commodity. This civilization gave rise to the kingdom of Zimbabwe in the 13th century—though it seems that a decrease in annual rainfall, with a concomitant decrease in pastoral activities, led to the eventual downfall of Mapungubwe. The area is now a World Heritage Site, as well as a national park; research continues here.

In modern times, the region is a melting pot of cultures; most people have rural livelihoods, and the area is characterized by relatively low infrastructure. The population density is high around the Limpopo River, however, at 35 people per 0.4 square mile (1 square kilometer). Poverty is severe, with many dependent on the variable flows of the Limpopo River for their subsistence-agriculture practices. Economic inequality across national borders here leads to tides of illegal immigrants, contributing to social instability and difficulties in policing environmental mandates.

The demand for water in the Limpopo River is unevenly spread. The agricultural sector taps about 50 percent, while the urban sector uses 30 percent. Livestock is very important, with an estimated 2.2 million animals in the area, 70 percent of which are cattle. South Africa is responsible for 60 percent of total water extraction, and Zimbabwe for 30 percent; Mozambique and Botswana combined account for only 10 percent of extractive use. Recognizing the need for collaborative efforts to manage this large watercourse, the governments of South Africa, Botswana, Zimbabwe, and Mozambique established the Limpopo Water Course Commission in 2003. It is mandated to foster equitable sharing of the resource of the Limpopo River, to promote sustainable development initiatives, and to help prevent a repeat of the type of climate-change-driven downfall that occurred here some 800 years ago.

Bernard W. T. Coetzee

Further Reading

Huffman, Thomas, N. *Mapungubwe: Ancient African Civilisation on the Limpopo*. Johannesburg, South Africa: Wits University Press, 2005.

Limpopo Basin Permanent Technical Committee. "Joint Limpopo River Basin Study Scoping Phase." January 2010. http://www.icp-confluence-sadc.org/project/docs/publicfile?id=190.

South African Development Community (SADC). "The Limpopo Watercourse Commission (LIMCOM)." SADC Water Sector, 2012. http://www.icp-confluence-sadc.org/rbo/60.

Logging Industry

Category: Forest Biomes.
Geographic Location: Global.
Summary: The logging industry, in its intensely mechanized form, is a major human impactor on ecosystems around the world.

Since the rise of civilization, humans have been logging forests and affecting forest ecosystems. Before the period when Europeans arrived in North America, logging around the world tended to exert a more local effect—although evidence indicates that the downfalls of various earlier dominant civilizations came about in part from the ecological effects of land-use practices, of which forest extraction and depletion was a component.

With the North American logging history as a clean-slate example, it was during the colonial period that the logging industry began to systematically harvest old-growth forests to supply the settlers with cleared land for agriculture, and with lumber for construction and their continued expansion by land and sea. The Industrial Revolution brought mechanization to extraction, and an increased rate of harvest, which began to allow transnational logging companies to supply the ever-increasing global market demands. This expansion has continued in regions around the world, and now has resulted in a reduction of nearly half of the Earth's original forest cover—with most of the known loss occurring since the late 1970s, mainly in the tropical forests.

The logging industry has become more regulated by governments keen to reduce forest ecosystem degradation and loss. Biotechnology has fueled the development of hybrid tree species grown in plantations in tropical zones; these have become a key part of the logging industry's approach to future harvests. Harvest approaches in some cases have been altered to improve sustainability, with third-party certification as part of the sustainable approach. The attention of governments, nongovernmental organizations, and the industry itself has turned to the problem of illegal logging, which continues to be an additional factor in deforestation and forest habitat degradation.

To ensure a robust future in the global economy, the logging industry is likely to continue to use tropical and temperate forest plantations, selective logging rotations, and to commoditize waste as renewable fuel; the industry will also negotiate new pathways for sustainability.

Early Logging

In the earliest of human times, forest extractions were minimal, with a small amount used for housing structures and early forms of work implements. Approximately 5,000 years ago, with the rise of cities and states, increased logging delivered needed wood supplies as civilizations grew. Large ecological effects on local forests were seen due to more concentrated and higher rates of logging. Evidence indicates that several earlier civilizations, such as the Mayans, rose and fell in part due to the destructive effects of localized logging.

Up until 500 years ago, most logged material was used within the region in which it was logged. This situation changed during the American colonial period, when North American settlers began to log the expansive stands of old-growth pine forests in their new homeland. From the 1700s, lumber exports outweighed domestic demand for products including construction materials, white pine for lumber, masts for ship building, and pulp for paper, all mainly to markets in Europe.

The technology of the logging industry during this time period was basic. Axes and the cross-cut saw were used to cut trees in the forests. The feller cut down the tree, the bucker cut the log into manageable sizes, and men with broad axes squared the timber at the stump. Oxen and horses pulled the skid with squared timbers to the water routes, and the water routes floated these squared timbers to ships, which transported them to colonial markets.

By the 1800s, sawmills became commonplace, and logs were no longer squared in the bush but were cut to the length dictated by the market. Waste in the logging industry during this period was high, often with two-thirds of each tree being left behind.

At the beginning of the logging industry, loggers were independent operators who worked in difficult conditions. They lived in remote camps, in small shanties constructed from logs, away from families, friends, and basic services. The hours were long, and the work was dangerous, with few safety measures put into place until the mid-20th century.

As capital costs increased, commercial operations moved into the hands of larger companies and out of the hands of individuals. By the late 18th century, the bigger logging companies were purchasing large tracts of land from the government and harvesting the standing timber. The work was carried out mainly by jobbers, or independent operators, on contract to the company. Some of the pioneers in this industry were Weyerhaeuser, Abitibi, McMillian Bloedel, and Domtar.

Industrial Revolution to Today

The Industrial Revolution brought heavier technology and mechanized methods to the logging industry, applied to cutting, extraction, skidding, milling, and transportation. By the early 20th century, railroads improved flow of product from the bush to sawmills and markets—which increased pressure on the forests. Today, mechanization includes the use of helicopters, tractors, cables, and specialized equipment such as feller-bunchers and forwarders to remove logs from the forest. Traditional approaches to logging, using axes, machetes, or chainsaws; non-mechanized skidding; and water and/or animal power make up a very small percentage of global logging today.

During the late 19th and early 20th centuries, the logging industry moved into full-scale mass production. Lumber and wood-based pulp and paper production was found in Russia, Scandinavia, Europe, and North America. In the 1950s, the established producer countries' products were sold on a wider world market, meeting demands for wood products, especially in developing countries in Asia and Latin America. Traditionally, logging has had global markets, but full globalization of the logging industry was pioneered after World War II by such national economies as Japan, which sought raw materials from around the globe. Today, the logging industry is important to the national economies in every quarter of the world.

With diminishing northern forests to harvest from and increasing regulations, many large logging companies turned their attention to logging transnationally in the 1950s. This initiative began with a move into Asian countries, which had lower labor costs. By the late 1990s, these transnationals and many of the large new Asian logging companies had depleted much of the Asian forests' stock and turned to logging tropical forests in Africa, the Amazon River basin, Burma, and Indonesia, where corruption makes forestry laws nearly unenforceable and a lack of transparency in commercial transactions gives little regard to the environment or local peoples.

The logging industry is a major global industrial sector; gross production accounted for $160 billion worldwide in 1998. The industry is projected to grow to $300 billion by 2020. Today, two-thirds of the world's forests are located in 10 countries: the Russian Federation, Brazil, Canada, the United States, China, Australia, the Democratic Republic of Congo, Indonesia, Angola, and Peru. There are 9.6 billion acres (3.9 billion hectares) of forests, of which 95 percent are natural forests and 5 percent are forest plantations. Forests cover about 30 percent of the world's land area. Nearly 44 percent of forests are located in the tropics, and about one-third in boreal regions, with temperate forests accounting for 13 percent, and subtropical forests amounting to nine percent.

Ecological Effects and Conservation

While the logging industry is economically important, the ecological effects of logging can be devastating. Almost half of the Earth's original forest cover is gone, much of it destroyed after 1970. Since 1990, about one-third of forest around the world has been lost, with the vast majority of this loss occurring in the tropics. Logging for wood products is responsible for one-third of total global deforestation; other factors are land-clearing for agriculture, urban development, and road-building.

The early industrial approach to harvesting forests mainly involved clear-cutting with no reforestation. This practice created ecological devastation leading to erosion, desertification and local droughts, and forest-composition changes, all of which resulted in reduction of habitat for many flora and fauna species. It became clear that the forests were in many ways worth more than the logging products that could be extracted from them. Stakeholders outside the logging industry began to see the value of forests beyond the extractable forest products.

A deeper understanding of key ecological services provided by forests included carbon storage, regulation of river flows and sedimentation, habitat support, non-timber forest products (mushrooms, medicinal plants, craft woods, etc.), and a variety of aesthetic and cultural benefits. In many counties, logging companies must now follow regulations and management practices that carry out integrated resource management (IRM) plans.

During the past 150 years, pioneering individuals brought key legislation and approaches to the logging industry to help reduce the devastating effects of the industry and aid in sustainability. Before the 1950s, the logging industry was concentrated in temperate and boreal forests, and coping with the effects of unsustainable approaches to logging in these forests was key to the sustainability of the industry and the ecoservices provided by these forests. The earliest adoption of forest regulations and conservation was in California in the 1850s.

Conservation of some forested areas was implemented when sets of protected areas, such as national parks in the United States and Canada, were established. The logging companies began to see that the seemingly vast expanses of forest were not everlasting. They began to buy up forested land and moved to using harvest rotations in these temperate northern forests. By the mid-20th century, they used reforestation, silviculture, and plantations to ensure that product was available to be logged in the future.

Unfortunately, strategies such as reforestation and plantations were slow in terms of growth in the north. Northern forests containing jack, pine, and black spruce take 80 to 100 years to grow to maturity. Thus, some of the largest logging companies turned their attention to tropical areas for harvest and for plantation. Plantations located in tropical countries, such as those in Brazil and Indonesia, are likely to become much more prominent, as hardwoods are easier to plant here and grow at rapid rates.

Along with the logging industry, commercial banks, the United Nations, and other international bodies have seen plantations as a possible antidote to deforestation; they have largely supported this approach to a conservation-and-industry partnership. Biotechnology and hybridization have brought diverse commercial species to the logging industry to replace northern pulp and commercial timbers. One example in pulp supply is plantation-grown eucalyptus. Most species take five to seven years to grow to commercial size, and coppiced growth allows three crops from one root. Opponents see plantations as monocultures that provide little habitat support for the high biodiversity found in tropical settings.

U.S. Fish and Wildlife Service staffers measuring selectively cut logs on a truck. Even though many countries now enforce logging regulations, about one-third of the world's forests have been lost since 1990. (U.S. Fish and Wildlife Service)

Harvesting Approaches

Logging companies, while historically built around clear-cutting, have to some measure changed their harvest approaches due to regulations and with an eye toward sustainability of forests. Harvest prescriptions are now set out to help mimic natural succession in different forest types. In even-aged stands, natural disasters such as fire and wind events are part of ecosystem cycling, in that most trees are killed off and grow up as one age class following the disaster. Newer forms of clear-cutting involve a reduction in the size of clear cuts, corridors for large terrestrial mammals, reserves on riparian areas, buffers around stick nests, and protection of forest reserves.

For temperate deciduous and most tropical forest types, which are mosaics of tree ages and species, single-tree selective logging can mimic natural gap formation.

Selective logging involves removing only a select proportion of the trees in a stand. Within selective logging, the approach has two main approaches. One involves a high grade or diameter cut. The loggers "take the best and leave the rest," resulting in leaving poor-genetic-legacy, diseased, and lower-valued trees in the stand. The second approach, also known as "best forest management practice," is the "worst first" approach. It removes lower-quality, diseased, and crooked trees. This method, in combination with attention to buffer zones around waterways and to preserving appropriate representation of all ages and classes of trees, including older trees with cavities for wildlife, allows for a more sustainable approach.

In tropical forests, harvest prescriptions and approaches can be controversial; it appears that logging in primary forests precipitates a significant loss of biodiversity. Selective logging is expanding rapidly throughout the tropics. In these settings, while selective logging is better than clear-cutting, selective logging may increase the vulnerability of forests to fire and future deforestation. Also, logging roads cut into remote forest areas typically grant easy access to developers, small farmers, hunters, and poachers. Selective logging in tropical forests is seen as degrading the forest, as the felling of a single large tree can bring down dozens of surrounding trees through the network of vines and lianas, creating large canopy openings and precipitating habitat changes by increased sunlight and drying winds. These types of effects are well-documented in the scientific study of forest fragmentation and edge effects.

One major problem in terms of regulating or creating sustainability within the industry is illegal logging. Its dramatic effects, especially in developing nations, include taking much-needed revenue away from governments, devastating effects on the livelihoods of forest-dependent people, and furthering corruption and civil conflict. At current rates of illegal logging, if all the timber involved had been harvested legally, governments could

have earned up to $6.5 billion. The World Bank estimates that the value of the illegal trade is $10 billion to $15 billion per year.

International Initiatives and Future Markets

Following the 1992 Earth Summit in Rio de Janeiro, Brazil, much-needed awareness was brought to the worldwide threats to forests, the implications of deforestation, and changing attitudes among the general public. As a result, attention was turned to furthering sustainability, keeping forests intact, and reducing the effects of climate change—as forests store 50 percent of the world's terrestrial carbon. The response by some of the largest companies within the logging industry was to partner with various groups, such as publicly recognized nongovernmental organizations, and to change their operations to be more sustainable through volunteer certification regimes.

There are more than 50 certification schemes in which the logging industry can participate, such as the Forest Stewardship Council and the Programme for the Endorsement of Forest Certification. These voluntary, third-party certifications are becoming more commonplace—but are not without criticism, as they are seen by some environmentalists as ways that the logging industry can manipulate the public by producing similarly named certifications with diluted standards.

Currently, the logging industry is moving into newer markets, such as turning waste into renewable energy. The industry also faces new challenges, such as the effect of invasive species and insects in the temperate forests of North America, urban sprawl, being called upon to mitigate various forms of pollution, and competition for land with mining and fossil-fuel extraction.

The logging industry has most recently turned its attention to the vast northern, or boreal, forests, where it is waging battles with government and environmental groups to obtain cutting rights. While both legal and illegal logging is found to be decreasing forest cover in tropical forest ecosystems, the boreal, temperate, and subtropical forests appear to be increasing in net forest cover due to reforestation, natural forest recovery, and establishments of forest plantations. The viability of these forest ecosystems remains at the heart of the logging industry's longevity—and that of forest habitats worldwide—into the 21st century and beyond.

KYMBERLEY A. SNARR

Further Reading

Bacher, John. *Two Billion Trees and Counting: The Legacy of Edmund Zavitz.* Toronto: Dundurn Press, 2011.

Bryant, Dick, et al. *The Last Frontier Forests: Ecosystems and Economies on the Edge.* Washington, DC: World Resources Institute, 1997.

Diamond, Jared. *Collapse: How Societies Choose to Fail or Succeed.* London: Penguin Group, 2005.

Food and Agriculture Organization of the United Nations. "Forestry." 2012. http://www.fao.org/forestry/en.

Lawson, Sam and Larry MacFaul. *Illegal Logging and Related Trade: Indicators of the Global Response.* London: Chatham House, 2010.

Marchak, M. Patricia. *Logging the Globe.* Montreal: McGill-Queen's University Press, 1995.

Loire River

Category: Inland Aquatic Biomes.
Geographic Location: Western Europe.
Summary: A braided river and one of the last wild rivers in Europe, the Loire supports a remarkable array of wildlife.

The Loire River is one of three rivers that rise from the Massif Central, a stark volcanic uplands region in south-central France, noted for its rugged landscape of forests and pasturelands. Several main tributaries feed into the Loire; these include the Allier, Cher, Indre, Vienne, Maine, and Seore Rivers. This freshwater system drains one-fourth of the lands of France. It is approximately 600 miles (1,000 kilometers) in length.

The catchment area, or watershed, of the Loire River encompasses about 45,000 square miles (177,000 square kilometers). It is considered one of the last wild rivers in Europe. The Loire River basin is remarkably rural and fecund; it supports two-thirds of the livestock and half of all grains produced in France. The Loire is used for navigation, for hydroelectric and nuclear power, and for fishing and tourism. Its many lakes and tributaries are linked by natural and manmade canals that have facilitated the production and transport of goods throughout western Europe for thousands of years.

Flowing north and then west from the central region near Orléans, the Loire passes through deep gorges and floodplains at its upper banks, through dikes and levees, across limestone plateaus, forested islands and sandy shores, and down through the Sologne wetlands to fill the brackish estuaries at the Atlantic coast. It transects the French landscape into three geographically distinct areas of urban, agricultural, and industrial development. The upper basin is bound by the Loire from its source in the Massif Central to its juncture with the Allier River. The Loire middle valley stretches from the Allier to the Maine River; it wanders in a broad circular pattern, creating what is known as the Garden of France. The lower Loire flows west across the granite terrain of Poitou and Brittany; it extends to the wetlands on the Atlantic coastline where it empties into the Bay of Biscay.

The climate of the Loire Valley is temperate, considered the most pleasant of northern France. There are hot summers, warm winters, and fewer extremes in temperatures here. Annual average minimum temperature is 45 degrees F (7 degrees C); annual average maximum temperature is 61 degrees F (16 degrees C). The climate is identified as temperate maritime, with rainfall of approximately 24–27 inches (63–69 centimeters) annually.

Biota

The Loire is a braided river, a definition describing the morphology of rivers that flow in multiple channels across a gravel floodplain. Braided rivers create unique and biologically diverse ecosystems including meadows, wetland marshes, forest vegetation, islands, and sedimentary banks. Because of the variability of riverine floodplain habitats, there is a constant redistribution of biological taxa. Along the Loire River, the result of these dynamics is an extraordinarily rich and complex array of habitats and their related biological species. The Loire River biome also encompasses part of an international flyway for migratory birds.

More than 100 algae species have been identified in the Loire, the richest phytoplankton diversity of all the French rivers. This abundance supports thriving communities of fish and migrating land birds. Every fish known to swim in French rivers can be identified in the Loire, including sea trout, shad, lamprey, flounder, mullet, sturgeon, and catfish. However, extensive damming, channel excavation, agricultural and industrial runoff, and overfishing have created migratory obstacles for the great schools of Atlantic salmon that come to these inland waters from as far as the waters off Greenland to spawn their eggs.

Wetlands are essential biofilters whose sediments, organic, and inorganic accretions act as important buffers for saltwater intrusion into freshwater environments; freshwater flooding is an essential dynamic of wetlands and estuaries. The Loire estuary provides habitat for 25 percent of the juvenile fish found in the Gulf of Biscay. The Sologne in the central Loire valley, just south of Orléans, is an internationally recognized inland wetland, a region whose marshes and forests provide valuable habitats for nesting birds and water fowl, including the goshawk, honey buzzard, the whiskered tern, and little bittern.

Similarly, growths of hawthorn, black thorn, and elder bush provide food and shelter for migrating flocks. Mallards, tits, house sparrows, woodpeckers, kites, larks, rails, and martins are common, as are great blue herons, terns, plovers, gulls, geese, and egrets that populate gravel and sand bank areas. Stands of riverine willows, elders, ferns, and climbers provide habitat for all of these bird species and others.

The Loire River valley's wet heaths, lakes, and flood meadows support a rich and varied number of moths, dragonflies, and butterflies. Common species include hawk moths, the peacock moth,

Grassland Butterflies

Considered an indicator species, grassland butterflies have a noted decline of nearly 70 percent in recent decades. Their disappearance is a direct result of intensive farming practices that favor monoculture crops over more variegated small farm land-use practices that included complex grasslands for pasturage. Noted species include the meadow brown (*Maniola jurtina*), the marbled white (*Melanargia galathea*), the small and large white (*Pieris rapae/napi*), gatekeeper (*Pyronia tithonus*), small heath (*Coenonympha pamphilus*), the chalkhill blue (*Polyommatus coridon*), adonis blue (*P. bellargus*), and the painted lady (*Vanessa*).

*A gatekeeper butterfly (*Pyronia tithonus*) found in Vendée, France, in 2006. (Wikimedia/ Guillaume Paumier)*

the common winter damselfly, clubtails, and pincertails. On warm spring days, the riverlands teem with insects (beetles, bees, flies, firebugs, and ants are common), an important food source for the migratory birds for which the area is noted internationally.

Finally, the lands contiguous to the Loire River are famous for their displays of orchids. The most common species include the tongue orchid, the violet Limodore, the early purple orchid, the lady orchid, the bee orchid, the early fly orchid, and the greater butterfly orchid. The Loire River is an important migratory system for plant life; species from the Central Massif, the Atlantic, and the Mediterranean have all made their way here, some taking centuries to migrate.

Human Impact

Civil engineering projects date back to antiquity and include dredging; flood control technologies; land reclamation; the extensive excavation of gravels and sediments; and the construction of levees, groins, bridges, dikes, canals, and dams for irrigation, transportation, and erosion control. For centuries, the habitual deposition of industrial, agricultural, and household wastes have contributed to chronic freshwater contamination. Protocols that now govern the management and protection of the Loire River and its internationally recognized habitats include the Ramsar Convention on Wetlands, adopted in 1971; the European Union (EU) Water Framework Directive; the Natura 2000 directives; and the policies of the EU Nature Legislation—its Birds Directive as adopted in 1979 and amended in 2009, and the Habitats Directive as adopted in 1992.

In 1986, the construction of four new dams in the Loire system was proposed by the French government and other private interests. That year, the Loire Vivante (Living Loire) network was established by the World Wildlife Fund and other nonprofit organizations to protest the new dam construction at Serre de la Fare on the upper Loire. As a result of the highly coordinated efforts of the Loire Vivante movement, the dams were either cancelled or substantially revised, and three other dams were demolished. In 1994, the Plan Loire Grandeur Nature initiative was introduced by the French government to implement sustainable programs of flood control and floodplain restoration. Today, the European Rivers Network is an outcome of the Loire Vivante movement.

In December 2000, the Loire Valley between Maine and Sully-sur-Loire was registered as a United Nations Environmental, Scientific, and Cultural Organization (UNESCO) World Heritage Site. This central Garden of France region includes a surface area of approximately 300 square miles (800 square kilometers). It is an exceptionally temperate, fertile area famous for its orchards and vineyards. As with other inland aquatic ecoregions, the lower areas of the river are susceptible to any changes in sea level. Hence, there is the present concern of long-term impact that climate change will have in this area, especially in the estuary habitats.

Victoria M. Breting-Garcia

Further Reading
Gray, Duncan and Jon. S. Harding. *Braided River Ecology: A Literature Review of Physical Habitats and Aquatic Invertebrate Communities.* Wellington, New Zealand: Science & Technical Publishing, Department of Conservation, 2007.
Hayes, Graeme. *Environmental Protest and the State in France.* New York: Palgrave Macmillan, 2002.
Hilbers, Dirk and Tony William. *Loire Valley: Loire, Brenne & Sologne (Crossbill Guides).* Philadelphia: Trans-Atlantic Publications, 2011.
Paskoff, Roland P. "Potential Implications of Sea-Level Rise for France." *Journal of Coastal Research* 20, no. 2 (Spring 2004).
Tockner, Klement, Urs Uehlinger, and Christopher T. Robinson. *Rivers of Europe: First Edition.* Waltham, MA: Academic Press, Elsevier, 2009.

Long Island Sound

Category: Marine and Oceanic Biomes.
Geographic Location: North America.
Summary: This estuary has a rich and constantly changing environment; it is under constant pressure from a high volume of human inputs.

Long Island Sound is an estuary located on the northeast coast of the United States, where the states of Connecticut and New York meet the North Atlantic Ocean. The watershed of Long Island Sound includes nearly all of Connecticut and western Massachusetts, large parts of Vermont, New Hampshire, and Rhode Island, and a small area of New York state. Here, saltwater from the Atlantic mixes with freshwater from the Connecticut, Thames, Housatonic, and a few smaller rivers. As these waters meet and mix, varying salinity levels are created, resulting in a wide range of flora and fauna that thrive in the marine and shoreline environments.

The sound supports a great spread of tidal marshes and wetlands, which play a huge role in the ecosystem, supporting open-water as well as seafloor communities. The sound also supports a very large human population: 21 million people, or between 6 and 7 percent of the population of the United States, live within 50 miles (80 kilometers) of the sound, which is why it is occasionally referred to as the Urban Sea. The heavy use of the sound and nearby coastal lands by humans for industrial, agricultural, municipal, commercial, and recreational activities has unfortunately resulted in significant levels of pollution. Both the federal government and local agencies are taking steps to reduce environmental threats and preserve the sound's ecosystem.

Long Island Sound is approximately 110 miles (177 kilometers) long and up to 21 miles (34 kilometers) wide, covering about 1,300 square miles (3,370 square kilometers). Its average depth is 65–79 feet (20–24 meters), and it is 300 feet (91 meters) at its deepest point. The sound's orientation is unique, as most estuaries along the east coast of North America have a north-south orientation, but Long Island Sound has an east-west configuration.

The eastern portion of the sound has higher saline levels than the western portion, as it is closer to the Atlantic Ocean, but, in general, salinity levels range between 27 and 32 parts per thousand. Tides play a very important role in the sound's ecosystem, as the incoming marine tides carry plant nutrients into the marshes to feed their inhabitants, and the outgoing tides carry excess nutrients back into the estuary. Average temperatures in the sound are about 69 degrees F (21 degrees C) in the summer and 37 degrees F (3 degrees C) in the winter.

Biota

Organisms that live in the Long Island Sound biome have had to develop the ability to adapt to a rapidly changing environment, as this water body has several ecological factors that are constantly in flux. The sound experiences dramatic changes in temperature from season to season. During the winter months, the temperature often drops below freezing, and in the summer it can be extremely hot. Salinity levels also experience periodic flux, due to the strong tides. The biota of the intertidal zone has had to adapt for the particularly harsh contrasts here. The sound is nevertheless home to more than 100 species of plants, 1,200 species of invertebrates, and 170 species of fish.

In general, coastal salt marshes are known as high-productivity environments due to the strong base of plant life within the favorable habitat of such marshes. Long Island Sound has three categories of plants: algae/phytoplankton, seaweeds, and vascular plants. Saltmarsh plants such as saltwater cordgrass grow along ditches, or slat pans, and on the seaside edges of marshes where high tides inundate daily.

Salt meadow cordgrass and amulet spikegrass grow in areas less frequently inundated by saltwater, typically closer to higher, drier land. Other plants found in the pans are sea lavender, saltmarsh aster, and seaside gerardia. In areas where the brackish water is more diluted with freshwater from rivers, cattail marshes replace saltmarshes. Various types of grasses, including wild rice and sedges, are found here. Eelgrass is typically found in protected bays, coves and other areas of brackish water.

All of these plants provide shelter to the mollusks, crustaceans, and other fauna breeding, hatching, feeding, and maturing here. Even as the plants decompose, the result is a rich organic soup that feeds the shellfish and smaller finfish that live within the marshes. The marshes host many types of snails, mussels, fiddler crabs, and species of minnows. Found on the sound proper's seafloor are shrimp; multiple species of crabs; lobsters; and benthic fish species such as flounders, skates, and monkfish. These crustaceans and smaller fish act as food for the larger fish that live within the open-water community, such as bluefish and dogfish, as well as migratory types such as striped bass, Atlantic salmon, eels, and turtles. At least 50 species of fish spawn here.

Throughout the year, the sound hosts more than 280 species of migratory birds that take advantage of the shellfish and smaller fish populations, including gulls, herons, egrets, and other waterfowl and shorebirds. Long Island Sound also has a harbor-seal population and is occasionally visited by whales and dolphins.

Human Impact

Long Island Sound supports the third-largest lobster fishery in the United States; oysters, clams, mussels, and many of the larger fish species are harvested by humans as well. At the same time, the sound has more than 40 sewage-treatment plants that empty about 1 billion gallons (3.8 billion liters) of treated sewage into its waters every day. Unfortunately, this activity has resulted in high levels of pollution, which has not only contaminated organisms that live in the sound, but also greatly depleted oxygen levels in its waters. The low oxygen levels have sometimes resulted in mass lobster and fish casualties, with one of the largest die-offs occurring in 1999.

Fortunately, in 1987, the Clean Water Act created the National Estuary Program, which eventually led Congress to pass the Long Island Sound Restoration Act of 2000. Millions of dollars per year are being used to clean up the sound and save the plant, animal, and human populations that depend on the sound to survive.

Studies are now being carried out and data is being collected on the potential effects of climate change for this biome. As yet, scientists feel a higher level of confidence in observed and predicted changes in global average temperature, while sea-level rise projections are more uncertain. Observations indicate that global sea-level rise is occurring and will continue to occur; indeed, the impact in the Long Island Sound biome is expected to occur faster than the global average. However, the magnitude of impact on coastal areas is complex, and will depend on rates of Arctic ice sheet melt, changes in ocean circulation due to additional freshwater inflow from melting ice, accompanying temperature fluctuations, altered salinity differentials, and naturally-cyclic climate patterns.

ELIZABETH STELLRECHT

Further Reading

Andersen, Tom. *This Fine Piece of Water: An Environmental History of Long Island Sound.* New Haven, CT: Yale University Press, 2002.

Dreyer, Glenn D. and William A. Niering, eds. *Tidal Marshes of Long Island Sound: Ecology, History and Restoration.* New London: Connecticut College Arboretum, 1995.

Wahle, Lisa and Nancy Balcom. *Living Treasures: The Plants and Animals of Long Island Sound.* Groton: Connecticut Sea Grant College Program, 2002

Weigold, Marilyn E. *The Long Island Sound: A History of Its People, Places, and Environment.* New York: New York University Press, 2004.

Lord Howe Island Subtropical Forests

Category: Forest Biomes.
Geographic Location: South Pacific Ocean.
Summary: Situated on this remote volcanic island, these forests have protected status, but introduced species threaten the endemic ones.

Lord Howe Island is located in the South Pacific Ocean, some 435 miles (700 kilometers) east of Australia and 840 miles (1,350 kilometers) northwest of New Zealand. Approximately 6 miles (10 kilometers) long, the crescent-shaped island is a remnant of a volcano that formed about 7 million years ago. Over the course of this time, the volcano eroded to create Lord Howe Island; a series of islets of the Admiralty Group; and Ball's Pyramid, an islet separated from Lord Howe Island by approximately 13 miles (21 kilometers) of water. There are two prominent peaks on the southern portion of the island: Mount Gover and Mount Lidgbird, with elevations of 2,870 feet (875 meters) and 2,550 feet (777 meters), respectively.

The climate is subtropical, with mean annual rainfall of just under 68 inches (173 centimeters). The average annual high temperature is 72 degrees F (22 degrees C); the average annual low temperature is 63 degrees F (17 degrees C). The warm temperatures and steady rainfall throughout the year create an optimal environment to support the lush rainforest that covers the vast majority of the island, along with some areas of scrub, grassland, and clearings in the lowlands and on the cliffs.

Biota

In 1982, 75 percent of the island was declared a Natural World Heritage Site, and the biological and cultural heritage was entrusted to the Lord Howe Island Board, which is responsible for eradicating the invasive goats from the island and for significantly decreasing the population of feral pigs. Efforts to control other introduced species, such as rats and plants, are ongoing. The removal of some feral animals has resulted in the recovery of the forest understory.

Due to the geographic isolation of the island, Lord Howe Island and its marine surrounds are home to a unique assemblage of birds, insects, plants, and marine animals, all of which reflect a high degree of endemism (species found nowhere else). Because the vast majority of Lord Howe Island's fauna and flora evolved in isolation, the biodiversity that exists here is amazing. Of the 241 native species of vascular plants, 105 are endemic species, of which at least 16 are considered to be rare, endangered, or vulnerable, and many of which have very restricted ranges. *Chionochloa conspicua*, for example, is known from a single clump located on Mount Lidgbird.

The forests here, mainly rainforest, can be assessed as either lowland, submontane, or montane. The lowland swaths are characterized by mixed evergreen fruit trees, laurels, fringetrees, tea trees, broadleaf types, palms, and the palm-like shrubs known as pandans, or ketakis. Laurels and palms are the hallmarks of the submontane. Ferns, mosses, and epiphytes abound in the understory of both these zones. The montane zone hosts cloud forests, so called because they are nearly continuously subject to very high moisture. Dominant

This aerial view of the 6-mile-long (10-kilometer) Lord Howe Island in the South Pacific shows its crescent shape and the two peaks, Mount Gover and Mount Lidgbird, at the southern end. The island formed from the remains of a volcano that dates back 7 million years. With a current population of fewer than 400 people, it is one of the few Pacific islands that has not been significantly altered by humans; only about 10 percent of the island has been cleared for human use. (Thinkstock)

here are fragrant flowering shrubs from the family *Winteraceae* and the heavily-fruited, but stunted trees of family *Eriaceae*.

There are a total of 202 bird species that have been identified on the island. Currently, 129 native and introduced bird species breed regularly here. Nine of the original 15 native land bird species have gone extinct, including seven endemic species. There are four extant endemic land birds, including two species and two subspecies: the abundant Lord Howe white-eye (*Zosterops tephropleurus*), Lord Howe Island woodhen (*Gallirallus sylvestris*), Lord Howe Island golden whistler (*Pachycephala pectoralis contempta*), and Lord Howe Island currawong (*Strepera graculina crissalis*). The woodhen was reduced to 25 individuals in the 1970s, but due to an aggressive management program, the population was well over 200 by the 1980s. Seabirds also use the island extensively. Lord Howe Island is one of the few known breeding grounds of the Providence petrel (*Pterodroma solandri*), located in the cloud forests of the island. Approximately half the world's population of the flesh-footed shearwater (*Puffinus carneipes hullianus*) breeds on the island, and the island has the greatest concentration of red-tailed tropicbird (*Phaethon rubricauda*). A total 12 additional species of seabirds have been documented breeding on the island.

The large forest bat (*Eptesicus sagittula*) is the only native mammal. As a result of introduced predators, there are only two identified native reptiles—a skink (*Oligosoma lichenigera*) and a gecko (*Phyllodactylus guentheri*)—but these both may have gone extinct by the mid to late 1990s. Many endemic invertebrates inhabit these subtropical forests, including snails, flies, earthworms, and spiders. Of the 100 spiders described on the island, up to 50 percent are thought to be endemic.

Human Impact

Due to the island's geographic isolation, the island was not settled until the early 19th century, and it has never been heavily populated. Consequently, it is unlike most other Pacific islands in that it has not been dramatically altered by people, and most of the island is still forested. It is estimated that only 10 percent of the island has been cleared for human use, and an additional 20 percent of the native vegetation has been disturbed by the introduction of animals, including cattle, goats, feral sheep and feral pigs, and rats.

In the late 1800s, trade began from the island in the form of the Kentia palm, endemic to the island. This palm, one of four endemic here, became very popular in Britain, Europe, and the United States by the early 1900s, and is now the most popular decorative palm in the world. The sale of seeds and seedlings remains today, along with tourism, the only form of outside income for the small population of the island, currently less than 400 people.

Although most of Lord Howe Island is protected, continued management is necessary to prevent the introduction of invasive species. Historically, introduced predators have had a devastating effect on some of the endemic fauna, and introduced plants have displaced native species. The long-term recovery of the island's many threatened species will be successful only through the continued control and management of these exotic species. Meanwhile, a more immediate threat to the ecosystems here may be in the form of climate change. Due to the small size of the island, any change in rising sea levels translates into potentially heavy loss of land mass.

JEFFREY C. HOWE

Further Reading

Davey, A. *Plan of Management: Lord Howe Island Permanent Park Preserve*. Sydney, Australia: New South Wales National Parks and Wildlife Service, 1986.

Department of Environment and Climate Change. *Lord Howe Island: Biodiversity Management Plan*. Sydney: Australian Government, 2007.

McDougall, I., B .J. J. Embleton, and D. B. Stone. "Origin and Evolution of Lord Howe Island, Southwest Pacific Ocean." *Journal of the Geological Society of Australia* 28 (1981).

Recher, H. F. and S. S. Clark. "A Biological Survey of Lord Howe Island With Recommendations for the Conservation of the Island's Wildlife." *Biological Conservation* 6, no. 4 (1974).

Luang Prabang Montane Rainforests

Category: Forest Biomes.
Geographic Location: Asia.
Summary: This rich and rugged biome in north-central Laos as well as parts of Thailand and Vietnam is under pressure from human activities.

The Luang Prabang Montane Rainforest biome is situated in the central massif or highlands of southeast Asia, comprising portions of Laos and Thailand. Its area in total extends across 27,700 square miles (71,800 square kilometers) mainly in north-central Laos and, depending on definition, also includes areas in the northeastern Isan region of Thailand and some mountains of Vietnam.

The biome consists of several ecological types, which merge with one another, as well as with the subtropical rainforest to the north, and the northern Annamite rainforest region of Vietnam to the east. It contains both open montane and open conifer forests, and also mixed conifer-hardwood forests. As part of a monsoon area, the region receives 7 to 10 feet (2 to 3 meters) of rain annually, but this is concentrated in a comparatively short season; most of the year, it is dry.

Flora and Fauna

Hardwood forests here are dominated by members of the beech family (*Fagaceae*) and the laurel family (*Lauraceae*). Where evergreen forests have been degraded above a height of 3,300 feet (1,000 meters), these species have been supplanted by the broadleafed evergreen *Castanopsis hystrix*, together with the Indian gooseberry or aamla (*Phyllanthus emblica*), which is attracting attention for potential medicinal uses.

Slightly lower, in the range of 2,600 feet (800 meters) elevation, another transitional phase is seen with the emergence of various palms, along with the Keruing (*Dipterocarpus turbinatus*) and wax trees (*Toxicodendron succedanea*), which are common across southeast Asia.

The nature of the tree cover depends on both the elevation of the land and the nature of the soil.

Research indicates that some mountain-dwelling agriculturalists have extensive knowledge of the soils present and their implications for plant growth, but much of this knowledge has yet to be captured and codified.

More than 500 species of birds are known to have ranges within the Luang Prabang montane rainforest, but the lack of systematic surveying means that it is not known whether any are unique to the area. The inaccessibility of the region contributes to the lack of research, as do the lack of domestic technical capacity, the low level of importance attached to such activities in the postrevolutionary period (after 1975), and the low status historically accorded to upland people.

To date, in the Luang Prabang montane region, the forests have been minimally exploited, and it is believed that this represents one of the best possibilities for the preservation of large animals in the area, including the Asian elephant (*Elephas maximus*), silvered leaf monkey (*Semnopithecus francoisi*), Himalayan black bear (*Ursus thibetanus*), and tiger (*Panthera tigris*). Extensive surveying of the region, which has not yet taken place, presumably would reveal the presence of other species, including some that have not yet been observed.

Human Impact

Luang Prabang is a province centered in north-central Laos. Its provincial capital is an ancient and sacred city with the same name; it is currently being developed as a significant tourism destination. Laos is a landlocked country with a sparse population, much of which is concentrated in valley areas.

Historically, people displaced by arriving migrants have been forced upslope and have adapted their agricultural patterns accordingly. They have been joined by migrating ethnic groups, which have not always been officially integrated into the state, and have continued with swidden or slash-and-burn agriculture in the forested mountains. Previously, some tribes grew opium in upland areas, but this rarely occurs now. Not all agriculture in the region is beneficial. There remain clear signs of past swiddening on an extensive scale that has led to changes in the permanent land cover.

In surrounding forests in the greater Mekong subregion, notably in Thailand and Vietnam, shifting agriculture has caused the degradation of rainforest areas; only some 30 percent remains in a recognizable state in those areas. Intensification of existing agricultural activities in the Luang Prabang region is likely to result in the same degree of degradation. Increasing competition for land in Laos, which may occur if current intentions for economic development, transportation infrastructure construction, and investment by Chinese corporations continue, is likely to result in the conversion of much existing montane rainforest to scrubland.

In addition to land degradation, threats include the construction of dams for the production of hydroelectricity; hunting for exotic animal products destined mainly for the Chinese market; and the general process of economic development represented by the building of the Asian Highway Network. Potential impact of climate change on forest lands remains unclear. Currently, protected areas in the Luang Prabang Montane Rainforests biome include the Nam Phouy National Biodiversity Conservation Area, the Phu Hin Rong Kla National Park, and the Phu Soi Dao National Park.

JOHN WALSH

Further Reading

Saito, Kazuki, Bruce Linquist, Bouanthanh Keobualapha, Tatsuhiko Shiraiwa, and Takeshi Horie. "Farmers' Knowledge of Soils in Relation to Cropping Practices: A Case Study of Farmers in Upland Rice-Based Slash-and-Burn Systems of Northern Laos." *Geoderma* 136, nos. 1–2 (2006).

Souvanthong, Pleng. *Shifting Cultivation in Lao PDR: An Overview of Land Use and Policy Initiatives.* London: International Institute for Economic Development, 1995.

Walker, Andrew. *The Legend of the Golden Boat: Regulation, Trade, and Traders in the Borderlands of Laos, Thailand, China, and Burma.* Honolulu: University of Hawaii Press, 1999.

Lukanga Swamp

Category: Inland Aquatic Biomes.
Geographic Location: Africa.
Summary: This Ramsar-recognized wetland supports diverse habitats and is home to over 300 bird species. Overfishing, poaching, deforestation, and water pollution have endangered its sustainability.

The Lukanga Swamp biome is a wetland located in the Central Province of Zambia. It is the largest permanent water body in the Kafue Basin, comprising generally shallow swamps that allow light penetration to the bottom, permitting high photosynthetic activity. While the permanent swampy area remains largely inaccessible, human encroachment within the surrounding floodplain has had significant environmental impacts, most notably deforestation, land erosion, deteriorating water quality, and species population reductions—all of these caused by or related to fishing, hunting, poaching, and logging for charcoal production. Poverty, limited resources, and lack of legislative enforcement have limited preservation efforts.

Geography and Climate

The permanently swampy area encompasses approximately 710 square miles (1,850 square kilometers), covering an area that is somewhat circular, with a diameter of about 30 miles (50 kilometers). It has an elevation of approximately 3,690 feet (1,125 meters) above sea level. The average depth is shallow, at about 5 feet (1.5 meters). Numerous rivers and streams discharge into the Swamp, most notably the Lukanga River and the Kafue River. Lagoons, such as Lake Chiposhye and Lake Suye; pools; reed beds; papyrus islands; and channels dot the area. Pockets of higher ground form permanently dry islands within the swamp area and its environs, such as Chilwa and Chiposha Islands.

The nearest urban areas include Mumbwa and Kabwe. The larger floodplain surrounding the permanent swamp falls within the Zambezian Flooded Grasslands ecoregion. It is inundated on an annual seasonal basis, and can increase the Lukanga Swamp's area to 2,300–3,100 square miles (6,000–8,000 square kilometers) during the rainy season. Areas of woodland savanna also surround the swamp.

Flora and Fauna

The Lukanga Swamp and its environs provide diverse grazing and aquatic habitats that support an array of plant and animal life, and host a number of threatened species, such as the wattled crane (*Bugeranus carunculatus*), the red lechwe, African python, and the sitatunga, an antelope adapted to walking and swimming in marshy environments.

The area is also an important breeding ground for fish, the most abundant of which is tilapia, with *T. rendalli* and *T. sparmani* the predominant species. Resident and visiting animals include the hippopotamus, crocodile, elephant, buffalo, roan antelope, oribi, marsh mongoose, bushbuck, python, and eland.

Over 300 species of resident and migratory birds are known to inhabit the Lukanga Swamp. Notable bird species include the coppery-tailed coucal (*Centropus cupreicaudus*), pallid harrier (*Circus macrourus*), Dickinson's kestrel (*Falco dickinsoni*), brown firefinch (*Lagonosticta nitidula*), white-bellied sunbird (*Nectarinia talatala*), Hartlaub's babbler (*Turoides hartlaubii*), kurrichane thrush (*Turdus libonyanus*), and broad-tailed paradise whydah (*Vidua obtusa*).

Human Impact

Human habitation has encroached upon most of the Lukanga Swamp's borders, with the exception of the swamp's largely pristine western portion. Approximately 60,000 people reside within the swamp's environs, along the floodplain's edges, and on the swamp's permanently dry islands. Other, temporary, residents inhabit fishing camps within the islands.

There are no roads or bridges offering access into the permanently swampy areas. Several seasonal dirt tracks offer access to the surrounding floodplain. The Great North Road crosses the country near the floodplain, but does not offer direct access to it.

Human uses of the swamp include fishing, hunting wildfowl, and gathering natural resources such as reeds, which are used for construction and for basketry. Slash-and-burn methods and other forms of agriculture are practiced along the outskirts of the swamp. Large parts of the swamp forests have been cleared for charcoal production and agricultural use. Ecological consequences include subsequent land erosion, water turbidity, silt suspension and poor water quality. Human consequences include threats to food security. Fishing is a significant form of both food and income in a country marked by significant and chronic poverty. Most fishing is by traditional methods, with only a small presence of larger commercial fishing operations.

A Zambian farmer walks in a field among maize stunted by drought in 2006; a normal crop would have been shoulder height. Unsustainable agricultural practices and land-clearing have affected the Lukanga Swamp region and threatened food security. (USAID/F. Sands)

The swamp's fish populations have suffered from overfishing; the use of harmful fishing methods; water pollution; and global warming, which has put pressure on temperature-sensitive species directly and through alterations to their floral habitats. Results include the diminishment of fish stocks and the diminishing size and quality of remaining fish.

Since 2006, seasonal outbreaks of the deadly fungal disease known as epizootic ulcerative syndrome during the rainy season have begun to decimate the country's fish populations, especially along the nearby Zambezi River.

Conservation

Regional, national, and international environmental preservationists have recognized the need to prevent further environmental destruction. The Lukanga Swamp is one of eight Zambian sites recognized by the Ramsar Convention, an international agreement dedicated to the preservation of globally significant wetlands regions. Key species targeted for conservation include the lechwe, marsh mongoose, sitatunga, python, bushbuck, and oribi. Nearby protected areas include the Miombo Woodland game management area. There has been some discussion of extending habitat and species protection with the Lukanga Swamp.

Although some steps have been taken to protect the Lukanga Swamp and its environs, a lack of adequate knowledge and coordination among governmental and wildlife authorities, such as the Department of National Parks and Wildlife Service (NPWS), and limited governmental resources present significant barriers to ongoing environmental protection. Animal poaching and overfishing remain significant problems, as existing legislation is not adequately enforced. Poverty and the push for economic development are key factors in determining the Swamp's future ecological health.

Some government officials have targeted improved access to the region and the development of an ecotourism industry as potential future sources of economic growth. Tourism often has mixed environmental consequences, promoting habitat and wildlife conservation to maintain interest in the area—while at the same time increasing some of the negative environmental impacts of human presence.

MARCELLA BUSH TREVINO

Further Reading

Bellani, Giovanni G. *Vanishing Wilderness of Africa.* Vercelli, Italy: Whitestar, 2008.

Dugan, Patrick. *Guide to Wetlands.* Buffalo, NY: Firefly, 2005.

McCartney, Matthew, Lisa-Maria Rebelo, Everisto Mapedza, Sanjiv de Silva, and C. Max Finlayson. "The Lukanga Swamps: Use, Conflicts, and Management." *Journal of International Wildlife Law & Policy* 14, nos. 3–4 (2011).

Stone, Roger D. *The Nature of Development: A Recent Report From the Rural Tropics on the Quest for Sustainable Economic Growth.* New York: Knopf, 1992.

Luzon Rainforest

Category: Forest Biomes.
Geographic Location: Philippines.
Summary: A tropical rainforest that is home to a highly diverse range of flora and fauna, this biome includes a large number of endemic species, but faces significant threats from human activities.

On the western flank of the Pacific Ocean lies the Philippine archipelago, a cluster of 7,107 islands. The largest of these is Luzon, a 40,400-square-mile (104,700-square-kilometer) mosaic of highly developed urban areas and primary rainforest. Geographic isolation and a tropical climate have resulted in high biota diversity levels, with a significant number of endemic (found nowhere else) plant and animal species. Illegal logging, development encroachment, and other human activities pose significant threats to this biome's conservation prospects.

Geography and Climate

Luzon, an irregularly-shaped island at the northern end of the Philippines, has an uneven terrain. Two mountain ranges, the Cordillera and the Sierra Madre, run north and south on both sides of the Cagayan River, the country's longest. Isolated volcanic mountains above 3,300 feet (1,000 meters) break up the flat areas on the southern portions of the island. These are Mt. Banahaw, Mt. Makiling, Mayon Volcano, Mt. Isarog, and Mt. Bulusan.

The island's tropical climate has a relatively high average temperature of 80 degrees F (27 degrees C), high humidity at 70–90 percent, and abundant rainfall of 40–200 inches (102–508 centimeters) yearly. Three main seasons are influenced by equatorial monsoons: hot and dry (March to May), rainy (June to November), and cool and dry (December to February). Although some areas may experience torrential rains during typhoons, variations in temperature and rainfall occur mainly due to elevation and proximity to bodies of water.

Flora and Fauna

The Luzon Rainforest biome is composed of tracts of moist dipterocarp forests in areas below 3,300 feet (1,000 meters), as well as on islets north of the main island and the handful of volcanoes in the south.

Dipterocarps are hardwoods named for their two-winged seeds, which spin like helicopter blades as the seeds fall to the ground. These broadleaf trees rise as high as 200 feet (61 meters), with straight trunks stabilized by wide buttresses, and sometimes stilt roots, for extra support in the shallow soil.

Additional adaptations to the environment include smooth thin barks that require no protection against freezing or water loss; and drip tips, grooved leaves, and oily coatings that allow leaves to quickly shed water, preventing mold and mildew growth. Common genera in the Luzon Rainforest include *Shorea* and *Dipterocarpus*.

The canopy tends to be uneven, making the emergent or highest layer practically non-existent, and the lower understory rich in flora such as rattans, woody lianas, saplings, large ferns, and strangler vines climbing toward sunlight. Mosses, orchids, bromeliads, and other epiphytes flourish on tree branches, making the most of the available light, water, and nutrients.

The unique flora of Luzon is exemplified by the Philippine jade vine, a woody climber that produces striking blue-green flowers. The jade vine is endangered in the wild, but has been successfully cultivated through cuttings in conservatories, botanical gardens, and private gardens all over the world. The difficulty in propagating the plant from seed is due to its need for bat pollinators, which are themselves sensitive to the effects of deforestation and habitat loss.

The forest floor, which receives as little as 1 percent of the sunlight that falls on the canopy, is host to plants and animals that thrive in the cool, shaded environment. Mushrooms and other fungi, worms, ant colonies, termites, and small animals facilitate the breakdown of leaf litter and other organic matter, releasing the nutrients for immediate use by surrounding plants. The short cycle of breakdown, release, and absorption results in low nutrient content and poor quality of the rainforest soil.

The Luzon Rainforest supports a number of highly diverse bird species, 40 of which are either nearly or strictly endemic to the area. Two species are both threatened and strictly endemic: the green racquet-tail and the Isabela oriole, once thought extinct. In addition to habitat destruction and deforestation caused by illegal logging, an unchecked pet trade has caused the decline not only of the Isabela oriole, but also a number of parrot species, such as the Philippine cockatoo.

The critically endangered Philippine eagle, the country's national bird and most famous conservation symbol, includes the Luzon rainforest in its habitat zone. The species requires large areas of primary forest to thrive, and depends on the protection of remaining rainforests to avoid extinction in the wild.

The Luzon Rainforest biome contains 15 species of strictly endemic and near-endemic mammals, including three shrew-like animals found only on Mt. Isarog, and one identified through a single specimen from Sierra Madre. Thirteen mammal species are considered threatened.

These terraced rice paddies reach only to the base of thickly forested mountains on Luzon Island in the Philippines. The inaccessibility of the Luzon Rainforests, which are mainly found in the region below 3,300 feet (1,000 meters), on volcanoes in the south, and on islets to the north of the main island, has helped protect them somewhat from human encroachment, including logging and development. (Thinkstock)

Among them is the golden-crowned flying fox, which is possibly the largest bat in the world. Five large mammals live in the biome, including a macaque, warty pig, and civets; none of which are listed as threatened, but all are still vulnerable to habitat destruction.

Human Impact

The highly diverse flora and fauna of the Luzon Rainforest ecosystem are threatened by human activities such as illegal logging, the pet trade, conversion of primary forests for agricultural use, and increasing urban sprawl. Although many areas that overlap the biome have been given protection as natural parks—totaling approximately 1,320 square miles (3,400 square kilometers)—these parks are scattered throughout the island, with some covering 10 square miles (26 square kilometers) or less.

In addition to natural park status, inaccessibility appears to be the main contributing factor for successful conservation of the rainforest, as in the case of the Palanan wilderness in the Northern Sierra Madre. Palanan is the largest swath of forested area in Luzon and a refuge for many threatened species, including the Philippine eagle. It is facing the prospect of greater encroachment amid the planned construction of roads within the park.

Strict enforcement of designated protected areas, community education and involvement, and reduced incentives for commercial logging and clearing of forests are needed for effective conservation of the few remaining primary forests in this highly diverse but also significantly threatened biome. Protection will also be needed from the looming pressures caused by warmer temperatures and sea-level rise due to global climate change. It is predicted that there will be more frequent and more severe storms in this region, which could erode soils and accelerate habitat fragmentation.

MARICAR MACALINCAG

Further Reading

Corlett, Richard and Richard Primack. "Dipterocarps: Trees That Dominate the Asian Rain Forest." *Arnoldia* 63, no. 3 (2005).

Newman, Arnold. *Tropical Rainforest.* New York: Checkmark Books, 2002.

Wikramanayake, Eric, Eric Dinerstein, Colby J. Loucks, et al. *Terrestrial Ecoregions of the Indo-Pacific: A Conservation Assessment.* Washington, DC: Island Press, 2002.

Mackenzie River

Category: Inland Aquatic Biomes.
Geographic Location: North America.
Summary: The Mackenzie River is the longest river and the largest watershed in Canada, with good ecological status and management, but subject to future climate and development challenges.

The Mackenzie River is named after the Scottish explorer Sir Alexander Mackenzie, while its indigenous Dene name is Deh Cho, meaning "big river." It is the longest river in Canada, with the largest watershed, some 695,000 square miles (1.8 million square kilometers) draining one-fifth of the country. The river has the largest delta and the second-largest conglomeration of wetlands area in the country. Flowing through sparsely populated territory, it is one of the last pristine watersheds and ecosystems of the hemisphere.

Vital Arctic Flow

Beginning in Great Slave Lake, the 1,080-mile (1,738-kilometer) Mackenzie River flows north through the Arctic Circle to the Beaufort Sea. The watershed for the river is even greater, being over 2,600 miles (4,200 kilometers) long. The watershed begins with the Peace and Athabasca Rivers, gathers further waters from Lake Athabasca and the Great Slave Lake, and the Liard, Nahanni and Arctic Red Rivers. The watershed spans two provinces, Alberta and British Columbia, and two territories, the Northwest and Yukon Territories. All along its route and in the final delta, the Mackenzie River erodes, transports, and deposits sediments, creating a unique environment for many terrestrial, aquatic, and marine species.

The Mackenzie River discharges more than 78 cubic miles (325 cubic kilometers) of water each year, accounting for 11 percent of the total river flow into the Arctic Ocean. Due to such large volumes and seasonal discharges, the river has a significant impact on the climate of the Arctic Ocean, with large amounts of warmer freshwater mixing with the cold seawater. The highest flow occurs during the annual melt and breakup of ice and snow in June. The lakes and rivers of the Mackenzie and its tributaries are open, and flow from mid-June to the beginning of November. Despite this seasonal influence, annual flow is also considerable because water comes from the flat, barren shrubland tundra east of the river and the many large lakes in the watershed.

As it flows north, the Mackenzie River is a broad and slow-moving river. Its elevation drops just 512 feet (156 meters) from source to mouth. It is a braided river for much of its length, characterized by numerous sandbars and side channels. The river ranges from 1.25 to 3 miles (2 to 5 kilometers) wide, and is 26 to 30 feet (8 to 9 meters) deep in most parts. It is easily navigable by boat. The river ends in a wide fan-shaped delta of channels and islands where it empties into the Beaufort Sea lobe of the Arctic Ocean. The treeline narrows along the river, but never completely disappears, following the Mackenzie all the way to the delta, and representing the northern-most extension of boreal forest in Canada.

The Mackenzie Delta is a vast spread of low-lying alluvial islands, some 50 miles (80 kilometers) across, bordered on the west by the Richardson Mountains of the Yukon Territory, and in the east by the Caribou Hills of the Northwest Territories. In the delta, the river splits into three main navigable channels: an east channel, the west or Peel channel, and the middle channel where most of the water flows to the Beaufort Sea. The harbor in Tuktoyaktuk is a transfer point for ocean cargo, open from July to September.

Biota

The Mackenzie River and watershed is home to one of the largest and most intact ecosystems in North America. It is covered mainly by permafrost and tundra, wetlands, and boreal forests—most of which have never been logged. Tree species include dwarf birch, willow, alder, black spruce, lodgepole pine, tamarack, white birch, paper birch, quaking aspen, and balsam poplar.

Grasses and sedge are abundant in wetland areas, some of them pockmarked by peat bogs. Wildflowers thrive in the near-24-hour daylight of the short-lived summers here; species include Indian paintbrush, monk's hood, sweet pea, yellow cinquefoil, purple crocus, blue Arctic lupine, and lousewort.

Fish species include the northern pike, lake whitefish, and coney among the freshwater types; and Pacific salmon, cisco and Arctic char among the anadromous ones.

Migratory birds use the two major deltas in the watershed basin—the Mackenzie Delta on the Beaufort Sea and the Peace-Athabasca Delta far inland—as important resting and breeding grounds. Tundra swan, snow goose, and black brant are frequently seen in great numbers at the 170-mile-long (270-kilometer-long) bird sanctuary based in the Mackenzie Delta. This area is also vital to beluga whales, many of which calve in these waters.

Barren-ground and boreal woodland caribou travel across the tundra and forests that surround the Mackenzie River, feeding mainly on lichen. Permafrost lies under as much as three-quarters of the watershed, which restricts its biological diversity somewhat. Other larger mammals here include grizzly and black bear, moose, musk-ox, and Arctic wolf. Among the well-established community of smaller mammals are muskrat, Arctic and red fox, and snowshoe hare.

Threats and Conservation

Although the Mackenzie River has not been dammed, many of its tributaries have been modified by dams for hydroelectricity, flood control and agriculture. Two major dams on the upper Peace River in British Columbia generate electricity and control flooding in that province, but also challenge the sustainability of wildlife and ecosystems along the entire Mackenzie River.

The Mackenzie Delta on the Beaufort Sea is an important breeding ground for beluga whales, and a marine protected area safeguards the whales' migration routes through the area. (Thinkstock)

Future challenges and opportunities for the Mackenzie River biome include climate and environmental changes; proposed hydrocarbon, hydroelectric, and mining developments in the watershed; upstream diversions and impoundments of water, affecting rate and seasonality of flows; and the impact of transboundary airborne contaminants on the ecosystem. Cooperative management among the residents along the river and governments of the watershed areas will be necessary to encourage economic development while preserving this valuable ecosystem. The Dene, Inuvialuit, Métis and non-aboriginal people inhabit the Mackenzie River and Delta, though it is quite sparsely populated with isolated settlements primarily linked by the river. With the resolution of northern land claims for the Northwest Territories in the 1980s and 1990s, aboriginal people in the regions have important rights to, and participation in, land and water management, environmental assessment, and economic development.

There are complex and multiple jurisdictions and water authorizations for the Mackenzie River and watershed under national, provincial, and territorial processes. Four land and water boards regulate the portion of the watershed located in the Northwest Territories: the Gwich'in Land and Water Board, the Sahtu Land and Water Board, the Wek'eezhii Land and Water Board, and the Mackenzie Valley Land and Water Board. The Mackenzie River Basin Board was created under the Transboundary Waters Master Agreement; it commits all authorities to manage watershed resources for ecological integrity in a sustainable manner for present and future generations. The Basin Board influences legislative and administrative decisions by providing reports on the state of the aquatic ecosystem.

Among reserves in the biome are three marine protected areas in the Mackenzie Delta area; these help to preserve beluga whale migration routes, bird breeding and molting areas, and freshwater and marine fish habitat.

Climate change effects in the Mackenzie River biome are already being recorded, particularly permafrost melt and land subsidence (settling of the ground after water is excessively withdrawn from an aquifer). Warmer sea-surface temperatures have contributed to an increase in air temperature, and the warmer air penetrates into the permafrost layer that underlies much of the land in the Mackenzie watershed. This can disrupt local habitats by collapsing the soil structure and altering water flow; more broadly, it is known that melting permafrost releases greenhouse gases that then contribute to accelerated global warming—a positive feedback loop with negative consequences. Researchers note that the Mackenzie provides extensive ecological services—water filtration, carbon sink, snow albedo enhancer—that help to counter global warming and keep the Arctic Ocean region more stable. This is another reason why its protection as an intact ecosystem is of vital concern.

MAGDALENA ARIADNE KIM MUIR

Further Reading

Alliston, W. George. *A Monitoring Study of the Species, Numbers, Habitat Use and Productivity of Waterfowl Populations on Two Study Areas in the Wooded Mackenzie Delta, Summer 1983.* Burlington, Canada: LGL Ltd., 1985.

Doyle, Alister and Janet Lawrence. "Canada's Mackenzie River Needs Aid as Climate 'Refrigerator.'" Reuters, September 3, 2012. http://www.reuters.com/article/2012/09/03/us-canada-idUSBRE8820M620120903.

Gummer, W. D., K. J. Cash, F. J. Wrona, and T. D. Prowse. "The Northern River Basins Study: Context and Design." *Journal of Aquatic Ecosystem Stress and Recovery* 8, no. 1 (2000).

Lesack, Lance F. W. "River-to-Lake Connectivities, Water Renewal, and Aquatic Habitat Diversity in the Mackenzie River Delta." *Water Resources Research* 46, no. 1 (2010).

Madagascar Lowland Forests

Category: Forest Biomes.
Geographic Location: Indian Ocean.

Summary: One of the foremost ecoregions for species endemism, the lowland forests of Madagascar also represent one of the world's most threatened terrestrial biodiversity hot spots.

Occupying a narrow slice of alluvial land running along the eastern edge of the world's fourth-largest island, the lowland forest of Madagascar is among the world's foremost regions for species endemism: A total of 80 percent of its flora and fauna exist here and nowhere else on Earth. The forest's remarkable number of endemic species is evidence of the island's unique evolutionary path, a product of its longtime isolation from the planet's major land masses.

Madagascar sits in the Indian Ocean not far south of the Equator, 250 miles (402 kilometers) across the Mozambique Channel off Africa's southeastern coast. One of the world's most biodiverse and threatened ecoregions, it is the focus of international conservation efforts.

Along with the subhumid forests and scrub *ericoid* thickets, the lowland forests make up one of Madagascar's three terrestrial ecoregions—and support more than half of the island's species, many of which are endangered.

The lowland forests here fill a narrow, 30-mile-wide (48-kilometer-wide) strip of land between the east coast and the mountainous central highlands of Madagascar. The strip stretches from the southeastern tip of the island to Marojejy National Forest beyond Antongil Bay and Masoala Peninsula in the northeastern corner, covering an area of 43,500 square miles (112,664 square kilometers). Altitude of the forests ranges from sea level to more than 2,600 feet (792 meters).

Warmed by the moist southeastern trade winds of the Indian Ocean anticyclone, and nourished by a steady rainfall that averages more than 100 inches (2,540 millimeters) per year—sometimes even reaching up to 340 inches (8,636 millimeters)—this subtropical ecosystem is marked by moist broadleaf rainforests; closely spaced evergreen trees with wide trunks; hanging vines and lianas; and sparse vegetation on the forest floor, a result of sunlight being blocked by the thick canopy above.

Madagascar has two seasons: rainy and dry. During the rainy season, it experiences frequent thunderstorms, lightning, and sometimes destructive cyclones coming primarily from the Mascarene Islands to the east.

Biodiversity

The lowland forests are perhaps best known for having a remarkable rate of endemism among their plant, insect, reptile, and fish species, and in particular for 66 terrestrial mammalian species, all of which are endemic. Madagascar split from the African continent 150 million to 180 million years ago, and from the Indian subcontinent 88 million years ago, creating what various observers and researchers have called an "alternate world," a "living laboratory," a "world apart," and the "seventh continent."

The dense canopy of leaves and branches provides shelter for a wide variety of species, with 70 to 90 percent of them living above the forest floor, some as high as the treetops, at around 210 feet (64 meters). Important plant species here include rosewood or pea (family *Dalbergia*), persimmon and ebony (genus *Diospyros*), bamboo, and epiphytic orchids.

Notable among the high diversity of mammals are 100 species of lemur, all endemic, and considered by many scientists as a parallel clade to anthropoid primates. The Madagascar Lowland Forests biome also supports the highest diversity among the island's birds, with 42 of the 165 breeding species recorded in these forests being endemic. Recently rediscovered species, which were previously thought

A three-month-old crowned sifaka (Propithecus coronatus), one of the 100 species of lemur that are endemic to Madagascar. (Thinkstock)

to be extinct, include two lemurs and one bird: the hairy-eared dwarf lemur (*Allocebus trichotis*), golden bamboo lemur (*Hapalemur aureus*), and Madagascar serpent eagle (*Eutriochis astur*).

Effects of Human Activity

Once virtually covered by rainforests and other dense vegetation, Madagascar has lost 90 percent of its forests since humans arrived some 2,000 years ago. Around 70 percent of the forests were destroyed under French rule from 1895 to 1925. Since the 1970s, a third of the remaining forest cover disappeared, primarily due to clearance for coffee plantations; illegal logging; and slash-and-burn activity known locally as *tavy*. These practices have led to habitat destruction, river siltation, and widespread land degradation and erosion. This deforestation, much of which is irreversible, has pushed many of the region's species to the brink of extinction.

Hunting also threatens the survival of many species. Today, virtually all of Madagascar's endemic animals and their unique habitats face serious threats. The nation has more critically endangered primates than any other nation in the world. Deforestation of the rainforests of Madagascar has potentially grave climate change impact, as the vegetation, when intact, very effectively helps absorb carbon dioxide. Cleared areas are slow to recover, however.

Conservation Efforts

In recent years, significant efforts have been made to expand the nation's protected areas. The Durban Vision, an initiative to increase the area of protection from approximately 6,500 square miles (16,835 square kilometers), or 3 percent of Madagascar's total area, to 23,000 square miles (59,570 square kilometers), or 10 percent of the total area, was announced at the 2003 International Union for Conservation of Nature (IUCN) World Parks Congress in Durban, South Africa. As of 2011, areas receiving official state protection included five strict nature reserves, 21 wildlife reserves, and 21 national parks.

REYNARD LOKI

Further Reading

Dufils, J. M. "Forest Ecology: Remaining Forest Cover." In Steven M. Goodman and Jonathan P. Benstead, eds., *The Natural History of Madagascar.* Chicago: University of Chicago Press, 2003.

Dumetz, Nicolas. "High Plant Diversity of Lowland Rainforest Vestiges in Eastern Madagascar." *Biodiversity and Conservation* 8, no. 2 (1999).

Goodman, Steven M. and Jonathan A. Benstead. "Updated Estimates of Biotic Diversity and Endemism for Madagascar." *Oryx* 39, no. 1 (2005).

Green, G. M. and R. W. Sussman. "Deforestation History of the Eastern Rain Forests of Madagascar from Satellite Images." *Science* 248, no. 1 (1990).

Myers, Norman. "Threatened Biotas: 'Hotspots' in Tropical Forests." *The Environmentalist* 8, no. 3 (1988).

Madeira Evergreen Forests

Category: Forest Biomes.
Geographic Location: North Atlantic Ocean.
Summary: This unique habitat has survived on part of an Atlantic archipelago, where the difficult geography has protected unusual species.

The evergreen forests called the Laurisilva, for *laurel forest*, cover great parts of the Atlantic Ocean archipelago of Madeira. Forests of this type once grew across large areas of northwest Africa and southern Europe, but are relatively rare now. Similar forests may be found in the Azores; the western Canary Islands; the Cape Verde Islands; and also on the islands of São Tomé and Príncipe. These Atlantic archipelagos are all volcanic in origin. Because of population pressure, the only significant Laurisilva forests to be found are on Madeira. These were designated a World Heritage Site by the United Nations Educational, Scientific and Cultural Organization (UNESCO) in 1999.

Although the islands were undoubtedly discovered around 1300 C.E, they were not inhabited until 1420, when Portuguese settlers arrived. They called the island *Ilha da Madeira*, or *Island of the Wood*, because this subtropical evergreen forest thrived there. This forest flourished on parts of the island where relatively stable and mild temperatures prevailed throughout the year, as well a high level of humidity. The rest of Madeira is dry, and in many of the areas where people live, there have been occasional water shortages.

There undoubtedly were similar forests around the Mediterranean basin, but they are thought to have disappeared around 8000 B.C.E. as those regions became drier. The unique subtropical and moist conditions in this biome create an unusual ecosystem with interesting flora and fauna. Such forests are typically found at 1,312–3,937 feet (400–1,200 meters) in elevation.

Among the small animals that inhabit the Madeiran forests is the Madeira wall-lizard (Teira dugesii), seen here. The archipelago's Laurisilva forests were designated a World Heritage Site in 1999. (Wikimedia/Iain and Sarah)

Flora and Fauna

The canopy found in the Madeira evergreen forests consists of tall trees in the laurel family. Several species of laurel here are endemic (found nowhere else), including dry laurel (*Apollonias barbuana*) and moist-laurel (*Laurus azorica*). Dry laurel stands tend to be found on southern slopes, while species that thrive in moist environments are found on north-facing slopes. There is a dense undergrowth of ferns and bryophytes in most areas, with many evergreen climbing plants (*Folhado* and *Barbusano*), as well as purple and white orchids. These flora led to Madeira's being called the Flower Garden of the Atlantic. Many of the plants that flourish on Madeira are Mediterranean species, including at least 62 endemics.

Although there are giant rats on the Canary Islands, the only mammals in the Madeira evergreen forests are various species of bats, especially the Madeira bat (*Pipistrellus maderensis*). There is also a wide range of small creatures, such as the Madeira wall-lizard (*Teira dugesii*). The various species of butterflies here were collected and studied in detail by some of the English settlers on Madeira in the early 19th century. The biome also hosts over 500 endemic invertebrates, including endemic spiders, insects, and mollusks.

Most notable among the fauna of Madeira are the bird species, which attract ornithologists from around the world. The Madeira firecrest (*Regulus madeirensis*) is now recognized as being a distinct species rather than a subspecies of the common firecrest (*Regulus ignicapillus*); morphology clearly shows the separate development of the Madeira firecrest since they settled on the island. Two species of birds are strictly endemic to Madeira: Zino's petrel (*Pterodroma madeira*) and the Trocaz pigeon (*Columba trocaz*).

Environmental Threats

Unfortunately for the Madeira Evergreen Forests ecosystem, on some parts of the island the trees were cut down for housing; fuel; and cropland, especially for growing bananas and sugar cane, and planting grapevines to make Madeira wine. The island is densely populated, and indeed some may say overpopulated. However, thanks to the rich volcanic soil, farmers can typically coax two crops per year in some parts of the country. Terraces have been constructed so that all possible land can be used for agriculture. In the early 19th century, with pressure on agricultural land in settlements such as Seixal, farmers sometimes had to climb ropes to get to isolated terraces. Many areas were inaccessible, allowing the forests there to remain untouched.

Stress now derives from overharvesting of timber, as well as continued intensive farming practices needed to sustain the large population. Centuries ago, climate change shifted the area covered by Laurisilva forests, so that no more remain in Europe; continuing threats from climate change may once again change the landscape of Madeira. Temperatures moderated by the heating and cooling of ocean waters may fluctuate, causing as yet unforeseen changes in this unique biome.

These evergreen forests have survived somewhat intact, mainly because they flourish in areas that are mountainous and unsuitable for traditional agricultural practices. To try to preserve these forests, there have been efforts to reduce the grazing of sheep and goats in these regions. In another effort, the biome was recognized by the Council of Europe in 1992; it gained its World Natural Heritage listing from UNESCO seven years later.

JUSTIN CORFIELD

Further Reading

Boyer, David S. "Portugal's Gem of the Ocean: Madeira." *National Geographic Magazine* 105, no. 3 (1959).

Goetz, Rolf. *Madeira.* Munich, Germany: Bergverlag Rother, 2011.

Heineken, Karl. "Notice of Residents and Migratory Birds of Madeira and the Canaries." *Edinburgh Journal of Science* 1, no. 2 (1829).

McCarry, John. "Madeira Toasts the Future." *National Geographic Magazine* 186, no. 5 (1994).

Press, J. R. and M. J. Short. *Flora of Madeira.* London: Natural History Museum, 1994.

Thomas, Veronica and Jonathan Blair. "Madeira, Like Its Wine, Improves with Age." *National Geographic Magazine* 143, no. 4 (1973).

Magdalena River

Category: Inland Aquatic Biomes.
Geographic Location: South America.
Summary: The Magdalena River is the main waterway and watershed in Colombia and runs through many of the country's unique and important ecosystems.

The Magdalena River, known as the mother river of Colombia, is located in this northwestern South American country; it is the continent's fifth-largest river basin. It runs through nearly 956 miles (1,540 kilometers), possessing a watershed of 99,397 square miles (257,438 square kilometers). The river covers almost a quarter of Colombia's territory, crossing 11 of the 32 departments (provinces) that are home to nearly 70 percent of the nation's human population.

The Magdalena River originates in the Central Andes range, within the Puracé National Natural Park, in a small lagoon of the same name located in the *Páramo de las Papas* (Potatoes Páramo). This name was coined by Rodrigo de Bastidas, a Spanish conqueror, who arrived at the river's mouth in 1501; he named the waterway after the Christian iconic figure Mary Magdalene. The fluvial system of the river includes three main water courses: the Magdalena River (736 miles or 1,185 kilometers), the Cauca River (116 miles or 187 kilometers), and the Canal del Dique (70 miles, 114 kilometers).

Biodiversity

Generally, the watershed is divided into three sections—upper, middle, and low—each with important ecological, economic, and socially distinctive features. As the main entrance and arterial waterway for the country since pre-Hispanic times, the river has suffered from intensive and extensive use, from transportation to agriculture, fisheries, and irrigation. All of these have seriously impacted the river and the natural systems of the watershed.

Biologically, this system is one of the most complex and diverse biomes in the tropics because of its wide range of elevational floors. From páramos—the tundra-like scrubland ecosystem above the continuous forest line and below the permanent snowline—and Andean oak and cloud forests—also called fog forests, a tropical or subtropical evergreen montane moist forest type—in the upper parts of the river, to the coastal wetlands, mangroves, and lowland lagoons in the lower parts, this river represents the heart of Colombian ecological history. It is the perfect representation of the enormous diversity found in the country.

The three sections of the river are easily distinguishable from one another. Both the upper and lower Magdalena basins are typically dry and naturally dominated by forested landscapes until the sub-alpine elevations, where they become páramo. The upper part of the basin is generally more dry; its high-elevation regions are mostly dominated by high Andean ecosystems with some areas influenced by moist winds and cloud forests, while some other parts have mixed stands of mid- and high-elevation forests. This is the most populated portion of the country, and it includes the capital city of Bogotá. The natural ecosystems here have been seriously affected by both traditional and contemporary agricultural activities.

The dry peri-Caribbean zones of northern Colombia are located in lower parts of the basin; these are some of the most interesting and complex wetland-lagoon-depression systems in South America. The river's mouth to the Caribbean Sea—the Ciénaga Grande de Santa Marta wetland complex—is among the most unusual and important wetlands in Colombia. This complex includes the largest wetland in the country and unique mangroves. It is a wetland-lagoon fresh-salty water complex classified as a Biosphere Reserve, Important Bird Area, and a Ramsar Wetland of International Importance. The area is protected by several preserves, including the Via Parque, Isla de Salamanca, and Ciénaga Grande de Santa Marta Wildlife Sanctuary.

Between the dry upper and lower sections, the middle basin is mostly piedmont and isolated ranges from the Andes. Climatically, it is considered a wet/moist region. This section includes highly differentiated forest ecosystems, such as the unique geological complex of the Serranía de San Lucas. It also includes wetlands in the Depresión Momposina, or Mompox Depression.

Biota

Biologically, the Magdalena River area is a diverse and complex habitat system that features more than 190 freshwater fish species, 200 types of mammals, and nearly 400 bird varieties. For some groups, such as dry forest vegetation, more than 300 species are found only in few dry forest fragments across the basin. Mammals also are important for the basin, including most of the Andean and non-Amazon species, from small rice mice and the enormous spectacled bear to the Andean tapir in the upper parts of the region; to jaguar, puma, deer, and manatee in the lowlands. There is a fairly high incidence of species that are endemic (found nowhere else). For birds, the Magdalena represents a vital refuge that includes several Important and Endemic Bird Areas, from coastal and marine species to those that depend on high Andean native and migrant plants.

The vegetation along the river features such dominant trees as Spanish cedar (*Cedrela odorata*), the conifer *Podocarpus oleifolius*, Spanish elm (*Cordia alliodora*), Andean oak (*Quercus humboldtii*), hemsl (*Aniba perutilis*), wild cashew (*Anacardium excelsum*), yellow lapacho (*Tabebuia serratifolia*), jacaranda (*Jacaranda caucana*), and palm trees such as *Ceroxylon quindiuense*, *C. alpinum*, *C. parvifrons*, and *C. sasaimae*. Several orchids are endemic here, including *Cattleya trianaei* to the upper Magdalena and *C. warscewickzii* to the San Lucas-Nechi region.

The upper Magdalena, in the Huila territory, has species of birds such as the burrowing owl (*Athene cunicularia tolimae*) and velvet-fronted euphonia (*Euphonia concinna*). Flora include the May flower or Christmas orchid (*Cattleya trianaei*), which is the Colombia national flower. In the hills along the cordilleras, there are other endemics such as *Ceroxylon sasaimae*, a flowering plant in the *Arecaceae* family; *Odontoglossum crispum*, an epiphytic orchid; honey tree; and others.

Large vertebrates in the forests surrounding the Magdalena include the mountain lion (*Puma concolor*), spotted cat (*Leopardus tigrina*), spectacled bear (*Tremarctos ornatus*), spider monkey (*Ateles geoffroyi*), Andean wolly monkey (*Logothrix lagothricha lugens*), Colombian tapir (*Tapirus terrestris colombianus*), mountain tapir (*T. pinchaque*), mountain paca (*Agouti taczanowskii*), red howler monkey (*Alouatta seniculus*), and several others. Among the birds, the blue-billed curassow (*Crax alberti*), huamán or mountain eagle (*Oroaetus isidori*), wattled guan (*Aburria aburri*), and yellow-eared parrot (*Ognorhynchus icterotis*) are on the verge of extinction. Also found here is the golden-headed quetzal (*Pharomachrus auriceps*), crested quetzal (*P. antisianus*), Andean cock-of-the-rock (*Rupicola peruviana*), and many more.

Migratory songbirds and raptors such as the rose-breasted grosbeak (*Pheucticus ludovicianus*), the broad winged hawk (*Buteo platypterus*), swainson's hawk (*B. swainsoni*), and the summer tanager (*Piranga rubra*) use many areas of the Magdalena.

Species of special concern include the yellow-eared parrot (*Ognorhynchus icterotis*), the Christmas orchid, Andean rosewood (*Aniba perutilis*), Andean woolly monkey, the mountain tapir, spectacled bear, and wax palms (*Ceroxylon* spp.).

Environmental Impacts

As a result of pollution, overexploitation, transformation, and deforestation, the Magdalena River basin is one of the most seriously threatened biomes in South America. As the central artery of the country, the Magdalena is vulnerable to threats including overfishing; deforestation; agrochemical pollution; interruption of natural flows due to water-management infrastructure; and the phenomena of El Niño and La Niña that, probably linked to related effects of global warming, have flooded the river's shores. Although the Magdalena River area receives a lot of public attention, there are numerous communities and unique ecosystems that remain unprotected. However, there are a growing number of conservation efforts and zones, including 12 National Natural Protected Areas.

Working with local Colombian authorities are several United States-based organizations. These include The Nature Conservancy, Great Rivers Partnership, and the U.S. Army Corps of Engineers. Colombian bodies involved in these cooperative efforts are led by the Ministry of Environment and the local environmental authority Cormagdalena. Their goals are to address the environmental needs of the basin, protect the Magdalena's biodiversity, and secure the prosperity of its people who rely on the river for their livelihood. Particular attention is being paid to the dams, since they are both a boon to human water needs and a culprit in the declining fish populations here.

José F. González-Maya
Diego Zárrate-Charry
Alexandra Pineda-Guerrero
Mauricio González

Further Reading

González-Maya, J. F., A. Cepeda, S. A. Balaguera-Reina, D. A. Zárrate-Charry, and C. Castaño-Uribe. *Ecological Integrity Analysis, Conservation Targets and Delimitation of the Ramsar Site Delta Estuary System of the Magdalena River, Ciénaga Grande de Santa Marta.* Bogotá, Colombia: Conservación Internacional Colombia-Ministerio de Ambiente Vivienda y Desarrollo Territorial, 2009.

Haffer J. "Speciation in Colombian Forest Birds West of the Andes." *American Museum Novitates*, 2294, no. 1 (1967).

Hernández-Camacho J., T. Walschburger, R. Ortiz-Quijano, and A. Hurtado-Guerra. *Origin and Distribution of South American and Colombian Biota.* Xalapa, Mexico: Instituto de Ecología A.C., 1992.

Herzog, Sebastian K., Rodney Martinez, Peter M. Jorgensen, and Holm Tiessen, eds. *Climate Change and Biodiversity in the Tropical Andes.* Sao Jose dos Campos, Brazil: Inter-American Institute for Global Change Research (IAI), and Scientific Committee on Problems of the Environment (SCOPE), 2011.

Restrepo J. D. and B. Kjerfve. *The Pacific and Caribbean Rivers of Colombia: Water Discharge, Sediment Transport, and Dissolved Loads.* Berlin, Germany: Springer Verlag, 2004.

Magdalena Valley Montane Forest

Category: Forest Biomes.
Geographic Location: South America.

Summary: The Magdalena Valley montane forest is one of the most unique and underprotected biomes in Colombia.

The Magdalena Valley Montane Forest biome is located along the northern foothills of Colombia adjacent to the central and western cordilleras. This independent mountain range of the Andes has an extension of 3,628 square miles (9,397 square kilometers), with limits at the Caribbean floodplains in the north, the Magdalena River in the west, the Cauca River in the east—these two being the most important rivers in Colombia—and with the central cordillera in the south. Its elevation ranges from sea level to 8,858 feet (2,700 meters), from the flooding areas of the Cauca and Magdalena rivers and the Mompox depression, to the highest peak, La Teta. The biome includes various types of habitats, from tropical dry and montane forests to large marshes and tropical rainforest. The region is a connector between the Amazon forests and the Caribbean forests and savannas.

The region retains extensive natural forests totaling 2,316 square miles (600,000 hectares), all considered native unique forests, especially the area known as the San Lucas *orobioma*, or elevation-oriented biome. Located in the highlands of the valley, the San Lucas is now listed in the national protected areas system. A priority region for conservation projects, it hosts many flora and fauna considered endemic (found nowhere else). Another related strategic ecosystem type occurs in the lowlands of the montane forest; these comprise the dry and wet *heliobiomas*, or sunlight-oriented biomes, of the Colombian Caribbean, characterized by warmer temperatures than the upland sections.

The climate in the Magdalena Valley is known as *Serrania de San Lucas* (SdSL), and includes three different weather patterns: dry warm, warm humid, and warm very humid. Each one has a dry season between December and March and a rainy season between April and November in the northern part of the region. The rainy season in the central and southern portions is a little longer, usually lasting into April. In the lowlands of the SdSL, the average temperature exceeds 75.2 degrees F (24 degrees C), with an average rainfall of 59 inches (1,500 millimeters). Temperatures in the highlands fluctuate in the range of 64.4–75.2 degrees F (18–24 degrees C), with average rainfall of 78–157 inches (2,000–4,000 millimeters).

The hydrography of the area is complex. In the southwestern zones, the Cauca River—one of the largest bodies of water in the country—flows north to empty into the Magdalena River. This union generates a complex series of wetlands and lagoons present most of the year, forming the Mompox depression, as the Magdalena continues flowing north. This river receives all the small tributaries that come from the SdSL and the drainage forms several micro-watersheds that connect the entire system. The eastern slope of the SdSL is covered by the Magdalena River basin, which runs across most of the country and empties into the Caribbean Sea.

Biodiversity

The Magdalena Valley Montane Forest biome is considered an endemic species center that includes the largest populations of the critically-endangered and endemic blue-billet curassow, known as the Paujil *(Crax alberti)*; the chestnut-bellied hummingbird *(Amazilia castaneiventeris)*; and the brown spider monkey *(Ateles hybridus brunneus)*, one of the rarest primates on Earth. Large species of vertebrates in the biome include vulnerable species such as the puma *(Puma concolor)*, spectacled bear *(Tremarctos ornatus)*, and Colombian tapir *(Tapirus terrestris colombianus)*, as well as the jaguar. At least 128 fish species reside in this region, as well as 42 species of amphibians and reptiles, 478 types of birds, and 212 different mammals.

The floristic composition of these montane forests is poorly described; there are no complete inventories for the area. The only available information was gathered from 1970 to 2009 in the rapid ecological surveys by the environmental authority of the area, resulting in 240 species of plants having been reported. Among the known endemic species are *Steriphoma colombiana, Amaria petiolata, Pithecellobium bogotense,* and in the transition forests between dry and moist regions, the national flower of Colombia, the highly endangered *Cattleya trianaei,* also called the May Flower or the Christmas Orchid.

The area also has thorny vegetation, including several cacti, such as *Opuntia* spp., *Melocactus* spp., *Armathocereus humilis, Stenocereus griseus, Acanthocereus pentagonus,* and *Pilosocereus colombianus.* There are woody species of bushes and trees such as *Pithecellobium bogotense, Capparis odoratissima, Bulnesia carrapo,* old fustic or Dyer's mulberry (*Maclura tinctoria*), and *Fagara pterota. Prakinsonia aculeta,* a perennial flowering tree in the pea family, *Prosopis juliflora,* and needle bush (*Acacia farnesiana*) are also found here.

Approximately 1,100 species of the area's known flora and fauna are reported from isolated forests of the Cordillera of the Andes, dispersed throughout the entire mountain range. Of those, 509 species are considered threatened; there are five species of fish identified as threatened or endangered. They are: categories of ray-finned fish (*Ichthyoelephas longirostris, Salminus affinis, Brycon labiatus*), tropical freshwater fish (*Prochilodus magdalenae*), and long-whiskered catfish (*Pseudoplatystoma fasciatum*).

There are three endemic bird species of the biome: the Colombian tinamou (*Crypturellus columbianus*), which is endangered; beautiful woodpecker *(Melanerpes pulche);* and sooty ant tanager (*Habia gutturalis*). Another six species are known as restricted-distribution birds. They are the northern screamer (*Chauna chavarria*), red-billed emerald (*Chlorostilbon Gibson*), white-whiskered spinetail (*Synallaxis candei*), yellow-browed shrike-vireo (*Vireolanius eximius*), white-shouldered tanager (*Heterospingus xanthopygiu*), and golden-winged warbler (*Vermivora chrysoptera*).

Among mammals, the biome includes eight threatened species: giant anteater (*Myrmecophaga tridactyla*), common woolly monkey (*Lagothrix lagothricha*),), brown spider monkey, gray-bellied night monkey (*Aotus lemurinus*), white-footed tamarin (*Saguinus leucopus*), jaguar (*Panthera onca*), speckled bear, and South American tapir (*Tapirus terrestris*). Some global initiatives con-

sider this biome a critical corridor for large, charismatic or umbrella species migrating in from Central America.

Environmental Protections

The Magdalena Valley montane forests are to some degree protected by a Colombian forest reserve law, Act 2 of 1959, which was created to protect soils, water, and wildlife from the main developmental processes. This category was designated for five vast ecoregions in the country belonging to the Magdalena River Forest Reserve. The large territories included in the category, coupled with expanding colonies and agricultural borders as the economy and social network evolved, quickly became obsolete and poorly enforced.

This young man was a participant in a forest protection project in Colombia in 2006 in which residents developed internal regulations for sustainable land use. (USAID)

In 1999, the government created the first Rural Reserve, which provided for a more sustainable land use program, allowing livestock farming communities and agricultural activities in the forest reserve. Currently—because these ecosystems are not represented in the national protected areas systems—the Colombian government, supported by international agencies and national nongovernmental organizations (NGOs), is working to establish a national park in the region to preserve this unique ecosystem for the long term.

Some of the major problems threatening the biome include industrialized production of cocoa (*Erythroxilum coca*), mining, social conflict, and resources overexploitation. Cocoa cultivation results in soil erosion and homogenization of the landscape, and it pollutes the rivers and creeks with runoff from the chemicals used in its processing. Mining activities have led to high rates of natural forests loss through deforestation and elevated pollution levels resulting from cyanide and mercury additives used in mining extraction.

The introduction of heavy machinery for soil removal also increases sediment in the rivers, which is seriously affecting the fauna. All these processes are aggravated by increased illegal activities by armed groups that threaten the safety and normal development of the region, and create a complex social situation that disrupts human activities and nature conservation. Overexploitation and illegal traffic of species have significantly affected the ecosystem, seriously reducing populations of some endemic and endangered species. These disruptions may lead to the extinction of some local species. Regional authorities expect that the creation of a national protected area in this biome will largely reduce the environmental problems, and will aid human and nature development as well as conservation long-term.

José F. González-Maya
Ivan Mauricio Vela-Vargas
Diego Zárrate-Charry
Alexandra Pineda-Guerrero
Jaime Murillo-Sanchez
Mauricio González

Further Reading

Cleveland, Cutler J. "Magdalena Valley Montane Forests." *Encyclopedia of Earth*. Washington, DC: Environmental Information Coalition, National Council for Science and the Environment, 2008.

Herzog, Sebastian K., Rodney Martinez, Peter M. Jorgensen, and Holm Tiessen, eds. *Climate Change and Biodiversity in the Tropical Andes*. Sao Jose dos Campos, Brazil: Inter-American Institute for Global Change Research (IAI), and Scientific Committee on Problems of the Environment (SCOPE), 2011.

Orejuela, J. E. "Tropical Forest Birds of Colombia: A Survey of Problems and a Plan for Their Conservation." In A.W. Diamon and T. E. Lovejoy, eds., *Conservation of Tropical Forest Birds*. Norwich, UK: Paston Press, 1985.

Magellanic Subpolar Forests

Category: Forest Biomes.
Geographic Location: South America.
Summary: This subpolar realm, also known as *Nothofagus* forests, extends from Tierra del Fuego at the southern tip of South America into Chile and Argentina; it is a refuge for Antarctic flora.

The subpolar *Nothofagus* forests extend along the base of the southern Andes Mountains and across the Chilean archipelago, bounded on the south and west by the ocean; on the east by the dry Patagonian steppe; and on the north by the higher-biodiversity Valdivian temperate forests. Genus *Nothofagus*, also known as southern beech, is the predominant tree type and is characteristic of the biome. In fact, the presence of *Nothofagus* across the austral continents (southern Pacific Rim) is evidence that Antarctica, Australia, and South America were connected in geological time.

This biome represents the southernmost forests in South America: It ranges through the southern Aisen and Magallanes regions of Chile; some western patches of Santa Cruz Province, Argentina; and to southern swaths of Tierra del Fuego, mainly from Lake Buenos Aires to Staten Island. The name refers to its proximity to the Strait of Magellan, which memorializes the global circumnavigation voyage of the Portuguese explorer Ferdinand Magellan in 1520.

Vegetation

The southern beech varieties found here include *Nothofagus betuloides*, *N. antarctica*, and *N. pumilio*. Generally, species richness is much lower than in the Valdivian forests to the north, but this unique biome supports relatively rich biodiversity in one of the harshest climates on the continent. Three plant community types are represented in the biome, mixed across ecotones and with neighboring biomes: moorlands, temperate forests, and deciduous forests. *N. antarctica* and *N. pumilio* are deciduous and typical of the eastern portion, while *N. betuloides* is evergreen and typical of the western zones.

At the highest elevations, this biome is dominated by permanent snow, ice caps, and glaciers. Generally speaking, it is characterized by cold temperatures, strong and permanent west winds, and high amounts of rain and snow. Landscapes are dramatic and often windswept, with few year-round human inhabitants.

The unique climate of the region has resulted in numerous endemic (found only here) plant species, but biodiversity is much lower than in surrounding biomes. Endemic plants include grasses such as *Deschamsia kingii*, *Festuca cirrosa*, *Poa darwiniana*, and *P. yaganica*; and a suite of herbs and shrubs, including *Onuris alismatifolia*, *Ourisia fuegiana*, *O. ruelloides*, *Senecio eightsii*, *S. humifusus*, *S. websteri*, *Nassauvia latissima*, *Acaena lucida*, *Perezia lactucoides*, and *Viola commersonii*. Many plants are stunted because of the high winds; some are confined to leeward rock and other formations that provide shelter.

Fauna

Endemism in animals here is much lower than in plants. Much of the fauna occur marginally, and are more predominant in neighboring biomes. Among the birds are the Magellanic woodpecker (*Campephilus magellanicus*), austral parakeet (*Enicognathus ferrrugineus*), and several near-endemic geese (*Chloephaga hybrida* and *C. rubiceps*).

Mammals have similar low diversity, but are notably visible in the landscape. Among the large mammals are puma (*Puma concolor*), two foxes (*Pseudalopex griseus* and *P. culpaeus*), guanaco (*Lama guanicoe*), Patagonian huemul (*Hippocamelus bisulcus*), southern pudu (*Pudu pudu*), and southern river otter (*Lontra provocax*). The only near-endemic mammals are grass mice such as *Abrothrix* (previously *Akodon*) *hershkovitzi*, *A. lanosus*, and *A. markhami*; currently, these are not considered to be threatened.

Environmental Threats

The Magellanic Subpolar Forests biome is seriously threatened by habitat conversion due to logging, human development in northern areas, grazing, and fires. In addition, the introduction of habitat-modifying exotic species such as the North American beaver (*Castor canadensis*) and the European rabbit (*Oryctolagus cuniculus*) have affected many plant communities. Beavers build dams and thus alter the habitat and soil characteristics, inhibiting forest restoration and damaging the original hydrological cycles.

Glaciers in the southern Patagonian region have been losing mass at a very rapid rate, one of the fastest glacier retreats recorded on Earth. Some of the permanent ice cover here has retreated up to 9 miles (14.5 kilometers) in the last 100 years—with some indications of recent acceleration. Besides more erratic spring runoff events, glacial shrinkage can lead to water shortages during drier months, affecting local habitat resilience and growth rates.

Climate change impacts upon this region have yet to be sharply predicted, but regional warming trends, precipitation disruption, or seasonal pattern shifts hold the potential to erode soils and challenge vegetation viability at different altitude gradients, and thus to fragment and otherwise stress the current habitat structure, food web, and fauna mix.

JAN SCHIPPER
JOSÉ F. GONZÁLEZ-MAYA

Further Reading

Brion, C. D., J. Puntieri Grigera, and E. Rapoport. "Plantas Exóticas en Bosques de Nothofagus. Comparaciones Preliminares Entre el Norte de la Patagonia y Tierra del Fuego." *Monograf Biologias de la Academia Nacional de Ciencias Exactas, Fisicas y Naturales* 4, no. 1 (1988).

De la Peña, M. R. and M. Rumboll. *Birds of Southern South America and Antarctica.* Princeton, NJ: Princeton University Press, 1998.

Gajardo, R. *La Vegetación Natural de Chile, Clasificación y Distribución Geográfica.* Santiago, Chile: Editorial Universitaria, 1994.

Lliboutry, Louis. "Glaciers of South America: Glaciers of Chile and Argentina." In Williams, R. S. and J. G. Ferrigno, eds. *Satellite Image Atlas of Glaciers of the World.* Washington, DC: U.S. Geological Survey, 1998.

Redford, K. H. and J. F. Eisenberg. *Mammals of the Neotropics 2, The Southern Cone.* Chicago: University of Chicago Press, 1992.

Veblen, T. T., C. Donoso, T. Kitzberger, and A. J. Rebertus. "Ecology of Southern Chilean and Argentinean Nothofagus Forests." In T. T. Veblen, R. S. Hill, and J. Read, eds., *The Ecology and Biogeography of Nothofagus Forests.* New Haven, CT: Yale University Press, 1996.

Malabar Coast Moist Forests

Category: Forest Biomes.
Geographic Location: India.
Summary: Once known for its species-rich forests, this biome is degrading from intensive grazing, mining, and submersion under reservoirs created by the construction of hydroelectric dams.

The Malabar Coast Moist Forests biome consists of tropical and subtropical deciduous broadleaf forests located in southwestern India. These evergreen forests lie in a relatively narrow, north-south strip between the Arabian Sea and the Western Ghats Mountains, along the Konkan and Malabar coasts. The Western Ghats, also known as the Sahayadri Mountains, are older than the Himalayas, cover 62,000 square miles (160,000 square kilometers), and form the catchment area for a complex of river systems. Rivers such as Tapti, Narmada, Mandovi, and Zuari originate from the Western Ghats and flow into the Arabian Sea, a marginal sea of the Indian Ocean.

Most of these rivers form estuaries at the coast. The Western Ghats intercept the southwestern monsoon rains to create a moist area with 98 inches (250 centimeters) of annual rainfall on the western side of the mountains.

The Malabar Coast moist forests extend from sea level to 820 feet (250 meters) high. The region includes the middle- and upper-elevation ecozones of the northwestern mountain range; a stretch of forest that extends north of Maharashtra State through Goa, Karnataka, and Kerala to Kanyakumari; and the southernmost point of the state of Tamil Nadu. The biome is bounded on the east by the northwestern Ghats moist deciduous forests in Maharashtra and Karnataka States, and the southwestern Ghats forests in Kerala State.

Neighboring habitats are similar, but with vegetation communities zoned by elevation. The southwestern Ghats, for example, cover the southern part of the Sahayadri and the Nilgiri Hills at the much higher altitude range of 820–3,280 feet (250–1,000 meters) in Kerala, Karnataka, and Tamil Nadu. The Agasthyamalai Biosphere Reserve here includes the moist deciduous forests of the southwestern Ghats montane rainforests and the Shola Grasslands complex. The Agastyamalai also is home to the Kanikkaran, one of the oldest surviving hunter tribes in the world, and is the habitat for 2,000 species of medicinal plants, of which at least 50 are rare and endangered.

Biodiversity

The Malabar Coast Moist Forests biome covers approximately 13,700 square miles (35,500 square kilometres), encompassing some sand dune and saline lake features. The area had once been a home to tigers, gaur, slender loris, Jerdon's palm civet, grizzled giant squirrel, Asian elephants, wild dogs, and sloth bear, and some 325 bird species, including the iconic Malabar grey hornbill, a near-endemic species, meaning one found almost nowhere else.

The distributions of the Nilgiri wood-pigeon, grey-headed bulbul, rufous babbler, and the Malabar parakeet extend throughout these moist deciduous forests. The white-bellied tree pie, white-bellied shortwing, and grey-breasted laughing thrush are primarily montane species. Both the white-bellied shortwing and the Wroughton's free-tailed bat are critically endangered. Other species, such as lesser florican, greater flamingo, and great hornbill, would benefit from conservation measures.

Human Impact

Forest commercialization has meant that teak, rosewood, and rubber plantations have replaced many swaths of the region's original northern forests. Human settlements and land used for livestock grazing have resulted in a change in the area's semi-deciduous vegetation. The forest area in the state of Karnataka has a huge herd of uncontrollable domestic livestock, and the Sanjay National Park in Maharashtra undergoes several anthropogenic fires every year.

Human activities have encroached on the natural habitat of such large animal species as the endangered Malabar civet. At present, the region contains fewer species of mammals than at any known time in its past; these include five near-endemic mammals and a single endemic rodent.

Hydroelectric power projects now dominate the region, with 13 dams built on Periyar River alone. One of the few perennial rivers, it also is considered a lifeline of Kerala State. The dam projects, with accompanying human encroachments, have fragmented the forests in many places. The paper pulp, plywood, sawmills, and fiber factories consume local timber and bamboo, and contribute to habitat degradation through erosion, water pollution and turbidity, blockage of fish and wildlife migration pathways, and other deleterious effects.

In Karnataka State, mineral extraction for vanadium, manganese, and iron contribute to habitat destruction. Human activities such as the collection of fuelwood and fodder, along with non-wood products and animal grazing, have intensified as rural populations grow and expand. The grasslands of the Malabar region, too, are highly vulnerable to fire, which have become frequent enough—and largely uncontrollable by the local population—to retard the growth and regeneration of the forests.

Climate change poses additional threats, such as shifts in the boundaries of the forests, unpredictable agricultural yields, drastically changed seasonal rainfall amounts, and ultimately disrupted habitats for both plants and animals. Scientific studies project that by 2050, most of India's forest biomes will be highly vulnerable due to the changing climate, and 70 percent of the country's vegetation susceptible to increased stress.

RHAMA PARTHASARATHY

Further Reading

Raman, T. R. Shankar and Divya Mudappa. "Hornbills—Giants Among the Forest Birds." *Resonance* 3, no. 8 (1998).

Sharma, Subodh. *Climate Change and India: A 4X4 Assessment.* New Delhi: Ministry of Environment and Forests, Government of India, 2010.

Wikramanayake, Eric, Eric Dinerstein, Colby J. Loucks, et al. *Terrestrial Ecoregions of the Indo-Pacific: A Conservation Assessment.* Washington, DC: Island Press. 2001.

Yahner, R. H. "Changes in Wildlife Communities Near Edges." *Conservation Biology* 2, no. 1 (1988).

Malawi, Lake

Category: Inland Aquatic Biomes.
Geographic Location: Africa.
Summary: Lake Malawi, the southernmost major lake in the Great Rift Valley of East Africa, is recognized as a global biodiversity hot spot.

Lake Malawi (known as Lake Nyasa in many countries) is located in East Africa at the juncture of the countries of Malawi, Mozambique, and Tanzania. It is in the Great Rift Valley, which began evolving at least 1 million years ago when the African tectonic plate split and created volcanic valleys that feature a string of great lakes: Malawi, Tanganyika, Victoria, and four others of lesser scale. The Rift Valley system includes some of the oldest, largest, and deepest lakes in the world; many of these freshwater ecosystems support vast biodiversity and specialized organisms.

The Rift Valley lakes are known for the evolution of hundreds of cichlid fish species. Few, if any, other bodies of water on the planet have as many species of freshwater fish as Lake Malawi. Because it and the other Rift Valley lakes represent ecosystems of such vast and specialized biodiversity, the World Wildlife Fund included the lakes in its Global 200 priority ecoregions for conservation.

Hydrology

Lake Malawi is the second-deepest lake in Africa, and is among the 10 largest lakes in the world. With a maximum depth of 2,316 feet (706 meters) and surface area of approximately 11,430 square miles (29,600 square kilometers), the total lake volume is estimated at 2,015 cubic miles (8,400 cubic kilometers). The water is generally warm, with temperatures near the surface ranging from 75 to 84 degrees F (24 to 29 degrees C). The deeper water runs several degrees cooler—averaging about 72 degrees F (22 degrees C) year-round. Due to the geological makeup of Lake Malawi, it is slightly alkaline (pH 7.7 to 8.6), with a carbonate hardness of 28.27 to 37.51 parts per million.

Located approximately 220 miles (350 kilometers) southeast of Lake Tanganyika, the largest portion of Lake Malawi is located in the nation of Malawi, with about one-quarter of the area belonging to Mozambique; these two countries, along with Tanzania, each claim some shoreline rights and occasionally air disputes over fishing rights in different areas of the lake.

Of hydrologic importance, the largest river that flows into Lake Malawi is the Tanzanian river Ruhuhu, flowing into the east side; the Shire River, a tributary of the Zambezi River, drains the lake from its southern shore. Only two islands—Likoma and Chizumulu—are inhabited. Characteristic of both islands is the abundance of baobab

Lake Malawi Cichlids

Lake Malawi is thought to contain approximately 30 percent of all known cichlid species in the world; approximately 350 species are endemic, or found only here. These cichlids initially entered Lake Malawi from its feeder rivers and speciation occurred, filling the many available biotypes. Because Lake Malawi is isolated from other bodies of water, these fish have developed impressive adaptive radiation into habitat niches. Many have evolved specialized diets and corresponding feeding apparatus. Some, for example, have evolved to forge on certain types of snails, or on specific plant leaves, zooplankton, algae, or other fish. In addition, most of these cichlids are maternal mouthbrooders, which makes the lake's ichthyofauna unique.

Hundreds of the endemic cichlid types found in Lake Malawi are suitable as commodities in the ornamental fish-trade business. Cichlids of the *Haplochromis* genus, often dubbed "happies" or "haplos" in the trade, are further divided into a clade of open-water and sand-dwelling species; and a branch of rock-dwelling species, or mbuna, which are smaller and specialized to prey on the biocover that grows on rocks, including algae, microorganisms, crustaceans, mites, snails, and zooplankton—collectively called aufwuchs. *Haplochromis* are mouthbrooders, that is, they carry their young in their mouths for protection at their earliest life stage.

Separate from the *Haplochromis* genus is the *Tilapii* genus. These latter are mainly substrate-spawning cichlids, although they do include four mouthbrooding species of chambo (*Nyasalapia* spp.). For the study of faunal speciation and evolution, Lake Malawi cichlids are considered to be as important to science as are the finches and honeycreepers, found in the Galapagos Islands and Hawaii, respectively.

A Malawi cichlid. Lake Malawi contains over 1,000 species of fish, including 30 percent of all cichlid species. (Thinkstock)

trees (*Adansonia digitata*), which are commonly referred to as "the upside-down tree" or the "tree of life." Both islets support several thousand fishers and farmers who grow a variety of food crops including cassavas, bananas, and mangoes.

Biodiversity

Lake Malawi is a rocky lake with few aquatic plant species; however, its banks feature wetlands and forests. The latter are a regionally valued source of timber for building and fuel. Notable trees include Mulanje cedar (*Widdringtonia cuppresoides*), mlombwa (*Pterocarpus angloensis*), and mbawa (*Khava anthotheca*).

Lake Malawi is home to a vast array of wildlife, including the painted hunting dog, which until recently was thought to be extinct; more prevalent are crocodiles, hippopotamuses, monkeys, and a large population of African fish eagles. Lake Malawi is probably best known for its diversity of fish. The tropical waters here are home to more than 1,000 species of fish, including unique types of cichlid, which is a family of which tilapia and sunfish are characteristic members.

In addition to cichlids, Lake Malawi is home to a plethora of other fish, including large populations of species belonging to families including *Clariidae, Claroteidae, Mochokidae, Poeciliidae, Mastacembelidae, Centropomidae, Cyprinidae,* and *Clupeidae*. While many such fish provide a major food source for other fauna, these populations are threatened by overfishing and pollution.

Preservation and Threats

Lake Malawi National Park was established specifically to conserve the lake's globally-recognized biodiversity among fish species and habitat niches. The park encompasses approximately 36.3 square miles (94 square kilometers) at the southern end of Lake Malawi, of which about 3 square miles (7 square kilometers) represents aquatic habitats, protecting some important vegetation types and breeding areas.

Also featured are rocky shorelines, sandy beaches, wooded hillsides, swamps, lagoons, granitic hills, and sandy bays. About half of all fish species within Lake Malawi are thought to occur in the park—including nearly all of the hundreds of mbuna cichlid species.

Unfortunately, because of the small area of the park in relation to the lake, many Lake Malawian animal and plant species are unprotected. The World Heritage Committee of the United Nations Educational, Scientific and Cultural Organization (UNESCO) has recommended that the scope of the park be extended. The integrity of the park relies on the conservation and management of the lake, which falls under the jurisdiction of three sovereign states—Malawi, Tanzania, and Mozambique—which in some ways complicates the matter.

Serious climate change impacts have already been felt in Lake Malawi. These include dry spells and droughts, many of which have seriously affected the spawning and feeding protocols of its denizens. Floods have also increased in frequency, duration, and severity. Dislocation of habitat due to these pressures is adding to the challenge of conserving the lake biome's richness against other anthropogenic factors.

Jeffrey C. Howe

Further Reading

Arnegard, Matthew E., et al. "Population Structure and Colour Variation of the Cichlid Fish *Labeotropheus fuelleborni* Ahl Along a Recently Formed Archipelago of Rocky Habitat Patches in Southern Lake Malawi." *Proceedings of the Royal Society of London* 266, no. 1 (1999).

Network of International Development Organisations (NIDOS). "Malawi Climate Change Fact Sheet." NIDOS in Scotland, 2009. http://www.nidos.org.uk/downloads/Malawifactsheet.pdf.

Olson, David M. and Eric Dinerstein. "The Global 200: Priority Ecoregions for Global Conservation." *Annals of the Missouri Botanical Gardens* 89, no. 2 (2002).

Maldives

Category: Marine and Oceanic Biomes.
Geographic Location: Indian Ocean.

Summary: One of the world's smallest nations, consisting of 1,192 islands, the Maldives is known for its rich coral ecosystem but is threatened by rising sea levels because of global warming.

The Maldives is one of the smallest nations in the world. The name derives from the Sanskrit word *Dhivehi*, which means "island." The Dutch called it the *Maldivische Eilanden*, and the British gave it the name Maldive Islands, which has become Maldives. As a nation, it consists of 1,192 islands spread over 332,000 square miles (859,000 square kilometers) in the Indian Ocean. The islands are spread across 26 atolls some 440 miles (708 kilometers) southwest of Sri Lanka and 250 miles (402 kilometers) southwest of India. Only 199 islands are inhabited, of which fewer than 30 are greater than 247 acres (100 hectares) in area.

Maldives is among the lowest-lying countries in the world, with an average ground altitude of only 5 feet (1.5 meters) above sea level. Because of this vulnerability, some of the islands are washed away regularly, even with a small rise in seawater. The total population of the country is about 300,000; about 70 percent of the inhabited islands each have a population of less than 1,000. The capital and largest city, Male, has a population of about 104,000. For administrative purposes, the islands are grouped into seven provinces.

Climate and Biota

The climate of the Maldive islands is tropical and regulated by the easterly and westerly monsoons. The daily mean temperature is about 82 degrees F (28 degrees C) with a consistently high humidity of around 80 percent. The Maldives has very little land or land-based natural resources; less than 10 percent of the land is arable. The inland soils are saline and in a very thin layer; only 150 islands, or 12 percent of the total, support mangrove vegetation.

The terrestrial environment includes 500 species of higher plants, two species of amphibians, 67 species of butterflies, six species of reptiles, five species of mammals, and 130 species of butterflies. Altogether about 200 species of birds have been reported, five of which were identified as endemic:

Maldivian pond heron, whiter tern, lesser frigate, black-naped tern, and red-footed booby. The most common trees are coconut, palms, dhigga, kaani, kandoo, midhili, and banyan. Along the shorelines, some 20 different species of grass grow. The common crops in Maldives are papaya, banana, tomato, watermelon, breadfruit, and chili.

The environment of Maldives is aquatic. The atolls are alive with coral reefs and sand bars that act as natural barriers against the sea. The coral biodiversity includes about 250 species of branching corals, five species of turtles, 51 species of echinoderms, some 5,000 species of mollusks, and 1,000 species of crustaceans; as well as five species of sea grasses, and 285 species of alga and sponges. The region also includes more than 2,000 species of fish, reef sharks, moray eels, sea turtles, and different varieties of rays.

The rich diversity in the reefs attracts thousands of tourists to the islands. The Maldives is considered as one of the top places for coral exploration. However, in recent years the coral reefs and the islands have faced several severe environmental challenges. The main damage to the islands is from climate change and human activities. Climate change has resulted in rising sea level and intensified frequency of cyclones and storms. An increasing sea level will continue to affect coral reefs, nutrient circulation, and fisheries. Impacts from human activities include coral degradation, water pollution, coastal habitat destruction, and over-fishing. Threatened by pervasive climate change, the government of Maldives has appealed to international institutions and governments to decrease greenhouse gases. Maldives was the first nation to sign the Kyoto Protocol.

Culture, Politics, and Economy

The culture of the Maldivians is very similar to the South Indian and Sri Lankan cultures. The official language, Dhivehi, is closely related to the Sinhalese language used in Sri Lanka. Historically, Maldivians shared similar religious beliefs as the neighboring cultures. However, in 1153, Arabs conquered Maldives and converted it into a Muslim nation. The Arabs also established a monarchy, under which the sultanate ruled the country from 1153 to 1968. During the latter period of the sultanate, from 1887 to 1965, the British colonized the island state under a contract to protect its sovereignty. In 1968, Maldives became a republic, replacing the monarchy; Ibrahim Nasir became the first president of the new republic. From its beginning, the new republic was mired in political instability. This resulted in Moumoon Abdul Gayoom ousting Nasir as the president in 1978. Moumoon then ruled the country for 30 years (1978–2008); Mohammed Nasheed (2008–12) replaced him and was in turn ousted, with vice president Mohammed Waheed Hassan Manik stepping in as president.

Economically, Maldives ranks 95th among the 182 countries in the United Nations Human Development Index. The literacy rate is 97 percent, and the per capita gross domestic product (GDP) was $4,400 in 2006. According to the Maldives government, the economy of the nation depends on four sectors: tourism, fisheries, agriculture, and the garment industry. Of them, tourism and fisheries are the main drivers. The government opened the first tourist resort in 1972 in an effort to diversify its economy away from its dependence on fisheries. Currently, tourism provides one third of the total revenue for Maldives. In order to attract more tourists, some 86 islands have been developed as tourist resorts—Maldivians are not allowed to settle in these tourist islands.

The fishery sector contributes about 15 percent of the GDP. The bulk of the revenue from fisheries (80 percent) comes from the export of tuna and tuna products. This dependence on just two sectors has made Maldives vulnerable socially and economically. Fluctuations in international trade and natural disasters have revealed this vulnerability. The 2004 tsunami devastated the country, all but destroying its tourism and fishing industries. The total damage was estimated to be more than $400 million.

Resettlement on the Horizon

The government of Maldives is discussing the possibility of buying land in neighboring nations and relocating the entire population. Recently, the government identified three nations as the potential destination: India, Sri Lanka, and Australia. How this transition would occur is not yet defined. There is also no information on the responses from these nations.

The government is also encouraging reforestation of its coastal areas, and cleaning and preventing litter in the sea. Environmental education is one of the top priorities of the government; various studies have been conducted to study the impact of climate change on the environment. In addition to these efforts, more actions are needed from developed countries to reduce carbon emissions and minimize the impacts from climate change. It is unfortunate that the existence of Maldives and its rich biodiversity depends on the actions of people living thousands of miles away.

Krishna Roka

Further Reading

Asian Development Bank (ADB). *Social Sector Profile, Maldives: Social Development Issues for the 21st Century*. Manila, Philippines: ADB, 2001.

International Business Publications (IBP), USA. *Maldives Ecology and Nature Protection Handbook*. Washington, DC: IBP, USA, 2009.

United Nations Development Programme (UNDP). *Assessment of Development Results: Maldives*. New York: UNDP, 2010.

Wilkinson, Clive, ed. *Status of Coral Reefs of the World: 2008*. Townsville, Australia: Global Coral Reef Monitoring Network, 2008.

Mana Pools

Category: Inland Aquatic Biomes.
Geographic Location: Africa.
Summary: A floodplain along the middle Zambezi River hosts a network of permanent and seasonal pools that provide water to a remarkable concentration of wildlife through the dry season.

Mana Pools is a complex habitat network of lakes, wetlands, and woodlands located along the northern border of Zimbabwe on the south bank of the Zambezi River. Not far south of Victoria Falls, the course of the Zambezi has gradually drifted northward across the floodplain, carving out channels and pools that fill with water during the rainy season—and retain much of that water throughout the dry season, supporting perhaps the greatest variety of large mammal herds on the African continent. There is concern, however, that changes in the natural flood regime of the river since the construction of two dams has altered the character of the floodplain and will put undue stress on its native species.

Geography and Habitat Types

Mana Pools derives its name from the indigenous Shona word for "four," referring to the four largest permanent pools created here by the meandering Zambezi. Many more seasonal pools and ponds dot the region, forming a complex network of watering holes flanked by woodlands. Three distinct habitats are found in the Mana Pools biome: jesse bush, mopane woodland, and the floodplain.

Jesse bush is a term encompassing a variety of shrubs, thorny bushes, and vines that grow so closely together that the bush is nearly impenetrable. A network of trails, created largely by elephants, rhinos, and buffaloes traveling between water holes create highways through the brush for smaller animals. The dense brush, in effect, creates a safe haven and refuge from the sun for many species.

Mopane trees (*Colophospermum mopane*) grow in poor soils in hot, low-lying areas; they tend to dominate where the soil is heavily clay-based. The clay in the soil retains pools of water for animals to drink, and mopane leaves provide valuable protein during the dry season. Mopane wood is extremely hard and termite-resistant, making it prized for construction and firewood among natives. Mopane woodland is usually open and easily walked, because the soil generally does not support dense undergrowth, while tree canopies are high with scant understory, and sparse in distribution.

The sandier, alluvial soils of the floodplain contain tracts of winterthorn, also favored fodder for herbivores, and more diverse woodlands including baobab, wild fig, sausage tree, dense albeida trees, and Natal mahogany. This area used to flood regularly during heavy rains before the construction of the Kariba Dam upstream on the Zambezi. It has been feared that damming the river would deprive the soil of new alluvial deposits.

Wildlife

Rich and varied mammal populations concentrate on the floodplains during the dry season, when water is scarce elsewhere. It is also the time when winterthorn trees drop their protein-rich pods. The Mana Pools biome has Zimbabwe's largest populations of hippopotamuses and Nile crocodiles. It also hosts huge herds of elephants, buffaloes, zebra, waterbuck, and several antelope species. Predators following the herds include lions, hyenas, leopards, and cheetahs.

Bird life is seasonally rich here, both in the bush and on the river. More than 450 species have been recorded. Spinetail, bee-eater, flycatcher, swift, sandpiper, plover, and redshank are among the sand-bank nesters and migrant waders that flock to the pools each year. Raptors such as Pel's fishing-owl (*Scotopelia peli*) and western banded snake-eagle (*Circaetus cinerascens*) use the tangled array of perches to good effect.

A hippopotamus surfacing in a bed of floating vegetation in the Zambezi River. The Mana Pools biome along the Zambezi is home to the largest concentration of hippopotamuses in Zimbabwe. (Thinkstock)

Effects of Human Activity

Since the construction of the Kariba Dam near Victoria Falls, and the Cabora Bassa Dam where the Lower Zambezi begins, the character of the floodplain has changed. The river is now regulated to the point that it rarely floods, and then only locally. Construction of the holding reservoirs Lake Kariba in 1959 and Lake Cabora Bassa in 1974 converted a once-fast-moving river into a sluggish body of water. This river now is unable to carry its silt load, depositing islands of sediment in the center of the riverbed. Blocking the flow of the water forces it to run faster on the edges, increasing erosion on the banks.

Mana Pools National Park, a United Nations Educational, Scientific and Cultural Organization (UNESCO) World Heritage Site since 1984, is located on the south bank of the Zambezi River, along the northern border of Zimbabwe. It is part of the Parks and Wildlife Estate that runs on both sides of the Middle and Lower Zambezi, from the Kariba Dam to the Mozambique border. There is no permanent human habitation in Mana Pools National Park. Strict regulation of visitor movements and vehicle traffic, combined with the remoteness of the site, has helped keep the park unspoiled. The presence of an array of tropical diseases, including sleeping sickness, malaria, and bilharzia, also discourages settlement.

In 1989, oil exploration within the preserve was proposed; many concerns were raised about impacts such as industrial littering, erosion, and improved access for poachers. Yet another dam was proposed for the Mapata Gorge, which would submerge much of the Middle Zambezi valley and halve the carrying capacity of the Mana Pools. Additional tourist facilities have also been suggested, which would overcrowd and strain the tourist resources currently in place. Environmental impact studies have shown that these and other development ideas would cause too much damage, and the ideas were abandoned. Currently, the greatest threat to the integrity of Mana Pools is the regulation of the flow of the Zambezi by the Kariba Dam.

In June 2010, Zimbabwe's Middle Zambezi Valley was granted enhanced conservation status as a Biosphere Reserve by UNESCO. This reserve extends from Lake Kariba down the river almost to the border of Mozambique, and includes Mana Pools and the adjoining Sapi and Chewore safari areas. Strictly controlled recreational hunting is permitted in the safari areas in the environmental management plan. While local operators are trying to increase tourism in the region, several private and public campaigns are working to protect Mana Pools from increased human interface and keep this portion of Zimbabwe's heritage intact for future generations to enjoy.

Climate change has already made its effects felt in the Zambezi River catchment area, as recent annual flows are down from persistent shortfall in annual precipitation. Scientists have predicted remorseless temperature increases for the region, which could act to impose drought by the action of increased evaporation. However, some areas are receiving more frequent storms to the point where the risk of flooding has become a constant. Deforestation and even desertification have been discussed in some parts of the region that depends upon the Mana Pools. Each of the pressures will make survival harder for plant and animal species here, and taken together may threaten even the current, somewhat degraded status quo of water management on the Zambezi River.

JILL M. CHURCH

Further Reading

Bone, Craig and Alistair Chambers. *African Seasons: Wildlife at the Waterhole.* London: Dragon's World, 1992.

Chenje, Munyaradzi. *State of the Environment 2000: Zambezi Basin, Vol 1.* Maseru, Lesotho: SADC, Environment and Land Management Sector, 2000.

Swain, Ashok, Ranjula Bali Swain, Anders Themner, and Florian Krampe. "Zambezi River Basin: A Risk Zone of Climate Change and Economic Vulnerability." *New Routes* 17, no. 3 (2012).

Timberlake, Jonathan. *Biodiversity of the Zambezi Basin Wetlands: Review and Preliminary Assessment of Available Information.* Harare, Zimbabwe: International Union for the Conservation of Nature, 2000.

Manchurian Mixed Forests

Category: Forest Biomes.
Geographic Location: Eurasia.
Summary: These diverse forests cover the low hills of Manchuria and the Russian Far East, home to ginseng and the Siberian tiger. Currently, they face jeopardy from overzealous logging.

The Manchurian Mixed Forests ecosystem is a lower-elevation biome that extends from hills in the northern part of the Korean peninsula, through China, to the Amur River region of the Russian Far East. It includes northeastern China's broad river valleys, the east slope of the Large Hinggan Mountains, the south and west slopes of the Small Hinggan Mountains, and the hill regions of the Changbai Mountains at the base of the Korean peninsula. The most direct neighboring biome is the Changbai Mountains Mixed Forests ecosystem, which comprises much of the higher-elevation sections in the same general area.

While considered mixed, this Manchurian sylvan system is dominated more by coniferous tree types than are the other, milder-climate forests to its south. Winters are the dry season here; they are long and cold, with mean low temperature range of minus 4 degrees F to 5 degrees F (minus 20 to minus 15 degrees C). Western Pacific monsoons influence the climate in the summer, with a regime of moist, warm air. Boosted by this seasonal rainfall, the annual precipitation range is approximately 20–40 inches (500–1,000 millimeters).

Forest Composition

At elevations of 1,640–3,280 feet (500–1,000 meters), forests are composed of both coniferous and broadleaf species. The conifer community here includes Korean pine (*Pinus koaiensis*), which frequently reaches 115 feet (35 meters); as well as fir (*Abies holophylla*) and spruce (*Picea obovata*). Deciduous types include oak (*Quercus mongolica*), ash (*Fraxinus mandshurica*), *Tila amurensis*, birch (*Betula schmidtii*), Manchurian elm (*Ulmus*

lacinata), maple (*Acer* spp.), and Manchurian walnut (*Juglans mandshurica*).

These forests also include shrubs such as *Lespedeza bicolor* and *Corylus mandshurica*, especially in the south. On the east slope of the Large Hinggan Mountains, the conifers are dominated by the *Pinus sylvestris* pine, and the broadleaf trees include birch, poplar (*Populus*), willow (*Salix rorida*), and oak.

The understory ecosystem includes *Oxalis acestosella, Phryma tenuifolia, Thalictrum filamentosum, Adiantum pedatum, Asarum sieboldii, Polystichum tripteron, Oplopanax elatus*, and *Kalopanax septemlobu*. The most famous native of the understory in this ecosystem is ginseng (*Panax ginseng*), valued as a medicinal and culinary rhizome. The genus name *Panax* means *all-heal*, related to *panacea*, and was applied by Swedish naturalist Carl Linnaeus to the plant because of its wide use in Chinese medicine. Today, ginseng continues to be widely used as a stimulant, aphrodisiac, and diabetes treatment, and is touted by adherents as having anticarcinogenic and antioxidant properties. Though the root is commonly used, the leaves actually contain the most ginsenosides, the compounds that distinguish the species.

Patches of grassland feature stipa, also known as spear grass or feather grass (*Stipa pennata*), and the drought-tolerant sheep's fescue (*Festuca ovina*), a grass that favors acidic mineral soil and is prized as a food by numerous species of butterflies and moths.

Wildlife

The brown-eared pheasant (*Crossoptilon mantchuricum*), native to the nearby grasslands, is often found here in small numbers. The steppe marmot (*Marmota bobak*) ranges across the steppes of all of eastern Europe and central Asia, and is common enough to be relied on as a food source by human populations in times of famine or following crop failures. Ungulates and large mammals here include deer, boars, moose, musk deer, bears, tigers, and ghorals.

The forests are home to four protected mammals: the Siberian tiger (*Panthera tigris altaica*), sable (*Martes zibellina*), Sika deer (*Cervus nip-*

pon), and leopard (*Panthera pardus*). The Siberian tiger, which once ranged through this entire region, is already extinct in South Korea and might be extinct throughout the Korean peninsula, though it is thought that some may survive in North Korea's Paektusan Mountains.

The tiger is known to inhabit the Changbai Mountains in China and the forests of Amur Oblast in the Russian Far East. The Amur tiger population is the best observed (and the largest), consisting of a population of fewer than 400 tigers, of which about 250 are breeding adults. Intensive conservation efforts have stabilized the population since the 1990s. The tiger is one of the largest felids that has ever lived. It preys primarily on Sika, red, Siberian roe, and musk deer; wild pigs; and smaller prey, including small mammals like rabbits and hares; and fish such as salmon.

When ungulate populations decrease, tigers are known to prey on brown and black bears, though this is unusual. Tigers attacking bears attack the throat and spine, and feed on the bear's abundant fat deposits. Though tigers do not prey on wolves—and do not consume them if they kill them—their presence will depress wolf population numbers, because they compete for the same prey.

The Siberian tiger suffers from low genetic diversity, with an extremely low effective population size relative to the census population size. The reduction in tiger numbers promises to compound this problem. Poaching of both tigers and the species they prey on is the most serious current threat to the species. In the past, deforestation and legalized tiger hunting were responsible for much of the decimation of the tiger population.

Threats

Substantial timber tracts in this area attract logging and related industries, and there is a danger of deforestation and overharvesting of timber. Habitat fragmentation is of high concern under such circumstances. Climate change impacts include the potential for increasing nitrogen in the soil, which could change the delicate balance of the forest ecosystem.

BILL KTE'PI

Further Reading

Guan, De-Xin, Jia-Bing Wu, Xiao-Song Zhao, Shi-Jie Han, Gui-Rui Yu, Xiao-Min Sun, and Chang-Jie Jin. "CO$_2$ Fluxes Over an Old, Temperate Mixed Forest in Northeastern China." *Agricultural and Forest Meteorology* 137, no. 1 (2006).

Kong, W. and D. Watts. *The Plant Geography of Korea, with an Emphasis on the Alpine Zones.* Boston: Kluwer Academic Publishers, 1993.

Zhao, J., ed. *The Natural History of China.* New York: McGraw-Hill, 1990.

Mannar, Gulf of, Coral Reefs

Category: Marine and Oceanic Biomes.
Geographic Location: Asia.
Summary: The Gulf of Mannar's coral reefs are among the richest marine zones of global significance—but with ever-increasing anthropogenic stress, its biodiversity is under threat.

Coral reefs in the Gulf of Mannar falling under the Indo-Malayan realm are formed around a chain of 21 islands that lie along an 87-mile (140-kilometer) coastline between the west coast of Sri Lanka and the southeastern tip of India. Their average distance from the mainland is 5 miles (8 kilometers). There are four island groups in the Gulf of Mannar, known as the Tuticorin, Vembar, Keezhakkarai, and Mandapam groups. These islands actually form the Pamban-to-Tuticorin barrier reef, which is about 87 miles (140 kilometers) long and 16 miles (25 kilometers) wide.

Other reef forms—shore, platform, patch, and fringing reefs—are also seen in the Gulf of Mannar. Narrow fringing reefs surround the islands, extending 33 feet (100 meters) from the shore. Patch reefs are also found and typically are 0.6 to 1.2 miles (1 to 2 kilometers) long, 17 feet (50 meters) wide, and 6.5 to 30 feet (2 to 9 meters) deep. Reef flats are extensive on all islands.

Biological Riches

The Gulf of Mannar coral reefs are one of the four major coral reefs in India. The Gulf of Mannar biome is comprised of three main ecosystems, namely coral reefs, mangroves, and sea grass beds. Approximately 3,600 flora and fauna species make it one of the richest coastal regions of India. The most commonly occurring genera of corals are *Acropora, Montipora,* and *Porites.* The coral fauna includes 117 species divided among 32 genera. Of these, seven species belonging to seven genera are ahermatypic (non-reef builders) and as such are considered of no economic value. Major reef-building corals (hermatypic) that are found in this area are *Poscillopora* (two species), *Acropora* (25 species), *Montipora* (20 species), the key reef-builders *Pavona* and *Goniopora, Porites,* and *Favia* (five species), *Favites* (six species), and *Goniastrea* (three species).

Ornamental coral fishes and associated marine fauna are found in the reefs. Some important ones belong to the family *Chaetodontidae* (butterfly fish), *Amphiprion* spp (clown fish), *Holocentrus* spp. (squirrel fish), *Scarus* spp. (parrot fish), *Lutjanus* spp. (snapper fish), *Abudefduf saxatilis* (Sergeant major), green turtles, Olive Ridley turtle, and the globally endangered marine mammal dugong (*Dugong dugong*). Of the 2,200 fish species in Indian waters, 450 species (20 percent) are found in the Gulf of Mannar, making it the single richest coastal area in the Indian subcontinent in terms of fish diversity. Other faunal composition is also very rich in the Gulf, including 168 migratory bird species, 79 species of crustaceans, 108 species of sponges, 260 species of mollusks, and 100 species of echinoderms. Fishery in the gulf is dominated by lesser sardine, silver belly, sciaenid, mackerel, anchovy, thread fin, bream, holothurian, lobster, mollusks, and prawns.

Nearly 20 different mangrove species occur and act as important nursery habitats for many faunal species. One mangrove species, *Pemphisacidula,*

The shell of a great rapa chank (Turbinella pyrum) found near Beruwela, Sri Lanka. (Wikimedia/H. Zell)

is endemic to the area; five others occur here and nowhere else in India. The main seaweeds are *Gracilaria, Gelidiella, Hypnea, Sarconema, Hydrodathrus, Caulerpa, Sargassum,* and *Turbinaria.* All six genera and 11 species of seagrass recorded in India occur in the Gulf of Mannar. Six of the world's 12 seagrass genera and 11 of the world's 50 species occur here. The biome also harbors 147 species of marine algae (seaweed). These sea grass and algal beds in turn support complex ecological communities and provide feeding grounds for many animals.

Conservation and Threats

To conserve the rich coastal biodiversity, the Gulf of Mannar Biosphere Reserve was established by the government of India with the assistance from the United Nations; the World Wildlife Fund; and the International Union for Conservation of Nature, through its Commission on National Parks and Protected Areas. It is the first marine Biosphere Reserve not only in India, but in all of south and southeast Asia. Krusadai Island within the Reserve is considered a "biologist's paradise" and harbors, for example, three species of seagrass that are found nowhere else in India. Representative members of every animal phylum known (except amphibians) are found on this island. Krusadai Island is also home to an endemic hemichordata called balanoglosus (*Ptychodera fluva*), a taxonomically unique living fossil that links vertebrates and invertebrates.

The major economic activities in the coral reefs of the Gulf of Mannar are fishing; coral mining for construction; harvesting of sacred chanks (*Turbinella pyrum*), sea cucumber, pipe fishes, sea horses, and seaweeds. Indiscriminate exploitation of reef-dwelling and reef-building corals, natural calamities and major bleaching events, excessive siltation, agricultural run-off, and sewage discharge are among the leading causes of the destruction and degradation of coral reefs in the Gulf of Mannar. Research shows that corals have been quarried here from the early 1960s.

Corals of the Gulf of Mannar have already faced several recent major natural calamities, including the tsunami of 2004. A recent study shows that after the tsunami the proportion of live coral—48.5 percent—was reduced to only 36 percent. The Gulf of Mannar appears to have had a barrier reef ecosystem originally; that has been lost over time largely due to anthropogenic causes. Legitimate and sustainable exploitation of the coral and associated organisms can conserve the rich biodiversity found here.

SABUJ KUMAR CHAUDHURI

Further Reading

Clarke, Arthur C. *The Treasure of the Great Reef—The Blue Planet Trilogy.* Jericho, NY: I Books, 2003.

George, Rani Mary and Sandhya Sukumaran. *A Systematic Appraisal of Hard Corals (Family* Acroporidae*) From the Gulf of Mannar Biosphere Reserve, Southeast India.* Kochi, India: Central Marine Fisheries Research Institute, 2007.

Linden, Olof, David Souter, Dan Wilhelmsson, and David Obura, eds. *Coral Reef Degradation in the Indian Ocean: Status Report 2002.* Kalmar, Sweden: Cordio and University of Kalmar, 2002.

United Nations Development Programme (UNDP) in India. "Conservation and Sustainable Use of the Gulf of Mannar Biosphere Reserve's Coastal Biodiversity." 2009. http://www.undp.org/content/india/en/home/operations/projects/environment_and_energy/conservation_andsustainableuseofgulfofmannarbiospherereservescoa.

Maracaibo, Lake

Category: Inland Aquatic Biomes.
Geographic Location: South America.
Summary: Lake Maracaibo, located in northwestern Venezuela, is the largest lake in South America. Its considerable biodiversity and unique combination of habitats is at grave risk from human activity and climate change.

Maracaibo is one of the most ancient lakes on Earth, created approximately 36 million years ago. It connects to the Gulf of Venezuela in the Caribbean Sea by a narrow strait, which was spanned in 1962 by one of the world's longest bridges. The north end of the lake is semi-arid, while the south end receives approximately 50 inches (1,270 millimeters) of rain annually. The Maracaibo dry forest realm that lies around the northern end of the lake is typical of neotropical dry forest, a forest type that globally has been degraded by agriculture and grazing.

Maracaibo is the largest lake in South America, covering approximately 5,000 square miles (13,000 square kilometers). The lake is slightly brackish in the north, while the Catatumbo River watershed—the main tributary to the lake—feeds freshwater into it from the southwest. Marshy lowlands surround the relatively shallow lake. Population explosions of duckweed, a small-leaved floating plant, become an issue whenever unusually heavy rains reduce the salt content of the lake and allow the plant to flourish—in some instances covering up to 15 percent of the lake surface.

Biota

Lake Maracaibo once featured extensive mangrove swamps. Even in their much-depleted state, they provide habitat for wetlands-evolved mammals, such as the common crab-eating raccoon (*Procyon cancrivorus*), crab-eating fox (*Cerdocyon thous*), cottontail rabbit (*Sylvilagus floridanus*), spotted paca (*Agouti paca*), kinkajou (*Potos flavus*), and the Orinoco agouti (*Dasyprocta guamara*). Jaguar (*Panthera onca*), South American tapir (*Tapirus terrestris*), giant anteater (*Myrmecophaga tridactyla*), ocelot (*Felis pardalis*), red howler monkey (*Alouatta seniculus*), and capuchin monkey (*Cebus* spp.) have also been sighted.

Aquatic mammals include the West Indian manatee (*Trichechus manatus*), river dolphin (*Sotalia fluviatilis*), Amazon River dolphin (*Ina geoffrensis*), and river otter (*Lutra* spp.).

Venezuelan mangrove bird species include herons, egrets, terns and gulls; the frigatebird

(*Fregata magnificens*); roseate spoonbill (*Ajaia ajaja*); anhinga (*Anhinga anhinga*); and jabiru (*Jabiru mycteria*). Winter migrant visitors include sandpipers (*Calidris mauri* and *Micropalama himantopus*) and the blue-winged teal (*Anas discors*). Non-aquatic species can include the orange-winged parrot (*Amazona amazonica*), yellow-headed parrot (*A. ochrocephala*), and macaws such as *Ara chloroptera* and *A. severa*.

Maracaibo dry forest vegetation along the northern reaches of the lake includes *Malpighia glabra, Acacia glamerosa, Gyrocarpus americanus, Myrospermum frutescens, Bulnesia arborea, Jacquinia pungens, Piptadenia flava, Copaifera venezuelana, Bourreria cumanensis,* and *Ritterocereus griseus*.

The Catatumbo moist forest habitat, sited to the southwest of Lake Maracaibo and extending into Colombia, includes a large block between the northern extension of the Andes (Cordillera Oriental) and the Cordillera de Merida to the east. This block is considered part of a northern refuge of species during the Pleistocene. Relict plant species include *Ochoterineae colombiana, Vochysia lehmannii, Miconia mocquerysii,* and *Palicourea buntingii*. Despite its species richness, relatively little is known about the botany or fauna here.

One of the rarest mammals in the Catatumbo moist forest is the brown hairy dwarf porcupine (*Coendu vestitus*), which inhabits lowlands up to Andean elevations. Diverse montane forests and alpine scrublands lie at higher elevations than the moist forest. The only protected area in the region, Catatumbo Bari National Park, mostly protects montane forest.

The southern range of many "northern" (Colombian and Central America) avian species ends around the Maracaibo basin. Such species include the russet-throated puffbird, northern screamer, citron-throated toucan, one-colored pecard, pygmy palm-swift, and crimson-backed tanager. Endemism, that is, the incidence of species that exist only in this biome, is not especially high among birds at Lake Maracaibo.

The area of the Catatumbo River mouth or delta is well-known for unusually spectacular lightning displays during thunderstorms. They occur 140–160 nights per year, last up to 10 hours, and include up to 280 strikes per hour. The phenomenon may occur from the collision of methane gases rising from decomposing Lake Maracaibo marshes, combining with dry winds moving away from the Andes Mountains to the southwest.

Human Impact

Lake Maracaibo basin contains some of the largest oil fields of one of the world's most productive petroleum exporters—as well as almost a quarter of Venezuela's population. Many of the rivers that feed the lake from the south are polluted by fertilizers and pesticides. The Lake's cultural features include *palafitos*, dwellings built on stilts above the water. The best known towns comprised of *palafitos* are on the east coast; these include San Timoteo, Ceuta, and Bachaquero.

Many of the raised boardwalks and dwellings are built from the wood of mangroves. The Lake originally contained the fourth-largest mangrove forest in Venezuela, which was reduced by 90 percent between the 1960s and 1990s. The mangroves that remain are greatly affected by shrimp farming, oil spills, pesticide use here, and siltation runoff from upstream. An ecosystem-wide watershed-conservation approach is needed to address these complex and intertwined issues.

Climate change impacts for the Lake Maracaibo biome include the potential results of increasing ambient temperature and lower rates of precipitation. Lower flow of freshwater into the lake, along with reduced rainfall and faster evaporation, could open the potential for considerable saltwater intrusion through the Tablazo Strait, altering habitat salinity in the lake proper. Increased outbreaks and severity of fires in the semiarid northern surrounds of the lake could also pose significant hazard to habitats there—and allow quicker penetration by nonnative species.

WILLIAM FORBES

Further Reading

Duker, L. and L. Borre. *Biodiversity Conservation of the World's Lakes: A Preliminary Framework for Identifying Priorities—LakeNet Report Series, No. 2.* Annapolis, MD: Monitor International, 2001.

Eisenberg, J. F. *Mammals of the Neotropics: The Northern Neotropics, Vol. 1.* Chicago: University of Chicago Press, 1989.

Harcourt, Caroline S. and Jeffrey A. Sayer, eds. *The Conservation Atlas of Tropical Forests.* New York: Macmillan, 1995.

Huber, Otto and Dawn Frame. "Venezuela." In David G. Campbell and H. David Hammond, eds., *Floristic Inventory of Tropical Countries.* Bronx, NY: New York Botanical Garden, 1988.

Marajó Varzea Forests

Category: Forest Biomes.
Geographic Location: South America.
Summary: These flooded forests line the mouth of the Amazon and cover Marajó and numerous other islands, as well as the interface of the Amazon River and the Atlantic Ocean.

The word *varzea* means flooded forest; it typically refers to the seasonally inundated forests that flank the Amazon River and its tributaries. In particular, *varzea* refers to those areas flooded by white-water rivers, contrasted with areas flooded by black-water rivers, which are called *igapo*.

The Marajó Varzea Forests biome is also tidal—and therefore quite different from the seasonal varzea of the middle and upper Amazon floodplains, which are inundated annually by flooded rivers. The Marajó varzea represents the flooded forests within the Amazon River delta, where the largest river system in the world meets the Atlantic Ocean. Similar to other varzea biomes, this area is a seasonally flooded, riparian forest ecosystem characterized by both terrestrial and freshwater elements.

However, the Marajó varzea has one additional complicating factor: the ocean. The Marajó varzea is therefore both seasonally flooded and tidally influenced, with a gradient of marine influence that gradually decreases moving westward (upstream). Aside from seasonal inundations, the landscape floods twice a day when the ocean tide pushes river water onto the landscape to a height of up to approximately 10 feet (3 meters).

At its mouth, the Amazon River splits into a braided network of islands as it becomes a river delta system. The largest of these is Marajó Island, at some 18,500 square miles (48,000 square kilometers), which lies roughly in the middle of the massive delta. The Marajó Varzea Forests biome is dominated increasingly by mangrove ecosystems along the coastline and around eastern Marajó Island.

Specially Adapted Species

Varzea biomes are harsh environments, where flood levels can reach 20 feet (6 meters), while at the same time they are unique, in that they are characterized by their mixture of terrestrial and aquatic elements. These include pacu fish that act as seed dispersers for trees, and pink river dolphins that hunt between the giant buttresses of riparian trees. It is also not unusual to see bull shark entering the flooded forest.

Because humans have aggregated along the river banks of the Amazon, much of this biome has been converted to agriculture and even to urban areas; however, the constant flooding ensures that large areas of varzea remain out of reach of development for now. This unique and threatened area supports habitats not only for a plethora of endangered species, such as manatee and giant otter, but it also has provided for many indigenous peoples who have become largely displaced.

Mammals are abundant and have very high diversity in this biome. Overlapping species from land, the canopy, the river, and the ocean all meet here and trade influences across habitats. In the treetops are the brown four-eyed opossum, Brazilian porcupine, Guianan squirrel, pale-throated three-toed sloth, and white-faced tree rat. Such monkeys as silvery marmoset, Midas tamarin, and red-handed howler are relatively common here.

The occurrence of mangroves and proximity to the river and ocean also make good habitat for crab-eating fox and crab-eating raccoon. During the dry season, the red brocket deer, jaguar, little-spotted cat, and southern naked-tailed armadillo make use of the understory. The short-tailed opossum (*Monodelphis maraxina*) is endemic to this biome (found nowhere else on Earth), and is considered endangered.

There are over 350 known species of birds on Marajó Island, and the entire biome contains an estimated 500 species. Colorful species with equally colorful names include black-throated mango (a hummingbird), blue-tailed emerald, blue-shinned sapphire, and blue-headed parrot. The proximity to the Atlantic Ocean means that some shorebirds are also frequent, if seasonal, residents, especially along the coastline and mangrove systems. These include American oystercatcher, least sandpiper, whimbrel, and ruddy turnstone. Swamps, flooded forests, and grasslands are common aggregation points for other waterbirds including numerous ducks, as well as snowy egret, little blue heron, cocoi heron, scarlet ibis, and wood stork.

Very large fish occupy this biome; they, too, often roam between the forest and the river system. Characteristic species here include tambaqui *(Colossoma macropomum)* and pirarucus *(Arapaima gigas).* The moriche palm (*Mauritia flexuosa*) is among the dominant species of tree along the swamps and river's edge. Among the amphibians is the Suriname toad, while reptiles include the tropical flat snake and Brazilian false water cobra.

Fragility and Threats

The varzea is an ever changing biome where the only constant is the annual flooding, which carries in nutrients from as far away as the Andes Mountains. Flooded forests are unique from nearby *terra firma* rainforests in many ways—but most notably in that seasonal inundations dramatically change tree recruitment, turnover, and seed dispersal on the forest floor when waters carry away seeds, leaf litter, and so forth. Tree diversity is therefore limited to those species that can survive frequent and prolonged inundations, and which have typically developed relationships with both terrestrial and aquatic fauna to ensure their dispersal and survival.

The effects of climate change on the Marajó Varzea Forests biome could thus be critical. The system is robust, but balanced in very calibrated ways. Sea-level rise, for example, could put just enough stress on less-salt-tolerant vegetation here to give way to radical changes in the plant community. Similarly, incremental temperature increases

Mangroves growing along the beach at Marajó Island in Para, Brazil, in 2011. Mangrove ecosystems have come to dominate much of the Marajó Varzea Forests biome along the coastline and around eastern Marajó Island. The 18,500-square-mile (48,000-square-kilometer) island is home to more than 350 species of birds. (Flickr/phogel)

could threaten the natural timetable for seed germination that has co-evolved with the capacity and seasonality of the unwitting efforts of various animals to aid in dispersal.

JAN SCHIPPER

Further Reading

Campbell, D. G. and H. D. Hammond, eds. *Floristic Inventory of Tropical Countries.* Bronx, NY: New York Botanical Garden, 1989.

Goulding, Michael. *The Fishes and the Forest: Explorations in Amazonian Natural History.* Berkeley: University of California Press, 1981.

Pires, Joao M. and Ghillean T. Prance. "The Vegetation Types of the Brazilian Amazon." In Ghillean T. Prance and Thomas E. Lovejoy, eds., *Key Environments: Amazonia.* New York: Pergamon, 1985.

Smith, Nigel J. H. *Amazon Sweet Sea: Land, Life, and Water at the River's Mouth.* Austin: University of Texas Press, 2002.

Marielandia Antarctic Tundra

Category: Grassland, Tundra, and Human Biomes.
Geographic Location: Antarctica.
Summary: One of the simplest and harshest terrestrial ecosystems on Earth is experiencing rapid environmental change.

At a distance, a visitor to Antarctica's rocky coast may see nothing but penguins against a lifeless-looking icy backdrop. Closer to shore, however, one of the world's simplest terrestrial ecosystems comes into view: tundra populated by low-lying plants, invertebrates, and microbes. This tundra ecosystem is at its most diverse and abundant in Marielandia, a region known most commonly as the Antarctic Peninsula and its nearby islands. Although Marielandia has a short history of direct human contact, the biome is changing rapidly because of human activities.

The Antarctic Peninsula is a curvy, roughly triangular piece of land jutting 777 miles (1,250 kilometers) from the western part of the continent toward South America. The moderating influence of the Southern Ocean helps keep the peninsula warmer than the rest of Antarctica. Average summer (December to February) air temperatures are just above 32 degrees F (0 degrees C), whereas average winter (June to August) air temperatures range from 21 degrees F (minus 6 degrees C) on the west coast to minus 9 degrees F (minus 23 degrees C) on the east coast. The peninsula is also wetter than Antarctica's dry interior. Palmer Station, on Anvers Island here, gets an average of 27 inches (682 millimeters) of liquid precipitation in the form of rain and melted snow each year.

Biodiversity

Marielandia has more than 300 species of lichens, more than 100 species of mosses and liverworts, one terrestrial alga, and just two species of flowering plants. There are even fewer fauna: under 100 species of free-living mites, springtails, nematode worms, rotifers, tardigrades, and flies. For comparison, at similar northern latitudes in Iceland, there are 329 species of flowering plants, and there are about 800 species of nonmarine invertebrates on the Svalbard Archipelago in Norway.

Terrestrial life in Marielandia tends to be restricted to ice-free coastal fringes where temperatures are the highest and liquid water is more freely available. Even so, the harsh environment poses multiple challenges. The short growing season is marked by low temperatures and rapid transitions between freezing and thawing. In the summer, 21 to 24 hours of daylight each day exposes organisms to high radiation. Although the air is cold, ground temperatures can spike to more than 77 degrees F (25 degrees C), increasing the risk of dehydration in an already dry environment.

Marine animals such as penguins and seals deposit nutrient-rich guano; this fertilizes plants and feeds invertebrates and microbial populations. But these animals also trample vegetation, creating large bare patches where moss and grass once grew in abundance. Organisms here must also deal with winds that fling salt spray and abrasive grit to shore.

Marielandia's tundra inhabitants have multiple adaptations that help them thrive despite harsh conditions. Lichens tend to be brilliantly colored because of natural sunscreens that protect their cells from ultraviolet (UV) radiation. Some moss species can temporarily shut down cellular activity when water is in short supply. Springtails, mites, and the Antarctic cushion plant (*Colobanthus quitensis*) avoid chilling damage by supercooling, meaning that their tissues remain unfrozen even when cooled well below 32 degrees F (0 degrees C). Antarctic hair grass (*Deschampsia antarctica*) and the wingless Antarctic midge (*Belgica antarctica*), on the other hand, can survive internal ice formation. Both freeze-tolerant species accumulate special proteins and large quantities of carbohydrates that reduce the size and number of potentially damaging ice crystals that form within their bodies.

One of the best protections against extreme cold is snow and ice. In the winter, a blanket of snow can provide an insulating microhabitat for both invertebrates and plants. Even if the air above the snow is minus 40 degrees F (minus 40 degrees C), the temperature underneath the frozen blanket stays a cozy 23 degrees F (minus 5 degrees C) or higher.

Conservation Efforts

Since 1961, Antarctica has been governed by an international treaty that sets aside the continent for scientific purposes. No one lives in Antarctica permanently, but each year about 4,000 scientists and support personnel and 30,000 tourists visit the region. More than 90 percent of tourism activities occur in western Marielandia. The Antarctic Treaty heavily regulates the activities of researchers and tourists alike. Hazardous waste, for example, must be removed from the continent, and visitors must keep their distance from wildlife.

Another international treaty is helping stabilize the ozone hole, a region of low ozone concentration that forms over Antarctica each spring. Earth's ozone layer, which absorbs harmful UV in sunlight, has been damaged by chemicals formerly used in refrigerators, aerosol cans, and other industrial and household items. Many ozone-depleting chemicals are being phased out under the Montreal Protocol, but Antarctic ozone concentrations are not expected to recover until late in the century.

Environmental Threats

One of the most pressing threats to the Marielandia tundra is climate change. The western Antarctic Peninsula is one of the fastest-warming regions on Earth, with winter temperatures increasing five times faster than the global average. The warming here has been accompanied by decreases in cloud cover, increased precipitation in some areas, and changes in storm patterns. The potential effects of these changes are still not well understood.

In the short term, warmer and wetter conditions likely will encourage plant growth and open new habitats by melting glacial ice. However, reduced snow and cloud cover may make freeze-thaw cycles more frequent, which early research suggests can damage both plants and invertebrates. And despite the restrictions of the Antarctic treaty, non-native species are more likely to invade Marielandia as conditions become milder.

JUANITA M. CONSTIBLE
RICHARD E. LEE, JR.

Further Reading

Convey, Peter. "Antarctic Terrestrial Biodiversity in a Changing World." *Polar Biology* 34, no. 11 (2011).

McGonigal, David, ed. *Antarctica: Secrets of the Southern Continent.* Buffalo, NY: Firefly Books, 2008.

Michaud, M. Robert, et al. "Metabolomics Reveals Unique and Shared Changes in Response to Heat Shock, Freezing, and Desiccation in the Antarctic Midge, *Belgica antarctica.*" *Journal of Insect Physiology* 54, no. 4 (2008).

Maudlandia Antarctic Desert

Category: Desert Biomes.
Geographic Location: Antarctica.

Summary: This tundra ecoregion is located in the coldest, windiest, driest, and most isolated place on Earth.

Located in the frozen environment of eastern Antarctica, the Maudlandia Antarctic Desert is a tundra ecoregion that is one of the harshest places on the planet. Despite the frigid and arid conditions, many species of flora and fauna have adapted to survive in this ice-covered ecosystem. The biome faces a variety of threats, such as global warming and human disturbance through tourism.

Antarctica is the coldest, windiest, driest, and most isolated of Earth's land masses. While 98 percent of it is permanently covered in ice, two regions have ice-free areas: Marielandia on the Antarctic Peninsula to the west, and Maudlandia on the continental coast to the east. Of the two biogeographical regions, which are separated by the 2,000-mile (3,219-kilometer) Transantarctic Mountains, Maudlandia spans a greater area: 81,700 square miles (211,602 square kilometers), or about two-thirds of the continent. It also has larger stretches of ice-free land; these are found inland and along coastal fringes. The Maudlandia Antarctic Desert has a harsh continental Antarctic climate: extremely cold, windy, and dry. Average monthly temperatures are below 23 degrees F (minus 5 degrees C).

Spanning the Victoria, Wright, and Taylor Valleys near McMurdo Sound, the Maudlandia Antarctic Desert biome is mostly situated in the Eastern Hemisphere, on the Indian Ocean side of the continent. It also contains the McMurdo Dry Valleys, parts of which are ice-free. This region receives little to no rainfall, with annual average precipitation around just 4 inches (102 millimeters).

Unlike the rest of the continent, this region's surface is kept ice-free due to strong winds that blow away snow cover and prevent ice buildup. Inland, where the central ice sheet is surrounded by the mountain and glacier zone, the temperature is low, and the environment is drier. Several substantial lakes occur in the valleys; some lakes are completely frozen.

Biodiversity

Maudlandia is a tundra ecosystem, marked by a semidesert environment and a lack of tree growth due to extremely low temperatures. But while this environment is perhaps the harshest in the world in terms of supporting life, there is a surprisingly rich terrestrial biodiversity living primarily along the ice-free coastal fringes. There are two native vascular flowering plants: the Antarctic hair grass (*Deschampsia antarctica*) and the Antarctic pearlwort (*Colobanthus quitensis*).

Vegetation is dominated, however, by lower plant groups, primarily mosses such as *Campylopus pyriformis*, liverworts such as *Cephaloziella exiliflora*, lichens, and fungi. The surfaces of rocks

The Maudlandia Antarctic Desert biome includes Wright Valley and its Don Juan Pond, shown here. The pond has been included in a National Science Foundation Long Term Ecological Research project, in part because some of the first effects of climate change in Antarctica are expected to be seen in the dry valleys. (National Science Foundation/Jenny Baeseman)

provide a habitat for a variety of mosses, algae, fungi, and bacteria. The soil can also support algae and nematodes.

Lichens in particular can form extensive communities. When water is not available, some lichens survive by entering cryptobiosis, a suspended metabolic state in which most biochemical processes stop. One lichen species, *Buellia frigida*, can have a life span of more than 1,000 years.

Some species of algae, such as *Hemichloris antarctica*, have adapted to withstand repeated cycles of freezing and thawing. Various species of blue-green algae form microbial mat communities at the bottom of some unfrozen lakes. In the Dry Valleys, cryptoendolithic microorganisms colonize the porous spaces near the surface of exposed semi-translucent sandstone rocks where they can acquire sunlight, moisture, and minerals from the rock. This area also supports a bacteria-eating nematode that enters a state of anhydrobiosis that allows it to survive for years of being frozen.

Fauna

Terrestrial invertebrates are limited to arthropods associated with coastal plant communities, such as springtails, midges, and mites such as *Nanorchestes antarcticus*. The coastal regions also support all the larger animals of the terrestrial ecosystem, including six native seal species: the Weddell seal (*Leptonychotes weddellii*); crab-eater seal (*Lobodon carcinophagus*); southern elephant seal (*Mirounga leonina*); Ross seal (*Omimatophoca rossii*); southern fur seal (*Arctocephalus gazella*); and the noted penguin predator, the leopard seal (*Hydrurga leptonyx*).

There are six native penguin species here: the chinstrap penguin (*Pygoscelis antarctica*), gentoo penguin (*P. papua*), emperor penguin (*Aptenodytes forsteri*), and Adélie penguin (*P. adeliae*)—all of which breed on the continent—as well as the king penguin (*A. patagonicus*), macaroni penguin (*Eudyptes chrysolophus*), and rockhopper penguin (*E. crestatus*). Of the 37 native flying seabird species, the ones that breed in Maudlandia include Wilson's storm petrel (*Oceanites oceanicus*), southern fulmar (*Fulmaras glacialoi-*

des), southern giant fulmar (*Macronectes giganteus*), south polar skua (*Catharacta maccormicki*), cape pigeon (*Daption capense*), snow petrel (*Pagodroma nivea*), and Antarctic petrel (*Thalassoica antarctica*).

Conservation Efforts

Due to their support of large breeding avian colonies, several areas have been designated as Specially Protected Areas (SPAs), such as Taylor Rookery in Mac Robertson Land, which supports one of the largest breeding colonies of emperor penguins. Because of the presence of unique undisturbed flora, Cryptogam Ridge on Mount Melbourne, and the Dry Valley known as Barwick Valley have also been designated as SPAs.

The Antarctic Treaty bans nuclear testing, radioactive-waste disposal, and oil and mineral exploration, but it does not protect against other activities that contribute to the region's environmental degradation. Tourism has become an increasing concern. Climate change remains a continuing threat to this environment, with looming glacial melt likely to stress these arid habitats. In addition, the persistence of chlorofluorocarbons in the Earth's atmosphere has created a loss of the ozone layer over Antarctica, which likely inhibits the robustness of some species communities.

REYNARD LOKI

Further Reading

Fogg, G. E. *The Biology of Polar Habitats.* New York: Oxford University Press, 1998.

National Science Foundation. *Final Supplemental Environmental Impact Statement (SEIS) on the U.S. Antarctic Program.* Washington, DC: National Science Foundation, 1991.

Parfit, M. "Timeless Valleys of the Antarctic Desert." *National Geographic* 194, no. 4 (October 1998).

Udvardy, M. D. F. "The Biogeographical Realm Antarctica: A Proposal." *Journal of the Royal Society of New Zealand* 17, no. 1 (1987).

Watson, G. E., J. P. Angle, and P. C. Harper. *Birds of the Antarctic and Sub-Antarctic.* Washington, DC: American Geophysical Union, 1975.

Mediterranean Conifer-Sclerophyllous-Broadleaf Forests, Eastern

Category: Forest Biomes.
Geographic Location: Middle East.
Summary: These forests feature a unique mosaic of plant formations adapted to harsh climatic conditions and are recognized as the southernmost limits for charismatic tree species—but they are threatened by human activity.

The Mediterranean Conifer-Sclerophyllous-Broadleaf Forests biome, located in the heart of the Middle East, lies across the Mediterranean Sea coasts of Turkey, Syria, Lebanon, Israel, and Palestine. The forests are home to diverse bird species, as many migratory pathways cross above the area. Eastern Mediterranean countries have a high rate of endemism, or incidence of unique species found nowhere else, with many of them presenting interconnected communities due to paleogeographical and historical patterns. The floristic richness and high endemism rate reflect the ancient origins of these forests. The vegetation cover is a mosaic of three groups of plants; the broadleaf sclerophyllous, also called maquis; the mixed coniferous forests and oak woodlands; and the steppe formations.

Broadleaf forests took hold in the Near East before stands of sclerophyllous (having hard, closely-spaced leaves) oak appeared in any significant numbers, except in Syria. Maquis formations comprise kermes oak, laurel, carob, Judas tree, terebenth, and lentisk, along with many species of shrubs, climbers, annuals, and perennials. Myrtle, broom, thorn bush, sage, oregano, ivy, and smilax are part of this formation. All of them are relatively long-lived and resistant, which no doubt explains why the ancient Greeks and Romans considered such evergreen plants as myrtle and laurel to be symbols of love, strength, and eternity.

Conifer and deciduous mixed forests are found in southwestern Anatolia and Taurus in Turkey; and in the Levantine uplands between Turkey, Syria, Lebanon, Palestine, Israel, and Jordan. They shelter a variety of tree species: Lebanese cedar; Cilician fir; juniper; Aleppo and Calabrian pine; Turkey oak; alder; storax; three-lobbed apple; hop-hornbeam and buckthorn.

While the Cilician fir, Lebanese cedar, Turkey oak, and hop-hornbeam have their southernmost limits in the Western Lebanese Mountain chain, Aleppo pine and Phoenician juniper are found sporadically in Jordan. Lebanese cedar grows in Turkey, Syria, and Lebanon. The current distribution of Aleppo pine, kermes oak, carob, and evergreen cypress is probably linked to human management rather than to strictly natural ecological factors.

Steppe formations, widespread in such arid bioclimates, constitute a transition zone in biogeographical and ecological terms between the Sahara and the Mediterranean region. Their vegetation, showing little variation, is composed of wormwood, shrubby horsetail, sunrose, and other scrubland species.

Wildlife

Eastern Mediterranean forests are home to various species of mammals, birds, reptiles, amphibians, and invertebrates. Resident animals tend to be small carnivores, such as the badger, stone marten, red fox, and Egyptian mongoose. Large mammals such as gazelles are mainly confined to southeastern Turkey; the caracal inhabits the arid, hilly steppe desert and mountain terrain to which it is adapted; and the wild boar is found in wooded hills and forests. Other large animals inhabiting these forests are the hyena, wolf, and golden jackal. In Lebanon, one may find rare mammals such as the swamp cat and common species such as Persian squirrel and the Indian porcupine, while a diminishing number of bears lives in the Syrian mountains.

Native birds include the flamingo and pelican, as well as various ducks, snipe, and other game birds. There is high diversity of insects and other invertebrates in the region. Many have developed close associations with specific plants, and are now entirely dependent on the presence of those plants for their survival.

Effects of Human Activity

Anthropogenic activities in the eastern regions of the Mediterranean Basin date back thousands of years. Throughout time, uninterrupted human pressure has led to the extinction of the majority of the large species, especially ungulates. Historically, timber extracted from the coniferous forests has been used for the production of all kinds of tools, as well as in the construction of temples and boats for the succeeding civilizations of the region.

Today, the Mediterranean forests here are vulnerable and fragile. Their sustainability is threatened by rapid population growth, overexploitation for fuelwood, clearing for agriculture, and overgrazing. Changes in communities' composition and extinction of local populations, along with the regressive evolution of Mediterranean forests, are threatening the biodiversity.

These forests are exposed to extreme climatic constraints. Variations in climatic conditions and global warming affects soil characteristics and makes the soil susceptible to biotic and nonbiotic risks. Changes in climate and land use have resulted in forest degradation, deforestation, soil erosion, desertification, and expansion of forest fires in Lebanon and Syria. The introduction of exotic species has caused biotic invasion, while altering the landscapes and decreasing native biodiversity.

Conservation Efforts

International conservation organizations have defined 10 hot spots in the Mediterranean basin with varying degrees of conservation priority. National conservation plans have targeted the forests with the declaration of Protected Areas, Important Bird Areas, and Biosphere Reserves. Biodiversity protection and management approaches followed strict conservation principles in their early stages; these principles were replaced at later stages by the *cosmovision* approach to sustainability, in which humans are considered to be part of the environment. This is crucial, as the long-term interaction between humans and nature in this area cannot be neglected, and has indeed been reflected by the mosaics of cultural landscapes.

ELSA SATTOUT

Further Reading

Blondel, J., and F. Medail. "Biodiversity and Conservation." In *The Physical Geography of the Mediterranean*, J. Woodward, ed. Oxford, UK: Oxford University Press, 2009.

Blondel, J., J.Aronson, J. Bodiou, and G. Boeuf. *The Mediterranean Region: Biological Diversity in Space and Time*. Oxford, UK: Oxford University Press, 2010.

Mazzoleni, S., G. di Pasquale, M. Mulligan, P. di Martino, and F. Rego. *Recent Dynamics of the Mediterranean Vegetation and Landscape.* Chichester, UK: John Wiley & Sons, 2004.

Medail, F. and P. Quezel. "Biodiversity Hotspots in the Mediterranean Basin: Setting Global Conservation Priorities." *Conservation Biology* 13, no. 6 (1999).

Mediterranean Sea

Category: Marine and Oceanic Biomes.
Geographic Location: Europe.
Summary: Bounded by three continents but connected to the Atlantic Ocean, the Mediterranean boasts a rich cultural history and contains great biodiversity.

The name *Mediterranean* originated in Roman times, when the sea was referred to as *Mediterraneus*, meaning "center of the world." The Mediterranean Sea is subdivided into various smaller lobes; these include: the Strait of Gibraltar, Alboran Sea, Balearic Sea, Ligurian Sea, Tyrrhenian Sea, Ionian Sea, Adriatic Sea, and the Aegean Sea. Smaller water-body divisions exist throughout, in the form of gulfs, bays, straits, and channels.

The Mediterranean Sea is bordered by the three continents of Europe, Asia, and Africa, with 21 countries along the coastline. The sea can count among its national borders the countries of Spain, France, Monaco, Italy, Malta, Slovenia, Croatia, Bosnia, Montenegro, Albania, Greece, Turkey, Cyprus, Syria, Lebanon, Israel, Egypt, Libya, Tunisia, Algeria, and Morocco. Water-based boundaries for the

Mediterranean include the Atlantic Ocean via the Strait of Gibraltar at the westernmost point; the Sea of Marmara by the Dardanelles; and beyond it the Black Sea, by the Bosphorus Strait to the east. The Sea of Marmara is often considered to be part of the Mediterranean Sea, whereas the Black Sea generally is not. The 101-mile (163-kilometer) human-made Suez Canal in the southeast connects the Mediterranean to the Red Sea. Because of its central location, the Mediterranean Sea has become one of the most studied water bodies in the world.

Biodiversity

This semi-enclosed sea is rich in islands and underwater beds. It is a critical area for wintering, reproduction, and migration. Although it covers less than 0.8 percent of the world's ocean area, this biome is one of the major reservoirs of marine and coastal biodiversity on the planet, with a great number of endemic (found nowhere else) species, as well as 7 percent of the world's marine fauna species and 18 percent of its marine flora represented here. The Mediterranean is a true crossroads of marine biodiversity.

The marine fauna and flora of the Mediterranean are derived from several biogeographical categories. More than 50 percent of Mediterranean species have their origins in the Atlantic Ocean; 17 percent originated in the Red Sea; and 4 percent

This Mediterranean jellyfish (Cotylorhiza tuberculata), which is also known as a fried egg jellyfish, is commonly found in the Mediterranean, Adriatic, and Aegean seas. (Wikimedia/Emanuele Ferraro)

are relict species, which are witnesses from earlier geologic periods, when the Mediterranean was subjected to a tropical climate. Recent estimates of Mediterranean marine species, taken from compilations of former biological studies, show that approximately 10,000 to 12,000 species are present in its waters. The presence of such a large number of distinct species qualifies the region as a biodiversity hot spot. The rate of endemism here is greater in the comparatively small Mediterranean than in the entirety of the Atlantic Ocean. About 19 percent of endemic Mediterranean species, however, are threatened with extinction; 7 percent are considered to be vulnerable, 7 percent are endangered, and 5 percent are critically endangered.

The marine biota of the Mediterranean Sea is mainly derived from origins in the Atlantic Ocean as a result of a flood more than 5 million years ago that introduced Atlantic biota to the Mediterranean through the Strait of Gibraltar. The marine life introduced to the Mediterranean Sea from the Atlantic Ocean has had to adapt to different climactic conditions. The North Atlantic was then and is now considerably colder and more nutrient-rich than the Mediterranean. Certain species form the foundations of the dense life in this area, and are consequently known as foundation species; they inhabit a multitude of habitats here.

Habitats and Biota

The Mediterranean Sea's diverse habitats together complete the landscapes and biological ranges of the Mediterranean coastal areas. Submarine meadows of endemic Neptune grass (*Posidonia oceanica*) and coral beds are prime examples. Many other sensitive habitats are found here, including deep-sea coral communities, underwater caves, dune areas, coastal forests, lagoons, and wetlands.

The Mediterranean Sea also has a vast array of deep-sea habitats, including hydrothermal vents, seamounts, and deep-sea coral reefs. Underwater canyons are of major importance for many species as places for reproduction and feeding for fish and marine mammals such as Risso's dolphin (*Grampus griseus*) and the sperm whale (*Physeter macrocephalus*). Such habitats also represent a remarkable reservoir of endemism among jellyfish and annelid worms. Chemosynthetic communities harboring essential microorganisms—hydrothermal vents, deep hypersaline habitats, and cold-water corals—are of great ecological value, but are threatened by deep-sea trawling by humans and by the direct seawater temperature-elevating effects of global warming.

The flora of the Mediterranean Sea has a high incidence of aquatic plant endemism, ranging as high as about 20 percent of the aquatic plants. More than 1,000 species of flora exist in the Mediterranean in all.

The Alboran Sea region of the Mediterranean is the most important feeding grounds in Europe for loggerhead sea turtle. In addition, the Alboran Sea hosts important commercial fisheries, including those for sardines and swordfish.

The most commonly seen marine mammals here include the fin whale (*Balaenoptera physalus*), sperm whale, striped dolphin (*Stenella coeruleoalba*), Risso's dolphin, long-finned pilot whale (*Globicephala melas*), bottlenose dolphin (*Tursiops truncates*), common dolphin (*Delphinus delphis*), and Cuvier's beaked whale (*Ziphius cavirostris*). Other, less-prevalent cetaceans here include: the minke whale (*Balaenoptera acutorostrata*), killer whale (*Orcinus orca*), false killer whale (*Pseudorca crassidens*), rough-toothed dolphin (*Steno bredanesis*), and humpback whale (*Megaptera novaeanglia*). There is also a small population of harbor porpoises (*Phocoena phocoena*) that range into the Black Sea.

The Mediterranean monk seal (*Monachus monachus*) is the only pinniped to be found within the Mediterranean Sea. It is now very rare and listed as an endangered species, with the only known colonies in the Alboran Basin and the Aegean Sea. The Alboran Sea, with the largest population of bottlenose dolphins in the western Mediterranean, is also home to the last population of the harbor porpoise in the Mediterranean.

Environmental Threats

Some of the greatest ecological concerns for the marine biodiversity of this area are human population growth that has come in tandem with habitat destruction, increased pollution, invasive species, and climate change. The coasts are under constant and growing human pressure caused by the activities of 150 million residents and the arrival of 200 million tourists every year, the consequences of which have for decades been nearly uncontrollable.

Unchecked urbanization and the overexploitation of resources, combined with the introduction of nonnative species; increased maritime transport; and increased runoff of sediments, nutrients, and heavy toxics have created devastating environmental consequences. Degradation of essential biodiversity, scarcity of the most sensitive species, and destruction of unique and irreplaceable coastal habitats, such as Port-Cros National Park and the Zembra Archipelago, are just a few of the most blatant effects of human activity across the region.

Worldwide, nonnative introduced species are of great ecological concern to regional habitats. Invasive species have become a major problem for the Mediterranean ecosystem in particular, and have endangered many local and endemic species. Analyzed populations of exotic species present in the Mediterranean Sea show that more than 70 percent of the nonindigenous decapods and about 63 percent of the exotic fish occurring in the Mediterranean are of Indo-Pacific origin, introduced into the Mediterranean through the Suez Canal. This pathway through the Red Sea is complicated by the differing salinity, nutrient, and climate regimes between the two—which has often allowed the somewhat hardier and more adaptable Red Sea species to colonize the Eastern Mediterranean basins, crowding out native taxa.

When the Suez Canal was initially constructed, the natural hypersaline condition of the Red Sea at first acted as a barrier, blocking the migration of Red Sea species into the Mediterranean for decades. Over time, however, the salinity in the channel roughly equalized with that of the Red Sea, opening the doors to invasion and colonization of the eastern area of the Mediterranean by species adapted to the lower-nutrient, higher-salinity, warmer-temperature conditions of the Red Sea, Arabian Sea, and Indian Ocean.

ALEXANDRA M. AVILA

Further Reading

Bianchi, C. and C. Morri. "Marine Biodiversity of the Mediterranean Sea: Situation, Problems and Prospects for Future Research." *Marine Pollution Bulletin* 40, no. 1 (2000).

CIESM (International Commission for Scientific Exploration of the Mediterranean Sea). "The Mediterranean Science Commission." http://www.ciesm.org.

International Union for Conservation of Nature (IUCN). "IUCN Guidelines for the Prevention of Biodiversity Loss Caused by Alien Invasive Species." (February 2000). http://www.issg.org/pdf/guidelines_iucn.pdf.

Myers, N., R. A. Mittermeier, C. G. Mittermeier, G. A. B. da Fonseca, and J. Kent. "Biodiversity Hotspots for Conservation Priorities." *Nature* 403, no. 1 (2000).

Pinet, Paul R. *Invitation to Oceanography.* Sudbury, MA: Jones & Barlett Learning Publishers, 2009.

Meghalaya Subtropical Forests

Category: Forest Biomes.
Geography Location: Asia.
Summary: The Meghalaya subtropical forests have shown shocking transformation in recent decades as forest reserves have been exploited for shifting cultivation, mining, and human infrastructure requirements.

The Meghalaya Subtropical Forests biome is one of the most species-rich ecoregions of India. This zone typically comprises montane subtropical moist broadleaf forests, and is spread over 16,100 square miles (41,700 square kilometers) of the Khasi, Garo, and Jaintia Hills. Based in Meghalaya state, it is the wettest ecoregion in India, receiving annual rainfall of more than 36 feet (11 meters). Cherrapunjee and Mawsynram, located in the southern part, receive perhaps the greatest amount of rainfall of anywhere in the world.

In biogeographic classification, these forests are located in the Indo-Malayan realm; they are known for remarkable range and number of mammals, plants, and birds. Throughout the forests are long trails of more than 300 species of orchids. The biome is also a dynamic resource base for timber, cane, bamboo, cane grass, medicinal plants, honey, wax, and mushrooms. Many rare plants found here are confined to sacred groves, which are remnants of past climate vegetation, and are almost untouched due to religious beliefs.

The land is largely owned by communities and tribes, and therefore regulated differently from other states in India. The rapid fall in the forest cover can be attributed to shifting cultivation, coal and limestone mining, urbanization, and industrialization. Due to expanding human population here, the pressure on the forest land for cultivation has increased, and consequently the Jhum cycle, or periodic shifting of cultivation, has been reduced to two to three years instead of the traditional 10–15 year period. The resulting damage to the region includes high levels of acidity in the soil and water.

Floral Biodiversity

The major forest types of this ecoregion are the Assam subtropical hill savanna, Khasi subtropical hill forest, Assam subtropical pine forest, and Assam subtropical pine savanna. The floral dimension is considerable, and distribution includes tropical evergreen forests, tropical semievergreen forests, tropical moist and deciduous forests, temperate forests, and subtropical pine forests. These forest types are located at different elevations, with high to low rainfall patterns.

The tropical evergreen forests are in low-lying areas with large amounts of rainfall, and are characterized by species such as *Bischofia javanica, Fermiana colorata, Pterygota alata, Mesua ferrea, Castonopsis indica, Talauma hodgsonii, Pterospermum acerifolium,* and *Acrocarpus fracinifolius.* The tropical semievergreen forests are at elevations up to 3,937 feet (1,200 meters), with annual rainfall of 49–66 feet (15–20 meters). Dominant species here include *Eleocarpus floribundus, Dillenia pentagyna, Dillenia indica,* and *Hovenia acerba.* Tropical moist and deciduous forests have less than 49 feet (15 meters) of annual rainfall; vegetation includes *Shorea robusta, Tectona grandis, Terminalia myriocarpa, Tetrameles nudiflora,* and *Schima wallichii.*

Though no typical grasslands or savannas are found here, some unique types have developed due to favorable interactions of climate, topography, and other biotic factors. The species of such areas appear across the tops of the Garo, Khasi, and Jaintia Hills, including *Saccarum spontanuem, Neyraudia reynaudiania, Chrysopogon aciculatus,* and *Setaria glauca.* Some patches of temperate forests—with species such as *Lethocarpus fenestratus, Castanopsis kurzi, Quercus griffithii, Quercus semiserrata,* and *Myrica esculenta*—are also present near the southern slopes of the Khasi and Jaintia Hills. Subtropical pine forests of *Pinus khesia* are strictly confined to the Shillong Plateau. Their forest cover is closely woven with a high-density growth of bamboos and canes. The most common species identified is *Dendrocalamus hamiltonii.*

As the ecoregion has suffered multiple disturbances in its vegetation cover, some species once popular in the state have become threatened. Ferns such as *Dipteris wallichii* and *Cyathea gigantea* have become rare in Meghalaya. *Llex embloides, Styrax hookerii,* and *Fissistigma verrucosum* are considered to be extremely rare, but were collected from sacred groves recently. Several orchid species such as *Dendrobium, Pleione,* and *Vanda* have ornamental value and have become rare. The most important example of endemic (not found elsewhere) flora in this ecoregion is the pitcher plant (*Nepenthes khasiana*), found in limited distribution in the Khasi Hills only.

Faunal Biodiversity

The Meghalaya Subtropical Forests ecoregion is reported to have over 110 species of mammals, but none are known to be endemic. Important species include the tiger (*Panthera tigris*), clouded leopard (*Pardofelis nebulosa*), Asian elephant (*Elephas maximus*), wild dog (*Cuon alpinus*), sloth bear (*Melursus ursinus*), Indian pangolin (*Manis crassicuadata),* Assamese macaque (*Macaca assamensis*), capped leaf monkey, and hoolock gibbon (*Hylobates hoolock*). Bats are ubiquitous.

The ecoregion has a large gathering of bird fauna. It coincides with two Endemic Bird Areas: the Assam Hills and the Eastern Himalayas. About 450 species of birds are found here. Almost every species is near-endemic, with five species completely endemic: the Manipur bush quail (*Perdicula manipurensis*), marsh babbler (*Pellorneum palustre*), brown-capped laughing thrush *(Garrulax austeni),* tawny-breasted wren babbler (*Spelaeronis longicaudatus*), and wedge-billed wren babbler (*Sphenocichla humei*).

Other globally threatened species spotted in the Meghalaya subtropical forests include the rufous-necked hornbill, white-winged duck, Pallas's fish eagle, Blyth's kingfisher, black-breasted parrotbill, dark-rumped swift, and beautiful nuthatch. The wreath hornbill (*Aceros undulates*), brown bill (*Anorrhinus tickelli*), and great hornbill (*Buceros bicornis*) are indicators of intact habitat. These birds are entirely dependent on mature trees for nesting and laying eggs. If the habitats are fragmented, their populations will decline.

Protected Areas

At present, nearly two-thirds of the ecoregion is fragmented. To conserve and enhance the natural resources, small protected areas have been delineated. Those owned by the state forest department are diverse and support a large number of species, which are important with respect to biodiversity. Also, patches of habitats are traditionally protected because of the high religious value placed on the area by the local tribes.

Two vital protected areas are the Baghmara pitcher-plant sanctuary for the conservation of *Nepenthes khasiana,* and Nokrek National Park in the Garo Hills for management of rare citrus taxa. The population of elephants from Assam migrate comfortably through the habitat areas of the ecoregion.

Environmental Threats and Existing Policies

The major threat to the Meghalaya subtropical forests is uncontrolled land use. Cultivation is unsustainably practiced; mining activities and overexploitation of ornamental plants and medicinal plants further accelerate damage to the environment. The aquatic vegetation of the ecoregion is compromised by refuse from mining processes. As the quality of water dips due to discharge of highly acidic or alkaline mining waste, the amount of dissolved oxygen for aquatic flora gets reduced.

Conservation and restoration practices should synchronize with traditional practices of this region. The major guiding policies are the 1988 forest policy and other state and tribal laws; the need is to merge them with urban growth strategies, so that the unique biodiversity of this ecoregion does not disappear. Climate change impacts include potential changes in rainfall amounts and seasonal timing, which would have a direct impact on the flora and fauna of this diverse ecosystem.

MONIKA VASHISTHA

Further Reading

Champion, H. G. and S. K. Seth. *A Revised Survey of the Forest Types of India.* New Delhi: Government of India Press, 1968.

Jamir, S. A. and H. N. Pandey. "Vascular Plant Diversity in Sacred Groves of Jaintia Hills in Northeast India." *Biodiversity and Conservation* 12, no. 2 (2003).

Ramachandra, T. V. and A. V. Nagarathna. *Biodiversity Characterization at the Landscape Level in North East India Using Remote Sensing and Geographic Information System.* Dehradun, India: Indian Institute of Remote Sensing, 2002.

Rodgers, W. A. and H. S. Panwar. *Planning a Wildlife Protected Area Network in India.* Dehradun, India: Department of Environment, Forests and Wildlife, Government of India, 1988.

Talukdar, G., S. Ghosh, and P. S. Roy. "Landscape Dynamics in North East Region of India (Meghalya State) Using Spatial Decision Tree Model." *Geocarto International* 19, no. 1 (2004).

Mekong Delta

Category: Inland Aquatic Biomes.
Geographic Location: Asia.

Summary: The richly diverse habitats of the vast Mekong Delta, heavily affected by past exploitation and destructive events, are newly threatened by rapid and uncontrolled development.

The Mekong Delta, the final downstream portion of one of the longest rivers of the world, extends to the southern tip of Vietnam and Cambodia on a triangular plain of approximately 21,236 square miles (approximately 55,000 square kilometers). It is the third-largest deltaic plain in the world, after the Amazon River delta and that of the Ganges-Brahmaputra system. The delta includes a wide variety of ecosystems. It faces several key environmental problems, mainly related to largely unsustainable economic activities and to its historical background.

The Mekong River flows for more than 2,485 miles (4,000 kilometers) from the Tibetan plateau

Boats on the Mekong Delta in Vietnam. Six reserves currently protect 77 square miles (200 square kilometers) of wetland ecosystems in the delta. (Thinkstock)

to the South China Sea, crossing six countries: China, Myanmar, Laos, Thailand, Cambodia, and Vietnam. The collision of the Indian tectonic plate with the Eurasian continent resulted in tectonic uplift and folding, which in this area shaped a variety of physical landscapes, ranging from mountains and highland in the northern regions, to broad floodplains in the southern ones.

In the vicinity of Phnom Penh in Cambodia, the main channel of the river divides into two distributaries: the Mekong and Bassac rivers, considered to be the northern extremity of the deltaic system. The water network subsequently splits into nine main channels, giving this area the name Nine Dragons River Delta. The triangular delta extends from Phnom Penh in the northwest to the southern Vietnamese province of Soc Trang in the southwest and to the mouth of the Saigon River in the east.

The delta is characterized by a complex hydrographic network that includes about 5,592 miles (9,000 kilometers) of large and small channels. The Gulf of Thailand and the South China Sea, characterized by different tidal regimes, affect the duration of flooding and the salinity of the brackish waters and soils. This hydrographic complexity determines the presence of different soil types that influence the associated ecosystems.

According to topography, properties of the sediment, and agro-ecology, researcher Nguyen Huu Chiem divided the delta into five different landform units: floodplains, a coastal intertidal complex, a broad depression, old alluvial terraces, and the hills-and-mountains district.

Biodiversity

Plants found in the Mekong Delta area tend to be types that thrive in the tropical moist, swampy environment. Tree communities are important to maintain the health of the ecosystem; for example, they prevent water runoff during the rainy season. Reed beds form important habitats here for waterfowl, including the eastern sarus crane (*Grus antigone*) and various ibis, such as the near-endemic *Pseudibis gigantea*, or giant ibis. Mammals commonly found in the area include such ungulates as Eld's deer (*Rucervus eldii sia-*

mensis) and the Indochinese hog deer (*Axis porcinus annamiticus*).

Threats

The two Indochina wars in the second half of the 20th century deeply affected both the social and the environmental frameworks. During the second war, largely known as the Vietnam War (1955–75), 20 million gallons (75 million liters) of toxic herbicides and defoliants such as the notorious Agent Orange were sprayed on large areas of the delta, targeting both forested areas and agricultural lands.

All the mangrove forests in the Dong Nai river system and 40 percent of the mangrove forests in the southern part of the delta were destroyed. About 3 million people were affected by the toxic chemicals, and a massive migration from rural to urban areas affected the whole socioeconomic structure of this region.

In spite of the dramatic human loss occurred in this period, the population in the delta increased by one order of magnitude, from 1.7 million people at the end of the 19th century to nearly 17 million people today. This growth has driven intensive development of primary agricultural activities, which in this region are constituted mainly of rice production and aquaculture.

After the end of the Vietnam War, an explosive expansion of the network of canals strongly affected the delta ecosystems in new ways, even as they provided access to previously undisturbed extensive wetland portions. Surface and subsurface waters were drained, decreasing the average period of flooding from year round to four to six months per year. The resulting dramatic changes in ecological conditions had widespread negative effects, such as soil acidification, saline intrusion, and pesticide pollution. In many cases, the exploitation of acid sulphate soils failed, with both socioeconomic and environmental fallouts: The number of landless farmers increased, and the original forests could not recover, leading to overall decline in biological diversity.

The fisheries industry in the delta is based on aquaculture farms that raise several species of shrimp and prawn (e.g., *Penaeus*, especially *P.*

monodon) and fish (*Pangasius bocourti* and *P. hypophthalmus*). In particular, cultured shrimp production has skyrocketed in the past few decades, from 220 tons (200 metric tons) per year in 1981 to about 165,347 tons (150,000 metric tons) per year in 2002.

Currently, shrimp aquaculture is considered to be the single most important factor responsible for the destruction of mangrove forests in the Mekong Delta, although these forests reportedly play a fundamental role in the maintenance of shrimp productivity. Despite some attempts of sustainable management, such as integrated mangrove-silviculture-prawn-aquaculture systems, most shrimp farming activities are conducted in an unsustainable way, with producers such as multinational companies rapidly exploiting local areas and moving to others, clear-cutting the mangrove forests and installing shrimp ponds.

Development projects that threaten the natural ecosystems of the delta range from the elimination of periodic flooding by damming and building upstream reservoirs, and diking river banks to the rapid increase of industrial activities, deeply affecting the hydrological regime and sedimentation rates. Recently, a damming project in the upper part of the Mekong River, part of the Xayaburi Dam in Laos, has been the object of debates between the Laotian government and the Mekong River Commission about possible consequences on the Mekong fisheries, including the migratory Mekong giant catfish (*Pangasianodon gigas*) and the river's giant freshwater stingray (*Dasyatis laosensis*).

Conservation Efforts

To cope with the rapid and deep alterations of the environment, several management and conservation measures have been taken. Six reserves have been established to protect parts of the wetland ecosystems of the delta, for a total area of about 77 square miles (200 square kilometers). In 1983, a joint research unit of the Mekong River Commission and the United Nations Environment Programme (UNEP) developed ecologically compatible measures to be adopted in the exploitation of water and land resources.

In 1986, Vietnam and Cambodia signed an international agreement on cooperation in wildlife conservation, attaching special importance to endangered waterfowl species, which are particularly abundant in this area. In a more recent review, 16 organizations, programs, and networks focused on the conservation of the Mekong River basin. Despite these attempts, sustainable approaches to better management of the economic activities of the area are considered to remain urgent priorities—and complex challenges.

Laura Ribero
Gianluca Polgar

Further Reading

Bucholtz, R. H., A. S. Meilvang, T. Cedhagen. J. T. Christensen, and D. Macintosh. "Biological Observations on the Mudskipper *Pseudapocryptes elongatus* in the Mekong Delta, Vietnam." *Journal of the World Aquaculture Society* 40, no. 6 (2009).

Chiem, Nguyen Huu. "Geo-Pedological Study of the Mekong Delta." *Southeast Asian Studies* 31, no. 2 (1993).

Christensen, S. M., D. J. Macintosh, and N. T. Phuong. "Pond Production of the Mud Crabs *Scylla paramamosain* (Estampador) and *S. olivacea* (Herbst) in the Mekong Delta, Vietnam, Using Two Different Supplementary Diets." *Aquaculture Research* 35, no. 2 (2004).

Ellison, Aaron M. "Managing Mangroves with Benthic Biodiversity in Mind: Moving Beyond Roving Banditry." *Journal of Sea Research* 59, nos. 1–2 (2008).

Stewart, Mart A. and Peter A. Coclanis. *Environmental Change and Agricultural Sustainability in the Mekong Delta*. New York: Springer, 2011.

Tong, Phuoc Hoang Son, et al. "Assessment From Space of Mangroves Evolution in the Mekong Delta, in Relation with Extensive Shrimp-Farming." *International Journal of Remote Sensing* 25, no. 21 (2004).

Torrell, Magnus, Albert M. Salamanca, and Blake D. Ratner. *Wetlands Management in Vietnam: Issues and Perspectives*. Penang, Malaysia: World Fish Center, 2003.

Mekong River

Category: Inland Aquatic Biomes.
Geographic Location: Southeast Asia.
Summary: Development along the 12th-longest river in the world has been slow, therefore sparing the region's extraordinary biodiversity, but global economic pressures loom.

The biodiversity of the Mekong River and its drainage basin is almost incomparable. Only the Amazon River's richness of plant and animal life outshines that of southeast Asia's longest river. The Mekong's eccentricity sprouts from the various ecosystems that it simultaneously creates and flows through. The river's name comes from *Mae Nam Khong*, meaning the "mother of water."

Up until the end of the 20th century, the river went mostly unhindered by human engineering, including the machinery of modern warfare, particularly that of the Second Indochina (or Vietnam) War, which exacted a heavy toll on the people and environment of mainland southeast Asia but largely spared the river's contours and bedrock, along with most of the basin's ecology. Today, as the world extols the virtues of the global marketplace, the economic dragons and tigers of Asia expect the Mekong to nurture regional development and generate financial wealth. While this is happening, the river and its surroundings are changing.

Biodiversity and Human Impact

Because the Mekong flows some 2,600 miles (4,200 kilometers) and generally from north to south, the Mekong region contains myriad ecosystems formed by an immense variety of flora and fauna: about 20,000 species of plants, 1,200 species of birds, 800 species of reptiles and amphibians, and 430 species of mammals, all specific to local types of terrain, temperature, precipitation, and drainage patterns.

The river joins five biomes that spread across parts of southeast Asia: Tropical Savanna, Tropical Savanna Altitudinal Zone, Tropical Rainforest, Humid Subtropical Altitudinal Zone, and Tropical/Subtropical Steppe Attitudinal Zone, all within the broader category of Humid Tropical. Although the classification of biomes depends on natural and geophysical circumstances, the river's massive fan-shaped delta, unfolding into the South China Sea, has been so transformed by culture that this particular area of the river could now be classified as an Anthropogenic biome.

Carved into the delta are canals and irrigation ditches controlling the flow of the river and dividing its fresh waters among the intensively cultivated paddies, plantations, and fish ponds that, when seen from above, create a mosaic of distinctly human layout. With intensive agriculture comes high population density, which in turn gives birth to cities. Small wonder that one of the first recorded empires of southeast Asia—the Funan—arose from the nutrient rich soils of this plain, which now reaches all the way to Phnom Penh, the capital city of Cambodia. Ho Chi Minh City, Vietnam, sits more toward the end of the lines where sand, silt, and other sediments that have broken away from the 309,000-square-mile (800,000-square-kilometer) drainage basin reach the sea. Here, where fresh water encounters salt, were once thick mangrove forests.

Moving upstream away from the coast (as sometimes seawater does with devastating consequences on agricultural land), the Mekong remains within the Humid Tropical Zone but leaves the Tropical Rainforest of evergreen broadleaf trees, often hosting epiphytes and embracing lianas, to enter a Tropical Savanna region. Unlike the forest—which teems with wildlife including the Asian elephant, Siamese crocodile, leopard cat, agile gibbon, king cobra, Asian emerald cuckoo, sun bear, and Indochinese tiger—the savanna has an annual wet and dry season, a climate regime referred to as monsoon. Less overall rainfall yields a landscape covered with tall grasses, including bamboo, and interspersed with deciduous trees.

Tonlé Sap, the Great Lake of Cambodia, straddles these biomes; and though the Mekong curves around the lake, its tributaries with their cargo of sediment and biota shape the intricate ecology of the area, much of which has been designated a United Nations Educational, Scientific and Cultural Organization (UNESCO) Biosphere Reserve.

Farmers and fisherman also rely on Mekong replenishment; they have done so here at least since the dawn of Angkorian civilization, the last of the great Indianized empires of southeast Asia.

Never-Tamed Stretches

More than 1,000 miles (1,600 kilometers) above the Great Lake, the Mekong begins its flow from the upper portion of a drainage seated in the Tibetan Plateau, the highest region in the world. It is the upper basin that makes the length of the Mekong notoriously un-navigable. The bedrock—raised, twisted, and folded by the tectonic foundation of the Indian subcontinent pushing toward the northeast—lines the river with rapids and waterfalls, and also supplies the contrastingly flat lower basin with half of its alluvium. On the basin plain live most of the 70 million people inhabiting the Mekong region, and in the lower course of the river, the largest collection of giant fish species in the world, including the giant Mekong catfish that can reach 9 feet (3 meters) in length. It is in the lower reaches, too, where the famous dugong or river dolphin is found.

Where the Mekong flows out of higher lands, the basin begins to narrow and biomes transition into Tropical Savanna Altitudinal Zone. This upper basin's biomes are altitudinal, meaning that differences in land elevation are prime influences on the biotic and abiotic characteristics. For instance, as elevation increases, average temperatures decrease—and generally so does biodiversity. The difference in elevation between the source of the Mekong in the Tibetan Plateau and the mouth

Rowboats navigating among thick vegetation alongside the Mekong River. The river flows for 2,600 miles (4,200 kilometers) and supports a wide variety of ecosystems that are home to 20,000 species of plants, 1,200 species of birds, 800 species of reptiles and amphibians, and 430 species of mammals. (Thinkstock)

of the river in the delta is approximately 15,000 feet (4,500 meters). At one end is an evergreen rainforest, or, rather, an anthropogenic biome 10 percent of which is covered by old growth forest; in the middle, deciduous and coniferous trees and bushes in the savanna; and at the other end, a rocky desert of ice sheets, alpine glaciers, and orographic (mountain-created) rainfall—the true mother of waters in the Mekong.

Through the rugged mountains of China's Yunnan province, the Mekong, Salween, and Yangtze Rivers etch parallel gorges that UNESCO has designated a natural heritage site. Below, the pristine gorges give way to hydroelectric dams. Since this region contributes only 15 to 20 percent to the Mekong's water volume, the dams' effect on water supply downstream is tolerable, except when the wet monsoon disappoints. Water held up by dams on many of the Mekong's tributaries also makes dry spells worse for those depending on the river for their livelihood. Fluctuating water levels, typical of dammed-up river systems, disrupt traditional trade and transport networks, and the corresponding increase in water temperature is responsible for a recent rapid decrease in aquatic species in the Tonlé Sap and in other internationally recognized eco-preserves.

The mainstream section of the Mekong in southeast Asia remains free of dams. Even so, river traffic tends to be localized, leaving long stretches of the river undisturbed by barges, boats, or shoreline development. Few sturdy roads line the banks and fewer bridges connect them; many backwater ports seem more like simple docks. But this may all change soon as China, India, Japan, and even the United States design a future for the region that will, they promise, expand river-based economies, increase standards of living, and integrate the nations of the region: China, Burma (Myanmar), Laos, Thailand, Cambodia, and Vietnam.

However, for such economic dreams to become reality, far more of the length of the Mekong will have to become navigable and also supply a source of electricity. Blasting away river banks, rapids, and waterfalls to make room for ships, dams, bridges, and ports would only be the beginning of a long process that may recast all of the Mekong as a biome. The Mekong River Commission, founded in 1995 to promote sustainable development and scientific management of the Basin, hopes to prevent the negative aspects of this from happening.

KEN WHALEN

Further Reading

Leinbach, Thomas R. and Richard Ulack. *Southeast Asia: Diversity and Development*. Upper Saddle River, NJ: Prentice Hall, 1999.

Rainboth, Walter. *Fishes of the Cambodian Mekong*. Bangkok, Thailand: Food and Agriculture Organization of the United Nations, 1997.

Stewart, Mart A. and Peter A. Coclanis, eds. *Environmental Change and Agriculture Sustainability in the Mekong Delta*. New York: Springer, 2011.

Weightman, Barbara A. *Dragons and Tigers: A Geography of South, East, and Southeast Asia, 3rd Ed.* Hoboken, NJ: John Wiley & Sons, 2011.

Mesoamerican Barrier Reef

Category: Marine and Oceanic Biomes.
Geographic Location: Central America.
Summary: The Mesoamerican Barrier Reef is the world's second-longest barrier reef and a biodiversity hot spot, yet it is highly threatened by anthropogenic activities in the region.

The Mesoamerican Barrier Reef is the Western Hemisphere's longest barrier reef, and one of the world's biodiversity hot spots, stretching more than 621 miles (1,000 kilometers) from the northern tip of Mexico's Yucatan Peninsula southward along the coasts of Belize, Guatemala, and Honduras; it extends more than 50 miles (80 kilometers) out from shore in places. Sited in the western Caribbean Sea, the Mesoamerican Barrier Reef (MAR) is affected by the Caribbean Current and the Loop Current of the Gulf of Mexico. There is a distinct wet season from July to October in the MAR region.

The MAR ecosystem is highly threatened yet biologically rich, encompassing long barrier reefs, near-shore fringing reefs, offshore atolls, and hundreds of patch reefs. Associated shallow and deep lagoons, mangrove forests, and seagrass beds provide habitats, foraging, and nursery grounds for fish and invertebrates. Seagrasses also help stabilize sediments, reduce beach erosion, and promote water clarity.

Biodiversity

The Mesoamerican Barrier Reef is home to more than 500 species of fish. Of particular interest is the recent discovery of major spawning aggregation sites at Gladden Spit, Half Moon Caye, and Caye Bokel for lutjanid, serranid, and carangid fish—important food fish in the Caribbean. Resident fish spawn close to their home reefs, but others, such as cubera and dog snappers, migrate over large distances to spawn in transient aggregations. This comes in response to cues that include location, reef morphology, season, temperature, sunlight period, and lunar and diurnal cycles.

Spawning aggregations are temporally consistent, and generally occur in shallow waters of about 82 feet (25 meters) along the shelf or in deeper waters of about 98–213 feet (30–65 meters) beyond the shelf break, between March and September, with a peak occurring in May each year. Peak abundances can reach 8,000 individuals, but generally there are about 4,000 to 6,000 individuals from March to July in each event.

Sharks and rays form a vital component of the fauna here. Southern stingray (*Dasyatis americana*), lemon shark (*Negaprion brevirostris*), and nurse shark (*Ginglymostoma cirratum*) frequent the lagoons and inter-reef areas for spawning, feeding, hunting, and shelter in both shallow and deep reaches.

In addition to fish, the reef system is home to 350 species of mollusk and 65 species of stony coral. Manatees, sea turtles, the American crocodile, and Morelet's crocodile are also found in this area. The whale shark, largest fish on the planet, may be found near Isla Contoy. Normally a solitary animal, whale sharks gather in this area to find mates and breed.

Threats and Conservation

Overfishing of fish, lobster, and conch is considered to be a persistent and extensive threat to the MAR, resulting in a reduction of fish sizes, spawning potential, and future catches. Other threats of great concern include unregulated fishing, deforestation in the watersheds, loss of habitat and nursery areas, marine and watershed pollution, coral disease and bleaching, boat groundings, mining and dredging, urban runoff, coastal development, and agriculture.

Global climate change effects such as ocean warming, stronger hurricanes, and ocean acidification are of grave and increasing concern. Extensive coral bleaching occurred in the MAR in 1995, 1998, and 2005, which were particularly warm years. Over time, severe bleaching and disease can lead to a reduction in species diversity, coral cover, and eventually loss of reef framework. High sedimentation rates and nutrient enrichment of coastal waters from fertilizers increase algal and sponge abundance at the expense of corals, and can be indicative of stressful conditions on the reef.

The region has about 894 square miles (2,315 square kilometers) of coral reefs in total, 23 percent average coral cover, 25 percent macro-algae abundance, and some 1,350 square miles (3,500 square kilometers) of mangrove forest; there are no estimations of seagrass cover. Seagrasses have also been severely damaged by dredging operations, prop scars, and poor water quality. The greatest threats to mangrove and seagrass habitat are direct losses associated with coastal development and freshwater flow.

More than 60 marine protected areas (MPAs) have been established in the MAR region. These MPAs often function better in theory than in reality, however. Several protected areas are "paper parks," in that they apparently fail to meet their management objectives.

The Mexican MPAs include Santuario del Manatí (including Chetumal Bay), Arrecifes de Xcalac Reserve, and Banco Chinchorro Biosphere

Reserve. Priority monitoring sites for these Mexican sites are Isla Contoy, Cancún, Puerto Morelos, Cozumel, Sian Ka'an, Akumal, and Majahual.

The MPAs in Belize are Glover's Reef Marine Reserve, South Water Caye Marine Reserve, Sapodilla Cayes Marine Reserve, Port Honduras-Deep River Forest Reserve, Sarstoon-Temash National Park, Gladden Spit, Bacalar Chico Marine Reserve and National Park, and Corazol Bay Wildlife Sanctuary. The priority monitoring sites include Hol Chan, Caye Caulker, Lighthouse Reef, Turneffe Atoll, and Laughing Bird Caye.

There are two protected areas within the MAR region in Guatemala: Punta de Manabique Wildlife Refuge and Sarstún National Park. Río Dulce and Santo Tomás Bay are considered to be priority monitoring sites.

In Honduras, the MPAs are Omoa-Baracoa Marine Reserve, and Turtle Harbor Wildlife Refuge and Marine Reserve. Priority monitoring sites include Cayos Cochinos, Roatán, Guanaja, Río Aguán, Río Plátano Biosphere, Laguna de Caratasca, Puerto Cortés, Tela, and La Ceiba.

Lucia M. Gutierrez

Further Reading

Coral Reef Alliance. "Mesoamerican Reef Alliance." http://www.coral.org/where_we_work/caribbean/mar.

Pikitch, Ellen K., et al. "Habitat Use and Demographic Population Structure of Elasmobranchs at a Caribbean Atoll (Glover's Reef, Belize)." *Marine Ecology Progress Series* 302, no. 1 (2005).

Spalding, Mark D., Corinna Ravilious, and Edmund P. Green. *World Atlas of Coral Reefs*. Berkeley: University of California Press, 2001.

Mexican Wetlands, Central

Category: Inland Aquatic Biomes.
Geographic Location: Central America.

Summary: Home to an exceptional number of endemic species and an important refuge for overwintering migratory birds, these wetlands are threatened by water extraction, buildup of sediment and salt, contamination, and invasive species.

Ancient volcanic activity in the center of Mexico fragmented the watershed of the Lerma and Santiago Rivers, resulting in the formation of many isolated lakes and wetlands. This isolated nature gave rise to an exceptionally high number of species that are unique to the ecoregion. These wetlands are also an important habitat for resident and migratory birds. The ecoregion is threatened by overextraction of water, pollution, and invasive species. Several sites within the Central Mexican wetlands have been designated as protected areas, and efforts are under way to conserve this important ecoregion.

The Central Mexican wetlands are associated with the many lakes in the hydrological basin of the Lerma and Santiago Rivers. This basin includes portions of the Central Mexican states of Guanajuato, Hidalgo, Jalisco, Michoacán, Morelos, Puebla, Queretaro, Tlaxcala, Veracruz, and the Federal District. It is located in the Trans-Mexican Volcanic Belt, a mountainous region that crosses the center of Mexico. Geological and biological evidence indicates that the basin was originally an open, interconnected hydrologic system, but volcanic activity in the Pleistocene fragmented the basin, resulting in the formation of many isolated, endorheic lakes (lakes that do not drain into the ocean).

The majority of Mexico's natural lakes are found in this area, including Chapala and Cuitzeo, the country's two largest lakes. Lake Chapala is also the third-largest lake in Latin America, and the second-highest in the Americas.

Biota

The vegetation of the wetlands is characterized by cattails (*Typha angustifolia, T. dominguensis,* and *T. latifolia*), bulrushes (*Scirpus* spp.), arrowheads (*Sagitaria* spp.), black flatsedge (*Cyperus niger*), and willow (*Salix bonplandiana*), as well as floating plants such as water lilies (*Nymphoides* spp.), pondweed (*Potamogeton* spp.), duckweeds, bogmat (*Wolffiella lingulata*), and water milfoil (*Myriophyllum* spp.). Mexico's national tree, the Montezuma cypress (*Taxodium mucronatum*), also grows along rivers and in marshes in the Central Mexican wetlands biome. Endemic (not found elsewhere) plants associated with the area include *Arenaria bourgaei*, *Panicum sucosum*, and *Sagittaria macrophylla*.

Approximately 100 species of fish are native to this ecoregion, 70 percent of which are endemic. Many of them depend on the wetlands at one or more life stages. Fish families with species and whole genera that are endemic to the Central Mexican Wetlands biome include the neotropical silversides (*Atherinopsidae*), sucker fish (*Castotomidae*), chubs and shiners (*Cyprinidae*), splitfins (*Goodeidae*), catfish (*Istaluridae*), and lampreys (*Petromyzontidae*). The wetlands are also home to endemic or near-endemic species of mollusks, crustaceans, and amphibians. The endemic frogs *Rana megapoda*, *R. montezumae*, and *R. neovolcanica* and the salamander *Ambystoma dumerilii* are endangered species.

The Central Mexican wetlands are an extremely important habitat for some 200 species of resident and migratory birds. During the winter months, hundreds of thousands of migratory waterfowl seek refuge, feed, hibernate, and breed here. The endemic black-polled yellowthroat (*Geothlypis speciosa*) is a globally endangered species. Two other endemic bird species, the yellow rail (*Coturnicops noveboracensis*) and slender-billed grackle (*Quiscalus palustris*), have not been observed recently; they are likely extinct. Approximately 40 species of mammals, including threatened ones such as the Mexican long-tongued bat (*Choeronycteris mexicana*), can be found in the wetlands.

Threats and Conservation

The areas surrounding the Central Mexican Wetlands biome support large populations of people, including some major urban areas. The wetlands provide critical ecosystem services, such as water treatment and aquifer replenishment. Many people depend directly on the wetlands for their livelihoods through fishing, aquaculture, and the manufacture of crafts.

Sites of archeological and cultural significance are also associated with the wetlands. The area surrounding Lake Patzcuaro is home to the

Fishermen on Lake Patzcuaro, which is home to the P'urhépecha indigenous group. The isolated lakes of the Central Mexican Wetlands biome have high numbers of endemic species, including such fish species as the neotropical silversides (Atherinopsidae), sucker fish (Castotomidae), chubs and shiners (Cyprinidae), splitfins (Goodeidae), catfish (Istaluridae), and lampreys (Petromyzontidae). (Wikimedia/Thelmadatter)

P'urhépecha indigenous group; each November, Day of the Dead celebrations are held on the shores and islands of Lake Patzcuaro, attracting tourists from around the world.

The wetlands have been highly affected by human activity. The water that feeds the wetlands is used intensively to support cities and for irrigation, power generation, and transportation. Due to diversion and overextraction of water, much of the marshes and bogs are now dry for much of the year.

Soil erosion associated with severe deforestation, agriculture, and other land clearing activity has caused sediment and salts to build up in water. Water in the wetlands is also contaminated with agricultural chemicals, pollutants from industrial areas, untreated wastewater, and solid wastes. Many native species are at risk of overexploitation. Some invasive species are displacing native species, such as the whitefish *Chirostoma estor*.

Climate change impacts in this area include higher average temperatures, which will tend to increase the evaporation rate of water in the wetlands. Increases in the frequency and severity of tropical storms are also predicted; this brings with it the threat of constant erosion problems and increased siltation. Habitat stresses will be the result.

Parts of the Central Mexican wetlands biome are protected, and some conservation activities are under way. Lago de Camécuaro is a Mexican national park; the Lerma marshes are protected at the national level as a nature reserve for flora and fauna, and internationally under the Ramsar Convention on wetlands. Other locations that are targeted under the Ramsar Convention include Lake Chapala, the Zapotlán and Sayula lagoons, and the wetlands of Lake Patzcuaro. There are plans to recover populations of some native species and to improve water management.

Melanie Bateman

Further Reading

Barbour, C. D. "A Biogeographical History of Chirostoma (Pisces: Atherinidae): A Species Flock from the Mexican Plateau." *Copeia* 3, no. 1 (1973).

Mitsch, William J. and James G. Gosselink. *Wetlands.* Hoboken, NJ: John Wiley & Sons, 2007.

Olson, D., E. Dinerstein, P. Canevari, I. Davidson, G. Castro, V. Morisset, et al., eds. *Freshwater Biodiversity of Latin America and the Caribbean: A Conservation Assessment.* Washington, DC: Biodiversity Support Program, 1998.

Mexico, Gulf of

Category: Marine and Oceanic Biomes.
Geographic Location: North and Central America.
Summary: The Gulf of Mexico contains great biodiversity, from shallow coral reefs to mysterious deeps to more than half of all the coastal wetlands of the United States. Rich in biodiversity, these waters are vulnerable to intensive human activity.

The Gulf of Mexico is an arm of the Atlantic Ocean that extends for about 700,000 square miles (1.8 million square kilometers), forming the marine border between the extreme southeastern section of the United States and eastern Mexico, and between the United States and northern Cuba. The gulf flows for more than 1,100 miles (1,770 kilometers) from west to east and for some 800 miles (1,287 kilometers) from north to south. The Gulf of Mexico flows into the North Atlantic Ocean past northern Cuba, by way of the Straits of Florida. Along southern Cuba, the Gulf of Mexico merges with the Caribbean Sea via the Yucatán Channel.

As warm water from the Caribbean flows into the gulf, it creates a loop current along the coasts. In the Florida Straits, that loop current turns into the Florida Current as it exits and joins with the Gulf Stream. The largest lobes of the Gulf of Mexico are the Bay of Campeche in Mexico, and Apalachee Bay in Florida. At 12,714 feet (3,875 meters), the deepest part of the gulf is Sigsbee Deep, 875 miles (1,408 kilometers) off the coast of Mexico.

The largest river, by far, that empties into the gulf is the Mississippi. Other major flows come from the Alabama, Brazos, and Rio Grande Rivers. The chief ports of the gulf are Tampa and Pensacola in Florida, Mobile in Alabama, New Orleans in Louisiana, Galveston and Corpus Christi in Texas, Tampico and Veracruz in Mexico, and Havana in Cuba. In the United States, the Intracoastal Waterway runs along the northern coast of the gulf. Along the shores of the Gulf of Mexico, the land tends to be low, sandy, and marshy.

More than half of all coastal wetlands in the United States are located in the Gulf of Mexico. Louisiana alone is home to 40 percent of all American coastal wetlands. According to a report issued by the National Oceanic and Atmospheric Administration (NOAA) in 2010, the loss of coastal wetlands in the United States has amounted to 25 square miles (65 square kilometers) each year for the past five decades. More than 1 million square miles (2.6 million square kilometers) have been lost over the past century. A total 90 percent of all U.S. coastal wetlands loss occurs in Louisiana.

Biodiversity

The complex ecosystem of the Gulf of Mexico includes open water columns, floating sargassum mats, deep-sea soft coral, hard coral reefs, sandy bottom, muddy bottom, marshes, submerged aquatic vegetation, bays, lagoons, and sandy beaches where sea turtles lay their eggs. Shrimp is the major commercial fishery in the gulf; the diverse aquatic habitats here support brown shrimp, white shrimp, pink shrimp, royal red shrimp, seabobs, and rock shrimp (although neither seabobs nor rock shrimp are much prized by commercial fishers).

Whale species include Bryde whales and sperm whales, an endangered species. Other cetaceans that frequent the gulf are the dwarf sperm whale, pygmy sperm whale, Blainville's beaked whale, Gervais's beaked whale, short-finned pilot whale, killer whale, false killer whale, and pygmy killer whale. Bottlenose dolphin, Atlantic spotted dolphin, Risso's dolphin, rough-toothed dolphin, Fraser's dolphin, pantropical spotted dolphin, striped dolphin, Clymene dolphin, and spinner dolphin are either permanent or seasonal residents here. All these species are protected under the Marine Mammal Protection Act. Five species of sea turtles also live in the gulf: the endangered Kemp's Ridley turtle, and the leatherback, loggerhead, green, and hawksbill. The olive Ridley, a threatened species, can sometimes, but only rarely, be found in the gulf.

Disasters: Natural and Human-Made

The most recent major impact event threatening Gulf of Mexico habitats and species was the April 2010 BP Deepwater Horizon oil blowout off the coast of Louisiana. By November 2010, the U.S. Fish and Wildlife Service had issued a report on damage, declaring that 6,000 birds, 600 sea turtles, and 100 mammals had died within the spill area. Experts believed that the actual count was much higher because so many dead animals had not been found. Researchers also located dozens of dead and dying coral deposits. An ecologist from the Lawrence Berkeley National Laboratory discovered a new species of bacteria that guzzled oil and was able to survive up to 3,600 feet (1,097 meters) below the surface. Great concern over the effects of the dispersant Corexit on the biota—BP pumped massive quantities of the toxic brew into Gulf waters—remains an open question.

Assessing damage to marine ecosystems in light of the disaster, the International Union for Conservation of Nature (IUCN) determined that there were 15,419 recorded species of marine life in the Gulf of Mexico, 10 percent of which were endemic, or found nowhere else on Earth. The IUCN Red List identified 53 threatened and 29 near-threatened species, including five of 28 mammals, all five species of sea turtles, 43 of 131 species of sharks and rays, more than half of the 22 species of groupers, and three of 40 species of seabirds. Also in jeopardy: two of nine seagrasses, and 11 of 60 reef-building corals.

Starting in the 1980s, a large dead zone—with insufficient oxygen to support marine life—was discovered some 60 miles (97 kilometers) off the coast of Louisiana and Texas. By 2011, scientists said it was approaching 9,500 square miles (25,000 square kilometers) in size. It is thought to be the result of nitrate- and phosphate-rich agricultural runoff, particularly from fertilizers used far up the Mississippi River basin in the Corn Belt.

There is evidence that some species of fish, such as the female croaker, are already showing physical abnormalities because of oxygen depletion. Conservationists cite the 1990s dead-zone reversal in the Black Sea as an example that could be a goal for here.

Anticipated climate change impacts in the Gulf of Mexico include rising sea levels and increased storm activity. Coral reefs are at risk from rising seawater temperature, which tends to drive away their symbiotic caretaker species, leading to bleaching events. Excess greenhouse gas absorption into the seas is also linked to lower reproductivity among crustaceans and mollusks; putting such pressure on native species in the Gulf of Mexico allows faster and broader penetration by invasive species, further destabilizing habitats and food webs.

ELIZABETH RHOLETTER PURDY

Further Reading

Blomberg, Lindsey E. "Dead in the Gulf." *The Environmental Magazine* 22, no. 5 (2011).

Dale, Virginia H., et al. *Hypoxia in the Northern Gulf of Mexico.* New York: Springer, 2010.

Felder, Darryl L. and David K. Camp, eds. *Gulf of Mexico—Origin, Waters, and Biota: Vol. 1, Biodiversity.* College Station: Texas A&M University Press, 2009.

Freudenburg, William R. and Robert Gramling. *Blowout in the Gulf: The BP Oil Spill Disaster and the Future of Energy in America.* Cambridge, MA: MIT Press, 2011.

Gore, Robert H. *The Gulf of Mexico: A Treasury of Resources in the American Mediterranean.* Sarasota, FL: Pineapple Press, 1992.

Tunnell, John W. Jr., et al., eds. *Coral Reefs of the Southern Gulf of Mexico.* College Station: Texas A&M University Press, 2007.

Michigan, Lake

Category: Inland Aquatic Biomes.
Geographic Location: North America.

Summary: The third-largest of the Great Lakes, Lake Michigan is abundant with aquatic life and shipping history, but is damaged from pollution, overfishing, and invasive species.

Lake Michigan is the only one of the five Great Lakes to lie entirely within the borders of the United States. Surrounded by rocky shores, sandy dunes, and beaches in Wisconsin and Illinois to the west, Indiana to the south, and Michigan to the east and north, Lake Michigan is the third largest of the Great Lakes. It is contiguous with Lake Huron via the Straits of Mackinac; it also has been connected to the Atlantic Ocean and the Mississippi River via canals. Once abundant with diverse aquatic life, Lake Michigan has been subjected to problems of overfishing, industrial and agricultural pollution, and invasive species.

It is the fifth-largest lake in the world and third largest of the Great Lakes by surface area at 22,400 square miles (58,000 square kilometers)—and the second largest of the Great Lakes in volume, at 1,180 cubic miles (4,918 cubic kilometers). It is 307 miles (494 kilometers) long and averages 75 miles (120 kilometers) across; it has a 1,640-mile (2,640-kilometer) shoreline. The lake's average and maximum depths are 279 feet (85 meters) and 923 feet (281 meters), respectively. Lake Michigan's mean surface level is 577 feet (176 meters) above sea level, the same as Lake Huron. Lake water levels are highest in autumn and lowest in winter. Highest and lowest recorded levels occurred in summer 1986 and winter 1964, respectively.

Hydrologically, Lakes Michigan and Huron are the same body of water (sometimes called Lake Michigan-Huron), but are geographically distinct. Counted together, it is the largest body of fresh water in the world by surface area. The Mackinac Bridge is generally considered the dividing line between them. Both lakes are part of the Great Lakes Waterway connecting the westernmost port of Duluth, on Lake Superior, to the Atlantic Ocean via the St. Lawrence Seaway and canal system.

Geology and Formation

The Great Lakes formed when the glaciers of the Laurentian, or Wisconsin, glaciation retreated north 12,000 to 7,000 years ago. Along with carving out the lake basins, the glaciers depressed the land with their tremendous weight. Meltwater initially filled Lake Michigan, which was originally much larger.

Along the shores of northwestern Michigan and the southern Upper Peninsula of Michigan, one can find unique Petoskey stones. These stones are fossilized coral from a reef that grew during the Devonian Period, around 350 million years ago. When wet or polished, they show the surface pattern of six-sided corals. They were moved and deposited during the last glaciation.

During the Silurian Period, over 400 million years ago, Wisconsin was located near the equator and the land was submerged under a massive inland sea. Over time, the sea disappeared but left behind some of the most extensive reefs in the world today. Several of those reefs are in evidence in Wisconsin today, and one runs through Lake Michigan, dividing the lake into a northern and a southern section. This reef occupies the lake bottom between Milwaukee and Racine on the Wisconsin side, and Grand Haven and Muskegon on the Michigan side. Because of winds, river water, and the coriolis effect, water both north and south of the reef flows in a clockwise direction, bringing moderating warmer surface water to Michigan, which experiences summer temperatures 5 to 10 degrees F (2 to 5 degrees C) higher than the Wisconsin side.

The beaches of the western coast of the lake and the northernmost part of the east coast are rocky, while the southern and eastern beaches are sandy and dune-covered. This is partly because of the prevailing winds from the west, which also cause thick layers of ice to build on the eastern shore in winter.

Along the southern and southeastern shorelines lies the largest freshwater dune system in the world. Large sand dunes rise hundreds of feet above the surface of the lake; the largest, Sleeping Bear Dune, is more than 400 feet (122 meters) high. These dunes are comprised of soft, off-white quartzite sand, sometimes called "singing sands" for the squeaky noise made when one walks through it. The dunes there and at other points

Invasive Species

Along with overfishing, the introduction of exotic species, which found their way to the lake via human-made canals and waterways, has severely impacted Lake Michigan fish populations. In particular, the invasive sea lamprey (*Petromyzon marinus*), smelt, and alewife have affected native species.

Sea lamprey are native to the Atlantic coast, Lake Champlain in Vermont, and the Finger Lakes region in New York. They entered Lake Michigan via the Welland Canal and Lake Huron. These parasites latch onto the bodies of other fish and rasp away at their flesh with sharp teeth and tongues. Victims often have multiple wounds and frequently die from resulting fluid loss and infection.

Alewives also invaded Lake Michigan via the Welland Canal. These plankton-eating fish compete with many other small- and medium-bodied fish in the lake. Around the time of their arrival in the 1950s, their primary predator, lake trout, were in steep decline due to overfishing and mortality from sea lamprey. This low predation pressure enabled the alewife population to increase rapidly.

Alewives have particular habitat needs and prefer the oxygen-rich, colder water of the lake depths. When waters warm in the summer, or during spawning times, alewife populations can go through huge die-offs, with millions of fish washing up on beaches. To control their populations, salmon were introduced to Lake Michigan.

Many other nonnative species have become established in Lake Michigan, such as smelt, round goby, and zebra mussels. Zebra mussels compete with fish for food and form thick colonies in the sediments, drainpipes, and other substrates. Control has proven to be difficult and costly.

One of the most recent invasive species to threaten Lake Michigan is the Asian carp. This aggressive species has migrated up the Mississippi River; the U.S. Army Corps of Engineers has barricades in place near Chicago in the effort to block Asian carp from reaching Lake Michigan.

around the lake can be visited at a variety of state and national parks.

Aquatic Life

Lake Michigan is home to a variety of species of fish and other organisms. The food web is based on the primary producers, the phytoplantkton, mainly green algae, blue-green algae, diatoms, and flagellates. These are consumed by a variety of filter-feeding animals, including zooplankton, shrimp, mollusks, and clams. These in turn are consumed by the smaller foraging fish, such as sculpin (various species) and perch (*Perca flavescens*), which are consumed by the predatory walleye (*Sander vitreus*), trout (*Salvelinus namaycush*), bass (*Micropterus* species), and salmon (*Onchorynchus* species).

Approximately 150 species of fish were once native to Lake Michigan. There are currently 134 species, including 17 non-natives, in Lake Michigan. At least eight native species have become extinct from the lake.

Native American populations thrived on Lake Michigan fish communities, but when European settlers imported industrial fishing techniques, populations began to decline. Fishing catches peaked in the late 1880s, with massive amounts of lake trout and whitefish caught. The most dramatic example of this is the lake whitefish (*Coregonus clupeaformis*), of which nearly 6,000 pounds (2,722 kilograms) were caught in 1947. By 1956, less than 60 pounds (27 kilograms) were caught.

Fisheries

Historically important commercial fish species include lake trout (*Salvelinus namaycush*), lake whitefish (*Coregonus clupeaformis*), walleye (*Sander vitreus*), white sucker (*Catostomus commersonii*), yellow perch (*Perca flavescens*), lake herring (*Coregonus artedi*), coho salmon (*Oncorhynchus kisutch*), chubs, and alewife (*Alosa pseudoharengus*).

One of the most productive fish to be caught in Lake Michigan was the lake trout. The commercial catch in 1878 was about 2,659 pounds (1,206 kilograms); it peaked in 1896 at just over 9,000 pounds (4,082 kilograms), and plummeted to 54 pounds

(24 kilograms) in 1950. In addition to overfishing, predation by the invasive sea lamprey contributed to the decline. In 1962, the fishery for lake trout was closed in Wisconsin, Illinois, and Indiana. Catches began increasing again in the late 1970s when sea lamprey control began; the most productive year in recent history was 1999 with just over 900 pounds (408 kilograms).

Lake whitefish catches were also highest in the early days of the fishery, with over 12,000 pounds (5,443 kilograms) caught in 1879. Whitefish catches reached their nadir in 1920, fluctuating after this but never regaining the historic levels. The Wisconsin Department of Natural Resources stocks about 13 million salmon and trout in Lake Michigan each year, for both the commercial and recreational fisheries.

Human Factor

Lake Michigan water quality has diminished due to multiple contaminants, including human waste water and industrial discharge. Some of the most serious pollutants include mercury, polychlorinated biphenyls (PCBs), and the chemicals and bacteria in human waste. Green Bay, along the northwestern shore of the lake, contains one of the highest concentrations of paper mills in the world; these are a major source of PCBs and other pollutants. In 1993, an outbreak of the bacterium *Cryptosporidium* sickened over 400,000 people via the Milwaukee drinking water system, among the largest public drinking water contaminations in U.S. history.

The climate of the northern part of the lake is colder; the coastlines are generally less developed. It is sparsely populated except for the Fox River Valley, which drains into Green Bay. The more temperate southern basin of Lake Michigan is among the most urbanized areas in the Great Lakes system and contains the Milwaukee and Chicago metropolitan areas. This region is home to about 8 million people.

Lake Michigan is connected to other waterways by way of several canal systems. The Saint Lawrence Seaway and Great Lakes Waterway opened Lake Michigan to ocean-going vessels. The Great Lakes are also connected by canal to the Gulf of

Mexico by way of the Illinois River (via Chicago) and then the Mississippi River.

Historical Notes

The name "Michigan" is derived from the Native American Ojibwa name, *Michi-gama,* which means "great water." The French explorer Samuel de Champlain called it the Grand Lac. It was later named Lake of the Stinking Water and also Lake of the Puans, after the Winnebago native people who occupied its shores. In 1679, the lake became known as Lac des Illinois because it gave access to the country of the Native Americans bearing that name. Three years before, the missionary Allouez called it Lac St. Joseph, by which it was often designated by early writers; others called it Lac Dauphin. Through the further explorations of Jolliet and Marquette, the lake received its current name of Michigan.

Before 800 c.e., Lake Michigan's shores were inhabited by the so-called Hopewell Indians. After their decline, the region was home for several hundred years to the Late Woodland Indians. The area is still inhabited by their descendants, including the Chippewa, Menominee, Sauk, Fox, Winnebago, Miami, Ottawa, and Potawatomi.

The first person to reach the deep bottom of Lake Michigan was J. Val Klump, a scientist at the University of Wisconsin–Milwaukee. Klump reached the bottom via submersible as part of a 1985 research expedition.

SUSAN MOEGENBURG

Further Reading

Ashworth, William. *Great Lakes Journey: A New Look at America's Freshwater Coast.* Detroit, MI: Wayne State University Press, 2000.

Brammeier, Joel, Irwin Polls, and Scudder Mackey. *Preliminary Feasibility of Ecological Separation of the Mississippi River and the Great Lakes to Prevent the Transfer of Aquatic Invasive Species.* Ann Arbor, MI: Great Lakes Fishery Commission, 2009.

Changnon, Stanley A. and Joyce M. Changnon. "History of the Chicago Diversion and Future Implications." *Journal of Great Lakes Research* 22, no. 1 (1996).

Farid, Claire, John Jackson, and Karen Clark. *Fate of the Great Lakes: Sustaining or Draining the Sweetwater Seas.* Darby, PA: Diane Publishing, 1997.

National Wildlife Federation. "Asian Carp Threat to Great Lakes." http://www.nwf.org/Wildlife/Wildlife-Conservation/Threats-to-Wildlife/Invasive-Species/Asian-Carp.aspx.

Noone, Michael. *Interbasin Water Transfer Projects in North America.* Bismark: North Dakota State Water Commission, 2006.

Middle East Steppe

Category: Grassland, Tundra, and Human Biomes.
Geographic Location: Middle East.
Summary: A study in contrasts that includes both arid, desert-like zones and wetlands, this biome hosts animal and plant life that is under threat from human development and from naturally occurring drought.

The arid, grass-covered plains that make up steppes are found on every continent on the planet except Australia and Antarctica. Steppes are generally found close to mountains and away from oceans. The Middle East Steppe biome encompasses the Tigris and Euphrates Rivers, which flow from Turkey southeastward into the Persian Gulf.

While steppes are not always hospitable to humans because of poor soils that make it difficult to grow crops, the Middle East steppe includes the Fertile Crescent, that greener area of western Asia that is surrounded by arid and semiarid land. Shrubs and grasses that are typical in steppes are found here, but the area is also home to riverine woodlands and extensive wetlands.

Like most steppes, the Middle East steppe experiences hot summer temperatures and cold winters. Average temperatures within the region range from January lows of 44.6 degrees F (7 degrees C) to July highs of 80.6 degrees F (27 degrees C). Violent winds are common on steppes, and the Middle East steppe is no exception. It experiences the hot, dusty *khamsin* wind that originates during summer in the eastern and southeastern regions. When the *khamsin* is blowing, temperatures frequently rise above 100 degrees F (38 degrees C). Less than 10 inches (254 millimeters) of rain falls here in a given year, and it can be even drier in the eastern and southern sections that are close to the Syrian Desert.

Within the Middle East steppe, the area known as the Fertile Crescent stretches from the Mediterranean Sea on the west to the Euphrates and Tigris River valleys on the east. It encompasses all or parts of Israel, the West Bank, Jordan, Lebanon, Syria, and Iraq. This area has a centuries-long bloody human history that has often made it impossible for nations within the area to agree on responsible use of its resources, whether water, oil, or the biota.

Vegetation and Animals

The open shrub of the Middle East steppe begins in western Jordan and reaches into southwestern and central Syria before crossing into northern Iraq, where it arrives at the Euphrates and Tigris rivers. From there, the steppe stretches out toward the east and the south, toward the foothills of the Zagros Mountains, which spread across northern Iran. Boulder fields of black basalt are ubiquitous in the southeastern portion of the steppe.

Within the valley surrounding the two rivers, the soils are alluvial-colluvial. Most of the vegetation within the Middle East steppe is herbaceous, and features dwarf shrub sage (*Artemesia herba-alba*) and grasses (*Poa bulbosa*). Soils may be stony outside riverine areas. Seasonal lakes and marshes, which may be either freshwater or saline, are permanently scattered throughout the steppe. Tamarix and poplars like *Populus euphratica* may be found near water sources. Many reed species grow in the wetlands areas.

In addition to the domestic animals that live in the more populated areas of the steppe, animal life tends to consist of grazing animals and predators that are able to survive the conditions in isolated areas. Unpopulated areas of the Middle East steppe are home to wolves, Ruppell sand foxes,

caracals, jungle cats, and wildcats. Badgers are common in more vegetated areas, and wild boars are found in reed thickets and in semidesert areas.

Because of the variety of land area encompassed in the Middle East steppe, bird life is varied. The semidesert areas are home to the houbara bustard, the great bustard, and the little bustard. Both the great bustard and the lesser kestrel have been listed as vulnerable. Other birds include the Eurasian griffon vulture and the lanner falcon.

The availability of wetlands that dot the river valley areas between southern and central Turkey and Iraq means that the Middle East steppe is a significant migration route for scores of waterbird types, such as the greater flamingo, the pygmy cormorant, and the marbled teal, and for terrestrial birds such as the turtledove. The pin-tailed sandgrouse, which lives on the nearby plains, also frequents the river waters.

Effects of Human Activity

There is still great biodiversity in the Middle East steppe, even though numbers of some species are declining. Efforts are being made to protect vulnerable species such as the Arabian goitered gazelle, the marbled teal, and the lesser kestrel.

While much of the steppe is not conducive to agriculture, arid areas are experiencing a backlash of problems associated with farming in the Fertile Crescent. Some areas of the steppe have been converted to farmland, and water from the riverine areas is used for irrigation. Chemical fertilizers used to grow cotton, wheat, barley, rice, olives, millet, sugar beets, and tobacco for export are polluting the waters.

Some of the lakes within the steppe have been turned into fish ponds. The construction of the Asad Dam on the Euphrates River has led to a decrease in steppe area, and officials plan to build more dams in the future. Farmers burn reed beds so that they can be regrown to feed livestock, and domestic animals are overgrazing the steppe. Hunting and taxidermy are depleting waterfowl and game birds. Woods in riverine areas are being used for firewood, and industrialization in the area increases pollution and places additional pressure on the environment.

In the early 21st century, the most serious environmental problems are persistent drought, the proliferation of dams, and ubiquitous oil pipelines. A 2007 study by Japanese and Israeli meteorologists predicted that the recent and current drought, which has been exacerbated by climate change, may become permanent. Despite evidence that the area cannot handle indiscriminate damming of its waters, new dams continue to be built. Tensions among the countries, particularly between Turkey and Iraq, and between Syria and all its neighbors, are becoming more strained—with water the leading flashpoint issue—making it highly unlikely that regional cooperation can be expected to provide a solution to ongoing problems.

ELIZABETH RHOLETTER PURDY

Further Reading

Bretschneider, Joachim. "Life and Death in Nabada." *Scientific American* 15, no. 1 (2005).

Hillstrom, Kevin. *Africa and the Middle East: A Continental Overview of Environmental Issues.* Santa Barbara, CA: ABC-CLIO, 2003.

Pearce, Fred. "Fertile Crescent Will Disappear This Century." *New Scientist* 203, no. 2719 (2009).

Schroeder, Bruce. "Shelter or Hunting Camp? Accounting for the Presence of a Deeply Stratified Cave Site in the Syrian Steppe." *Near Eastern Archaeology* 69, no. 2 (2006).

Stattersfield, Alison J., et al. *Endemic Bird Areas of the World: Priorities for Diversity.* Cambridge, UK: BirdLife International, 1998.

Mid-Pacific Gyre

Category: Marine and Oceanic Biomes.
Geographic Location: Pacific Ocean.
Summary: The Mid-Pacific Gyre contains a vast floating mass of plastic and other human debris that creates a major environmental hazard for seabirds, sea turtles, and sea mammals—and potentially for humans through the oceanic food chain.

One of the five vast oceanic gyres, or roughly circular current systems, the structure of the Mid-Pacific Gyre is formed by four major currents. Together, they rotate through the entire North Pacific Ocean in a generally clockwise motion. The currents—North Equatorial, Kuroshio, North Pacific, and California—stretch for thousands of miles, influenced by and influencing air and water temperature, humidity, air pressure, and the biota that inhabit the Pacific, its islands, and continental shorelines.

In more detail: The California Current moves toward Central America to meet the North Equatorial Current on the Mexican Coast before heading to Asia as the North Equatorial Current. From these equatorial waters and along the coast of Japan, the system picks up warmth as it becomes the Kuroshio Current. This segment flows northeastward to confront the cold subarctic Oyashio

Thousands of pieces of plastic washed up along a beach in Hawaii in August 2008. The nearby North Pacific gyre helps bring this debris to Hawaii. (NOAA/Eric Johnson)

current; their merger fuels the warm North Pacific Current, which flows across to its encounter with North America, where it splits into the northbound Alaska Current and the California Current, completing the gyre.

With few major island chains or continental masses, and with generally low winds, the surface water of the gyre tends to push floating matter toward its low-energy center. In recent decades, this trend has precipitated a phenomenon that has been variously dubbed the Trash Vortex, the Great Pacific Garbage Patch, the Eastern Garbage Patch, and other monikers.

This vast extent of mostly human-generated debris is composed of up to 90 percent polymer or plastic materials. Generally the mundane products of consumer society, the leading recorded items include plastic bottles, cups, bags, and packing materials. Some estimates are that up to 10 million tons of plastic materials enter the world's oceans each year; perhaps one-quarter to one-third of this flotsam finds its way ultimately to this vortex. While much of the material degrades into smaller particles, virtually none of it is naturally recyclable into the environment. Instead of decomposing, then, it remains a threat for marine organisms of many sizes and scales.

Although this diffuse Pacific Ocean garbage patch was first observed in the 1970s, it came to wide public attention in 1997 when Captain Charles Moore, founder of the Algalita Marine Research Foundation of Long Beach, California, observed the mass and began to take measurements. In many samples scooped out of the ocean, Moore discovered that there were six times more plastic in the water than plankton.

Aquatic Danger

The garbage gyre presents a dire threat—beyond the disturbing appearance of floating trash such as fishnets, clothing, bottles, plastic bags, and cigarette lighters. Believing that the larger items are prey or plants, many seabirds and sea turtles eat the plastic—which in many cases has absorbed toxins during its manufacture, its intended use, and/or after disposal. Digestive processes can release many such toxins. The chips of plastic also can cause digestive, pulmonary, and/or respiratory blockage that eventually kills the creature that ingests it.

In this way, dead fish, birds, and marine mammals also become caught up in the vortex. Scientists have estimated that around the world, more than 1 million seabirds and in excess of 100,000 mammals and sea turtles die each year from ingesting plastic that floats out to sea. Jellyfish, too, may eat the poisoned debris; they in turn are eaten by fish. When fish containing the concentrated and released toxins are caught by commercial fishers, those fish can then taint the human food chain.

In the summer of 2010, a team of scientists from Woods Hole, Massachusetts, recovered 20,000 bits of floating plastic per square mile (2.6 square kilometers) from the Great Garbage Patch, giving some idea of the scope and intensity of the problem.

While plastics and polymers are the chief component of the Great Garbage Patch, this is by no means the only problem. Commercial fishers who use 20-mile (32-kilometer) castaway fishnets snare everything in their nets from dolphins to sea turtles, which after being discarded as worthless by-catch, then in turn become debris heading toward the patch.

The problem of floating garbage also exists in the North Atlantic Ocean, to a somewhat lesser extent. Between 1986 and 2008, scientists collected 64,000 tiny particles of plastic from the Atlantic. Another large area known as the Western Garbage Patch lies off the coast of Japan. Other researchers believe that the biggest patch of all may be located in the South Pacific, but no dedicated scientific examination of that area has been undertaken.

In addition to the plastic and debris that end up in various surface garbage patches of the world's oceans, researchers believe up to 70 percent of ocean debris sinks to the bottom, winding up on the ocean floor. No one is sure what havoc that debris may be creating for the ecosystems deep in the oceans.

Mitigation and Prevention

Many scientists believe that the only solution to the issue of floating garbage vortices is to prevent plastic and other debris from ending up in the ocean in the first place. Even if that task were accomplished, it is possible that current damage could persist for thousands of years, because plastic does not decompose.

Raising and activating public awareness will be crucial to get anything meaningful done about this threat. Environmentalist Moore, still active more than 15 years after his first discoveries, has concentrated his efforts on investigating the phenomenon, and informing both the scientific community and the public about the extent of the problem. Moore and his Algalita Foundation team, always seeking to better map the garbage gyre, think that it may be 100 feet (30 meters) deep or more.

Various suggestions for keeping plastic out of the ocean include recycling and reusing plastic, opting for items that use fewer or no plastic materials in either content or packaging, and initiating local cleanup activities on all the beaches, estuaries, lagoons, and marine coasts and waterways of the world. In one such local effort, for example, Mary Crowley, who owns a California-based boat-chartering business, in 2010 launched Project Kaisei to call attention to the Great Garbage Patch by assembling a team of volunteers to begin hauling debris away from the patch one boatload at a time.

Elizabeth Rholetter Purdy

Further Reading

Blomberg, Lindsey E. "The Great Pacific Garbage Patch." *Environmental Magazine* 22, no. 3 (2011).

Coe, James M. *Marine Debris: Sources, Impacts, and Solutions.* New York: Springer, 1997.

Greenpeace. "The Trash Vortex." http://www .greenpeace.org/international/en/campaigns/ oceans/pollution/trash-vortex.

Kostigen, Thomas M. "The World's Largest Dump: The Great Pacific Garbage Patch." *Discover*, July 2008. http://discovermagazine.com/2008/jul/10 -the-worlds-largest-dump.

Perkins, Sid. "Oceans Yield Huge Haul of Plastic." *Science News* 177, no. 7 (2010).

Mingo Swamp

Category: Inland Aquatic Biomes.
Geographic Location: North America.
Summary: In an epic story of environmental ruin and renewal, conservationists have revived the Mingo Swamp ecosystem that was once reduced to a burnt and eroded wasteland.

Lying within an abandoned channel of the Mississippi River, and nestled between the Ozark Escarpment on the west and the limestone bluffs of Crowley's Ridge on the east, the Mingo Swamp sits squarely within the Boot Heel of southeastern Missouri, approximately 150 miles (240 kilometers) south of St. Louis. Alluvial fans formed by tributaries to the old Mississippi River channel act as natural levees here, slowing drainage through the basin that also is interrupted by several small sand ridges. These may be riverine deposits or, alternatively, may have been forced to the surface by earthquakes. Indeed, the region lies within the New Madrid seismic zone, the source of some of the most powerful earthquakes recorded in North America.

The Mingo Swamp was once part of a 25-million-acre (10.1-million-hectare) complex of forested wetlands extending along both sides of the Mississippi River from Illinois to Louisiana. The seasonal flooding of the Mississippi created a shifting mosaic of dynamic habitats that supported a diverse and abundant array of fish and wildlife species. Native Americans used the area seasonally for hunting and fishing, attracted by its abundant wildlife. The area was later included in the 1803 Louisiana Purchase, but remained relatively unpopulated until the 1880s, when lumber companies began intensively harvesting its vast forests of cypress and tupelo. By 1935, the majority of the venerable trees had fallen—mainly for railroad ties and building materials—and the lumber companies were forced to look elsewhere for supplies.

Seeking to reap additional revenues through agricultural production, landowners constructed a system of drainage ditches to divert water from the swamp and into the nearby St. Francis River.

However, periodic flooding of the river and the region's poorly-drained acidic soils, combined with falling land prices during the Great Depression, dashed any remaining hopes of commercially farming the Mingo.

What followed was a free-for-all of unregulated land uses. Timber was cut indiscriminately without regard to ownership, wildlife and waterfowl were rampantly over-harvested, and lands were burned with impunity to maintain open grassland conditions for cattle and hogs. Wildlife populations were decimated. The region's once-abundant deer and beaver populations were locally extirpated; wild turkey nearly so.

Nevertheless, deforested, drained, and over-exploited, the Mingo was designated a National Wildlife Refuge (NWR) in 1944, with a primary mandate to preserve Mississippi Lowland Forest habitats and to provide resting, feeding, nesting, and brooding habitat for waterfowl and other migratory birds traveling the Mississippi Flyway. The western portion of the refuge was designated a Wilderness Area in 1976. Together, the Mingo NWR and Mingo Wilderness Area preserve and provide opportunities to study the unique ecosystems and natural features of the region.

Biodiversity

Nearly 70 percent of the 21,592-acre (8,737-hectare) Mingo NWR is composed of Mississippi bottomland hardwood forest, complemented by marsh and open water habitats (23 percent) and grassy openings (3 percent). Flanked by bluffs, the swamp floods seasonally. Vegetation communities vary along a narrow elevational gradient that corresponds to the degree of inundation. Three upland community types are recognized, as well as four bottomland community types, the most extensive being Oak Hardwood Bottoms that occupy shallowly inundated areas between drainage ditch levees and floodplains.

Bottomland hardwood forests are one of the lowest and wettest types of hardwood forests. They serve as a transition area between the drier upland hardwood forest, and wet floodplains and forested wetlands. The trees located in this area are cherry-bark oak, swamp white oak, swamp chestnut oak, Shumard oak, pin oak, willow oak, overcup oak, shagbark hickory, and water hickory. Also in the swamp forest ecosystem are bald cypress and water tupelo, species that have adapted to the flooded conditions of their environment.

A total of 279 resident and migratory bird species use the Mingo's refuge habitats throughout the year, including tens of thousands of migrating waterfowl. Hundreds of thousands of migrating ducks can be seen at any one time, as can masses of herons, ibises, swans, grebes, geese, gulls, loons, rails, and terns. Whooping crane, an endangered species, finds haven in the Mingo Swamp, as do the bald eagle and peregrine falcon.

A spider lily in the Mingo Swamp National Wildlife Reserve (NWR) in 2008. Today comprising only 21,592 acres (8,737 hectares), the Mingo Swamp was once part of a larger 25 million acre (10.1 million hectare) wetland complex in the heart of North America. (U.S. Fish and Wildlife Service)

Thirty-eight mammal species occur here, including the swamp rabbit, a relative of the eastern cottontail that regularly takes to water to move about and escape predators. Other mammals include deer, bobcat, and beaver.

More than 30 species of amphibians and reptiles can be found, including three species of venomous snakes, such as the cottonmouth, which hibernate in the cracks and crevices of the bluffs surrounding the refuge. While a complete list of fish species has not yet been compiled, at least 46 species are known to occur in the area's ponds and ditches.

Environmental Protection

Preservation and restoration of Mississippi Bottomland Forest remnants is a top priority for conservation planners, and the Mingo NWR is an important part of this strategy. Areas of concern include enhancing the natural productivity of the swamp and its surroundings; improving drainage through the use of water control structures, ditches, and dikes, providing habitat, and producing food for wildlife, including preserving open marsh areas that nurture aquatic invertebrates and high-energy seeds.

Wildfires are managed with prescribed fire where appropriate to restore natural ecological processes and minimize greenhouse gas emissions. Ongoing reforestation using native plants helps increase carbon sequestration, a natural way to ward off climate change.

Various natural populations here—including groups as diverse as mussels, moths, and mushrooms—are monitored as part of local and national surveys. Careful monitoring and management of invasive plants and animals is an ongoing need, as is maintaining environmental controls over contaminants, such as mercury, in order to minimize their potential effects on wildlife and human health.

Through careful management, the majority of plant and animal species historically present in the Mingo Swamp are being restored to the biome, and ecological succession to a complex mosaic of mature forested wetlands is occurring. Ongoing challenges include keeping swamp forces in balance in the face of climate change-

induced alterations in rainfall patterns, temperature, and humidity.

MELANIE L. TRUAN

Further Reading

Conservation Fund. "Restoring a Forest Legacy at Mingo National Wildlife Refuge." http://climate-standards.org/projects/files/mingo_missouri/The_Conservation_Fund_CCBA_Application_Mingo_NWR.pdf.

Heitmeyer, Mickey E., et al. *Water and Habitat Dynamics of the Mingo Swamp in Southeastern Missouri.* Washington, DC: U.S. Fish and Wildlife Service, 1989.

Rundle, W. Dean and Leigh H. Fredrickson. "Managing Seasonally Flooded Impoundments for Migrant Rails and Shorebirds." *Wildlife Society Bulletin* 9, no. 2 (1981).

U.S. Fish & Wildlife Service. "Mingo, Pilot Knob, and Ozark Cavefish National Wildlife Refuges Comprehensive Conservation Plan." http://www.fws.gov/midwest/planning/mingo.

Miombo Woodlands

Category: Forest Biomes.
Geographic Location: Tropical Africa.
Summary: Africa's most common vegetation type is experiencing the effects of climate change.

The most widely distributed vegetation type in Africa is the savanna, which is also the most extensive vegetation stratum in Africa south of the Equator. The most common of these savannas is the miombo woodlands, which predominantly comprise closely related genera from the legume family (*Fabaceae*, subfamily *Caesalpinioideae*) in particular *Brachystegia*, *Julbernardia*, and *Isoberlinia*. There are reportedly 21 species of *Brachystegia* in the Miombo Woodlands biome, and three species of each from *Julbernardia* and *Isoberlinia*.

Vegetation types associated with the miombo woodlands are the Acacia savannah, Mopane woodland, and other dry savanna woodlands. The ecology of the miombo woodlands make it the most extensive tropical seasonal woodland type and the most extensive dry forest type in Africa, perhaps even globally.

When undisturbed, the miombo woodlands is composed of a closed deciduous non-spinescent woodland type that occurs on geologically old, rather infertile soils at an average annual rainfall around 28 inches (700 millimeters). Mature stands boast trees ranging to 33–66 feet (10–20 meters) high. Small trees and shrubs in the scrub layer vary in density and composition, while the understory ranges from dense course grass to scattered herbs or forbs, small sedges, and short grasses. Fires are a typical feature of the miombo woodlands ecosystem.

Miombo woodlands are relatively homogenous in community composition internally, yet extremely rich in plant species. Consequently, the Miombo Woodlands ecosystem has an estimated 8,500 species of higher plants; out of this, 334 are tree species. More than half of these are thought to be unique to this biome, in other words, endemic. Across the biome, it is in Zambia where the highest species diversity for trees is found, with, for instance 17 species of *Brachystegia* that are endemic. Comparatively, there is one such species in Kenya, six in southeastern Tanzania, and 11 in western Tanzania.

However, generic endemism is found in less than 15 percent of the genera; such species are linked to the Sudanian and coastal formations. In many herbaceous plant genera, like *Crotalaria* and *Indigofera,* there is high species diversity and some localized endemism. In Zimbabwe, where there are serpentine soils, there are additional localized sites of speciation and endemism.

Plant Life Cycle Stages

There are basically five main phenological, or plant life cycle, seasons for the miombo woodlands. These include: the warm, dry pre-rain season (September to October), the early rainy season (November to December), the mid-rainy season (January to February), the late rainy season (March to April), and the cool and dry season (May to August). Phe-nology of miombo plants is also determined by seasonal variations in soil moisture. Leaf fall and flush occurs during the dry season in most trees and shrubs, while flowering takes place during the warm, dry pre-rains season (September–October).

However, fruit production varies from year to year. Sometimes, fruiting failure may occur due to lack of flowering and/or flower abortion, which may be caused by bud infestation by the *Curculionid* beetle larvae and imagoes. Fruits normally take six months to mature in the miombo, with a few exceptions that take under six months. Fruit dispersal occurs in the late dry season; in rare cases, it overlaps into the early rainy season. There is huge diversity in fruit types and dispersal modes in the miombo.

Seed germination of most trees and shrubs takes place soon after dispersal, as long as optimum germination conditions exist. In terms of regeneration, miombo regrows naturally from stumps, suckers, shoots, old stunted seedlings or wildlings, and from coppices. A more significant number of seedlings and wildlings may be concentrated under tree canopies than in the open. Stem height increments in regrowth miombo woodlands are accelerated within the first and second years, and later on decline. Success of seedling establishment is low, due to various factors including moisture and heat stress.

Ecological System

Miombo woodlands are involved directly and indirectly in global change, in terms of land-use, atmospheric composition, and climate. Conversion of miombo woodlands to short-duration croplands has resulted in accelerated carbon emissions, as well as disturbances in the rainfall formation processes. This, in other words, is a positive feedback loop contributing to accelerated global warming effects.

Ecosystem services provided by the miombo woodlands fall into two broad categories: material products, which include fertilizer, foods, fiber, medicines, energy, browse fodder, construction and craft material; and ecological services, comprising climate regulation, erosion control, and hydrological control, as well as cultural and spiritual values.

The greatest concern regarding this suite of products and services revolves around the dramatic recent, current, and projected future change in the Miombo Woodlands biome. These must be interpreted in view of the changing demands of various African communities, the large portions of miombo woodlands that are cleared each year for different reasons, and the looming effects of climate change.

Alterations in the coverage and quality of miombo woodlands have been rising gradually in response to several underlying causes, such as the pre-colonial, long-distance caravan trade; the Ngoni penetration; the rinderpest epidemic; colonial intervention; crafting of ploughs as animal drought implements; the market economy; growing rural populations; and post-independence political, economic, and land-use restructuring. To address these concerns, it is perhaps best that the Miombo Woodlands biome be examined and analyzed multi-dimensionally as a complex, interactive ecological-social-economic system.

Cliff S. Dlamini

Further Reading
Campbell, Bruce, ed. *The Miombo in Transition: Woodlands and Welfare in Africa.* Bogor, Indonesia: Center for International Forestry Research, 1996.

Chipeta, M. E. and Godwin S. Kowero. "Valuation of Indigenous Forests and Woodlands. An International Perspective." In Michael John Lawes, et al., eds., *Indigenous Forests and Woodlands in South Africa: Policy, People and Practice.* Scottsville, South Africa: University of KwaZulu-Natal Press, 2004.

Nhantumbo, Isilda and Godwin Kowero. *Modeling Method for Policy Analysis of Miombo Woodlands.* Bogor, Indonesia: Center for International Forestry Research, 2001.

Rodgers, W. A. "The Miombo Woodlands." In T. Mclanahan and T. Young, eds., *The Habitats of East Africa.* Nairobi, Kenya: Longmans Harare, 1996.

Sumaila, U. R., A. Angelsen, and G. Kowero. *A System Dynamics Model for Management of Miombo Woodlands.* Bogor, Indonesia: Center for International Forestry Research, 2001.

Mississippi Lowland Forests

Category: Forest Biomes.
Geographic Location: North America.
Summary: This biodiverse region is threatened by land conversion, but has attracted forces willing to battle to preserve it. The region is home to vast numbers of wintering waterfowl, endangered species like freshwater mussels, and the possibly extinct ivory-billed woodpecker.

The Mississippi Lowland Forest ecoregion—also called the Mississippi Alluvial Valley—encompasses the historic floodplain of the lower Mississippi River that was formed by receding glaciers more than 12,000 years ago. Extending from the southern tip of Illinois to the Gulf of Mexico, a distance of approximately 500 miles (800 kilometers), this region drains nearly 40 percent of the North American continent.

While much reduced today, this former 38,610-square-mile (10-million-hectare) complex of old-growth forested wetlands interspersed with swamps, cypress-tupelo brakes, shrub-scrub wetlands, and emergent wetlands once provided wetland functions and wildlife values of inestimable worth.

These habitats supported a stunning variety of forest interior species, including the ivory-billed woodpecker (*Campephilus principalis*), one of the largest woodpeckers known and the biggest ever to inhabit the United States. An intensive search effort is currently underway for this species following unconfirmed sightings in recent years.

Biodiversity
Historically, seasonal flooding was the primary ecosystem driver for the Mississippi Lowland Forest, creating a dynamic patchwork of diverse habitat types that supported a wide variety of plant and animal species. The historical climax plant community (a steady-state community composed of species best adapted to average conditions in that area) consisted of more than 70 species of bottomland hardwood trees, dominated by hard

(nuts) and soft (berries) mast-producing trees. These featured several species of oak (e.g., *Quercus nuttallii*, *Q. lyrata*, *Q. phellos*, *Q. nigra*), hackberry (*Celtis occidentalis*), and green ash (*Fraxinus pennsylvanica*). Mast and other foodstuffs produced by these trees provided forage for huge populations of migrating and wintering waterfowl.

Frequency and duration of periodic flooding interacted to determine plant community composition and species distribution. For example, *Q. pagoda* and *Q. phellos* tended to occur on higher, less flood-prone sites, while *Q. lyrata* occurred on low sites that flooded frequently and for longer periods of time. Bald cypress (*Taxodium distichum*) and tupelo (*Nyssa aquatica*) dominated the permanently-flooded sloughs.

Today, however, an elaborate and highly-managed flood control system acts to constrain the Mississippi River within a relatively narrow floodplain; natural flooding has been reduced by 50–90 percent. As a result, fully 80 percent of the Mississippi Lowland Forest biome has been lost, making the Mississippi Alluvial Valley (MAV) one of the most heavily-converted ecoregions in the United States. What does remain comprises tens of thousands of tiny fragmented patches, fewer than 100 of which are believed to be large enough to support self-sustaining populations of forest-breeding birds and other area-sensitive wildlife. Yet, despite this attrition, scientists estimate that 372 animal species still utilize this ecoregion for all or part of their life cycle.

Avian and Fish Activity

Mississippi lowland forests currently serve as the continent's most important wintering habitat for mallard (*Anas platyrhynchos*) and wood duck (*Aix sponsa*). Other waterfowl species, such as gadwall (*Anas strepera*) and green-winged teal (*Anas crecca*), are also common. Catahoula Lake and the lower Mississippi River delta together also support 10–25 percent of the continent's population of canvasback (*Aythya valisineria*), the largest concentration of these ducks in the world.

Unfortunately, while the region's wetlands still support good numbers of waterfowl, habitat loss and fragmentation has impacted feeding areas and

forested sites that provide needed isolation for pair bonding and thermal refuges during cold snaps. Ironically, while land conversion has decimated habitats for some species, it has benefited others—such as northern pintail (*Anas acuta*), green-winged teal, northern shoveler (*Anas clypeata*), greater white-fronted goose (*Anser albifrons*), and snow goose (*Chen caerulescens*), which all respond favorably to open, agricultural habitats.

In addition to waterfowl, Mississippi lowland forests once provided habitat for some of the most diverse and abundant freshwater mussel populations in North America. Today, as a result of alterations in hydrology and increased agricultural sediment loads, many of those species are endangered or extinct.

Moreover, more than 60 species of fish, many of sport or commercial value, are found here. Many of these have life history strategies that are intricately tied to flooding and the natural hydrology of the system. For example, some fish species rely on peak flood flows to stimulate their spawning activities; others use flooded forest as nursery habitat.

Historically, several species of neotropical migrant passerine birds, woodpeckers, and raptors most likely had source populations in the MAV. Though current populations are much reduced and/or restricted in distribution, certain areas still support a diverse resident and migrant avifauna. Species such as the cerulean warbler (*Dendroica cerulea*) and Swainson's warbler (*Limnothlypis swainsonii*) currently persist in a few large, isolated forest remnants. The swallow-tailed kite (*Elanoides forficatus*), once widespread in the mid-MAV, is now restricted as a breeding bird to the Atchafalaya Delta in the extreme southern portion of the ecoregion. The future of the MAV's remaining bird populations is uncertain, particularly because the ultimate effects of relatively recent large-scale clearing and resultant fragmentation have not been fully realized.

Mammal Habitats

Like birds, several species of mammals have suffered population declines or extirpation as a result of landscape-scale changes and human encroach-

Close-up of a canvasback duck (Aythya valisineria). Catahoula Lake and the lower Mississippi River delta are home to the world's largest concentration of these ducks. (U.S. Fish and Wildlife Service/Gary Kramer)

ment. Panthers (*Puma concolor*) once ranged throughout the region; remnants of this population still exist in Florida. The endangered Louisiana black bear (*Ursus americanus luteolus*) hangs on in small numbers; this habitat generalist often overwinters in hollow cypress trees, a rare commodity, in or along sloughs. While poorly studied, it is likely that the region's several species of bats also depend on cypress trees for nesting and roosting.

Environmental Conservation

The MAV's largest and least-spoiled forested tracts generally remain in the wettest areas: the Cache River Wetlands in Illinois, Crowley's Ridge and Mingo National Wildlife Refuge in Missouri, White River National Wildlife Refuge in Arkansas, and the Atchafalaya Basin and Big Woods Conservation Area in Louisiana. These forests provide important ecosystem services, such as groundwater recharge, flood attenuation, and water quality enhancement through sediment filtration. They also support a high degree of biodiversity, much higher than in corresponding upland forests.

Conservation partnerships between federal, state, and nongovernmental organizations such as Ducks Unlimited are working to restore or enhance public lands in the MAV. They are also working to return cultivated private lands to wildland through the application of conservation easements. The possible rediscovery of the ivory-billed woodpecker has led to a larger effort by such organizations, hunters, and landowners to conserve and restore the ecosystem. In the meantime, the conservation of large hardwoods—home to the beetles that are a favorite meal of this species—is contributing to the resurgence of other native creatures, such as the Louisiana black bear.

The impact of climate change on these forests will post additional challenges to such conservation and species protection programs. Higher temperatures and drought stress may disrupt native breeding and spawning schedules, and facilitate the introduction and spread of nonnative flora and fauna in the region.

Melanie L. Truan

Further Reading

Cooperrider, Allen Y., Raymond J. Boyd, and Hanson P. Stuart. *Inventory and Monitoring of Wildlife Habitat.* Washington, DC: U.S. Bureau of Land Management, 1986.

Ducks Unlimited. "Mississippi Alluvial Valley—Level I Ducks Unlimited Conservation Priority Area, the Most Significant Winter Habitat Area for Mallards in North America." http://www.ducks.org/conservation/mississippi-alluvial-valley.

Luneau, David. "Ivory-Billed Woodpecker—News and Info From Arkansas." http://www.ibwo.org/index/php.

Schwartz, Charles W. and Elizabeth R. Schwartz. *The Wild Mammals of Missouri, 2nd rev. ed.* Columbia: University of Missouri Press, 2001.

Mississippi River

Category: Inland Aquatic Biomes.
Geographic Location: North America.
Summary: An expansive area that drains one-third of the United States, the natural landscape

of the Mississippi River varies widely, including floodplains, wetlands, forests, and glacial lakes that support vast numbers of aquatic, avian, and terrestrial species.

The Mississippi River is the largest river system in North America, and the longest when combined with its largest tributary, the Missouri River. Globally, the Mississippi is the fourth-longest river. It begins in Minnesota and flows south for 2,515 miles (4,048 kilometers) to drain into the Gulf of Mexico near New Orleans, Louisiana. The river basin covers nearly one-third of the United States, impacting 31 states. It has played an important role in shaping the culture, economy, and environment of the country since long before its founding.

Tribal groups like the Choctaw, Chippewa, Koroa, Chicksaw, Tunica, Yahzoo, Pascagoula, Natchez, and Alibamu inhabited the basin centuries before the arrival of Europeans. During the period of colonial rule, France and Spain controlled much of the river basin before the country's Louisiana Purchase in 1803. The expansion of settlements along the river increased with the advent of the paddlewheeler boat, which brought new immigrants interested in farming the fertile land. Today, the river maintains a very significant role in the transportation of goods by barge from Louisiana to Minnesota and elsewhere. The ecological nature of the river and its surrounds have changed profoundly through each of these phases.

Upper Mississippi River

The upper Mississippi River begins at Lake Itasca, Minnesota, and flows south about 1,248 miles (2,008 kilometers) to join the Ohio River at Cairo, Illinois. This part of the river includes areas in Minnesota, South Dakota, Wisconsin, Illinois, Indiana, Iowa, and Missouri that drain water from a set of other major rivers: the Missouri, Ohio, Minnesota, St. Croix, Wisconsin, and Illinois. The upper Mississippi flows through boreal forests, glacial lakes, sand plains, and bog and spruce swamps.

Forest composition in this segment includes paper birch, black spruce, jack pine, white pine, sugar maple, red maple, oak, and balsam fir. Evidence suggests that in the past, prairies and savan-

nas may have been integral parts of these floodplain communities. However, in recent years, urban development, intensive agriculture, and fire suppression practices have led to the replacement of both prairie and savanna lands by flood-tolerant and fire-intolerant species.

The upper Mississippi River, along with its tributaries, maintains an annual discharge rate of 126,285 cubic feet (3,576 cubic meters) per second from headwaters to the confluence with the Ohio River. When the discharge from the Missouri River is added, the rate increases to well above that average. The discharge is highest in April and May, from snowmelt and spring rains. The annual spring flood inundates the floodplain, which ranges from less than 1 to 6 miles (1 to 10 kilometers) wide here. This action changes the physical characteristics of the floodplain, along with islands, bank areas, and channels. This annual flood cycle is vital for nutrient cycling and for the reproduction of fish that are dependent on floodplain spawning habitats.

The upper Mississippi is rich in both invertebrate and vertebrate populations, with an estimated 430 taxa of invertebrates and 145 fish species inhabiting the river system. Invertebrates include 51 species of freshwater mussels, 13 endangered species listed by one or more states, and three species of federally endangered invertebrates: the winged mapleleaf, Higging's leaf, and fat pocketbook.

The most common fish species in the upper Mississippi River are the northern pike, common shiner, bluegill, bullhead minnow, largemouth bass, black crappie, channel catfish, white bass, river shiner, and emerald shiner. Five fish species are listed as endangered by at least one state, including the skipjack herring, whose migratory path has been cut off by dam construction. Other vertebrates include 11 species of salamanders, 14 types each of turtles and frogs, six kinds of snakes, and three species of mammals.

Because of its location, the upper Mississippi River has been a center of commerce and civilization since prehistoric times. The river has been affected by human actions intended to alter the river for transportation and to irrigate the surrounding floodplain for agricultural and urban development. Major human effects include the construction of 11 dams in the headwaters, and 26 locks and dams constructed between St. Paul, Minnesota, and the confluence with the Missouri River. These dams have increased the sedimentation rate, decreased the quality of habitats, reduced the zones of primary production, and formed a large slow-flow area before the river reaches the next dam.

Agriculture in the floodplain is probably the most important water-quality issue in the river. It has led to increases in the concentration of nitrogen and phosphorus as the river flows to the confluence with the Ohio River. Similarly, the presence of nonnative species in the river has become an ecological concern. The common carp is one such species threatening native species in the river.

Lower Mississippi River

The lower Mississippi River begins at Cairo, Illinois, and flows southward to Baton Rouge, Louisiana, before draining into the Gulf of Mexico. This part of the river basin drains from the Arkansas, White, Red, and Yazoo Rivers, and three distinct physiographic provinces: coastal plain, Ouachita province, and Ozark plateau. The coastal plain is large, with alluvial deposits; the Ouachita province is characterized by ridges and valleys; and the Ozark plateau consists of highlands composed of limestone, sandstones, and shales.

The basin area here covers 339,770 square miles (880,000 square kilometers) in Arkansas, Mississippi, Missouri, Louisiana, Tennessee, and Kentucky. This lower basin is highly productive land that has been cleared for growing soybeans, rice, cotton, and corn. Most of the agricultural land is maintained as cleared fields for row-crop farming. In addition, confined animal production facilities for poultry or swine are combined with pasture for beef production. Although the manure from the facilities is spread in the pasture, it contributes to nonpoint source (non-discernable source) river pollution as organic matter.

The lower Mississippi River is rich in biodiversity, and it is mostly forested, unlike the upper section of the river. The lower river contains parts of six terrestrial ecoregions: Central U.S. Hardwood Forests, Ozark Mountain Forests, Piney Woods Forests, Southeastern Mixed Forests, Western Gulf Coastal Grasslands, and Mississippi Lowland Forests. The forests are mainly composed of oaks, hickories, gums, hackberries, birches, loblolly pines, and willows. The floodplain of the lower Mississippi River is one of the largest in the world; it contains 13,900 square miles (36,000 square kilometers) of wetland habitats.

The channels of the river are abundant in several species of caddisflies, chironomids, oligochaetes, and clams. Some 91 species of fish have been reported, including channel catfish, common carp, freshwater drum, threadfin shad, smallmouth buffalo, blue catfish, yellow bullhead, minnows, and pirate perch.

Likewise, amphibians and reptiles are diverse and abundant, especially toward the southern coastal zone. Species include American alligators, turtles (such as the common snapping turtle, read-eared turtle, and alligator snapping turtle), snakes (such as the diamondback water snake and western cottonmouth), and frogs (such as the bullfrog, pig frog, spring peeper, and cricket frog). Large animals include beavers, muskrats, raccoons, minks, river otters, white-tailed deer, swamp rabbits, black bears, and endangered red wolves. The majority of the fish and animal species are hunted and fished for their commercial value.

Human Impact

Human activities have extensively altered the physical, chemical, and biological integrity of the lower Mississippi River. The major sources of these alterations are engineering projects to improve navigation and control flooding; pollution from industrial, urban, and agricultural activities in 40 percent of the nation's contiguous states; and the introduction of nonnative species.

With the advent of powerful boats, the Mississippi River became an important avenue for transportation of materials from the North American heartland to the ports of Baton Rouge and New Orleans. Several dams and levees were constructed to maintain the flow of water for transportation. One such project involved reservoir and lock-and-dam construction on tributaries as part of the Mississippi River and Tributaries Project. In addition, the input of municipal wastes, sedimentation, industrial effluents, and agricultural runoff has created a dead zone downstream in the estuary and out into the Gulf of Mexico. It is estimated that the Mississippi River carries 331–441 billion tons (300–400 billion metric tons) of sediment past New Orleans each year.

Despite a long history of human activities, the upper part of the Mississippi River retains much of its natural condition, with more than 80 percent of the floodplain connected to the river, and its hydrology still resembles natural conditions. The point-source pollution has been significantly reduced because of the Clean Water Act, better industrial waste management, and sewage treatment around urban areas. However, various human activities have significantly altered the region. Infrastructure construction for transportation of materials and inefficient use of resources by communities have resulted in decreased biodiversity and led to severe flooding in recent years. Climate change is increasing the chance of flooding from shifting weather patterns and more severe storms and erosion—at the same time that it is extending and deepening drought effects on neighboring lands and applying stress to riverine habitats dependent on year-round moisture.

KRISHNA ROKA

Further Reading

Benke, Arthur C. and Colbert E. Cushing. *Rivers of North America.* New York: Elsevier Academic Press, 2005.

Morris, Christopher. *The Big Muddy: An Environmental History of the Mississippi and Its Peoples From Hernando de Soto to Hurricane Katrina.* New York: Oxford University Press, 2012.

National Oceanic and Atmospheric Administration (NOAA). "Major Flooding on the Mississippi River Predicted to Cause Largest Gulf of Mexico Dead Zone Ever Recorded." NOAA. http://www.noaanews.noaa.gov/stories2011/20110614_deadzone.html.

Ruth, Maria Mudd. *The Mississippi River.* New York: Benchmark Books, 2001.

Missouri River

Category: Inland Aquatic Biomes.
Geographic Location: North America.
Summary: Once known as the Big Muddy, the Missouri River is the longest river in the United States. It has been altered by dams and its meandering channel has been straightened to accommodate modern life—affecting the ecosystem and stressing local wildlife.

The Missouri River is the longest river in the United States, flowing 2,341 miles (3,767 kilometers) from its headwaters near Three Forks, Montana, to just north of St. Louis, Missouri, where it joins the Mississippi River. The Missouri River basin, the area of land drained by the Missouri River and its tributaries, covers one-sixth of the continental United States. Once a meandering river that flooded regularly, the Missouri's natural channel was dammed and straightened in the 20th century, greatly impacting its ecosystems and the basin's plant and animal communities.

Biodiversity Before Dam Construction

Prior to the intervention of the last century, the Missouri River had high flows in spring and summer that were fed by spring rains and regional snowmelt, especially from the Rocky Mountains. During this time of year, the river would erode sediment from both the riverbed (river bottom) and riverbanks (sides of the river). When the river flooded, it would become connected with backwaters, chutes, oxbow lakes, and wetlands. During high flows, the river also would carve out new side channels and sweep riparian vegetation such as trees into the main channel. The high water would replenish the groundwater table, which helped plants in floodplain cottonwood forests, wetlands, and wet meadows grow and reproduce. As river flows declined later in the year, sediment in the slower-moving water would drop to the riverbed, creating sandbars and isolating some side channels and meanders.

Fish, wildlife, and plants were adapted to these wide seasonal changes in water levels. For example, floods were a reproductive cue for many fish species. Floodwaters transferred organic matter and nutrients from the main river channel to floodplain habitats, nourishing fish, aquatic invertebrates, and plants. Fish species in the Missouri included goldeye, blue catfish, channel catfish, and interior cutthroat trout. Riparian tree and shrub species included cottonwood, bur oak, dogwood, green ash, box elder, slippery elm, eastern red cedar, and willow.

Raptors, particularly bald eagles, depended on the sturdy branches of cottonwoods to build their large nests, while birds that were year-round residents used riparian forests for shelter during the winter. Migratory waterfowl and songbirds depended on riparian forests and wetlands as habitat during migration. Reptiles and amphibians were common in wetland and riparian forest areas. Mammals such as mink, beaver, muskrat, bobcat, fox, white-tailed deer, mule deer, raccoons, possums, skunks, shrews, and mice also lived along the river.

A side channel of the Missouri River suffering from the effects of erosion; downed trees are visible on the far bank. The destruction of side channels and backwaters like this have affected spawning by the endangered pallid sturgeon. (U.S. Fish and Wildlife Service/Steve Hillebrand)

Some wildlife and plants were adapted to using sandbars and other areas of freshly deposited sediment. For example, shorebirds such as least terns and piping plovers built their nests on bare sandbars to avoid predators and forage for insects. Turtles and snakes used the bare sandbars for basking in the sun and building up energy.

Impact of Dam Construction

The Pick-Sloan Missouri Basin Plan was crafted in 1944 by Colonel Lewis Pick of the U.S. Army Corps of Engineers and William G. Sloan of the Bureau of Reclamation to address many of the transportation, irrigation, hydropower, and flood control demands of people all across the upper and lower Missouri River states. Five dam sites were chosen in North Dakota and South Dakota, with the locations primarily determined by their topography, minimal impact to major cities and towns, and cost-to-benefit ratios. Because the Missouri River States Committee had no Native American members, impacts to reservations were not a large consideration.

The Flood Control Act of 1944 authorizing construction directed that the Missouri River be managed for eight primary purposes: flood control, water supply, water quality, irrigation, navigation, hydropower, fish and wildlife, and recreation. The flood of 1952 helped secure political support and funding for these five Missouri River dams—all were constructed by 1963.

The completion of these major dams brought into being reservoirs that inundated some of the most productive reservation land, including cottonwood forests that provided year-round habitat for wild game such as deer, beaver, rabbits, and raccoons, and winter habitat for pheasants and other game birds. Also heavily impacted were habitat for many wild fruits, herbs, and legumes—consumed both by animals in the areas and used by the indigenous peoples for food and medicine.

Nearly 1,000 Native American families were displaced as more than 100 million acres (40.5 million hectares), much of it belonging to the tribes, were put underwater forever. The dams significantly impacted the Missouri River ecosystem. Many of the large meanders that once characterized the river were eliminated, shortening it by more than 200 miles (320 kilometers), or nearly 10 percent of its original length.

The modern river does not flood as often as it once did and consequently deposits fewer areas of bare sand along the riverbanks, which are sorely needed for the growth of new willow and cotton-wood trees. There also are fewer sandbars in the river channel. Two bird species that depend on sandbars, the least tern and piping plover, were placed on the endangered species list because of declining populations.

Because reservoirs trap sediment, the flowing stretches of the river downstream of reservoirs carry less sediment to the Mississippi River, which then lessens the amount of sediment deposited in the Gulf of Mexico. The chain reaction continues with fewer barrier islands being formed to protect the Louisiana coast from storm surges. The clearer, deeper, and colder water now present in the Missouri River is not appropriate habitat for many native fish species that are used to shallower and more turbid conditions. There are fewer side channels or backwaters along the river for fish to spawn, affecting the now-endangered pallid sturgeon.

During the 1980s, scientists became increasingly concerned about these changes. In 1986, the U.S. Congress authorized the U.S. Army Corps of Engineers to implement the Missouri River Fish and Wildlife Mitigation Project, with the goal of restoring 166,750 acres (67,480 hectares) of aquatic and terrestrial habitat along the lower Missouri River.

Habitat is restored by planting native vegetation and excavating aquatic habitats such as backwaters, chutes, and wetlands on land purchased from willing sellers. In 1989, the Corps and the U.S. Fish and Wildlife Service began consultation under the Endangered Species Act to reduce impacts of Missouri River water management on threatened and endangered species. The Water Resources Development Act of 2007 further expanded the geographic reach of restoration projects within the upper river states.

All of these efforts currently fall under the Missouri River Recovery Program. The Corps collaborates with the service, tribes, federal, state, and local agencies, non-governmental organizations, and other stakeholders within the Missouri River basin to plan and implement the program. The main goals of the program include habitat creation for native species, flow modification, and the incorporation of science and public involvement to restore ecosystem form and function to the river.

Climate Issues

The Missouri River dams and reservoirs have provided social and economic benefits to the country, such as recreational opportunities, electricity, irrigation, navigation, and improved flood control. However, large floods can still occur. A rainy autumn in 1992, heavy snowfall during the winter of 1992–93, and stormy weather in early spring 1993 led to saturated soils across the Missouri River basin and widespread flooding. Many privately built levees along the Missouri failed during the summer of 1993. In 2011, snowmelt from record snowfall in the Rocky Mountains and heavy spring rainfall combined to trigger large-scale flooding along the Missouri River during the spring and summer.

Climate change is impacting the Missouri River basin. Scientists are convinced that warming temperatures are increasing precipitation along the river's banks, enhancing flooding incidents, which could alter life and land use in the future. These floods reflect the complicated and uncertain nature of river management. In coming years, agencies and Missouri River basin inhabitants will need to consider and balance a diverse array of economic, social, and ecological issues when managing the Missouri River ecosystem and planning for its future.

Kristine Nemec

Further Reading

Committee on Missouri River Ecosystem Science and National Research Council. *The Missouri River Ecosystem: Exploring the Prospects for Recovery.* Washington, DC: National Academies Press, 2002.

Lawson, Michael L. *Dammed Indians Revisited: The Continuing History of the Pick-Sloan Plan and the Missouri River Sioux*. Pierre, SD: South Dakota State Historical Society, 2009.

Mullen, Tom. *Rivers of Change: Trailing the Waterways of Lewis and Clark*. Malibu, CA: Roundwood Press, 2004.

Schneiders, Robert Kelley. *Unruly River: Two Centuries of Change Along the Missouri*. Lawrence: University Press of Kansas, 1999.

U.S. Army Corps of Engineers. "Missouri River Recovery Program." http://www.moriverrecovery .org.

Mojave Desert

Category: Desert Biomes.
Geographic Location: North America.
Summary: As California's largest desert, the Mojave is a scenic and biologically diverse ecoregion, threatened by recent urbanization.

The Mojave Desert landscape offers many dramatic geologic features, including peaks, dry washes, salt pans, cliffs, canyons, sand dunes, and alluvial fans. It is richly diverse in plant and animal species that have adapted to its extreme climate. The recent sprawl of cities in southern California and Nevada has had a significant effect on the ecology of the region. As a result, the U.S. Bureau of Land Management, National Park Service, and other public agencies have moved to protect its biological diversity and to ensure sustainability of this uniquely scenic region.

Transitional Desert

The natural boundaries that define the wedge-shaped Mojave Desert can easily be identified in satellite photographs and on topographical maps. The Mojave is considered to be a transitional desert, lying between the cooler Great Basin Desert to the north and the hot Sonoran Desert to the south. To the east, the Mojave Desert stretches across four states: California, Nevada, Arizona,

and Utah. On the west, it is bordered by the intersection of two major fault systems: the Garlock at the northern edge of the San Gabriel Mountains, and the San Andreas at the southern boundary of the Sierra Nevada range.

Known for its distinctive typography, the Mojave is part of a geologic area known as Basin and Range Province, caused by an expansion of the Earth's crust 5 million to 23 million years ago. The result was a series of parallel mountain ranges punctuated by broad, flat valleys. At the point where the mountains and valleys meet, alluvial fans spread like giant feet onto the floor of the desert. Phreatophyte plants (those with deep roots constantly in touch with water) such as the creosote bush often grow here, extending their roots to water 50 feet (15 meters) or more below the surface.

Climate

The Mojave Desert rests in the rain shadow of three mountain ranges: the southern Sierra Nevada, and California's Transverse and Peninsular ranges. It receives less than 13 inches (330 millimeters) of precipitation a year, most of it during the winter months and often in the form of snow, with an occasional summer thunderstorm. Most winters see less than 3 inches (76 millimeters) of rain; prevailing winds from the Pacific Ocean drop most of their moisture in the mountains; the rest often evaporates before ever hitting the desert floor.

Though its elevation generally ranges from 2,000 to 5,000 feet (610 to 1,524 meters), the Mojave Desert is guarded by the towering Telescope Peak in the Panamint Range, which features one of the steepest vertical ascents of any mountain in the contiguous United States. The peak rises 11,331 feet (3,454 meters) above the lowest point in North America, Badwater Basin in Death Valley, at 282 feet (86 meters) below sea level. Death Valley broke the record for the highest temperature ever recorded in the Western Hemisphere when Furnace Creek reported 134 degrees F (57 degrees C) on July 10, 1913.

Biodiversity

With such extremes of elevation and temperature, it is not surprising that there is notable biological diversity among the plant and animal populations of the Mojave Desert. Telescope Peak, for example, supports a variety of trees, including single-leaf pinyon; limber pine; and, at the highest elevations, bristlecone pine (*Pinus longaeva*), believed to be one of the oldest living species on Earth. By contrast, Badwater Basin features a pool of brackish water that looks as though it could not support life at all. The pool is undrinkable by humans, leaching toxic salts from the surrounding area, but it doesn't deter pickleweed, aquatic insects, and badwater snails from thriving.

The climate of the Mojave Desert varies from west to east. The western desert is more typical of California's mediterranean climate—with hot, dry summers and cool, wet winters—while the eastern Mojave has a more balanced pattern of summer and winter moisture. Utah agave, Spanish bayonet, Mojave yucca, and grasses such as galleta are found more often in the east, where summer rain falls, while winter flowers like desert coreopsis, goldfields, and California poppy are found in the western portion of the desert during the wet winter season. The Mojave's dune systems also support unique plant and animal species such as Eureka dune grass and the Mojave fringe-toed lizard.

Some animals living in the Mojave have developed physical attributes that allow them to survive in the extreme desert climate. The Mojave ground squirrel and jackrabbit have light-colored coats that reflect the heat. The multicolored coat of the coyote allows it to stalk prey among rocks and in sandy washes. The venomous Mojave green rattlesnake has a neurotoxin that is considered to be 10 times as strong as that of other North American rattlesnakes, allowing it to hunt quickly and efficiently. Most desert animals hunt or forage for food at dawn or after the sun goes down.

The desert tortoise has a domed brown shell that can easily be mistaken for a rock. It burrows 3 to 6 feet (1 to 2 meters) deep in the desert soil to escape the heat, and can live for years without water. Desert tortoises living in the Mojave Desert are federally listed as a threatened species. They are vulnerable to illegal collection by humans and to habitat disruption from urban expansion. In recent years, ravens have played a unique role in the dilemma of the desert tortoise. Ravens are both predators and scavengers. They love human trash, and at the same time, they prey on young tortoises, whose tender shells have not hardened. Attracted in artificially high numbers to areas where human activity has spiked, the ravens then turn on the young tortoises that are seeking new shelter. The human-raven-tortoise relationship provides a good example of how human activities can disrupt ecological balance.

Joshua Trees

The best-known plant in the Mojave Desert is the endemic Joshua tree (*Yucca brevifolia*). Joshua trees are seen in abundance at 1,300–5,900 feet (400–1,800 meters) elevation and can reach a height of 50 feet (15 meters). They develop a deep, extensive root system and have been known to live for hundreds and perhaps thousands of years.

Flowers bloom on Joshua trees from February to April, but fruit will develop only after a winter freeze. The soft flesh contains many flat seeds that provided nutritious food for the indigenous Cahuilla tribe, who also used the leaves to weave sandals and baskets. Today, the plant is still an important part of traditional Cahuilla culture.

One of the most alarming projected impacts of climate change is its potential effect upon the Joshua tree. The U.S. Geological Survey examined how the Joshua tree reacted to climate changes over 12,000 years ago; based on these findings, they project that global warming could reduce their population by as much as 90 percent.

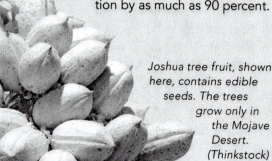

Joshua tree fruit, shown here, contains edible seeds. The trees grow only in the Mojave Desert. (Thinkstock)

Effects of Human Activity

The Mojave indigenous peoples here centered their activities on the Colorado River, using its annual overflow to grow melon, corn, pumpkin, and bean crops. They used nets made from grasses to catch fish, and gathered seeds and pods to supplement their diet. The Chemehuevi, Cahuilla, and Serrano were native tribes who lived in this subsistence manner in and around the Mojave Desert.

During the Gold Rush period of 1848–54, it is estimated that more than 250,000 people crossed the Mojave to find their fortunes in the mountains and cities of California. The effect on desert ecosystems was minimal until the rapid urban expansion of Las Vegas, Nevada, and Lancaster and Barstow, California, more than a century later. Between 1990 and 2000, for example, Las Vegas showed an increase of more than 710,000 people.

As early as the 1970s, public and private agencies recognized the need to protect desert lands, but it was not until 1994 that Congress passed the California Desert Protection Act, designating large areas of the Mojave Desert as wilderness. About 80 percent of the Mojave in California is now managed by federal agencies—not all of them focused on conservation. The Bureau of Land Management is the largest land manager of the region, overseeing 8 million acres (3 million hectares) of land. The National Park Service manages the Death Valley and Joshua Tree national parks, as well as the Mojave National Preserve. The Department of Defense manages five military bases covering about 13 percent of the region. State parks and fish and game agencies manage the rest.

Even with extensive management, Mojave Desert plants and wildlife are still at risk. Urban sprawl, off-highway vehicle activity, overuse of water resources, and overgrazing of cattle and sheep continue to cause problems. Of the 439 vertebrate species that inhabit the Mojave Desert, 135 species of birds, mammals, reptiles, amphibians, and fish are included on a special watch list for at-risk species. Of these species, 14 are endemic to the Mojave region. Only with continued commitment and strategic conservation efforts will the extraordinary Mojave Desert ecoregion be restored, protected, and preserved for future generations.

DEBORAH FOSS

Further Reading

Abella, Scott R. "Disturbance and Plant Succession in the Mojave and Sonoran Deserts of the American Southwest." *International Journal of Environmental Research and Public Health* 7, no. 4 (2010).

Aitchison, Stewart. *Death Valley National Park: Splendid Desolation.* Mariposa, CA: Sierra Press, 2002.

Bakker, Elna. *An Island Called California: An Ecological Introduction to Its Natural Communities.* Berkeley: University of California Press, 1971.

Nevada Fish and Wildlife Office. "Threats to Desert Tortoises." U.S. Fish and Wildlife Service. http://www.fws.gov/nevada/desert_tortoise/dt_threats.html.

Schoenherr, Allan A. *A Natural History of California.* Berkeley: University of California Press, 1992.

Mongolian-Manchurian Grassland

Category: Grassland, Tundra, and Human Biomes.
Geographic Location: Asia.
Summary: This large ecosystem once was a home for large populations of wild ungulates. Today, intensive pastoralism and global warming threaten its ecological integrity.

The Mongolian-Manchurian Grassland biome is the largest remaining block of the once-endless grasslands of Asia. It is found in the eastern third of the Eurasian steppe belt. This area is covered with diverse grassland communities, supporting small populations of wild ungulates such as the Mongolian gazelle and the reintroduced Przewalski's horse. At the same time, species such as the Bactrian camel and Przewalski's gazelle have become extinct because of human activity. Hunting and grazing by large numbers of domestic sheep and goats threaten this ecosystem.

The Mongolian-Manchurian steppe spreads from the inland side of northeastern China's coastal hills toward the boreal forests of southern Siberia and to the desert regions of southern Mongolia and north-central China. It forms a large crescent around the Gobi Desert, extending across central and eastern Mongolia into the eastern portion of Inner Mongolia, and eastern and central Manchuria. The main part of this ecosystem, which covers more than 386,102 square miles (1 million square kilometers), consists of nearly flat or undulating grasslands and river basins. The climate is extreme continental, with warm summers and cold, wind-blown winters.

Biodiversity

The dominant flora consists of medium to tall grasslands. Grassland communities include feather grass and sheep's fescue grass. Areas closer to the Gobi Desert are desert steppes that have lower productivity. Dominant plant species on the steppes include drought-resistant grasses; forbs; and small, spiny shrubs that are well adapted to arid conditions. The southwestern slopes of the Greater Khingan range support pockets of broadleaf deciduous forest as well.

Several threatened mammal species may be found on the Mongolian-Manchurian grasslands; these are remnant populations, small and fragmented. The Asiatic wild ass, or kulan, is still found along the Mongolian border regions, but other species are extinct as a result of overhunting and competition with domestic ungulates. Przewalski's horse may be found on the steppes today, thanks to several international efforts to reintroduce this species to its historic homeland; small groups of these wild horses can be seen again grazing in the Mongolian turf. Other mammals characteristic of the biome include the Mongolian and goitered gazelle, and Hangai and Gobi wild sheep.

Carnivores also roam the steppes; marbled polecats, Pallas's cat, red fox, corsac, wolf, and brown bear are found in the Mongolian-Manchurian grasslands. As usual in steppe ecosystems, many rodent species, such as ground squirrels, marmots, pikas, hamsters, numerous species of voles, gerbils or jirds, and jerboas, dwell among the grasses. Subterranean rodents, including numerous species of zokors, also are typical members of these steppe communities.

The brown-eared pheasant is an endemic bird of the Mongolian marshes, and reed beds provide breeding shelter for the great-crested grebe, Oriental white stork, Japanese crane, and relict gull. Many rare birds breed on the plains of the Mongolian-Manchurian grassland, such as the great bustard and Oriental plover.

Environmental Threats

Sheep and goat grazing are the dominant activities in the Mongolian-Manchurian Grassland biome. In the past decade, the number of livestock has increased dramatically, and overgrazing is the greatest danger for the remaining natural values of this ecosystem. Factors that are exacerbating this danger include mining activity and global warming. Water scarcity, lower productivity of various plant communities, soil loss, and desertification are some problems already being measured here as impacts of climate change—largely driven by faster evapotranspiration due to higher temperatures. Solutions will entail cooperative approaches to sustainable grazing, water conservation, and related practices.

ATTILA NÉMETH

Further Reading

Finch, C., ed. *Mongolia's Wild Heritage.* Boulder, CO: Avery Press, 1999.

Girvetz, Evan H., et al. "Eastern Mongolian Grassland Steppe." In Jodi A. Hilty, Charles C. Chester, and Molly S. Cross, eds., *Climate and Conservation.* Washington, DC: Island Press, 2012.

Laidler, L. and K. Laidler. *China's Threatened Wildlife.* London: Blandford, 1996.

MacKinnon, J. *Wild China.* Cambridge, MA: MIT Press, 1996.

Zhao, J., Z. Guangmei, W. Huadong, and X. Jialin. *The Natural History of China.* New York: McGraw-Hill, 1990.

Montana Valley and Foothill Grasslands

Category: Grassland, Tundra, and Human Biomes.
Geographic Location: North America.
Summary: In addition to the unique biodiversity it harbors, this biome is vital because its valleys provide habitat connectivity across the foothills of the Rocky Mountains.

Located in Montana in the United States and Alberta in Canada, the Montana Valley and Foothill Grasslands biome runs along the foothills and high valleys of the central Rocky Mountains. The ecosystem covers some 9,300 square miles (24,000 square kilometers), and ranges between 1,800 and 5,400 feet (550 and 1,650 meters) in altitude. The grasslands extends from the uppermost flatland reaches of the Missouri River watershed on the Rocky Mountain Front, into the Clark Fork-Bitterroot catchment of the Columbia River system within Montana and parts of Idaho; and in Alberta, along the Oldman, Little Bow, and Bow Rivers, as well as other tributaries of the Peace and Athabasca Rivers.

Climate, Topography, and Vegetation

The climate in the Montana Valley and Foothill Grasslands biome is generally cool and dry. However, there is a mix of moisture microclimates here, with one major factor being the shape, orientation, and elevation of the valleys as they either capture, deflect, or are themselves deprived of precipitation that arrives mainly from the prevailing Pacific Ocean-generated Chinook wind pattern. There are valleys and foothills in the rain shadow, where semiarid conditions maintain; others effectively capture moisture and have habitats that make good use of rainfall. Most areas experience seasonal runoff from snowmelt. The result is broad biological diversity.

Riparian corridors are important because they allow the passage of species across valley bottoms, opening critical habitat linkages for plant and especially animal types between separate moun-

tain ranges. The stream beds and rivers are also key corridors for species that descend seasonally from sheltering elevations; specifically, the wet and high-productivity valleys represent ideal sites for the grizzly bear (*Ursus arctos*) on its path to annual recolonization of the grassland habitats.

Overall, the mean annual temperature across this biome is 26 degrees F (3.5 degrees C). The summer mean is 57 degrees F (14 degrees C); winter mean hovers at a frosty 18 degrees F (minus 8 degrees C). Mean annual precipitation is approximately 17 inches (425 millimeters).

The dominant grass species here include fescue (*Festuca* spp.) at higher and wetter elevations, interspersed with bluebunch (*Pseudoroegneria spicata*) on very steep slopes. Western wheatgrass (*Pascopyrum smithii*) is most abundant on lower slopes. Mixed in are likely to be such shrubs as creeping juniper (*Juniperus horizontalis*), shrubby cinquefoil (*Dasiphora floribunda*), and kinnikinnik (*Arctostaphylos uva-ursi*). Sagebrush (*Artemisia* spp.) is a dominant species in many of the semiarid swaths.

While the grasslands are ubiquitous across the drier zones here, shrubland habitat is often seen in niches of higher water availability. Closest to the riparian corridors are found considerable stands of Douglas fir (*Pseudotsuga menziesii*) and ponderosa pine (*Pinus ponderosa*).

At elevations above the limits of the grassland, forests of quaking aspen (*Populus tremuloides*) and woodlands of limber pine (*Pinus flexilis*) are common, often with an understory cover of chokeberry (*Prunus virginiana*), Woods's rose (*Rosa woodsii*), or Saskatoon serviceberry (*Amelanchier alnifolia*). Below the grassland, where wetter conditions tend to prevail, are often found stands of willow (*Salix* spp.) and red-osier dogwood (*Cornus sericea*).

Fauna

The traditional wildlife of this biome once included the remarkably vast herds of the American bison (*Bison bison*), the elk (*Cervus canadensis*) and the bighorn shell (*Ovis canadensis);* the populations of each remain in small numbers. The most iconic mammal today is the grizzly bear, which makes its seasonal descent from the mountainsides in warmer months to forage along the rivers and in grassland habitats. Characteristic species here include white-tailed deer (*Odocoilecus virginianus*), pronghorn antelope (*Antilocapra americana*), coyote (*Canis latrans*), mountain cottontail (*Sylvilagus nuttallii*), red-tailed hawk (*Buteo jamaicensis*), and dusky grouse (*Dendragapus obscurus*).

The Montana Valley and Foothill Grasslands biome supports some fish species and sub-species that are endemic, meaning found nowhere else, or near-endemic. These include the west slope cutthroat trout (*Oncorhynchus clarkii lewisi*), Yellowstone cutthroat (*Oncorhynchus clarkii bouvieri*), and the fluvial arctic grayling (*Thymallus arcticus*), a relict fish species still present from the last glaciation.

The reptile suite features prairie rattlesnake (*Crotalus viridis*), western skink (*Eumeces skitonianus*), and greater short-horned lizard (*Phyrnosoma hernandesi*). The plains spadefoot (*Spea bombifrons*), boreal chorus frog (*Pseudacris maculata*), and barred salamander (*Ambystoma mavortium*) are among the most frequently noted amphibians here.

Human Impact

Historically, the heavy grazing by native, free-ranging herbivores such as buffalo was a continuous influence on the grasslands ecosystem here. This changed rapidly to more intensive stock grazing by domesticated fauna. Draining the wetland areas downhill from the grasslands has been a common practice as the land has attracted more human settlements, with conversion of grasslands to row crops, in effect replacing native species with exotic grasses. Such practices have also opened the door to noxious weed invasion.

The biome is considered a critical one, and generally thought of as in an endangered status, with approximately one-fourth of its area remaining as relatively original, intact habitat, mainly along the East Front of the Rockies from Great Falls, Montana, to near Calgary, Alberta. The most prevalent current threat is the conversion of surviving native habitats into urban and semi-urban areas, which sharply reduces the habitat connectivity for wild-life species. Building projects are concentrated in the Paradise Valley on the Yellowstone River, in the Bitterroot Valley, and in the Gallatin River Valley. Another ongoing threat is groundwater pollution from several Superfund toxic waste sites and various large mining and expanding hydrocarbon drilling operations.

About one-quarter of its area is protected under various conservation schemes; these include the National Bison Range, Pine Butte Swamp, and Red Rock Lakes National Wildlife Refuge in Montana; and in Canada, Banff National Park and Fish Creek Provincial Park, the landmark large urban natural preserve in Calgary.

Lower rainfall, higher temperatures, and more damage from wildfires are among the projected impacts of climate change to the Montana Valley and Foothill Grasslands biome. The results of such trends could include a gradual loss of grasslands to encroaching woodlands and forest—or the opposite could occur, with grasslands mounting somewhat higher elevations as the trees give way.

Interactions at the forest-grassland ecotone (overlap area) are quite complex, and are further complicated by animal behavior, both wild and domestic. Livestock grazing patterns, for example, will be a factor, as many herd animals are quite

The arctic grayling (Thymallus arcticus), shown here, is considered a relict fish species that dates back to the time of the last glaciation. (U.S. Fish and Wildlife Service)

effective at eradicating young tree shoots along forest fringe areas. During warmer periods in the past, some trees, such as larch, have tended to regenerate faster, and to more successfully move downhill, colonizing areas of grassland.

Certain species adapt by migration far more easily than others to temperature changes in their habitats. Research has shown, for example, that the long-tailed vole, happiest in aspen forests here, will readily move downhill into nearby grassland habitat when such areas are vacated by their montane vole rivals. The montane vole, on the other hand, has shown little if any propensity for moving uphill into the forest zone when the long-tailed vole moves out, and when warming temperatures or other factors make its own grassland habitat untenable.

José F. González-Maya
Marylin Bejarano

Further Reading

Foresman, Kerry R. and Alexander V. Badyaev. *Mammals of Montana.* Missoula, MT: Mountain Press Publishing, 2012.

Kudray, Gregory M. and Stephen V. Cooper. *Montana's Rocky Mountain Front: Vegetation Map and Type Description.* Helena: Montana Natural Heritage Program and U.S. Fish and Wildlife Service, 2006.

Restani, Marco. "Resource Partitioning Among Three Buteo Species in the Centennial Valley, Montana." *The Condor* 93, no. 4 (1991).

Vance, Linda K. and Tara Luna. "Montana Field Guide—Rocky Mountain Lower Montane, Foothill, and Valley Grassland." Montana Natural Heritage Program. http://fieldguide.mt.gov/displayES_Detail .aspx?ES=7112.

Monte Alegre Varzea Forests

Category: Forest Biomes.
Geographic Location: South America.

Summary: These periodically flooded forests line the Amazon River along its central reaches and contain many oddly adapted animal species.

The Monte Alegre Varzea Forests biome represents the flooded forests within the central and lower Amazon River Basin, roughly from the confluence of the Tapajos River westward (upstream) to the confluence of the Purus River—including nearly the entire length of the Madeira River Basin. The city of Manaus, Brazil, lies roughly in the middle of the biome, which also includes the confluence of the Solimões River ("white water") and Rio Negro ("black water") near Manaus.

The Monte Alegre section of the region is characterized in particular by white-water rivers—tributaries that are clouded by sediments brought downstream from the Andes. This biome comprises a vast amount of the central Amazonian region, as much as 25,000 square miles (65,000 square kilometers) in total area, although it is in many cases not contiguous but rather contorted or distended. The topographic relief of the multiple channels, rivulets, meanders, and gorges that riddle the central stretches of the Amazon form the eccentric contours of this biome, defined by the very nature of its relation between dry gradient and seasonal stream overflow.

Flora

Generally speaking, varzea biomes represent a complex and seasonal interface between freshwater and terrestrial ecosystems—where the river systems flood their banks and become the understory of a flooded forest. Varzea biomes are harsh environments, where flood levels can reach 20 to 40 feet (6 to 12 meters), and they are unusual in that they are characterized by their intimate mixture of terrestrial and aquatic elements. These include fish that act as seed dispersers for trees, and the region's iconic pink river dolphins that hunt between the giant buttresses of riparian trees.

Tropical and subtropical evergreen and moist broadleaf forest types share the landscape here, depending on factors including elevation, slope alignment, and inundation patterns. The canopy generally reaches to about 80 feet (24 meters), topped by such members as capirona, cumala or wild nutmeg, acai palm, and *aguajé* or Mauriti palm. The understory features such river-friendly shrubs as eugenia, the evergreen *chaparro de agua* or symmeria, and isabgol or psyllium. Grasses abounding here include bitter cane, millet, and water paspalum.

While large portions of this biome have faced development pressure from logging, monoculture farming and plantation forest cultivation, ranching, mining, and other human activities, the periodic inundations have effectively barred large areas from such impacts.

Fauna

The Monte Alegre varzea, as part of the network of biomes that make up the lowland Amazon Basin, has very high biodiversity but low endemism, that is, few species that uniquely evolved here. Many of the flooded-forest-adapted species occur along much of the Amazon River and its tributaries, and during the flood season there is a massive turnover in animal species composition as water replaces the forest understory. During flood stages, it is not uncommon to have complete overlap of terrestrial and aquatic elements in virtually the same space. For example, the brown bearded saki monkey and toco toucans feed in the rainforest canopy, while arawana, arapaima, and several types of piranha feed among the inundated lower branches of the same trees.

Mammals that typify the biome are both permanent residents and seasonal visitors. Aquatic mammals including the giant otter, manatee, and Amazon River dolphin are present all year in the rivers and streams, but enter the forest only during flood stages. On the other hand, terrestrial mammals like the short-ear dog, jaguar, and ocelot tend to stick to dry land at all times. Arboreal mammals, such as the brown-throated three-toed-sloth, a variety of primates, and bats such as the Amazonia sac-winged bat, remain in the forest throughout the year, as they have little trouble staying dry in the canopy during flood season. The capybara, however, is a highly adapted large rodent perfectly at home in and out of the water in the varzea—as is the yapok, or water opossum.

Among the richest of the vertebrate fauna here are birds, with many species well adapted for land, air, or water. Along the river banks and into the forest, it is not uncommon to see the neotropical river warbler and fork-tailed wood nymph, while Amazon kingfisher typically perch on overhanging branches looking for small fish. The harpy eagle, among the largest raptors in the Americas, is a well adapted and opportunistic hunter of small mammals and other prey.

The ancient hoatzin, however, is the most characteristic and perhaps best-adapted bird of the flooded forests. Hoatzin chicks feature claws on two of their wing digits, which help them crawl up and down trees to and from the water. The claws are retained until the chicks mature enough to use flight as their preferred mode of movement within the varzea. More colorful are the red-and-green macaw, toco toucan, and numerous hummingbirds.

The flooded nature of these forests makes ideal habitat for many amphibians and reptiles. Among the amphibians are a number of species whose behavior is as curious as their names imply; examples include the Rio Mamore robber frog, nauta mushroom-tongue salamander, and Manaus slender-legged treefrog. Among the reptiles are a vast number of turtles and tortoises who take advantage of the freshwater, while green and tropical thornytail iguana lizards bask on the branches above. Along the river bank, it is not uncommon to see both dwarf and black caiman.

Threats and Conservation

Among the ongoing effects of the combined anthropogenic impacts noted above, the Monte Alegre Varzea Forests biome is vulnerable to habitat fragmentation. While many of the species here are robustly adapted to a natural environment with inherently dramatic changes on a seasonal basis, there are severe disruptions of the soil, water, and vegetation base that can quickly undermine many of these finely-tuned adaptations. Gross sedimentation from the erosion and runoff released by clear-cut logging, clearing for roads, mining, and monoculture plantation farming can cause immediate habitat destruction by blocking oxygen availability, upsetting nutrient mix, and inputting toxic materials.

Climate change looms as a complex challenge to this ecosystem. Precipitation, wind currents, temperatures, and air pressure are all subject to long-term variance under different models for predicting the effects of global warming on the Amazon. Combinations of pronounced changes in some or all of these vectors are expected to put stress on many of the unique local habitat niches that are already under pressure from aggressive human activities. Accelerated—or delayed—onset of seasonal flooding or of its drawdown is a core pattern to be monitored as these coming changes become more clear in their general trends and relative severity.

JAN SCHIPPER

Further Reading

Goulding, Michael. *The Fishes and the Forest: Explorations in Amazonian Natural History.* Berkeley: University of California Press, 1981.

Prance, Ghillean T. "Notes on the Vegetation of Amazonia III: The Terminology of Amazonian Forest Types Subject to Inundation." *Brittonia* 31, no. 1 (1979).

Sioli, Harald. *The Amazon: Limnology and Landscape Ecology of a Mighty Tropical River and its Basin.* Dordrecht, Netherlands: W. Junk, 1984.

Smith, Nigel J. H. *The Amazon River Forest: A Natural History of Plants, Animals, and People.* Oxford, UK: Oxford University Press, 1999.

Monterey Bay Intertidal Zone

Category: Marine and Oceanic Biomes.
Geographic Location: North America.
Summary: The intertidal region in the Monterey Bay region on the central coast of California owes its biodiversity to a mild climate, a ready mix of freshwater and seawater, and rich natural nutrients.

Consisting of rocky shorelines interspersed with sandy beaches along a semicircular bay and above a deep underwater canyon, the Monterey Bay Intertidal Zone biome hosts a vast array of organisms. Monterey Bay is an ecosystem built around the interface of the Pacific Ocean and the mouths of the Pajaro River and Salinas River. The bay is ensconced in the rugged central coast of California at a point just south of the Santa Cruz Mountains and just north of the Santa Lucia Mountains, both segments of the coastal ranges here.

Over the course of the year, the average surface temperature range of the bay waters nearest the intertidal area is approximately 52–56 degrees F (11–13.5 degrees C).

The rocky-shore-and-beach habitat, particularly in central California, is one of the most varied, most studied, and best understood biological regions of the world. Traditionally, such environments across the world are closely linked to the movements of ocean tides, in what is called the intertidal zone. In the intertidal zone, the most common organisms are small, and most are very resilient to the harsh environmental conditions. This occurs for a variety of reasons. First, the supply of water which marine organisms require to survive is intermittent. Second, the wave action around the shore can wash away or dislodge poorly suited or adapted organisms. Third, because of the intertidal zone's high exposure to the sun, the daily temperature range can be extreme from very hot to near freezing in frigid climates (with cold seas). Lastly, the salinity is much higher in the intertidal zone, because salt water trapped in rock pools evaporates, leaving behind salt deposits. These four factors make the intertidal zone an extreme environment in which to live.

Zonation of the Intertidal

According to the manner of classification by the U.S. National Oceanic and Atmospheric Administration (NOAA), there are four intertidal zones in general: the splash zone (Zone 1), the high intertidal (Zone 2), middle intertidal (Zone 3), and low intertidal (Zone 4).

Zone 1, the splash zone, also known as the supratidal or supralittoral zone, is the highest and the driest of these gradients, and can be found above the spring high-tide line. It tends to be exposed to the air most of the time, as it receives no more than the occasional ocean spray during high tides and is only actually covered by water during storms.

The splash zone has relatively few species. Mollusks such as limpet and periwinkle sea snails, crustaceans in the form of tiny barnacles, and algae survive in this hot, dry zone. In fact, the periwinkle snail (*Littorina keenae, L. scutulata*) is sometimes used as an indicator of this zone. Microscopic algae are common in winter months, when large waves produce consistent spray on the upper portions of the rocky shore.

Zone 2, the high intertidal zone, is exposed to air twice daily. It is here that the tide pools form during low tides. Tide pools are a harsh environment because the high temperature and repeated exposure drives evaporation, which in turn increases salinity, making it necessary for organisms living here to have an elevated tolerance both for high salinity and high temperatures.

The common *Balanus glandula* barnacle, along with several types of red algae (*Endocladia muricata, Mastocarpus papillatus*), are used as indicators of this zone, although these species are also found in other areas of the rocky shore. The other most common animal types in the intertidal are tube worms, wooly sculpin fish, lined shore crabs, and mussels. Giant green sea anemones and ochre sea star live in the deeper end of zone 2.

Zone 3, the middle intertidal zone, features temperature fluctuations that are less extreme, due to briefer direct exposure to the sun. Therefore, salinity is only marginally higher than regular ocean levels. However, wave action is generally more extreme than the high intertidal and spray zones. This mid-intertidal zone is exposed to air once or twice a day for relatively brief periods. Many common organisms are found here; at wave-exposed sites, the mussel (*Mytilus californianus*) often dominates the available attachment substratum.

Zone 4, the low intertidal zone, is exposed only during the lowest tides. The presence of the Scouler's surfgrass (*Phyllospadix scouleri*) and Torrey's surfgrass (*P. torreyi*) are good indicators of the mean low-water tide level. This zone is where sponges and tunicates—anchored filter feeders sometimes called sea squirts—are most common. The sea lemons, a range of nudibranches or sea slugs; the toadfish; sculpin fish; reef surfperch (*Micrometrus aurora*); and monkey-faced eel (*Cebidichthys violaceus*) all thrive in this zone.

The aggregating anemone (*Anthopleura elegantissima*) is a key link in the lower food web here, as it hosts zooxanthellae—much as corals do—thereby helping to buttress the algae population of the intertidal reaches of the bay.

In Monterey Bay, an additional type of intertidal environment exists in the estuarine area known as Elkhorn Slough, near the mouth of the Salinas River. This considerable-sized tidal salt marsh area—one of the largest on the west coast of North America—includes mud flats, sand bars, and swaths of eelgrass that all harbor the types of algae, snails, crustaceans, and small fish that make the site a haven for great numbers of resident and migratory birds across hundreds of species. Sea otters are among the most recognizable denizens of the Elkhorn.

Elkhorn Slough, one of the largest tidal salt marsh areas on the west coast of North America, is home to large numbers of sea otters (Enhydra lutris) like this one. (Thinkstock)

Research and Conservation

Intertidal zones have received a large amount of scientific attention, for several reasons. Foremost among those reasons is the highly structured, or zoned, biological diversity present in such habitats. Furthermore, complex relationships among species here are readily studied because of the distinct boundaries between zones and inhabitants. Also, relatively large numbers of species can be present within the zones. For example, upper intertidal *Endocladia* algal colonies may contain as many as 93 different species of the algae within the colony, while some mussel beds have been shown to contain more than 300 associated taxa.

These factors are accentuated in the Monterey Bay Intertidal Zone biome, in part due to its considerable tidal range, typically as much as eight feet (2.4 meters). The upwelling of nutrient-rich water is another factor, along with the cool fog prevalent nearly every warm day around the bay, which helps prevent desiccation during low tides in otherwise dry summer months.

There are several very small nature preserves around the bay; all are benefited by the presence of the Monterey Bay National Marine Sanctuary (MBNMS), a major, federally-protected marine area that encompasses a shoreline length of 276 miles (444 kilometers) and covers nearly 6,100 square miles (15,800 square kilometers) of the ocean and bay. Supporting one of the world's most diverse marine ecosystems, the MBNMS helps protect marine mammals, seabirds, fish, many invertebrates, kelp stands, and seagrass beds in this highly productive coastal environment.

The MBNMS rocky shores have historically been, and will continue to be, used to detect human impacts on a global scale. The effects of human-derived ecological impacts such as oil spills, disturbance from ship groundings, habitat destruction, and real estate development are some of the disturbances that have been studied here. By measuring levels of radioactive iodine–131 in the intertidal brown algae *Fucus* here, scientists were able to detect nuclear fallout from the 1986 Chernobyl nuclear power plant disaster in Russia.

Various researchers have detected increasing numbers of warmer-water species in Monterey Bay, and have concluded that the pattern is consistent with the anticipated trends of global warming. Continuing and improved monitoring efforts in this area will facilitate studies that impact global comprehension of both human-induced and natural changes and threats to intertidal ecosystems.

ALEXANDRA M. AVILA

Further Reading

Addessi, Loana. "Human Disturbance and Long-term Changes on a Rocky Intertidal Community." *Ecological Applications* 4, no. 1 (1994).

Barry, J. P., C. H. Baxter, R. D. Sagarin, and S. E. Gilman. "Climate-Related, Long-Term Faunal Changes in a California Rocky Intertidal Community." *Science* 267, no. 1 (1995).

Dawson, E. Y. and M. S. Foster. *Seashore Plants of California.* Berkeley: University of California Press, 1982.

Hewatt, Willis G. "Ecological Studies on Selected Marine Intertidal Communities of Monterey Bay, California." *American Midland Naturalist* 18, no. 2 (1937).

Light, S. F., Ralph I. Smith, and James T. Carlton, eds. *Light's Manual: Intertidal Invertebrates of the Central California Coast, 3rd ed.* Berkeley: University of California Press, 1975.

McDonald, Gary. "Intertidal Invertebrates of the Monterey Bay Area, California." Long Marine Laboratory, University of California, Santa Cruz. http://seymourcenter.ucsc.edu/Inverts.

Ricketts, Edward Flanders, Jack Calvin, and Joel W. Hedgpeth. *Between Pacific Tides, 5th ed.* Palo Alto, CA: Stanford University Press, 1985.

Mount Desert Island Intertidal Zone

Category: Marine and Oceanic Biomes.
Geographic Location: North America.

Summary: This unique and important intertidal biome is important for the variety of habitats it supports.

Mount Desert Island, a large island off the coast of Maine in the far-northeastern United States, is best known for its dramatic confluence of glacially formed mountains, windswept ridges, and rocky coastline in the Gulf of Maine on the North Atlantic Ocean. The island consists of nearly 60,000 acres (24,000 hectares), approximately half of which are included within Acadia National Park, with the remainder under private ownership. The preservation of this significant resource was an important milestone in the conservation movement in the United States, as this national park was the first one created east of the Mississippi River.

Geography and Climate

Mt. Desert Island falls in a transitional zone along the Maine coast, as well as being ecologically distinct from inland Maine ecosystems. The characteristics which typify the southwestern portions and the more northerly coastal regions of Maine meet at this area. The overlap between the southerly ecosystems—such as pitch pine woodlands—and the characteristic northern spruce-fir forests, is the result of climate conditions in which a variety of ecological systems coexist within a relatively small geographic footprint.

Mt. Desert Island experiences more moderate temperatures than inland Maine, thanks to the moderating effect of the surrounding waters. In a ranking of continental U.S. microclimates, the island was found to be second only to the Pacific Northwest as far as its level of annual precipitation. Temperatures vary widely by season here, from below zero degrees F (minus 17 degrees C) in the winter to a high of approximately 90 degrees F (32 degrees C) in the summer, with average temperatures around 70 degrees F (21 degrees C).

Geographically, glacial and post-glacial activities have shaped the island into a north-south series of ridges separated by deep U-shaped valleys. The ridges also are separated by crests that rise sharply from the forest cover well below the snowline. These geologic features make for an abundance of streams and rivers that carve out a variety of microhabitats along the rocky shores as they enter the sea. Somes Sound, the only real fjord on the eastern seaboard of the United States, is also found at Mt. Desert Island. The rugged surface topography extends through the intertidal zone and beneath the waters here.

Habitats and Biota

In and above its intertidal zones, Mt. Desert Island features many wetland types. Big Heath is a coastal plateau bog ecosystem, notable for its presence this far south. Deer-hair sedge and black crowberry are hallmark flora here. Bass Harbor Marsh, a streamshore ecosystem, has gradients of pure freshwater that transition to brackish tidal marsh and then to reed-rush-forb saltmarsh. Great Meadow contains a red maple and fern swamp, as well as a shrub-reed marsh; Fresh Meadow features tidal creeks amid raised bog areas; while patches of dune grassland are scattered over the largely rocky shorelines.

A man stands at the high tide mark on a beach of large boulders in Acadia National Park on Mount Desert Island. About half of the island is part of the national park, which receives more than 3 million visitors a year. (NOAA)

Robust eelgrass beds are found in the intertidal zones around most points of the island. This habitat is vital as a fish nursery, rich feeding zone for waterfowl, nesting shelter for waders including some of the many migratory species that stop on Mt. Desert Island each year, and the most diverse base for the crustaceans and mollusks so important to the food web here.

These habitats provide for both terrestrial and aquatic fauna as diverse as the river otter, horseshoe crab, and bald eagle. Sea birds such as the Atlantic puffin and common eider also are sighted in these waters. The peregrine falcon, harlequin duck, purple sandpiper, and least bittern are among vulnerable species that depend on the Mt. Desert Island Intertidal Zone biome. Approximately 40 species of mammals are residents on the island; among those most drawn to the intertidal areas are beaver and muskrat.

Threats

Despite National Park status, threats continue to call into question the long-term vitality of the island and its ecosystems. Most severe are those impacts associated with the recreational over-use of Mt. Desert Island. With over 3 million visitors a year, Acadia National Park is one of the most heavily used national parks, and mitigating ecosystem degradation is an ongoing challenge for park managers and residents alike. Development pressures on the remaining privately owned land also remain a constant; such land is valuable and sought after.

Other threats common to the Atlantic seaboard include algae bloom, or red tides, and periodic invasions of nonnative species. The potential impacts of sea-level rise associated with global warming also presents the challenges of protecting against saltwater intrusion, and beach and bank erosion.

JESS PHELPS

Further Reading

Collier, Sargent F. and G. W. Helfrich. *Mt. Desert Island and Acadia National Park: An Informal History.* Rockport, ME: Down East Books, 1978.

Dubuc, Leslie J., William B. Krohn, and Ray B. Owen Jr. "Predicting Occurrence of River Otters by Habitat on Mount Desert Island, Maine." *Journal of Wildlife Management* 54, no. 4 (1990).

Maine Department of Inland Fisheries and Wildlife. "Focus Areas of Statewide Ecological Significance—Acadia East and West." http://maine.gov/doc/nrimc/mnap/focusarea/acadia_eastwest_focus_area.pdf.

Morison, Samuel E. *The Story of Mount Desert Island,* New York: Little, Brown, and Co., 1960.

Murray River

Category: Inland Aquatic Biomes.
Geographic Location: Australia.
Summary: The Murray River supports many rich habitats in a variety of climate and elevation zones, but is under great stress from human activity.

The Murray River is the largest of three rivers in the Murray-Darling Basin, and among the longest in Australia, at 1,565 miles (2,520 kilometers). The climate and habitats along the Murray River change considerably, beginning as subtropical in the far north; changing to cool and humid in the eastern uplands, Snowy Mountains, and temperate southeast; and ending as semiarid to arid in the western plains. The Murray begins in the Australian Alps, drains the western side of the range, then meanders northwest over the inland plains as the border between Victoria and New South Wales, finally turning south for 310 miles (500 kilometers) to Lake Alexandrina and the ocean.

The Murray-Darling river system drains inland Victoria, New South Wales, and southern Queensland. The system as a whole extends 2,330 miles (3,750 kilometers). The prehistoric Lake Bungunia formerly was the terminus of the Murray River; at maximum, it stretched over 13,000 square miles (33,000 square kilometers). But the lake long ago drained, as the wet period of the Murray-Darling Basin ended a half-million years ago. Since then, the region has been mostly arid, and species long ago common to the Murray are now restricted to Queensland.

The Murray's volume of water is minuscule compared to other rivers of its length in the world. On rare occasions, extreme drought has dried up the river entirely. The river system as of 2010 received 58 percent of its natural flow; the region is Australia's breadbasket and the river is heavily drawn upon. Irrigation using the Murray's waters began in the 1850s, with the first large pumping station built in 1887, allowing farming along the river to rapidly expand. In the 20th century, the river was the site of four reservoirs, as well as locks and weirs for a proposed but obsolete system for river traffic.

Biodiversity

Notable flora in the Murray River basin encompasses one of the world's largest river red gum forests, a flood-fed, wetland-type ecosystem. One 70-mile (112-kilometer) segment of the river features the Barmah Forest, a Ramsar-recognized Wetland of International Importance, along one bank, and the Millewa Forest on the other. Forests of red gum appear both upland and right along the river fringe. These river red gum forests are a product of the Cadell Fault, which altered the route of the Murray 25,000 years ago.

These river red gum forest floodplains are vital in their ability to support a broad range of threatened and vulnerable species of every type. Observers here have documented the Barmah-Millewah's value to such species as the flowering slender darling pea, Mueller daisy, and Moira grass among flora; the trout cod, silver perch, southern bell frog, giant banjo frog, inland carpet python, broad-shelled turtle, and squirrel glider among terrestrial and aquatic fauna; and the intermediate egret, plumed egret, blue-billed duck, superb parrot, and white-bellied sea-eagle among avians.

The Murray River basin as a whole is home to identified endangered fauna including at least 35 bird species, 16 mammal species, and 35 native fish species. With some 30,000 wetlands, about 7 percent of the Murray River basin has been set aside in reserves particularly to support migratory birds.

Fish species that are relatively widespread here include the Murray cod, trout cod, golden perch, silver perch, Australian smelt, and the eel-tailed catfish. The Murray short-necked turtle and Murray River crayfish are among the species unique to the river, but it also has species that appear elsewhere in the southeast, including long-necked turtles, and varieties of shrimp, rat, and platypus.

Threats

River regulation since Europeans settled the area has adversely affected native fish and aquatic life, causing some species to decline until they become endangered or rare. The Murray's natural flow changed, benefiting agriculture but disrupting ecosystems inside and outside the river. Agricultural runoff and pest species have also damaged the river, and increased salinity of both the water and the land, potentially making it unusable in the medium to long term. The last major flood of the Murray was in 1956; it left river towns under water for longer than six months.

Weirs and dams negatively impact the river biome. They prevent migrations of fish and have in many cases displaced the river-adapted Murray crayfish with a floodplain species, the yabbie. Dams and weirs alter seasonal flood cycles, disrupting recruitment and spawning. Since 1950, the endemic Murray cod and river blackfish have declined as water control systems have proliferated. Human-built fishways along the Murray in some cases are poorly designed and tend to inhibit the seasonal movement of such species as golden perch, Australian bass, and barramundi.

Water removal from the Murray River takes up to 80 percent of the available flows, resulting in insufficient water for habitats at most segments, including its estuary and the floodplain forests. The Murray system includes several lakes whose salinity now varies, although until recently they were commonly fresh. This may be the result of human-created changes in the environment. Droughts in 2000–07 stressed the red gum forests, raising concern about their survival. Increased droughts in Australia, thought to be the result of climate change, are likely to continue to stress the red gum forests.

JOHN H. BARNHILL

Further Reading

Mallen-Cooper, Martin. *Fishways and Freshwater Fish Migration on South-Eastern Australia.* Sydney, Australia: University of Technology Sydney, 1996.

Pittock, Jamie. "Challenges of Freshwater Protected Areas." World Wildlife Federation. http://www.wwf.or.th/en/publications/feature_stories/?17772/Challenges-of-freshwater-protected-areas.

Roberts, Nick, et al. *13 Thirsty Species.* Victoria, Australia: Victorian National Parks Association and Environment Victoria, 2010.

Weir, Jessica K. *Murray River Country: An Ecological Dialogue With Traditional Owners.* Brisbane: Australian Institute of Aboriginal and Torres Studies Press, 2009.

Murrumbidgee River

Category: Inland Aquatic Biomes.
Geographic Location: Australia.
Summary: The Murrumbidgee is the second-largest river in the Murray-Darling Basin, but lower water levels due to river regulation and irrigation have resulted in a decline in the health of the river, a loss of biodiversity, and an increase in salinity.

The Murrumbidgee River has its source in Kosciuszko National Park in the state of New South Wales, Australia. From this alpine region, it runs southeast, winding through the Monaro High Plains before flowing north through the Australian Capital Territory. Then it heads west down the South West Slopes and across low-lying plains to flow into the Murray River, which meets the sea at Encounter Bay, South Australia.

Covering a distance of 1,050 miles (1,690 kilometers) from its source near Kiandra in New South Wales to its confluence with the Murray River, the Murrumbidgee is the third-longest river in the expansive Murray-Darling drainage system. This system covers a land area in excess of 386,102 square miles (1 million square kilo-

meters), collecting the water draining west of the Great Dividing Range.

The Murrumbidgee is fed by several tributaries: the Cotter, Tumut, Molonglo, Yass, and Lachlan rivers. The catchment has one of the most diverse climates in New South Wales, ranging from the alpine areas of Kosciuszko National Park and the Monaro plains, to the rich grazing and grain belts of the South West Slopes and plains, and to the shrublands and grasslands of the semiarid western Riverina.

Biodiversity

Within the Murrumbidgee catchment are a number of wetlands. Those of international ecological significance include Fivebough Swamp, Tuckerbil Swamp, and the Lowbidgee Floodplain. These wetlands support a diversity of plant, bird, fish, and animal life, some of which are considered to be "icons" because of their profile or endemic (found nowhere else on Earth) nature. Many are now endangered. High-profile species include the plains-wanderer, superb parrot, southern bell frog, trout bluenose cod, and the grey-crowned babbler. The glossy black cockatoo that once thrived in the lush Murrumbidgee wetlands is now considered endangered. Endangered species with a population entirely within the Murrumbidgee Catchment include the white-browed treecreeper and squirrel glider. Other endemic fauna include the southern corroboree frog and northern corroboree frog.

Endemic plant species found in the biome include the bethungra spider orchid, burrinjuck spider orchid, Mueller's eyebright, bredbo gentian, wee jasper grevillea, tumut grevillea, lemon zieira, yaas daisy, slender darling pea and kiandra leek orchid. Many of these species are considered threatened and of high conservation value.

Threats

Settlement of land along the Murrumbidgee took place largely from the 1830s. It is an area that has been transformed by irrigation. Initial investigations into irrigation were made in the 1890s, partly in response to drier years, but also in an attempt to more closely settle the land. For sheep and cattle grazing to give way to farming, a more reliable source of water was needed rather than the periodic inundations in wet years.

The development of a virtual monoculture of rice in areas such as the Coleambally highlights the problems that have developed from the addition of large quantities of water to the land. The Coleambally Irrigation Area was the last major component that used water diverted via the Snowy Mountains Scheme. A rise in the water tables here became evident after 10 years. Before irrigation, the water table was about 66 feet (20 meters) below the surface, but by 1983, it was estimated to be rising by 5.2 feet (1.6 meters) a year. By 1985, shallow water tables within 7 feet (2 meters) of the surface were a problem in at least two districts, resulting in waterlogging and salt deposits.

The flow of the Murrumbidgee and many of its tributaries is now highly regulated, and they are thus lined with stressed habitats. In the 1960s, another major storage, the Blowering Dam, was constructed in the headwaters of the Murrumbidgee, on the Tumut River, to regulate the water made available from the Snowy Mountains Scheme. The Tantangara Reservoir was constructed from 1958 to 1960. Conservatively, this reservoir reduced by 50 percent the annual flow of the Murrumbidgee. The consequent damage to the river system became a cause for concern from the mid-1960s, when years of drought further reduced river flows.

Salinization has become one of the most serious problems in the Murray-Darling. Before river regulation, the periodic floods due to the cycles of El Niño Southern Oscillation flushed the river system. The reduced flow in the Murrumbidgee in particular has been a major contributor to increased salinization in the basin. Before extensive irrigation, the natural flows of the Murrumbidgee and its tributaries provided water to dilute the salinity of the lower reaches of the Murray River. By the mid-1970s, virtually all of the water in the Murrumbidgee had been allocated to irrigation. Climate change impacts upon this area will add to the challenges; they are likely to be felt in increased brushfires, intense storms, and prolonged droughts.

Julia Miller

Further Reading

Phillips, Peter J. *River Boat Days: On the Murray, Darling, Murrumbidgee.* Melbourne, Australia: Landsdowne, 1974.

White, Mary E. *Listen ... Our Land is Crying. Australia's Environment: Problems and Solutions.* Kenthurst, Australia: Kangaroo Press, 1997.

Young, W. J. *Rivers as Ecological Systems: The Murray Darling Basin.* Canberra, Australia: Murray Darling Basin Commission, 2001.

Muskwa-Slave Lake Forests

Category: Grassland, Tundra, and Human Biomes.
Geographic Location: North America.
Summary: The Muskwa-Slave Lake forests, which are home to a large population of the remaining free-ranging wood bison, are a collection of ecologically diverse plateaus and the narrow boreal forest corridor of the Mackenzie River plain.

The Muskwa-Slave Lake Forests biome of Canada is located in the southwestern section of the Northwest Territories (NWT), the northwestern section of Alberta, and the northeastern section of British Columbia (BC), with a small corridor stretching up most of the Mackenzie River valley. This boreal-taiga plain corresponds to several ecozones: the Mackenzie River plain, Hay River lowland, Northern Alberta uplands, and the Horn and Muskwa plateaus. The Mackenzie River plain is a narrow boreal forest corridor that runs between the Franklin and Mackenzie Mountains and parallels the Yukon River, stretching just north of Fort Good Hope significantly into northern NWT.

The Horn Plateau stretches from the Horn River to the Mackenzie River, and runs west along the Willow-Lake River. This plateau extends over the flat Hay River Lowland and the Great Slave Lake Plain regions in the south and northeast.

The Muskwa Plateau runs along the foothills of the northeastern BC Rocky Mountains, spans the border of the NWT, and makes up a portion of the Fort Nelson Lowland. The Northern Alberta Uplands encompasses the northern Alberta Caribou Mountains, as well as the Cameron Hills uplands that extend over the NWT and BC borders. The human population of the entire ecoregion is estimated at 18,000.

Climate, Geography, and Soil

As a biome, the Muskwa-Slave Lake forests are classified as boreal-taiga forest; they span 100,300 square miles (259,780 square kilometers). The climate of the region ranges from mid to high sub-humid boreal, depending on the location. This type of climate has very cold winters, cool to mild summers, and low precipitation. The average annual temperature of the Muskwa-Slave Lake forests ecoregion is around 26 degrees F (minus 3.3 degrees C). Winter temperatures range from 5 degrees F (minus 15 degrees C) to minus 12 degrees F (minus 24 degrees C), while the average summer temperature is 53 degrees F (12 degrees C). Annual precipitation is in the range of 10–20 inches (250–500 millimeters).

The Muskwa-Slave Lake forests are composed of a series of plateaus: flats; gentle slopes; scarps; and wide, undulating plains along Mackenzie River. The ecoregion is interspersed with wetlands. Most of the terrain of the forests lies at 980–3,280 feet (300–1,000 meters) elevation. The region has discontinuous permafrost that ranges from sporadic to extensive, with low to moderate ice content and sparse ice wedges in the Mackenzie River Plain.

The dominant soils here include organic and turbic cryosols (mixed frozen soils), with some silty loam soil covering glacial till. The characteristic lowlands soils are well supplied with mixed organic material but with some permafrost-affected soils, while the dominant soils of the upland areas are luvisols typical of forest floors. Overall, the soil regime is fertile, especially relative to the high northern latitude of the biome. Wildfires in the forest and shrublands here are common, helping to replenish nutrients and minerals.

Flora

The Muskwa-Slave Lake Forests biome primarily consists of subalpine coniferous forest and wetlands. The plateaus in general house a diverse collection of grassland communities, such as feathermoss, bog cranberry, blueberry, Labrador tea, sedges, shrubs, juniper, cottongrass, sagewort, wild rose, sedge tussocks, shrubby cinquefoil, cowberry, lichens, and various peat mosses (*sphagnum*) that dominate the wetlands areas. In the wetlands of the Horn Plateau, the vegetation is characterized by perennially frozen peat bogs, low oval-shaped frost heaves called palsas, northern ribbed fens, and horizontal fens.

In both the Horn and Muskwa Plateau regions, spruce trees are the primary species: black spruce in the Horn, and white and black spruce in the Muskwa. The forest cover landscape includes open, fragmented and closed areas. Many swaths are mixed-wood forest. Characteristic trees throughout the ecoregion are quaking aspen (*Populus tremloides*), white spruce (*Picea glauca*), balsam fir (*Abies balsamea*), and black spruce (*Picea mariana*).

Fauna

Several key species of animal are found in the Muskwa-Slave Lake Forests biome. Common mammals include woodland and barren ground caribou, moose, mountain goats, Dall sheep, deer, and elk. Common predators are brown and black

A wood bison calf. The Muskwa-Slave Lake forests' population of 5,000 wood bison represents half of all wood bison remaining in Canada. (U.S. Fish and Wildlife Service)

bears, red foxes, wolves, wolverines, lynxes, cougars, and martens. Small animals here include Arctic ground, red, and northern flying squirrels; beavers; snowshoe hares; muskrats; short-tailed weasels; minks; river otters; porcupines; deer mice; least chipmunks; singing voles; northern red-backed voles; and little brown myotis bats.

The ecoregion has a population of 5,000 free-ranging wood bison; the total population in Canada is about 10,000. These bison are mostly found in Wood Buffalo National Park. The wood bison is considered to be a threatened species in Canada.

There are abundant avian species; the Mackenzie Valley is a major migratory corridor for waterfowl. Examples of the waterfowl are northern pintails; Canada geese; Pacific and common loons; and horned, pied-billed, and red-necked grebes. Shorebirds include common snipe, great blue heron, American golden-plover, semipalmated plover, lesser yellowlegs, semipalmated and pectoral sandpipers, whooping cranes, and American coots. Game birds include willow ptarmigan and spruce, sharp-tailed, and ruffed grouse.

Other common birds are black-billed magpies, gray jays, ruffled and spruce grouse, boreal chickadees, common ravens, rusty blackbirds, three-toed woodpeckers, red-breasted nuthatches, Lapland longspurs, and Wilson's warblers. The abundance of waterfowl, songbirds, shorebirds, and game birds attract numerous winged predators, such as great horned owls, ospreys, bald and golden eagles, hawk owls, Merlin's and peregrine falcons, and gyrfalcons. The amphibian suite features frogs such as the northern leopard, wood, and boreal chorus; and Canadian toads. The one reptile reliably found here is the red-sided garter snake.

Land Use, Threats, and Conservation

Land-use issues are mainly various forestry activities such as logging. Logging is limited in some areas, but extensive in certain watersheds. There are small-scale pulpwood and sawmill factories. Other industrial activities include oil and gas extraction and exploration, as well as oil and gas pipeline corridors. Recreational activities are hunting, trapping, water sports, and tourism. The area is considered to be relatively stable, with 75 percent of the ecoregion remaining intact.

There are two main parks in the ecozone: Wood Buffalo National Park and Maxhamish Lake Provincial Park and Protected Area. Wood Buffalo is just south of Hay River and is the largest national park in Canada, at 11 million acres (4.5 million hectares). It was established in 1922 as a refuge for the free-roaming wood bison. Maxhamish Lake Provincial Park and Protected Area is approximately 68,000 acres (27,500 hectares).

Climate change appears to be the greatest threat to the Muskwa-Slave Lake boreal forests. This area may be one of the most vulnerable to the effects of rising temperatures. Disease, insect pests, and nonnative species will likely take advantage of the longer warm seasons. Methane released from permafrost will accelerate greenhouse gas concentrations in the atmosphere. Warming trends may result in the loss of important habitats here, and some fear that various native and endemic species will not be able to effectively move to cooler areas before becoming extinct.

ANDREW HUND

Further Reading

Beaumont, Linda J., et al. "Impacts of Climate Change on the World's Most Exceptional Ecoregions." *Proceedings of the National Academy of Sciences* 109, no. 40 (2010).

Malcolm, Jay. "Canada Rates Among the Most Vulnerable to Climate Change Impacts." Eco-Week, 2002. http://www.ecoweek.ca/issues/ISarticle.asp?aid=1000113335.

Perry, David A., Ram Oren, and Stephen C. Hart. *Forest Ecosystems, 2nd Ed.* Baltimore, MD: Johns Hopkins University Press, 2008.

Ricketts, T., E. Dinerstein, D. Olson, C. Loucks, W. Eichbaum, D. DellaSalla, et al. *Terrestrial Ecoregions of North America: A Conservation Assessment.* Washington, DC: Island Press, 1999.

Namib Desert

Category: Desert Biomes.
Geographic Location: Africa.
Summary: This Atlantic coastal desert has the second-largest dunes in the world and an ecosystem uniquely adapted to its fog belt.

The Namib Desert (the term *Namib* means *vast place* in the language of the Nama people), one of the oldest deserts in the world, stretches for 1,200 miles (1,931 kilometers) along the South Atlantic Ocean coasts of Angola, Namibia, and South Africa. It is a neighbor to the Kalahari Desert in the south, which is semiarid compared to the Namib, a traditional arid desert. In the north, from the Angola-Namibia border, the Namib is known as the Mocamedes Desert and includes the Kunene River, which flows from the Angola highlands to the border—one of the few perennial rivers in the Namib.

As a coastal desert, the Namib enjoys a unique climate and ecosystem. While the Namib's climate is arid, it is close enough to an ocean to receive moisture from advective fog, which forms above the ocean and drifts over the land. The ocean is thus a partial remedy for the desert's aridity—but also one of its causes. The Atlantic currents desic- cated the coast over time by being too cold to con- tribute much atmospheric moisture; the biome receives extremely limited coastal precipitation of less than 1 inch (25 millimeters) per year.

The Namib has some unique features, including: grand spreads of high dunes; the 700-million-year- old granite peaks of Spitzkoppe; the Sperrgebiet diamond mine in the southwest; and Swakop- mund, one of the nearest cities to the desert.

There are only a few small settlements in the Namib, and most of the human population belongs to indigenous pastoral groups like the Himba, who raise cattle and goats, moving their herds from place to place through the desert and protecting them from predation by lions. The Himba follow rainfall patterns to bring their herds to the places most likely to have food, and they use herbs from the omuzumba shrub mixed with butter to protect themselves from the sun.

Heat, Fog, and Rain

Though the South Atlantic is cold here, its pres- ence is a moderating influence on the Namib's temperature, and the diurnal range—the differ- ence between the cold night and the hot after- noon—is considerably foreshortened in the Namib compared with other deserts. This is less true in

865

The Namib Desert's treacherous "Skeleton Coast," a portion of which is shown here, is now under the protection of the Skeleton Coast National Park. While the Namib's proximity to the ocean allows it to receive moisture from ocean fog that drifts inland, the biome still receives less than 1 inch (25 millimeters) per year of coastal precipitation. (Thinkstock)

the Namib's furthest inland regions, where summer nights can drop to 32 degrees F (0 degrees C), while afternoons reach a high of 113 degrees F (45 degrees C).

The northern stretch of coastline is known alternately as Skeleton Coast and Gates of Hell, names that speak to the difficulty of safe navigation. Until the advent of motor-powered boats, it was impossible to launch boats from the shore, and the ones that did successfully navigate the surf to land on the coasts were forced to cross both marsh and desert to leave again.

The skeleton part comes from the whale and seal bones left behind on the beaches when the whaling industry was an ongoing concern. The wooden and metal skeletons remaining from numerous shipwrecks justify the name equally well. Today, a portion of the coast is a designated wilderness area and part of Skeleton Coast National Park. The inland riverbeds are home to baboons, giraffes, lions, springboks, and rhinoceroses. The coast's bird populations have been

the subject of numerous research studies and documentaries.

The Namib's coast is foggy even by the standards of other coastal deserts, with thick mist hanging over the area approximately 180 days a year. The fog occurs when the cold Benguela Current meets the warm air from the Hadley Cell. Much of the coast receives twice as much fog precipitation every year as rain. Long known as a hazard for sailing, the fog is vital to the life of the sand seas along the coast—broad, flat areas with little to no plant cover, where the sand has been formed by the wind.

The Namib's soil is fine-textured and porous, with little organic material. Plants like salt bush, buckwheat bush, and rice grass have adapted to the environment, with strong root systems to take advantage of the rains that do occur, as well as fleshy leaves and stems that store water. In the north, rain typically falls in the summer, and in the south, it falls in the winter. At the transition belt between the two, there is no seasonal pattern;

rains are even more erratic and less predictable than in most desert areas.

Most rivers in the Namib are underground or ephemeral streams: waterways that follow consistent paths, but flow only immediately after precipitation. One of the best-established of these rivers is the Tsauchab, which flows from the Naukluft Mountains when it rains—because the rock and soil of the mountain cannot accommodate seeping precipitation—resulting in flash floods. Within hours of rainfall, the Tsauchab emerges from the dust as a fast-running river with a strong flow through the Sesriem Canyon until reaching the Sossusvlei salt pan.

Vegetation

Lichens and succulents comprise most of the plant life near the coast, and grasses and shrubs predominate along the escarpment. Because of the dearth of plants, there is little to hold the soil in place and less microbial activity. When rain does occur, it can have a much greater erosive effect than it would elsewhere.

A thin layer of organisms forms on the surface of rocks and in their pore spaces, including bacteria, funguses, and cryptogams—organisms that reproduce by spores, in this case mainly lichens such as wreath lichen (*Phaeophyscia orbicularis*) and golden hair lichen (*Teleoschistes flavicans*). This biofilm excretes acids that contribute to the breakdown of rocks and the further desertification of the soil. Microbes and cryptogams also thrive in the fog zones, and the minerals they pick up from the fog help form "desert varnish," a paper-thin coating of mineral clay and iron on exposed rocks.

The Namib is part of the Succulent Karoo ecoregion, which includes the deserts of Namibia and South Africa, and contains the world's largest concentration of succulent plants—about one-third of the world's species, more than 3,000 in total. Like the rest of the Succulent Karoo region, the Namib is home to monkey beetles, masarine wasps, and colletid and melittid bees.

For centuries, people have marveled at the "fairy rings" of the Namib, though they are most striking from the air. The fairy rings are circles on the ground caused by shifts in the soil and sudden changes in the pattern of vegetation. Recent studies have suggested that these rings are the result of termite activity—*Hodotermes mossambicus*, *Psammotermes allocerus*, and *Baucalioterms hainsei* all live in the desert—but this is the most recent of numerous theories and may yet be disproved.

Fauna

Along the coast, brown fur seals and shorebirds feed on the many fish; lions feed on the seals and shorebirds. Away from the coast, much of the fauna obtain water through diet. Small mammals like the rock mouse (*Aethomys namaquensis*), cape short-eared gerbil (*Desmodillus aricularis*), dune hairy-footed gerbil (*Gerbillurus tytonis*), and Grant's golden mole (*Eremitalpa granti*) may feed mainly on plants and insects that retain water from the last rainfall. The same is true of reptiles like the dwarf puff adder (*Bitis peringueyi*), wedge-snouted desert lizard (*Meroles cuneirostris*), and Koch's chirping gecko (*Ptenopus kochi*).

Much of the desert fauna here consists of arthropods. There are 13 species of darkling beetles (*Onymacris*) in the Namib, which have longer legs than other beetles, along similar proportions to those of many spiders. These extra-long legs allow them to lift their bodies above the hot layer of air clinging to the desert floor—an adaptation known as stilting. Just as interesting is fog basking, a behavior shown by two darklings: *O. unguicularis* and *O. bicolor*. These darklings crawl to the tops of the coastal sand seas in large numbers when the fog settles in, letting it condense on their abdomens and drip into their mouths.

While large mammals cannot rely on such methods, ungulates like springboks have adapted by ceasing to sweat, making their water consumption more efficient while also requiring careful use of energy to avoid overheating.

The black widow spider (*Latrodectus indistinctus*) also thrives in the Namib Desert, feeding primarily on detrital-algae-feeding flies that have flown over the coast from the Atlantic. The presence of the black widow helps protect the dune vegetation, as it poisons and reduces the local herbivore population.

Welwitschia mirabilis

Welwitschia mirabilis, one of the world's most unusual plants, is found only in the Namib. Found at least 60 feet (18 meters) apart in broad shallow channels that are barely discernible as such, the plants are positioned to receive floodwaters on the rare occasion of rainfall. The plants are 1–4 feet (0.3–1 meter) tall, with two leaves that curl into unusual shapes as they reach the ground. Unlike some plants, *Welwitschia* doesn't lose and regrow leaves; it retains the same two leaves for its entire life. This makes its leaves the longest-lived leaves in the world, as the eldest *Welwitschia* plants, eking out their existence from occasionally damp soil, are 2,500 years old.

Ripening seed cones emerging from a female Welwitschia mirabilis *plant photographed in Namibia in 2007. (Wikimedia/Hans Hillewaert)*

Local Features

Namib-Naukluft National Park is a national game park encompassing part of the desert near the Naukluft Mountains. Located in the fog zone, the park is home to hyenas, jackals, gemsboks, and reptiles, as well as endemic (found nowhere else) insect species. The mountains are home to mountain zebras and leopards. The dunes are among the largest in the world, and are a striking orange color as a result of the oxidation, over millions of years, of the soil's iron content.

In the southern part of the Namib is the Sossusvlei, a salt and clay pan surrounded by reddish dunes that is a drainage basin without outflows for the Tsauchab River. Like the dunes of the Namib-Naukluft Park, the dunes here are highly oxidized. The marshes are seasonally flooded by underground rivers; they turn nearly white when dry from the high levels of salt.

Rodents, jackals, antelopes, and ostriches live in the area, and migrant birds arrive with the floodwaters. The nearby Deadvlei was an oasis once, but the river that fed it has since changed course, and it is now an empty salt pan filled with long-dead acacia trees. Some of the oasis vegetation has managed to survive without the river—notably clusters of saltwort and the nara melon (*Acanthosicyos horridus*), endemic to Namibia. The nara melon is a vital foodstuff for the Topnaar tribe, as well as the bush cricket (*Acanthoproctus diadematus*), which lives in the southern part of the desert, moving ceaselessly from one melon plant to the next.

In the southwest is the Sperrgebiet, a series of diamond mines jointly owned by the Namibian government and the De Beers corporation. Due to the limited human presence in the area in the 100 years since the German government first established a colony here for the purpose of operating the mines, the wildlife remain very diverse and largely unchanged from before the arrival of Europeans.

The Orange River is the only permanent water supply to the area, but there are 800 species of plants in the area, 234 of which are endemic to the Namib, primarily succulents. The Namba padloper, a rare tortoise, is found almost exclusively in this area.

Climate Change

Temperatures are rising globally and in the Namib Desert. In fact, temperatures have been recorded as rising in this desert since the 16th century. Scientists have found that the weather is getting hotter and the desert is growing drier. Rainfall seasons are shorter and starting later, which has pushed back the planting season and depleted some water supplies. The flora and fauna of this biome are hardened to arid conditions, and it remains to be seen how much the Atlantic fogs may counteract the rainfall drought.

BILL KTE'PI

Further Reading

Armstrong, S. "Fog, Wind, and Heat: Life in the Namib Desert." *New Scientist* 127 (1990).

Barnard, P., ed. *Biological Diversity in Namibia.* Windhoek: Namibian National Biodiversity Task Force, 1998.

Barnard, P., C. J. Brown, A. M. Jarvis, and A. Robertson. "Extending the Namibian Protected Areas Network to Safeguard Hotspots of Endemism and Diversity." *Biodiversity and Conservation* 7 (1998).

Griffin, M. "The Species Diversity, Distribution, and Conservation of Namibian Mammals." *Biodiversity and Conservation* 7 (1998).

Maggs, G. L., P. Craven, and H. H. Kolberg. "Plant Species Richness, Endemism, and Genetic Resources in Namibia." *Biodiversity and Conservation* 7 (1998).

Seely, Mary. *The Namib: Natural History of an Ancient Desert.* Windhoek: Desert Research Foundation of Namibia, 2004.

Wessels, D. C. J. "Lichens of the Namib Desert." *Dintera* 20 (1989).

Namibian Savanna Woodlands

Category: Grassland, Tundra, and Human Biomes.
Geographic Location: Africa.
Summary: A rich wildlife habitat, this savanna is bounded by deserts and shaped by climate extremes.

The Namibian savanna woodlands occur on Africa's Great Escarpment, a high ridge that separates the Kaokoveld Desert and Namib Desert from other interior biomes of southwestern Africa. The woodlands extend from north of the Fish River Canyon through Windhoek, the capital of Namibia, and on into southwestern Angola, spreading some 1,200 miles (1,940 kilometers) south to north. The biome is never more than about 200 miles (320 kilometers) wide.

There is a vast difference in rainfall across the region, with an average of 2 inches (60 millimeters) in the west, to 8 inches (200 millimeters) in the east—driven by the unpredictable thunderstorms during the October-to-March austral summer. While rainfall varies, the evaporation rate in the region exceeds 10 feet (3 meters) overall. Temperatures are predictable, and can range from 16 degrees F (minus 9 degrees C) on cold winter mornings to well above 104 degrees F (40 degrees C) on summer afternoons.

The stony flatlands of the highland plateau, at about 3,281 feet (1,000 meters) above sea level, are intersected by the broken and dissected remnants of the escarpment. This yields massifs such as Brandberg—6,686 feet (2,038 meters)—with its ancient White Lady rock artwork, and also the red rocks of Spitzkoppe. The Kunene River is the only perennial river in the ecoregion as the Swakop and Kuiseb rivers usually are dry.

Flora

The diverse topography and associated soils yield diverse flora. The ecoregion contains three vegetation types: mopane savanna, semidesert and savanna transition, and dwarf shrub savanna. Mopane trees (*Colophospermum mopane*) dominate the north, while the broad range of *Euphorbia* (or spurge) plants are common in the semiarid zone to the west. To the southeast is the Kalahari Xeric Savanna biome. In each of these areas, the mopane grows together with *Balanites welwitschii* in depressions and riverbeds, forming woodland stands. Also prominent in the mopane savanna are two species of genus *Sesamothamnus*, part of the sesame family *Pedaliaceae*. *S. benguellensis* is found along the Kunene River, and Herero sesame tree *(S. guerichii)* is spread across the mopane savanna. The succulent shrub *Ceraria longipedunculata* is typically abundant in the higher-altitude areas here.

The mopane savanna extends to the Brandberg massif in the south, which has an extremely high level of endemism (species found nowhere else) as an isolated, remnant inselberg (island mountain) supporting 90 species that are endemic to Namibia, such as eight that are found only at Brandberb, including quiver trees *(Aloe dichotoma)*, *Euphorbia guerichiana*, *Cyphostemma* spp., *Adenolobus* spp., Brandberg acacia *(Acacia montis-ustii)* and *A. robynsiana.*

Acacia senegal and *A. tortilis* are found mainly in the alluvial sands and silts along ephemeral rivers in the biome.

High diversity of the genus *Commiphora* of aromatic flowering plants—related to frankincense and myrrh—is particularly characteristic of both the mopane savanna to the north and the semi-desert and savanna transition zone. To the south are karoo shrubs and grasses such as *Rhigozum trichotomum*, *Parkinsonia africana*, *Acacia nebrownii*, *Boscia foetida*, *B. albitrunca*, and *Catophractes alexandri,* as well as smaller karoo bushes including *Pentzia* spp. and *Eriocephalus* spp. Tufted grasses, mainly *Stipagrostis* spp., are found scattered between the woody plants.

Fauna

The region is a center of faunal endemism, with seven endemic reptiles, including Alberts burrowing skink and the Nama padloper tortoise. The Brandberg thick-toed gecko is found only on the Brandberg massif, and the Kaokoland escarpment has the highest levels of bird endemism in Namibia, with 297 species present, including the Herero chat. Among the mammals, Hartmann's mountain zebra roams the western sections of the Namibian savanna woodlands and is virtually endemic to Namibia.

Common but iconic African wildlife also occur here. Etosha National Park protects many of them in more than 8,880 square miles (23,000 square kilometers) of Namibian savanna woodland surrounding the vast Etosha Pan. Desert-dwelling elephants move vast distances between water sources and foraging grounds in the biome. These are the largest elephants in Africa, but their tusks are invariably broken and stunted due to a mineral deficiency in the soils that makes the ivory brittle.

The region also supports the last unfenced black rhinoceros population in the world; individuals range over 965 square miles (2,500 square kilometers) and can go without water for up to four days. These behemoths have a flexible diet in this plant-deficient region but have a passion for euphorbias, despite their toxicity to most animals.

Black-rhinoceros numbers plummeted in the 1970s through illegal poaching for their horns, which are used for Middle Eastern dagger handles and Asian semitraditional medicines. The population fell from 70,000 in the 1960s to 2,500 individuals in 1995. Northeastern Namibia faced some poaching, and conservation biologists removed the horns from some rhinos in an effort to reduce their attraction to poachers.

The region supports a unique black-faced form of the common impala that appears to be genetically distinct. Springboks, kudus, gemsboks, zebras, and wildebeest also all provide meals for the abundant large predator community that includes lions, leopards, spotted hyaenas, and cheetahs.

Human Settlement

The Khoisan people lived around Etosha for centuries. The Owambo named it *Etosha*, meaning *big white place*, in honor of the salt pan, which is 80 by 45 miles (129 by 72 kilometers) in scope. The pan is hypersaline, which excludes plant growth, but soaks around the edges provide valuable fresh water for animals. The pan is generally empty, but can fill in places during the summer; it then supports flocks of flamingos and pelicans.

The Khoisan have left their mark on the landscape through their majestic rock art and petroglyphs that reflected their spiritual beliefs. Places in the Namibian savanna woodlands like Twyfelfontein have galleries of rock carvings of all manner of creatures, including giraffes, lions, and eland. The rock-art galleries in the Brandberg depict similar standard images, but the meaning of the White Lady has confounded anthropologists for decades.

Europeans did not enter the region until Francis Galton (Charles Darwin's cousin) and Charles Andersson passed through in 1851. The rinderpest plague in the 1890s devastated wildlife and livestock numbers in southern Africa, and created an impetus for German authorities to become more influential in the region. They created a livestock-free zone on the southern edge of the Etosha Pan to minimize the risk of transmission of the disease. Governor Friedrich von Lindquist created Etosha National Park in 1907. At one stage, Etosha was the largest conservation area in the world, at more than 38,610 square miles (100,000 square kilometers), but it was reduced in 1967 to its current size.

Boreholes were sunk to open the land for pastoralism, but the names of the boreholes around Etosha attest to the difficulties the early explorers faced in trying to open the region: Stinkwater and Bitterwater are among these. Today, each of the major rest camps in Etosha occupies a former German fortified outpost, and many camps are named after the early explorers and the park's creator. The waterholes at the Okauhuejo, Namatoni, and Halali camps are lit at night; they are among the best places in the world to see black rhinos, elephants, and lions.

Environmental Threats

The region faces a range of threats, irrespective of the protection offered by Etosha National Park, Brandberg National Monument, Damaraland Wilderness Reserve, and the surrounding community conservation areas, such as ecotourism destinations like the Ongava and Onguma game lodges. Fencing along the 528-mile (850-kilometer) border of Etosha limits animal access to areas of fresh plant growth after localized rainfall, while fire and elephants are thought to have contributed to the alteration of the fauna-community dynamics such that the blue wildebeest population declined from more than 25,000 to just 2,300 by the late 1990s. While the ecotourism benefits created via the communal conservancies are large, they also can cause problems of poaching.

Climate change could affect migration patterns of animals, forcing them to areas they don't usually travel to, and stranding them in areas unsuited to their dietary needs. Rising temperatures will further dry the arid areas, which could lead to species loss. It is likely that climate change will exacerbate the weather extremes of the region, reflected in both increased temperature and rainfall, making all species vulnerable.

MATT W. HAYWARD

Further Reading

Berger, J. "Science, Conservation, and Black Rhinos." *Journal of Mammalogy* 75 (1994).

Berger, J. and C. Cunningham. "Active Intervention and Conservation: Africa's Pachyderm Problem." *Science* 263 (1994).

Leggett, K. "Home Range and Seasonal Movement of Elephants in the Kunene Region, Northwestern Namibia." *African Zoology* 41 (2006).

Maggs, G. L., P. Craven, and H. H. Kolberg. "Plant Species Richness, Endemism, and Genetic Resources in Namibia." *Biodiversity and Conservation* 7 (1998).

Simmons, R. E., M. Griffin, R. E. Griffin, E. Marais, and H. H. Kolberg. "Endemism in Namibia: Patterns, Processes and Predictions." *Biodiversity and Conservation* 7 (1998).

Nariva Swamp

Category: Marine and Oceanic Biomes.
Geographic Location: Caribbean Sea.
Summary: A jewel nestled on the southeast coast of Trinidad and Tobago that straddles the line between conserving biodiversity and supporting livelihoods for the dependent human populations.

The Nariva Swamp, the largest freshwater wetland in Trinidad and Tobago, lies on the southeast coast of Trinidad and covers an area of approximately 23 square miles (6,000 hectares). Surrounded by a multitude of Moriche Palms and coconut trees, the swamp presents a myriad of opportunities for exploring both the natural beauty and built environments. The swamp has a forest reserve, Bush Bush Wildlife Sanctuary. The Forests Act, Conservation Act, the State Lands Act, and the National Wetlands Policy protect the entire biome. As one of the first protected wetlands in Trinidad and Tobago (the Caroni Swamp and Bon Accord Lagoon are also protected), the Nariva Swamp has both ecological and anthropogenic significance for the wider Trinidadian public as well as for the communities that dwell in and around the resource.

A view of the Nariva Swamp in 2006, with the mouth of the Nariva River on the left flowing into the Atlantic Ocean on the east coast of Trinidad. The swamp is among the first protected wetlands in Trinidad and Tobago. (April Karen Baptiste)

The land use in this area is diverse; it consists of a mixture of natural and human landscapes. The vegetation includes stands of palms next to swamps, mangroves, forested areas, and agricultural plots. It has rich flora and fauna, many threatened species, including the globally threatened West Indian manatee, *Trichechus manatus*. The swamp is natural habitat for several species in Trinidad and Tobago, and is important for maintaining the biological diversity in the Caribbean region. It is also known for the red howler monkey (*Alouatta macconnelli*) and the blue and gold macaw (*Ara ararauna*), the latter of which has been only recently repatriated into its natural habitat.

Reliance and Protection

The Nariva Swamp is almost entirely state or government owned. There are some surroundings that are private property; however, in many cases these lands are either leased or are occupied by squatters seeking redistribution letters from the state. Several human communities in the wetland are dependent on the resources for their social and economic needs, which include hunting, subsistence fishing, rice cultivation (which is no longer practiced in large volume), catching of conch and crabs, and the gathering of firewood and plant products for use in the crafts industry.

Several species of birds are caught mainly for the pet trade, and herds of water buffalo are kept in the marsh. With this relatively high human dependence on the Nariva Swamp for livelihoods, the area has been threatened by these stressors, and as such has led to a governmental and nongovernmental push to protect the ecosystem for further damage.

Trinidad and Tobago took action to protect the Nariva Swamp formally in 1992, when it signed the Ramsar Convention on Wetlands. This international policy, which was created in 1971, reflects a commitment by countries to provide guidelines for wise use and protection of wetlands. By making a commitment to this convention, Trinidad and Tobago began to prioritize the sustainability of Nariva Swamp, and launched programs that attempted to reduce and rectify the damage that had been sustained by the ecosystem.

One of these initiatives involved an extensive study conducted in 1999 and updated in 2005 by the international conservation group Ducks Unlimited—in collaboration with the Trinidad and Tobago national government and the United States Department of Agriculture Forest Service—to outline areas that needed extra management in the swamp. This initiative recognized several issues, including fire protection during the dry season; reduction in rice farming, which was damaging the soils; and reducing or managing deforestation.

Through its recommendations, there is now an ongoing reforestation project along both the entire coastal strip and a major portion of the inland margin of the swamp, where illegal rice farming once had taken place, as well as a volunteer fire patrol system where villagers monitor fires to ensure that they do not get out of control. In addition, rice cultivation has been banned, even though villagers still engage in small-scale, short-term cash crop farming in order to maintain their families. Several nongovernmental organizations, including the Nariva Environmental Trust and the Manatee Conservation Trust, along with nature-based tourism providers engage in eco-tours of the swamp to promote awareness and increase the education of the general public—local, regional, and international—about the value of this precious resource.

APRIL KAREN BAPTISTE

Further Reading

Baptiste, April K. *Evaluating Environmental Awareness: A Case Study of the Nariva Swamp.* Albany: State University of New York Press, 2008.

Mohlenbrock, Robert H. "Nariva Swamp, Trinidad." *Natural History* 102, no. 10 (1993).

Nariva Swamp Development Project. *Studies on the Biological Resources of Nariva Swamp, Trinidad, Vol. 1.* St. Augustine, Trinidad and Tobago: University of the West Indies, 1979.

Sletto, Bjorn. "An Alluring Course for Trinidad's Wetlands." *Americas* 50, no. 5 (1998).

Surhone, Lambert M., Mariam T. Tennoe, and Susan F. Henssonow. *Nariva Swamp.* Beau Bassin, Mauritius: Betascript Publishing, 2011.

Trinidad and Tobago Environmental Management Authority (EMA). *Nariva Swamp Restoration Initiative, Trinidad and Tobago—Terms of Reference for the Development of a Reforestation Scheme for the Nariva Swamp.* Port of Spain: Trinidad and Tobago EMA, 2006.

Narragansett Bay

Category: Marine and Oceanic Biomes.
Geographic Location: North America.
Summary: A good stewardship plan is in place to help Rhode Island's premiere ecological jewel recover from its rough industrial history.

Narragansett Bay, situated on the eastern side of Rhode Island, comprises about 15 percent of the state's total land mass. Ninety-five percent of the bay's surface area is in Rhode Island, with the remainder in southeastern Massachusetts; about 60 percent of the bay's watershed is in Massachusetts, as well. At the head of Narragansett Bay lies the city of Providence. This area of Narragansett Bay, called the Providence River, derives its major freshwater from four rivers: Blackstone, Moshassuck, Woonasquatucket, and Pawtuxet. The bay then widens into a more typical estuary that traverses suburbs and rural areas, and past Patience, Prudence, Conanicut, and Aquidneck Islands.

The mouth of Narragansett Bay opens into Rhode Island Sound, where the North Atlantic Ocean brushes up against Block Island, the state's largest offshore island. The bay has numerous smaller side embayments, the largest of which is Mount Hope Bay to the east; it receives fresh water from the Taunton River. Narragansett Bay's watershed areas have a legacy that combines indigenous, colonial, nautical, and industrial heritage manifest in its villages, ponds, beaches, forests, and rivers.

Narragansett Bay is a classic temperate drowned-river-valley estuary, formed from the action of glacial scouring and interglacial sea-level rise. It has two passages split by the islands in its center. The average depth is 26 feet (8 meters), and the bay deepens to more than 180 feet (55 meters) at its mouth. The average salinity is about 30 parts per thousand; it is saltier at the mouth and fresher at the head of the bay, where 90 percent of the freshwater flows in. The Providence River is highly stratified with a discernible freshwater layer above more saline waters, whereas the lower portion of the bay is well mixed by tidal and wind-induced circulation.

These coastal habitats have been identified in and around the bay: beaches, brackish marshes, common reed stands, sand dunes, eelgrass beds, oyster reefs, rocky shores, subtidal sand and mud sediments, salt marshes, shrub wetlands, streambeds, and tidal flats.

Biodiversity

Because of the rich assembly of habitats, many important fauna have been observed throughout the bay. The base of the food web is phytoplankton, the tiny free-floating plants that are the principal primary producers consumed by fish, crustaceans, and shellfish. These tiny plants use the sunlight, carbon dioxide, and dissolved nutrients to bloom in the late winter and early spring, mainly as diatoms. As the season progresses, other phytoplankton bloom, and tiny zooplankton and other animals—such as jellylike animals (comb jelly), clams (filter feeders), and small fish—actively consume them.

Numerous species of aquatic invertebrates—such as the hard-shell clam, bay scallop, horseshoe crab, and American lobster—inhabit the bay. Larger, more conspicuous animals, such as the northern diamondback terrapin and the harbor seal, dwell in the near-shore environment. Myriad benthic and pelagic fin fish—including black sea bass, bluefish, and winter flounder, as well as the American eel—are common throughout the bay. Shorebirds such as the American oystercatcher, black-crowned night-heron, common tern, double-crested cormorant, glossy ibis, great blue heron, herring gull, least tern, piping plover, and little blue heron thrive in and around this biome.

Human Activity

Narragansett Bay played a large role in the industrial development of the United States. It is often considered the birthplace of the American Industrial Revolution. The calm and sheltered waters of the bay provided easy transport of both raw materials and products of the industrial period, including textiles and jewelry. Due to these activities, the shoreline of the northern portion of Narragansett Bay has been significantly altered from its natural state.

Nowadays, the industrial sector is almost nonexistent, as economic changes have transformed commerce into the service and tourist industries of today. The population of the watershed increased markedly during the 20th century, reaching nearly 1.8 million. The bay now is the centerpiece of Rhode Island's tourist activity, with nearly 6 million people visiting for bathing, boating, fishing, and other recreational activities each year.

Threats and Conservation

The historical legacy of the region has left its mark on the environmental makeup of the bay, evidenced by sediments that were contaminated by both heavy metals and toxic organics. Since the exodus of the old industrial activities and effective regulation under the Clean Water Act, cleaner sediments have been deposited on top of contaminated layers. Over recent decades, bacterial and nutrient pollution, along with climate-related changes, have overtaken toxics as the main environmental concerns for the estuary.

Major planned and current upgrades to wastewater and storm water infrastructure are expected to improve the quality of the Bay. In 1987, Narragansett Bay was designated an "estuary of national significance," which marked it, along with 27 other estuaries, for special consideration, protection, and restoration under the U.S. Environmental Protection Agency's National Estuary Program (NEP). Each NEP site has developed a Comprehensive Conservation and Management Plan (CCMP) spelling out goals for the estuary and the steps needed to achieve them.

The goals of the CCMP for Narragansett Bay reflect the priorities that were current when the plan was written in 1992. Today, the state of Rhode Island, federal partners, and regional stakeholders are updating the management objectives through a visioning process that will articulate current priorities in light of what has been learned and achieved over the past 20 years.

The future ecological health of Narragansett Bay is not clear. The fisheries of the bay have changed over the years, with early data indicating an abundance of alewife, shad, and smelt. Recent data show declines in flounder species, scup, and menhaden; losses of soft-shell clams, scallops, and oysters also are apparent. By contrast, late-20th-century increases in hard-shell clams and lobster had been documented. These species, however, have shown more recent declines that seem due to increased fishing pressure, as well as to environmental changes such as water-temperature increases.

Narragansett Bay and its watershed have benefited from the cultural shift toward environmental stewardship that began in the 1970s. There have been clear success stories, including reduced toxic inputs resulting in lower sediment contamination, reduced shellfish harvesting, and beach-closing restrictions from better stormwater management. Today, however, the combination of larger-scale issues such as climate change, suburbanization, and competition for scarce resources pose a risk to continued improvements. Nevertheless, there is a strong environmental commitment by the citizens of southern New England, which, coupled with a concentration of educational and research institutions in the Narragansett Bay watershed, makes for a strong possibility of improvement in ecological and human well-being in the years to come.

James S. Latimer

Further Reading

Burroughs, R. H. and V. Lee. "Narragansett Bay Pollution Control: An Evaluation of Program Outcome." *Coastal Management* 16 (1988).

Desbonnet, A. and B. A. Costa-Pierce, eds. *Science for Ecosystem-Based Management: Narragansett Bay in the 21st Century.* New York: Springer, 2008.

Desbonnet, A. and V. Lee. *Historical Trends: Water Quality and Fisheries: Narragansett Bay.* Narragansett, RI: University of Rhode Island Coastal Resources Center, 1991.

Narragansett Bay Estuary Program (NBEP). NBEP Website. http://www.nbep.org.

Nasser, Lake

Category: Inland Aquatic Biomes.
Geographic Location: Africa.

Summary: Lake Nasser and the Aswan High Dam have produced great benefits to the people and life in Egypt, enhancing certain aquatic ecosystems and increasing the presence of migratory birds and mammals. These projects also have upset traditional ecological patterns and led to new environmental problems.

One of the largest human-made lakes in the world, Lake Nasser was created on the Nile River as the brainchild of former Egyptian President Gamal Abdel Nasser. Eighty-three percent of the lake is located in Egypt, in the southeastern portion of the country, with the remaining area in northern Sudan. The lake came into being in 1960 in conjunction with the construction of the Aswan High Dam. Egypt has used the lake and dam to generate hydroelectric power, provide fishing and irrigation for local residents, and prevent areas below Aswan from flooding uncontrollably, as they had done in the past. (The Sudanese reject the name Lake Nasser, preferring to call it Lake Nubia for the people who populate the area.)

The lake, which extends for some 1,550 square miles (4,014 square kilometers), is 600 feet (183 meters) deep in some areas. Throughout its history, Egypt has continued to view the Nile River as its key to survival. Fully 95 percent of Egypt's population still resides around the Nile, an area that comprises only 5 percent of the country's total land area. Outside experts have been highly critical of Egypt's stewardship of the river, considering the nation's handling as a misuse of the Nile's resources; they point out that Egypt continues to depend on imported food products for survival.

The lands around the Nile have historically been extremely fertile, but damming the waters has caused silt to be trapped rather than deposited downstream, and farmers have begun using human-made fertilizers that pollute waters of the surrounding area. The fertilizers have helped the bilharzia parasite, spread by snails attracted to standing waters, to infect people who live nearby.

Flooding has always been a major problem in the Nile basin. Intentional flooding during the building of the dam forced tens of thousands of Nubians who lived along the banks of the lake to relocate to higher ground. The flooding also destroyed numerous ecosystems, along with ancient sites of major historical significance. Historically, the shores of the lake had been dotted with Nubian temples, ruins, and monuments. Because of flooding, the United Nations Educational, Scientific, and Cultural Organization (UNESCO) stepped in to help in the 1960s, assisting Egyptian officials in relocating historical artifacts to safer, inland sites.

The Nile River has been a major reason for ongoing tension between Egypt and the Sudan, and those tensions mounted during negotiations for the construction of Aswan High Dam. The two countries did reach agreement in 1959, allotting the Sudan 653 billion cubic feet (18.5 billion cubic meters) of water from the Nile, rather than the 141 billion cubic feet (4 billion cubic meters) that had been in effect from 1929 to that date. Additionally, more than 2,510 square miles (6,500 square kilometers) of northern Sudan were flooded, including some of the most fertile land in the world. The Sudanese government agreed to engage in Nile development projects of its own, and in the 1990s, a new dam built along the border of the Sudan and Ethiopia displaced some of the same people who lost their homes when Lake Nasser and the building of the Aswan High Dam were under development.

Nasser Lake is thought to be the only breeding ground in the world for the Egyptian goose (Alopochen aegyptiacus). (Wikimedia/Andreas Trepte)

Climate

The climate around Lake Nasser is typical of extreme desert biomes: Residents may go for several years without ever seeing rain. In 1997, at a total cost of approximately $17 billion, then-president Hosni Mubarak of Egypt decided to use some of the water from Nasser Lake to create a "Nile clone" in the western desert. Mubarak accomplished this goal by transferring 3.6 billion gallons (13 billion liters) of water a day out of the lake into the New Valley Canal, to help irrigate a fertile valley among a string of ancient oases. The ambitious plan, which is expected to stretch for 192 miles (309 kilometers) when completed, involves creating 1 million acres (404,686 hectares) of new agricultural lands that would become home to evolving animal and plant habitats.

The initial plan was to entice some 7 million Egyptians to relocate by offering the land at low prices, promising an abundance of water, and granting tax breaks. In 2005, the National Project for the Development of Upper Egypt, or the Toshka Project, was in place on the northern section of Lake Nasser. It was considered a major engineering feat, pumping water from Lake Nasser into the Sheik Zayed Canal, which was named after the president of the United Arab Emirates, who donated $100 million to the project. The end stage of the project, which is set to provide 100,000 acres (40,469 hectares) of irrigated farmland, and to be home to 3 million people, is set to be completed in the 2020s.

Erosion is a major problem along the shores of Lake Nasser, and fossils of ancient plants and animals are often clearly visible. Climate change is projected to have both positive and negative effects on the lake. Greater swings in rainfall amounts and wider temperature fluctuations could mean increased evaporation and shore erosion, and strained water sources if rain eases too much. Since Lake Nasser is fed by the Nile River, a changing water table could affect everything from fish communities and local plant life to recreational activities. In addition, the farmland surrounding the lake could increase the incidence of pollution from pesticides, and the additional population could strain the resources of the lake community.

Biodiversity

The shores of Lake Nasser are composed of 85 major khors, or desert wadis—narrow, meandering shallows that provide an ideal environment for aquatic flora and vegetation, and serve as a breeding ground for fish. There are 48 khors on the eastern side of the lake and 37 on the west, the three largest being Allaqi, Kalabsha, and Tushka. These khors offer the richest habitats found in the Lake Nasser biome.

Fish have been introduced into the lake, and many species have thrived in the freshwater environment, including tilapia, Nile perch (*Lates niloticus*) or sangara, mputa or capitaine, tiger fish, and large vundu catfish.

Lake Nasser is considered one of the best spots in the world for freshwater fishing. Although commercial fishing is banned, the fish from the lake account for 25–40 percent of all inland fishing production in the country, and local peasants who live in temporary camps depend on those fish for their survival.

Drawn by the abundance of available food, birds flock to the western bank of Lake Nasser, and approximately 100 species have been recorded, including the tufted duck (*Aythya fuligula*). In the first two months of 1995, more than 200,000 birds were counted on the lake, which is now considered one of the most important wetlands in Egypt. Migrant and overwintering birds use the shores of the lake as a staging ground, and Lake Nasser is the only known breeding ground for the Egyptian goose (*Alopochen aegyptiacus*).

In addition to the tufted duck, among the most abundant species are black-necked grebe (*Podiceps nigricollis*), great white pelican (*Pelecanus onocrotalus*), common pochard (*Aythya ferina*), northern shoveller (*Anas clypeata*), Eurasian wigeon (*Anas penelope*), and black-headed gull (*Larus ridibundus*).

Characteristic breeding birds include Egyptian goose, black kite (*Milvus migrans*), Senegal thick-knee (*Burhinus senegalensis*), Kittlitz's plover (*Charadrius pecuarius*), spur-winged lapwing or spur-winged plover (*Vanellus spinosus*), crested lark (*Galerida cristata*), and the graceful prinia warbler (*Prinia gracilis*).

This is the only area where African skimmer (*Rynchops flavirostris*) and African pied wagtail (*Motacilla aguimp*) are known to breed in Egypt. During the summer months, there is a significant influx of yellow-billed stork (*Mycteria ibis*) and pink-backed pelican (*Pelecanus rufescens*) into Lake Nasser.

Large numbers of animals and reptiles are drawn to Lake Nasser, including the Nile crocodile, soft-shell turtle, monitor lizard, Dorcas gazelle, jackal, sand cat, and desert fox. Bedouins often bring their camels and sheep to graze on the sparse vegetation growing along the shore, which is largely dominated by salt cedar (*Tamarix*) growing in thin bands. Dominant aquatic vegetation in the shallow margins of the lake includes water-lilies (*Najas*). Also along the lake's banks are palm trees, including the type common throughout the Nile and Egypt, the date palm.

Elizabeth Rholetter Purdy

Further Reading

Belal, Ahmed, et al. *Bedouins by the Lake: Environment, Change, and Sustainability in Southern Egypt.* Cairo and New York: American University in Cairo Press, 2009.

Cooperman, Alan. "Egypt Clones a Nile." *U.S. News and World Report* 122, no. 19 (1997).

Crisman, Thomas L., et al. *Conservation, Ecology, and Management of African Fresh Waters.* Gainesville: University Press of Florida, 2003.

Mange, Maria A. and David T. Wright, eds. *Heavy Metals in Use.* London: Elsevier, 2007.

Wohl, Ellen E. *A World of Rivers: Environmental Change on Ten of the World's Greatest Rivers.* Chicago: University of Chicago Press, 2011.

Nebraska Sand Hills Desert

Category: Grassland, Tundra, and Human Biomes.
Geographic Location: North America.
Summary: A largely intact native temperate grassland habitat that features wetlands and small lakes scattered throughout a mixed-grass prairie ecosystem.

Located almost entirely within the state of Nebraska, near the geographic center of the United States, the Sand Hills Desert biome encompasses one of the largest sand dune formations in the Western Hemisphere. It also is one of the largest contiguous and least-disturbed prairies, and among the largest and most complex wetland ecosystems in the United States. Today, it faces several environmental challenges, such as the overgrazing of cattle, expansion of agriculture, climate change, and drought.

The Nebraska Sand Hills Desert is a relatively stable, mostly intact ecoregion that spans the Great Plains region and High Plains subregion; it covers about one quarter of the state of Nebraska. The Sand Hills have a more intact natural habitat—about 85 percent—than most other grasslands in the Great Plains. This is due primarily to the absence of agriculture, as very little of the Sand Hills has been plowed.

The Sand Hills are classified as a semiarid region, with average annual rainfall from 23 inches (580 millimeters) in the east, to less than 17 inches (430 millimeters) in the west. Temperatures range from minus 30 degrees F (minus 34 degrees C) in winter to 100 degrees F (38 degrees C) in summer.

The Sandy Formations

One of the smallest of the Great Plains ecoregions, the Sand Hills represents a relatively stable and intact biome of mixed grasslands covering an area variously defined as 19,600 or 23,600 square miles (50,764 or 61,124 square kilometers). Its topography is characterized by mixed-grass and tallgrass prairie, and a wide variety of sand formations, most prominently sand dunes, some of which are stabilized by vegetation. Some of the dunes can reach heights of 340–400 feet (104–122 meters) and stretch up to 2 miles (3 kilometers) in length.

Other formations include blowouts—holes in the surface of the sand created by rapid wind erosion that can reach several hundred feet (meters) in diameter—and widely spaced barchans, which are arc-shaped ridges of sand with downwind-pointing "horns." The average elevation of the region increases gradually from 1,800 feet (549 meters) in the east to 3,600 feet (1,097 meters) in the west.

The Sand Hills sit atop the vast Ogallala Aquifer, one of the world's largest aquifers. When the water table of the massive but shallow Ogallala, also known as the High Plains Aquifer, rises above interdunal valleys, then wetlands, small ponds, or shallow lakes are formed, with a typically average area of approximately 10 acres (4 hectares). These can be seen scattered throughout the dry and sandy grassland dunes in the western and northern regions.

Creeks, rivers, and subirrigated meadows also are nourished by Ogallala's precipitation-charged groundwater, giving rise to rich and varied vegetation that provides ideal grazing ground for both wild ruminants and farm-raised cattle. The western part of the Sand Hills is drained by small interior drainage basins, while the eastern and central areas are drained by tributaries of the Niobrara and Loup Rivers. The Snake and Dismal Rivers also run through the region.

The sands are generally stabilized by the root systems of local vegetation spread throughout the sandy soil, comprised primarily of entisols, the second-most abundant soil order in the world after inceptisols. The Sand Hills represent the most intact natural and continuous habitat of all the ecoregions in the Great Plains, supporting a diverse and extensive array of flora and fauna.

Flora

While the Sand Hills ecoregion falls into the mixed-grass prairie classification, its singular landscape of dry and sandy upland dunes interspersed with thousands of lakes and wetlands becomes a biodiverse community of sand-tolerant vegetation from various prairie types. These include some species found in northern boreal forest ecosystems. Consequently, the Sand Hills are considered a distinct grassland association, supporting more than 700 species of plants, the majority of which are native. Only about 7 percent of Sand Hills flora are non-native species. Half of them are found in most other prairie ecoregions, due in part to the difficulty that foreign plant species have in negotiating the sandy soil.

While supporting a variety of short-grass, mixed-grass, and tallgrass species, the Sand Hills are dominated by a few hardy perennial monocots that are good for grazing wildlife and cattle, such as sand bluestem (*Andropogon hallii*), a bunchgrass known as turkey-foot that is high in crude protein and ideal for sandy-soil stabilization; prairie sandreed (*Calamovilfa longifolia*), an abundant, drought-tolerant, warm-season tallgrass, the seeds of which feed songbirds and rodents; and needle-and-thread (*Stipa comata*), a cold-winter bunchgrass that provides excellent forage in the fall and winter months.

Found only in the Sand Hills and central Wyoming, the endangered Hayden's penstemon (*Penstemon haydenii*), a stout perennial flowering herb, has adapted to survive within blowouts. Hence, it is also known as blowout penstemon, and it stabilizes the sandy soil. However, penstemon dies off when other species begin to recolonize the newly stabilized surface. It may take years for a blowout to return to its original stabilized state.

Additionally, ranchers are generally careful not to let cattle overgraze, fearing that a reduction of vegetation will lead to more blowouts. The prevention of naturally occurring fires has led to fewer blowouts. Such land management practices have increased the frequency of many prairie plants and significantly reduced the habitat of the Hayden's penstemon. Today, it remains Nebraska's only endangered plant.

Another species that thrives in blowouts is the sandhill muhly (*Muhlenbergia pungens*), a drought-tolerant bunchgrass well adapted to coarse-textured soil.

Fauna

The sandy-bottomed ponds, lakes, rivers, and streams that exists throughout the Sand Hills provide an ample freshwater source for wildlife and cattle, and also support many fish species. Some of the lakes are alkaline, and support several phyllopod shrimp species.

Game fish, primarily yellow perch, northern pike, largemouth bass, bluegill, and carp, were introduced into many lakes in the Sand Hills, and trout was added to a number of coldwater streams. In addition, there are 27 species of amphibians and reptiles in the region, including one salamander, three toads, four each of frogs and lizards, six turtles, and nine snakes.

The Nebraska Sand Hills Desert biome supports numerous insect species, including spiders, dragonflies, grasshoppers, and mosquitoes, the populations of which increases during the summer when bodies of still water increase across the region.

Limited human development and sprawling virgin prairie make this ecoregion a long-term stable habitat for varieties of animals, and an important habitat for more than 300 vertebrate species. Birds including waterfowl, shorebirds, songbirds, and birds of prey are particularly well represented in this area. Because of urban and agricultural development elsewhere in United States, the Sand Hills ecoregion has increased in importance as a last refuge for some bird populations.

The natural range of the greater prairie chicken (*Tympanuchus cupido*), a large bird in the grouse family, for instance, has been decreasing across central North America. Its territory has been impacted by human expansion, primarily agricultural development, and the prairie chicken has become endangered or extinct in some parts. Consequently, the species is listed as vulnerable by the International Union for Conservation of Nature (IUCN).

The Sand Hills sit along the Central Flyway, a bird migration route connecting Canada and Mexico through the Great Plains. Favored because of its lack of mountains and ample refuge, availability of food, and freshwater sources, the Central Flyway is used by some migratory birds traveling from the Arctic Circle to the southern tip of South America.

The ponds and lakes of the Sand Hills provide temporary refuge for several migratory bids, including ducks, geese, and cranes. Birds that make the Sand Hills their primary habitat include the wild turkey and Nebraska's state bird, the western meadowlark (*Sturnella neglecta*), a ground-foraging icterid.

Some of the many other birds that rely on the Sand Hills ecoregion include hunting birds such as Swainson's hawk (*Buteo swainsoni*) and burrowing owl (*Athene cunicularia*); shorebirds such as the upland sandpiper (*Bartramia longicauda*), long-billed curlew (*Numenius americanus*), and willet (*Catoptrophorus semipalmatus*); and seabirds such as the trumpeter swan (*Cygnus buccinator*), American white pelican (*Pelecanus erythrorhynchos*), and double-crested cormorant (*Phalacrocorax auritus*).

Also featured here are turkey vulture (*Cathartes aura*), sharp-tailed grouse (*Tympanuchus phasianellus*), barn swallow (*Hirundo rustica*), great crested flycatcher (*Myiarchus crinitus*), horned lark (*Eremophila alpestris*), loggerhead shrike (*Lanius ludovicianus*), western kingbird (*Tyrannus verticalis*), eastern kingbird (*T. tyrannus*), common nighthawk (*Chordeiles minor*), and lark sparrow (*Chondestes grammacus*).

Native mammalian species represented in the Nebraska Sand Hills include pronghorn antelopes, mule deer, white-tail deer, coyotes, red foxes, ground squirrels, bats, kangaroo rats, and porcupines.

The millions of wild bison that once roamed the Sand Hills have been replaced by more than 500,000 cattle. Today, one of America's largest bison herds lives on protected land at the Nature Conservancy's Niobrara Valley Preserve. Elk, which once roamed freely, are now found only at the Fort Niobrara Wildlife Refuge. Other protected areas include Valentine National Wildlife Refuge and Crescent Lake.

A view of the Sand Hills in Whitman, Nebraska. The Sand Hills cover 25 percent of the state and offer habitat that is 85 percent intact. (U.S. Geological Survey/Byron W. Sellers)

Environmental Threats

Environmental challenges to the Nebraska Sand Hills Desert biome are related primarily to human activity and include overgrazing by cattle, prevention of naturally occurring brush fires, and the development of agriculture using center-pivot irrigation methods.

Invasive species also are a threat, from the carp that have degraded many lakes, wetlands, and streams to the purple loosestrife and reed canary grass that threaten riparian areas and wetlands. The musk thistle and leafy spurge are negatively impacting prairie communities, as are Eastern red cedars.

Other concerns include interbasin water transfers that lower the water table, impact wetlands, and reduce flows in streams and rivers, as well as oil pipeline construction that could pollute the Ogallala Aquifer. Climate change poses a challenge to this desert biome, as warming average temperatures may make the grasslands here more unstable, increasing the chance of fires, mild drought and erosion.

REYNARD LOKI

Further Reading

Bleed, Ann. *An Atlas of the Sand Hills*. Lincoln: University of Nebraska at Lincoln, 1990.

Jones, Stephen. *The Last Prairie*. Camden, ME: Ragged Mountain Press, McGraw-Hill, 2000.

Ricketts, Taylor H., et al. *Terrestrial Ecoregions of North America: A Conservation Assessment*. Washington, DC: Island Press, 1999.

Nelson River

Category: Inland Aquatic Biomes.
Geographic Location: North America.
Summary: The ecosystem of Manitoba's Nelson River has experienced environmental changes ranging from rising water levels to coastal erosion in the wake of hydroelectric development.

Spanning much of the central Canadian province of Manitoba, the Nelson River drains from Lake Winnipeg and runs a length of 400 miles (644 kilometers), with its mouth providing freshwater inflow to Hudson Bay. The river's considerable volume and long drop have historically made it useful in hydroelectric generation, which has caused the river to be dammed repeatedly. This has led to controversy with First Nations (aboriginal Canadians) and concerns among environmentalists about the changes wrought on the region's ecosystem. The river's drainage basin affects a huge area, including four Canadian provinces and two American states.

Hydrology and Climate

The total length of the Nelson River, including the Saskatchewan River system, is about 1,600 miles (2,575 kilometers), and includes water from the Red, Grass, Burntwood, and Winnipeg Rivers, as well as from Lakes Cross, Sipwesk, Split, and Stephens. The Nelson River discharge rate is 84,000 cubic feet (2,379 cubic meters) per second. Its drainage basin of more than 892,000 square miles (2.3 million square kilometers) includes about 70,000 square miles (181,299 square kilometers) in the United States. This basin includes the Canadian provinces of Alberta, Saskatchewan, Manitoba, and Ontario; and the U.S. states of Montana, Minnesota, North Dakota, and South Dakota.

Located in an ecosystem characterized by a sub-Arctic climate, the region surrounding the Nelson River experiences short, cool summers and long, cold winters. Daily mean temperatures in January are minus 30 degrees F (minus 34 degrees C), while temperatures in July are in the range of 55–65 degrees F (13–18 degrees C). Annual precipitation averages 16– 20 inches (406–508 millimeters), of which approximately one-third falls as snow.

Traditionally, the riverine nature of many of the lakes in the Nelson River region are high in nutrients. The area has an ion concentration much higher than that of the nearby Churchill River, with elevated levels of chloride, magnesium, sodium, sulphate, and suspended sediments. Scientists believe this characteristic is due to the Nelson River draining primarily through weathered, fine-grained prairie soils, while the Churchill primarily drains the granitic Canadian Shield.

Fauna and Flora

Algal biomass is lower in the cloudy Nelson River than in surrounding lakes. The benthic macroinvertebrate fauna is dominated by midge flies (*chironomidae*), worms (*oligochaeta*), and fingernail clams (*sphaeriidae*). The fish found in lakes along the Nelson River are typical of relatively shallow boreal Canadian lakes, and diversity of fish is low. A small number of cool-water benthivores (bottom-feeders) such as lake whitefish (*Coregonus clupeaformis*), white suckers (*Catostomus commersoni*), walleye (*Stizostedion vitreum*), and northern pike (*Exos lucius*) are common, as is sturgeon. Commercial and domestic fisheries are present, although sport fishing is rare due to limited access, and the cost of production has kept the area from being overfished.

Mammals found along the Nelson River include polar bear, woodland caribou, timber wolf, moose, and black bear. Birds that frequent the area include North American ruddy turnstones, semipalmated sandpiper, dunlin, and least sandpiper. Among the waterbirds are white-winged scoter, surf scoter, sandhill crane, Bonaparte's gull, and bald eagles.

Vegetation in the river basin features tracts of grasses and willows interspersed with black spruce. The land behind the Nelson River estuary is boggy and has extensive expanses of stunted spruce.

Effects of Human Activity

Beginning in the mid-1950s, Manitoban power planners made the decision to build a series of hydroelectric generating stations on the northern part of the Nelson River. This process included constructing generating stations along the lower portions of the river at Kettle, Limestone, and Long Spruce, as well as the installation of a high-voltage direct-current transmission system to carry electricity to southern population centers. These changes diverted most of the flow of the Churchill River into that of the Nelson, causing certain ecological effects. Water levels of adjoining lakes reached 20-year highs shortly after the 1974 construction of a dam near Southern Indian Lake, raising its level by nearly 10 feet (3 meters).

Shoreline erosion rates and forms were altered by impoundment, especially since rising water levels swept over long-established granite beaches and began affecting land that was formerly permafrost. The change in water levels has caused shore erosion of nearly 30 feet (9 meters) per year in some areas, dramatically shifting the ecosystem. Sediment in the Nelson River, already high, has increased fivefold since the damming began.

The local flooding originally prompted increased crop production, due to the additions of organic matter and other nutrients, but subsequently this declined due to the depletion of the flooded organic matter. This caused a rapid response of macroinvertebrates in many of the reservoirs along the Nelson River. Zooplankton levels increased dramatically in some places, changes that coincided with the other effects of impoundment. Higher flow rates in certain areas of the Nelson River caused a decrease in measurable zooplankton levels elsewhere. Similarly, crustaceans responded in different ways to the damming of the river. Cyclopoid populations declined, while calanoids remained relatively stable.

Hydroelectric development had a negative effect on fisheries along the Nelson; the grade of the whitefish caught by commercial fisheries declined from export (A grade) to continental (B grade). Overall fish populations are down as well, and mercury levels in fish are up. Mitigation efforts have been undertaken to restore the Nelson River's ecosystem, but with limited effectiveness to date.

Climate change may further stress the area in terms of lower runoff by earlier seasonal snowmelt from diminished snowfall, resulting from higher temperatures, which will have ripple effects throughout the ecosystem.

Stephen T. Schroth
Jason A. Helfer

Further Reading

Bodaly, R. A., D. M. Rosenberg, M. N. Gaboury, R. E. Hecky, R. W. Newbury, and K. Patalas. "Ecological Effects of Hydroelectric Development in Northern Manitoba, Canada: The Churchill-Nelson River Diversion." In Patrick J. Sheehan, Donald R. Miller, Gordon C. Butler, and Philippe Bourdeau, eds., *Effects of Pollutants at the Ecosystem Level.* New York: John Wiley & Sons, 1984.

Coward, H. and A. J. Weaver. *Hard Choices: Climate Change in Canada.* Waterloo, Canada: Wilfrid Laurier University Press, 2004.

Minns, Charles K. "Factors Affecting Fish Species Richness in Ontario Lakes." *Transactions of the American Fisheries Society* 118, no. 5 (1989).

Nenjiang River Grassland

Category: Grassland, Tundra, and Human Biomes.
Geographic Location: Asia.
Summary: The Nenjiang River Grassland is an ecologically significant region under threat from agriculture and other human activities.

The Nen River, or Nen Jiang, rises in the mountains that form the borderland between China's northeast and the Russian Far East. It is a tributary of the Sungari River, which is in turn a tributary of the mighty Amur. It descends across the Songhua-Nenjiang, or Song-Nen, Plain, where it meanders through grasslands that have considerable ecological importance. Numerous tributaries flow into the Nen in its course across the grassland from both branches of the Khingan Ranges; these contribute to flooding in the region. Many soil types have been deposited by the different water flows, and the region is important for agriculture. As a result, there is more farming and fishing activity here, which threatens the sustainability of several species.

The Nen River is frozen for up to four months of the year. Flooding is a regular issue at the time of the thaw and during the summer rainy season, particularly at the confluences of the Nen and important tributaries. The river is navigable, by small craft at least, for extensive distances along the plain. River transportation was an important means of spreading populations, and developing economy and society in the premodern era; the Nen River consequently was instrumental in the early settlement of Heilongjiang Province in China.

The Nen River region has a continental monsoon climate, and is warmer and drier than the surrounding mountains. The mean annual precipitation is 15–17 inches (400–450 millimeters).

Biodiversity and Conservation

The grassland region as a whole consists of a complex mix of flowing and still water, interspersed with marshes and areas of reeds. The diverse environments represent space for fish, waterfowl, insects, and amphibians; the protected Siberian wood frog *Rana amurensis* is one of several species to depend on the area.

The extensive marshy and wetland habitats are protected to some extent by the Zhalong and Momoge nature reserves. The Zhalong reserve, known as Home of the Cranes, contains many permanently flooded areas, resulting in part from the overflowing Wuyu'er River, and is home to a variety of species, most notable of which are rare or vulnerable crane species. The 811 square miles (2,100 square kilometers) of the reserve are home to 46 fish species, more than 150 types of birds, 277 different kinds of insects, and 21 other animals.

Approximately 700 of the world's 1,000 endangered cranes are residents of the Nen, including the threatened red-crowned crane (*Grus japonensis*). Some 15 crane species are found in the Zhalong, which is an internationally important wetland area. Resident cranes include the white-naped crane (*G. vipio*), demoiselle crane (*Anthropoides virgo*), common crane (*G. grus*), the endangered Siberian crane (*G. leucogeranus*), and hooded crane (*G. monacha*), as well as other rare birds such as the endangered oriental white stork (*Ciconia boyciana*), black-headed ibis (*Threskiornis aethiopicus melanocephalus*), mandarin duck (*Aix galericulata*), and Eurasian spoonbill (*Platalea leucorodia*).

Cranes also are present in the Momoge Nature Reserve, which is known as Heaven for Birds, and is located mainly in Jilin Province. Its 355,832 acres (144,000 hectares) are 80 percent flooded. Cranes represented in this grassland region include hooded, demoiselle, and white-crowned cranes. Many cranes use this area as a stopping point while migrating to and from Siberia.

Typical vegetation in the lowlands of the Nen river basin consists of a distinctive coniferous swamp forest interspersed among meadows dominated by grasses and sedges. Forests are dominated by the larch (*Larix gmelini* spp. *olgensis*) and birch (*Betula japonica*). Meadows are filled with such grasses as *Calamagrostis epigeios* and *C. langsdorfii* that are adapted to flooded soils. These grasses often grow as dense tussocks. Lakes are often filled or lined at the margin by the salt-tolerant reed (*Phragmites communis*). Upland areas are dominated by grasslands, forest-grasslands with crooked elms, and shrub groves of wild apricot.

Common amphibians include toad (*Bufo raddei*), tree frog (*Hyla arborea*), and the frogs *Rana nigromaculata* and *R. amurenss*.

Environmental Threats

Unfortunately, salinity has begun to threaten parts of the grasslands, significantly impacting the area's habitat sustainability. Demand for water for agricultural irrigation can place pressure on the ability of the system to maintain its freshwater resources, which can intensify if the currently low level of tourism and economic development begins to increase. There is a further long-term threat represented by the possible presence of exploitable oil and gas reserves in the region.

River salinity concern is accompanied by overfishing and egg collecting, as well as pollution linked to agricultural and other human-based activities. The loss of soil may have long-term climate-change effects, which could alter the vegetation of the region. Wildfire incidents have grown to almost uncontrollable outbreaks when excessive reed clearance and low water levels have coincided, as occurred in 2000 and 2001, when more than 32,124 acres (13,000 hectares) of reedbeds were destroyed.

Experts are aware of the threats facing the region, and able stewardship has resulted in the creation of an incipient Asian Wetland Network project, which has brought together representatives from China with others from Iran, Russia, and Kazakhstan. Together, they are working to gather information, write manuals, and raise awareness.

Climate change poses additional threats. Rising temperatures and rainfall amounts could vary enough to impact the vegetation and the environment for local wildlife. Warmer temperatures could expand the agricultural lands and increase the accompanying chemical runoff that affects water levels and quality.

John Walsh

Further Reading

MacKinnon, John Ramsay, et al. *A Biodiversity Review of China.* Hong Kong: Worldwide Fund for Nature International, 1996.

Wang, Zhi-chaing, Jian-chun Fu, Chengyuan Hao, and Zhichao Chen. "The Spatial-Temporal Pattern Changes of the Red Crowned Crane (*Grus Japonensis*) Population in Zhalong NNR and the Related Driving Forces." *Acta Ecologica Sinica* 29, no. 6 (2009).

Xia, Xueqi, Zhongfang Yang, Yan Liao, Yujun Cui, and Yansheng Li. "Temporal Variation of Soil Carbon Stock and Its Controlling Factors Over the Last Two Decades on the Southern Song-Nen Plain, Heilongjiang Province." *Geoscience Frontiers* 1, no. 1 (2010).

New Caledonia Barrier Reef

Category: Marine and Oceanic Biomes.
Geographic Location: Oceania.
Summary: This large reef supports a diverse marine ecosystem but is vulnerable to the effects of mining, destruction of surrounding mangrove forests, overfishing, and climate change.

The New Caledonia Barrier Reef is located in New Caledonia in the South Pacific near Australia's eastern coast, and surrounds Grande Terre, New Caledonia's largest island. This barrier reef is the second-longest double-barrier coral reef in the world, after Australia's Great Barrier Reef. It surrounds a 9,300-square-mile (24,000-square-kilometer) lagoon that is 82 feet (25 meters) deep. The reef is located approximately 19 miles (30 kilometers) from the shore.

New Caledonia has a temperate climate. The average temperature is 75 degrees F (24 degrees C) during the winter and 59 degrees F (15 degrees C) at night, although during the hot season, temperatures range to 79–86 degrees F (26–30 degrees C). The cyclone season is January to March.

Biodiversity

The islands of New Caledonia were formerly a part of the Gondwana supercontinent and, as a result, the reef is populated with many ancient species, including varieties of sponge and mollusk. Also making a home in the reef are algae, plankton, crustaceans, cetaceans, sharks, and other fish. Seagrass beds comprise one-third of the reef, providing nesting and feeding grounds for several species of endangered turtles and dugongs.

The soils of the islands are heavily laden with nickel and magnesium, and mining on the islands causes sedimentation, which clouds the water and can bury the reefs, posing a serious threat to the ecosystems they support. Sedimentation also is caused by damage to the mangrove forests, which not only prevent soil erosion in the archipelago but also serve as a breeding ground for many fish, and filter out coastal pollution.

The barrier reef comprises more than 300 species of coral, which are supported by their symbiotic relationship with the photosynthetic algae called zooxanthellae. Elevated sea-surface temperatures in 1998 resulting from the El Nino weather pattern caused coral bleaching, which caused zooxanthellae to be expelled from the corals here. The corals of the reef are still considered relatively healthy, with the exception of those on the eastern coast of Grande Terre, which have been affected by effluent from nickel mining.

There are concerns that global warming will affect the growth of the corals, because of the predicted increase in both sea-surface temperature

This small island called Amédée sits about 15 miles (24 kilometers) off of Grand Terre in New Caledonia and is surrounded by coral reefs. New Caledonia's double-barrier reef is the second longest in the world and is home to thousands of species of flora and fauna, including close to 2,000 species of fish. (Thinkstock)

and salinity. The coral also is threatened by infestations of the carnivorous crown-of-thorns starfish, which feeds on polyps, and eutrophication due to shrimp and oyster aquaculture on the islands.

More than 350 species of algae are present in the barrier reef, including phytoplankton, on which coral can feed. Also feeding on algae are two species of giant clams (*Tridacna gigas* and *Hippopus hippopus*) that live among broken coral and are supported by symbiosis with zooxanthellae. Cyanobacteria, which can exist in symbiosis with coral, is also present on the reef. Though helpful because of their ability to fix nitrogen, some cyanobacteria secrete harmful toxins or form suffocating blooms, especially close to aquaculture farms that flood the waters with bloom-promoting nutrients. Small crustaceans known as copepods feed on such bacteria and possibly keep their populations in check.

The reef supports a wider benthic (bottom-dwelling) community of 600 sponge, 5,500 mollusk, and 5,000 crustacean species, as well as species of worms, arachnids, and scorpions; all

of these invertebrates supplying food to the fish population here.

There are more than 1,950 species of fish, some of which are endemic (found nowhere else) to the reef, including sea bass (*Luzonichthys williamsi*) and numerous species within the brightly colored *Labridae*, *Pomacentridae*, *Gobidae*, *Serrandiae*, *Chaetodontidae*, and *Apogonidae* families.

Notable commercial fish types include emperor and goatfish, which feed on benthic species such as worms, crustaceans, and mollusks. Snapper and wrasse are known to feed on other invertebrates, as well as small fish. Parrotfish, a type of wrasse, feeds on the coral polyps themselves, breaking down their skeletons and excreting them as sand, as well as preventing overgrowth of potentially polyp-choking algae. Some species of butterfly fish live almost exclusively on specific polyp species, and their survival is thought to be completely dependent on the survival of the coral they feed on.

New species of fish and invertebrates are being documented continuously, including a new species of extremely rare amphipod (*Didymochelia ledoyerisp*), and a *Palaemonidae* shrimp (*Brachycarpus crosnieri*).

Seagrass covers a considerable 347 square miles (900 square kilometers) of the 5,513-square-mile (14,280-square-kilometer) reef area. The seagrass meadows are spawning areas for fish and invertebrates, as well as feeding grounds for dugongs and marine turtles, which generally subsist on seagrass. Seagrass productivity is actually increased by such grazing, because it prevents overgrowth that promotes sedimentation and obscures light.

Notable turtle species of the New Caledonia Barrier Reef biome include: the hawksbill turtle, which feeds on sea sponges living in the reef; green sea turtle, which uses the region as an important nesting site; the invertebrate-eating loggerhead turtle; and leatherback turtle, which feeds almost exclusively on jellyfish, keeping their populations in check. Many of these turtles are prey only to sharks, although human pilfering of their nests for eggs and hunting for turtle meat has had dire effects on their populations.

A number of seabirds breed in the area surrounding the reef. They include: both the lesser and brown noddy; the red-footed booby, which dives into shallow water to catch small fish and squid living among the reef; and the sooty tern, known to breed on coral islands and pick fish. Large marine mammals such as blue fin and the sei whales have been known to pass through the reef.

Environmental Concerns

Most of the coral reefs here are in good health, with the exception of the eastern reefs that are threatened by erosion during cyclone flood surges from nickel mining and bush fires; heavy sedimentation in lagoon areas resulting from the destruction of sediment-retaining mangroves; coastal development; industrial, domestic, and marine pollution; overfishing; small-scale infestations of the coral-feeding crown-of-thorns starfish; and climate change.

Climate change poses numerous and varied threats to the barrier reefs, from warming temperatures that affect the chemical and species balance of the ecosystems, to greater incidence of droughts that could damage subsistence agri-culture and livestock farming, leading to heavier fertilizer use and harsher runoff into the sea. More severe cyclone activity could directly damage coral structures, and degrade habitat complexity.

YASMIN M. TAYAG

Further Reading

Charpy, L., et al. "Cyanobacteria in Coral Reef Ecosystems: A Review." *Journal of Marine Biology* 2012 (2012).

Payri, Claude E., et al. "Vulnerability of Mangroves, Seagrasses and Intertidal Flats in the Tropical Pacific to Climate Change." In *Vulnerability of Tropical Pacific Fisheries and Aquaculture to Climate Change*, edited by Johann D. Bell, Johanna E. Johnson, and Alistair J. Hobday. Nouméa, New Caledonia: Secretariat of the Pacific Community, 2011.

World Heritage Convention. "Lagoons of New Caledonia: Reef Diversity and Associated Ecosystems." United Nations Educational, Scientific, and Cultural Organization, 2012. http://whc.unesco.org/en/list/1115.

New Caledonia Dry Forests

Category: Forest Biomes.
Geographic Location: Pacific Ocean.
Summary: Supporting great biodiversity, this biome is subject to impacts from climactic changes that could imperil the habitats for native species.

New Caledonia is a French-controlled archipelago in Melanesia, about 930 miles (1,497 kilometers) east of Australia, and includes a land area of some 7,172 square miles (18,575 square kilometers). The climate is tropical, featuring a hot, humid summer from November to March with temperatures 80–86 degrees F (27–30 degrees C) and a cool, dry winter with temperatures of 68–74 degrees F (20–23 degrees C) from June to August.

Rainfall amounts vary widely, with 118 inches (3,000 millimeters) recorded in Galarino, the Northern Province, three times the average total of the west coast. Between December and April, tropical depressions and cyclones can cause winds of 62 miles per hour (100 kilometers per hour) with gusts of 155 miles per hour (250 kilometers per hour) and abundant rainfall. A quite violent cyclone hit New Caledonia in January 2011.

The main island, Grand Terre, has some of the most concentrated species diversity in the world. While the surrounding islands are recent and volcanic in origin, Grand Terre originally was a piece of the supercontinent Gondwanaland that separated from Australia 85 million years ago. This isolation has driven much of the ecoregion's diversity. The dry forests of New Caledonia are found only on Grand Terre, on the western side of the island. They harbor 59 endemic plant species, that is, species found only here. The present-day forests represent a surviving remnant of the vast, prehistoric ecosystem that once existed. Because they are so often cleared for agricultural land or logging purposes, tropical dry forests are among the most threatened types of forest worldwide.

Biodiversity

The best-known of those endemic plants is *Captaincookia margaretae*, a flowering variety of the madder family (*Rubiaceae*), with brilliant red flowers. Because the dry forests have been so severely reduced to create agricultural land and settlements since Europeans colonized the island in the 19th century, *C. margaretae* today is found in only six locations. It is the only member of its genus. The introduction of the Rusa deer (*Cervus timorensis russa*) to the island posed another threat, as the deer adapted well to the habitat and became a major consumer of plant life; it damages trees by rubbing its antlers against tree stems.

The only gymnosperm common in the ecoregion is queen sago (*Cycas circinalis*), a cycad with poisonous seeds that grows near the western coast. Other common plants include the verbena family (*Verbenaceae*), flowering plants of the *Premna* genus, shrubs and small trees of the *Dodonaea* genus, flowering plants of a *Pittosporum* genus that is shared with Australia, gardenias, and *Canthium* shrubs.

Seasonal wildfires have begun turning patches of dry forest into shrubland, where *Acacia spirorbis* and white popinac (*Leucaena leucocephala*) dominate. White popinac is highly invasive; it continues to spread whenever surrounding vegetation has been cleared, allowing it entry to the ecosystem. *Pittosporum tanianum*, a tree species discovered in the 1980s, was recently declared extinct after a wildfire left no surviving specimens. Fire poses a danger both in the form of naturally occurring wildfires and through the use of fire-setting as a means of protest by young unemployed men in the rural areas.

Newly introduced species also pose significant threats, from the neotropical ant *Wassmannia auropunctata*, which has diminished native lizard and invertebrate populations through competition; to the introduction of cats, rats, dogs, and pigs, which have taken serious tolls on various native flora and fauna.

Several species of bat in the *Nyctophilus* genus live on the island, including *N. nebulosus*, the New Caledonian long-eared bat, a species discovered in 2010 that may be the island's only endemic vertebrate. Near-endemic wildlife found in the dry forests and few other places includes the New Caledonia wattled bat (*Chalinolobus neocaledonicus*), ornate flying fox (*Pteropus ornatus*), long-tailed fruit bat (*Notopteris macdonaldi*), and New Caledonia flying fox (*P. vetulus*).

There are 23 near-endemic bird species, including the white-bellied goshawk (*Accipiter haplochrous*), red-bellied fruit dove (*Ptilinopus greyii*), New Caledonian imperial pigeon (*Ducula goliath*), horned parakeet (*Eunymphicus cornutus*), New Caledonian owlet-nightjar (*Aegotheles savesi*), Melanesian cuckoo-shrike (*Coracina analis*), long-tailed triller (*Lalage leucopyga*), New Caledonian grassbird (*Megalurulus mariei*), fan-tailed gerygone (*Gerygone flavolateralis*), yellow-bellied robin (*Eopsaltria flaviventris*), New Caledonian flycatcher (*Myiagra caledonica*), southern shrike-bill (*Clytorhynchus pachycephaloides*), streaked fantail (*Rhipidura spilodera*), New Caledonian whistler (*Pachycephala caledonica*), green-backed white-eye (*Zosterops xanthochrous*), New Caledonian myzomela (*Myzomela caledonica*), dark-brown honeyeater (*Lichmera incana*), New Caledonian friarbird (*Philemon diemenensis*), barred honeyeater (*Phylidonyris undulata*), red-throated parrotfinch (*Erythrura psittacea*), striated starling (*Aplonis striata*), and New Caledonian crow (*Corvus moneduloides*).

The New Caledonian crow is found only here and in the Loyalty Islands, and has increasingly been the subject of studies of animal intelligence. The crow is a tool-using bird species and perhaps the only nonhuman one with rather advanced tool skill. Scientists have observed the crow innovate by modifying existing tools to create new ones, and then pass the tool-making method on to other members of its group. The crow also has been observed making new copies of previously used tools from new materials, to take advantage of what is available. Again, it is the only nonhuman species that has been observed doing so.

Scientists have seen New Caledonian crows bending wires to create hooks to retrieve pieces of food. In the inhabited parts of New Caledonia, crows have been observed placing nuts in the street, waiting for a car to drive over them to crack the shells, and then waiting at the crosswalk before crossing with pedestrians to safely retrieve the cracked nut.

Environmental Threats

New Caledonia's biodiversity is threatened by introduced species, deforestation from logging, mining, uncontrolled fires, agricultural and urban development, and tourism, which all impact the fragile ecosystem. The New Caledonian rail and the New Caledonian lorikeet have not been seen for more than 100 years and are considered critically endangered if not actually extinct; the New Caledonian owlet-nightjar has been pushed to remote areas; and the New Caledonian crested gecko was considered extinct until it was rediscovered in 1994. Native grasses are being outcompeted by introduced species, such as molasses grass (*Melinis minutiflora*)—used for livestock feed—and native trees that are now part of government-created protective parks and reserves.

New Caledonia is a conservation priority for many environmental organizations, which have lobbied to preserve the archipelago's unique ecosystems, but so far governmental support has not stepped in to provide the desired protections. Efforts to secure United Nations Educational, Science, and Cultural Organization (UNESCO) World Heritage Site status to certain areas has failed, as environmental interests conflict with those of regional governments, and mining and development interests.

The impact of climate change has not been fully studied for this biome, but temperature and rainfall variations are known to imperil the survival of dry forest ecosystems. A change in rainfall totals could alter the habitat, and more fires could exacerbate damage. Climate change also could diminish the resilience of these natural habitats to recover, and instead provide a more favorable environment to advance the spread of invasive species.

BILL KTE'PI

Further Reading

Bouchet, P. and Y. I. Kantor. "New Caledonia: The Major Centre of Biodiversity for Volutomitrid Molluscs." *Systematics and Biodiversity* 1, no. 4 (2003).

Gillespie, Thomas W. and Tanquy Jaffe. "Tropical Dry Forests in New Caledonia." *Biodiversity and Conservation* 12, no. 8 (2003).

Pennington, Toby. *Neotropical Savannas and Seasonally Dry Forests*. Boca Raton, FL: CRC Press, 2006.

Steadman, D. *Extinction and Biogeography in Tropical Pacific Birds*. Chicago: University of Chicago Press, 2006.

New England-Acadian Forests

Category: Forest Biomes.
Geographic Location: North America.

Summary: The New England-Acadian Forests ecoregion is noted for encompassing a transition zone between two distinct vegetation types across a geologically complex area. It also is home to diverse, unique ecosystems, including cranberry bogs and glacial-remnant lakes.

The New England-Acadian forests are coastal forest communities along the North Atlantic Ocean that are best characterized as the transition zone between boreal forests to the north and temperate deciduous forests to the south. The biome consists of temperate broadleaf and mixed forests with an array of habitats on the hills, mountains, and plateaus of New England in the northeastern United States, along with the Maritime Provinces of eastern Canada.

Encompassing most of Maine down to northwestern Massachusetts, through Vermont and New Hampshire, as well as most of Nova Scotia, and parts of Quebec and New Brunswick in Canada, the region was once covered by glaciers 9,843 feet (3,000 meters) thick. The landscape was left with a visually striking variety of topographic features. Among these are broad sweeping valleys, glacial erratics, striations, glacial potholes and kettle ponds, cobblestone beaches, and marine-deposited clay soils.

A newly hatched bog turtle, which is a threatened species because of habitat loss. The wetlands of the New England-Acadian Forests provide an important home for these turtles. (U.S. Fish and Wildlife Service/Rosie Walunas)

The climate of this region consists of humid summers and cool winters, and is characterized as a cool humid continental-marine climate.

Natural Features

The New England-Acadian Forests biome contains a vast array of habitats and natural features, including beaches and shorelines, intertidal and subtidal zones along the Atlantic, mountains, marshes, and swamps. Each of these has associated animals and plants, which sometimes also give structure to the ecosystems they inhabit. The Acadian forest system also is known for its distinct assortment of unique ecological features, including serpentine rocks, raised peat bogs, ribbed fens, and raised coastal peatlands. The forest is home to Acadia National Park, a region noted for rugged, breathtaking landscapes.

The Atlantic Ocean can be a powerful driver of forest ecosystems, especially in New England, where the maritime currents provide a significantly moist and humid environment. Trees along the coast are subjected to higher concentrations of salts, which can lead to interesting physical changes on plant communities. Beaches and shorelines provide a transition from marine to terrestrial ecosystems, fundamentally bridging two major ecotypes.

Intertidal and subtidal zones are defining ecosystems within the New England-Acadian forests because they support numerous marine, estuarine, and freshwater organisms. The interactions between these organisms and their environments are important to understand, especially as their diversity is at risk as ocean levels rise and coastal habitats are exploited. These zones are interesting biologically, as estuarine environments (where salty and freshwater mix) can be crucial habitats for commercially important fish and other aquatic species. Estuaries serve as buffers for the transition from aquatic and marine to terrestrial environments.

Acadian forests are found in the White Mountains of New Hampshire, the Green Mountains of Vermont, and north of Mount Katahdin in Maine. Mountains provide topographic relief, and with this variation, the transitions among different forest community types become more apparent. At increased elevations, more boreal forest species appear, while deciduous species dominate the lower elevations.

Wetland communities feature soil that is inundated with water, including such wetland types here as ribbed fens and peat bogs. Marshes have areas of low-lying land that are flooded in wet seasons or at high tide, and typically remain waterlogged. The dominant vegetation in these habitats consists of nonwoody plants and grasses.

Swamps are flooded low-lying areas usually containing woody plants and trees such as tamarack and black spruce. Bogs have wet, spongy ground, usually comprising various sphagnum moss species, that is highly acidic and contains dead plant material that is decaying slowly. This soil is compressed over time into peat. Bogs can support a variety of plant species, including cranberries (*Vaccinum macrocarpon*), bog laurel (*Kalmia polifolia*), Labrador tea (*Ledum groenlandicum*), and sweet gale (*Myrica gale*).

Vegetation

The New England-Acadian Forests biome is home to more than 30 tree species distributed across this varied topographic backdrop. It will be an important migration zone as climate changes and tree populations effectively migrate poleward in an effort to track their ideal climates as global warming accelerates. Many species are currently at their northern range limit in the Acadian forests, including sugar maple (*Acer saccharum*) and American beech (*Fagus grandifolia*).

The forests are composed of several main forest cover types: northern hardwoods, spruce-fir, boreal, and old-growth forests. Each of these supports a different community of plants and animals.

Northern hardwoods trees of the Acadian forest include sugar maple, yellow birch (*Betula alleghaniensis*), American beech, balsam fir (*Abies balsamea*), and eastern hemlock (*Tsuga canadensis*). Associated and less common species include red spruce (*Picea rubens*), red oak (*Quercus rubra*), white ash (*Fraxinus americana*), white pine (*Pinus strobus*), ironwood (*Ostrya virginiana*), and striped maple (*Acer pensylvanicum*).

The understory is largely comprised of spring ephemerals, yellow trout lily (*Erythronium americanum*), and Carolina spring beauty (*Claytonia caroliniana*), as well as club mosses (*Lycopodium* spp.), Bush honeysuckle (*Diervilla lonicera*), and blueberry species (*Vaccinium* spp.). Soils within northern hardwoods forests are generally nutrient-rich loams that promote diverse understory species, especially the spring ephemeral flowers, which disappear as soil quality declines.

The spruce-fir forests are the most abundant forest cover type throughout the biome. The overstory of spruce-fir forests is dominated by red spruce, white spruce (*Picea glauca*), and balsam fir. Common associated tree species include black spruce (*Picea mariana*), eastern hemlock (*Tsuga canadensis*), American larch (*Larix laricina*), and birch species (*Betula* spp.). The understory in spruce-fir forests can be fairly sparse, but includes herbaceous vegetation and shrubs such as trillium (*Trillium* spp.), blue-bead lily (*Clintonia borealis*), hobblebush (*Viburnum latanoides*), trailing arbutus (*Epigaea repens*), and Canada mayflower (*Maianthemum canadense*). Soils within spruce-fir forests tend to decompose more slowly, and are more acidic and nutrient-poor than the soils of northern hardwoods communities. Spruce-fir also is commonly associated with bog communities, which are based on sphagnum moss as the dominant ground cover.

While boreal forests generally are not common in this biome, they are the forest type at one edge of the greater transition zones. Known as the taiga biome, the boreal forests extend throughout the northern reaches of North America and similar latitudes in Eurasia. The overstory is largely low-diversity and centers on black spruce, white spruce, and American larch. Common associated species include pines (*Pinus* spp.), balsam fir, aspens (*Populus* spp.), and eastern hemlock (*Tsuga canadensis*). The understory community depends largely on the nutrient-poor soils of the boreal region, but generally consists of hardy evergreens, especially mosses and bryophytes, although some ferns are also found there. The boreal forest overstory is common in bog settings, similar to the spruce-fir forest type.

The New England-Acadian forests also contain old-growth forest communities. These can vary greatly in composition and structure, but generally present complex structural features indicative of old age. New England old-growth forests generally have large standing dead trees called snags and great biological diversity. The Acadian forests ecoregion contains old-growth eastern hemlock, red spruce, and Atlantic white cedar forests, among other species. These habitats are largely recognized for their importance in preserving biological diversity and their unique structural features; they recently gained attention for their potential to sequester more carbon than second-growth forests, which has important implications in combatting global climate change. Some scientists consider the lack of human-based disturbances critical for old-growth forests, but with global climate change, most ecosystems will be indirectly affected by human actions, even if not through the use of a chainsaw.

Biodiversity

The biome supports diverse animals, from well-known megafauna to rare butterfly populations. Common mammal species include the white-tailed deer (*Ocodoileus virginianus*), American black bear (*Ursus americana*), moose (*Alces alces*), fisher (*Martes americana*), and muskrat (*Ondatra zibethica*).

Common bird species include the red-breasted nuthatch (*Sitta canadensis*), golden-crowned kinglet (*Regulus satrapa*), and several warbler species. Two species of interest here include bog turtles and bald eagles (*Haliaeetus leucocephalus*), which tend to nest in the tall white pines. This forest type supports more than 200 bird species overall, and is considered one of the 20 most diverse ecoregions in North America.

Threats and Conservation

One of the greatest threats to the New England-Acadian forests is global warming. Land fragmentation and construction practices also have huge effects on the biome. Compartmentalizing of land parcels breaks up the connectivity and flow within ecosystems, and can lead to species and population decline for many of them by limiting habitat, resources, and gene flow among populations.

Anthropogenic needs for paper and forest products continue to put additional pressures on forest ecosystems around the globe, and the forests in New England are no exception. The biggest threats to this ecosystem are those of human actions. Additionally, nonnative species (such as the invasive purple loosestrife) and pests (such as the hemlock woolly adelgid) threaten native forest communities, and can have long-term effects on structure and community composition within the forests.

Relena R. Ribbons

Further Reading

Bateson, Emily M. "Two Countries, One Forest—Deux Pays, Une Forêt: Launching a Landscape-Scale Conservation Collaborative in the Northern Appalachian Region of the United States and Canada." *Conservation Practice at the Landscape Scale* 22, no. 1, (2005).
Braun, E. Lucy. *Deciduous Forests of Eastern North America.* Caldwell, NJ: Blackburn Press, 1950.
Eyre, F. H., ed. *Forest Cover Types of the United States.* Bethesda, MD: Society of American Foresters, 1980.

New Guinea Freshwater Swamp Forests

Category: Inland Aquatic Biomes.
Geographic Location: South Pacific Ocean.
Summary: The New Guinea freshwater swamp forests are filled with a wide array of flora and fauna, including some species indigenous to this inland aquatic biome. Ecosystems important to the survival of indigenous people are threatened by mining wastes and logging.

New Guinea, the second-biggest island in the world, was once attached to Australia and features a range of lowland, freshwater, and peat swamp forests across its Irian Jaya and Papua New Guinea sectors. The area of the Freshwater Swamp Forests biome mainly encompasses the foothills on the northern side of the central cordillera and down to the north coastal areas.

The region's climate is tropical, with the temperature of the coastal plains averaging 82 degrees F (28 degrees C), 79 degrees (26 degrees C) in the inland and lower mountain areas, and 73 degrees (23 degrees C) in the higher mountain regions. The area's relative humidity is a high 70–90 percent. There is a somewhat more dry season, June to September, while the rainy season is December to March. The western and northern parts of New Guinea get the most rainfall, which averages 400 inches (10,160 millimeters) per year.

Flora and Fauna

The island of New Guinea represents approximately 1 percent of the world's landmass—but is home to nearly 10 percent of its vertebrate species and 7 percent of its vascular plant types. It also is home to the southern crowned pigeon (*Goura scheepmakeri*), considered the largest known pigeon, as well as the smallest parrot, the red-breasted pygmy parrot (*Micropsitta finschii*), and the longest lizard, Salvadore's monitor lizard (*Varanus salvadorii*). New Guinea is home to the world's largest butterfly, the Queen Alexandra birdwing (*Ornithoptera alexandrae*), and the conifer *Araucaria*, which can grow up to 230 feet (70 meters) and is thought to be the tallest tropical tree species on the planet.

Lowland swamp forests are located extensively in the northern portion of the island and particularly near the Sepik River. This area contains numerous habitats, including herbaceous swamp, Leersia grass swamp, *Saccharum-phragmites* grass swamp, *Pseudoraphis* grass swamp, mixed swamp savanna, *Melaleuca* swamp savanna, mixed swamp woodland, sago swamp woodland, pandan swamp woodland, mixed swamp forest, *Campnosperma* swamp forest, *Teminalia* swamp forest, and *Melaleuca* swamp forest.

The swamp forests include small- to medium-crowned dense to open areas, with 65–98-foot (20–30-meter) canopy. Sago palm (*Metroxylon sagu*) and *Pandanus* spp. palm generally are located in this subcanopy, which also features *Campnosperma brevipetiolata*, *C. auriculata*, *Terminalia canaliculata*, *Nauclea coadunata*, and *Syzygium* spp., with *Myristica hollrungii* in delta areas.

The canopy along the Mamberamo River is approximately 147 feet (45 meters) high; it includes ficus and *Pittosporum ramiflorum*, as well as *Timonius*, *Dillenia*, and *Nauclea*.

The most abundant forest type in this ecoregion is lowland broadleaf evergreen, which is split into alluvial and hill varieties. The lowland alluvial forest has a multitiered and irregular canopy with many emergents, and a shrub and herb layer understory with climbers, epiphytes, and ferns. Palms are most common in the shrub layer. The canopy trees include *Pometia pinnata*, *Octomeles sumatrana*, *Ficus* spp., *Alstonia scholaris*, and *Terminalia* spp. Additional important genera include *Pterocarpous*, *Artocarpus*, *Planchonella*, *Canarium*, *Elaeocarpus*, *Cryptocarya*, *Celtis*, *Dracontomelum*, *Sysoxylum*, *Syzygium*, *Vitex*, *Spondias*, and *Intsia*.

The lowland hill forest contains a more open shrub layer than the lowland broadleaf evergreen forest, and a denser herb layer. The dominant canopy trees are *Pometia*, *Canarium*, *Anisoptera*, *Cryptocarya*, *Terminalia*, *Syzygium*, *Ficus*, *Celtis*, *Dysoxylum*, *Buchanania*, *Koompassia*, *Dillenia*, *Eucalyptopsis*, *Vatica*, and *Hopea*. Dense stands of *Araucaria* are scattered throughout this area.

New Guinea is home to more than 2,000 species of orchids and an equal number of ferns. Eight out of 10 of these plants are found nowhere else in the world, that is, they are endemic here. Of the climbing plants in the forests here, there are climbing palms or rattans that can grow to 787 feet (240 meters).

Biodiversity

There are 76 mammal species in this ecoregion, including 13 that are endemic or near-endemic, such as tube-nosed bat (*Nyctimene draconilla*) and greater sheath-tailed bat (*Emballonura furax*). The western part of the ecoregion is the only known site in New Guinea for the western ringtail possum (*Pseudocheirus albertisi*). Others include

Tree kangaroos like this Dendrolagus goodfellowi *live in the New Guinea Freshwater Swamp Forests. New Guinea is only 1 percent of the world's landmass but contains 10 percent of all vertebrate species. (Thinkstock)*

Echymipera clara, Dorcopsis muelleri, Dorcopsis hageni, Emballonura furax, Hipposideros wollastoni, and *Paraleptomys rufilatus.*

The area is home to two types of tree kangaroo (*Dendrolagus dorianus notatus* and *D. goodfellowi*), long-beaked echidna (*Zaglossus bruijni*), and the mountain cuscus *(Phalanger carmelitae).* New Guinea has a diverse animal population, including monotremes (mammals that lay eggs). These include the short-beaked echidna or spiny anteater *(Tachyglossus aculeatus)* and the long-beaked echidna. Other animals include the forest wallaby *(Dorcopsis macleayi),* which are not found anywhere else, and 75 bat species and rodents.

Among the region's many reptiles are Papuan olive python (*Apodora papuana*); New Guinea snapping turtle (*Elseya novaeguineae*); lined gecko, which also is known as the striped gecko (*Gekko vittatus*); carpet python (*Morelia spilota*), and spotted tree monitor (*Varanus timorensis*).

There are more than 750 bird species, approximately half of which are endemic to New Guinea. Local bird families include *Ptilonorhynchidae, Eopsaltriidae, Meliphagidae, and Paradisaeidae.* Among the 16 endemic or near-endemic species is the vulnerable Salvadori's fig-parrot (*Psittaculirostris salvadorii*). Others include Edwards's fig-parrot (*P. edwardsii*), Papuan swiftlet (*Aerodramus papuensis*), brown-headed crow (*Corvus fuscicapillus*), white-bellied whistler (*Pachycephala leucogastra*), brown-capped jewel-babbler (*Ptilorrhoa geislerorum*), green-backed robin (*Pachycephalopsis hattamensi*), and pale-billed sicklebill (*Epimachus bruijnii*). There are more than 40 different kinds of bird of paradise, the landmark species of the island.

Many of the fish in freshwater here are migratory, coming from or going to the sea to spawn, such as the tarpon (*Megalops cyprinoides*) and barramundi (*Lates calcarifer*). Others are permanent inhabitants of freshwater habitats, such as fork-tailed catfish (*Ariidae*), gudgeons (*Cyprinidae*), gobies (*Gobiidae*), grunters (*Terapontidae*), and jacks (*Carangidae*). In the lowland rivers, where the waters are turbid and silty, are catfish, croakers (*Sciaenidae*), silver biddies (*Gerreidae*), and ponyfish (*Leiognathidae*). In the floodplain lakes, swamps, and backwaters are rainbow fish, gobies, gudgeons, and catfish.

Environmental Issues

One of the largest protected areas in all of Southeast Asia-Oceania is Lorentz National Park in West Papua. At 5.8 million acres (2.35 million hectares), this World Heritage Site stretches 93 miles (150 kilometers) to the Arafura Sea, crossing a large area of freshwater swamp forest.

Because population density is low in the swampy areas of lowland New Guinea, the indigenous people have presented little threat to biodiversity, except for their overhunting. Logging also has been a threat, particularly in better-drained areas; this mostly impacts bird habitat. Hunting for bird meat and feathers may also be contributing to the decline of several bird species, such as the southern and northern cassowary *(Casuarius casuarius johnsonii* and *C. unappendiculatus),* Salvadori's teal (*Salvadorina waigiuensis*), the black honey buzzard (*Henicopernis infuscatus*), and the New Guinea harpy eagle (*Harpyopsis novaeguineae*).

Other threats to the area come from open-pit copper and gold mines in the mountains upstream from large expanses of wetlands. Three of the world's largest mines were allowed by the Indonesian and Papua New Guinea governments to operate without a system to retain mine tailings, which are often discharged into local waterways.

Climate change could unsettle the balance in this ecosystem, with the possibility of decreased rainfall increasing shore erosion and affecting the salinity of the waterways. Any change in precipitation could have an ripple effect on freshwater ecosystems; and rising sea levels could interact with increased flow from the rivers, creating more back pressure at river mouths. In either case, the water flow changes can impact both the flora and fauna of the biome.

PATRICIA K. TOWNSEND

Further Reading

Paijmans, K. "Wooded Swamps in New Guinea." *Ecosystems of the World.* Amsterdam, Netherlands: Elsevier, 1990.

Ruddle, K., et al. *Palm Sago: A Tropical Starch from Marginal Lands.* Honolulu: University of Hawaii Press, 1978.

Townsend, Patricia K. "Palm Sago: Further Thoughts on a Tropical Starch From Marginal Lands." Australian National University Open Access Research. https://digitalcollections.anu.edu .au/handle/1885/39954.

Wikramanayake, Eric, Eric Dinerstein, and Colby Loucks. *Terrestrial Ecoregions of the Indo-Pacific.* Washington, DC: Island Press, 2002.

New Jersey Pine Barrens

Category: Forest Biomes.
Geographic Location: North America.
Summary: The New Jersey Pine Barrens is located in the southern portion of New Jersey, and is dominated by expansive stands of pitch pines over sandy soils.

The New Jersey Pine Barrens, also referred to as NJPB or the Pinelands, consists of approximately 1.25–1.4 million acres (505,857–566,560 hectares) of heavily forested coastal plains in southeastern New Jersey. This is an area consisting of sandy acidic soil where the dominant tree types are pines, oaks, and, in wetland areas, cedars. Despite being located in a densely populated state, the Pinelands remain mostly rural and sparsely populated. On the east, they are bounded primarily by salt marshes along bays connecting to the North Atlantic Ocean. Other boundaries are adjacent to farmlands and deciduous forests.

The NJPB is the largest section of an unconnected ecoregion known as the Atlantic Coastal Pine Barrens, which includes the southern half of Long Island in New York; Cape Cod, Massachusetts; and a small section of barrens near Albany, New York. Pinelands National Reserve in New Jersey occupies 22 percent of the state, and is underlain by aquifers containing some of the country's purest water. The Pinelands Comprehensive Management Plan restricts development on 934,000 acres (377,976 hectares), or about 70 percent of the original NJPB.

The mean temperature in the Pinelands is 33 degrees F (0.3 degrees C) in January and 75 degrees F (23.8 degrees C) in June, with annual precipitation totaling 44 inches (1,123 millimeters). The Pine Barrens, positioned on the Outer Coastal Plain of eastern North America, has terrain that includes globally rare pine plains. They are comprised of dwarf, 3–6 foot (1–2 meter) pitch pines, low-angle slopes dominated by dry, fire-swept pitch pine-shrub; and oak forests with a maximum elevation of 205 feet (62.5 meters). These pine plains are supplemented in dark swamp areas by Atlantic white cedar.

The Pinelands occurs over coarse, unconsolidated quartz sand of the Cohansey and Kirkwood formations. Relict sand wedges measure up to 8 feet (2.5 meters) deep and 1.3 feet (0.4 meters) wide, indicating previous existence of permafrost, likely during late Pleistocene times. Today, deposits contain an expansive, uncontained aquifer of uncontaminated but acidic, nutrient-poor water. The water table is within 23 feet (7 meters) of the surface in the uplands and can reach the surface in the lowlands and swamps. Swamp soils contain up to 12 inches (300 millimeters) of decaying organic material, compared to upland soils, which typically do not have an organic horizon more than 2–4 inches (50–100 millimeters) deep.

Vegetation

More than 800 species and varieties of plants have been documented for the NJPB, including 92 which are considered threatened or endangered. Of these, several species are orchids, such as pink lady's slipper, which is native to the Pine Barrens.

Upland forests constitute 62 percent of the forested area, and most of that space is dominated by homogeneous stands of pine-oak forest, with pitch pine (*Pinus rigida*) interspersed with oak (*Quercus* spp.), including white, black, post, and scarlet varieties. The understory, dominated by ericaceous shrubs, contains huckleberry (*Gaylussacia* spp.) and blueberry (*Vaccinium* spp.), with shrub oaks (*Q. ilicifolia* and *Q. marilandica*).

The lowland swamps of the NJPB generally have a closed canopy composed of red maple (*Acer rubrum*), black gum (*Nyssa sylvatica*), and Atlantic white cedar (*Chamaecyparis thyoides*). Many rare, endemic (found only here) plants such as Knieskern's beaked rush (*Rhynchospora knieskernii*) and Pine Barrens livid sedge (*Carex livida*), as well as many orchids and carnivorous plants, are found in the NJPB, which also provides habitat to the Pine Barrens treefrog (*Hyla andersoni*) and the northern pine snake (*Pituophis melanoleucus*).

Fauna

The Pinelands is home to many species of mammals, birds, reptiles, amphibians, and fish. Approximately 35 species of mammals inhabit the expansive area, including white-tailed deer, coyotes, the rare bobcat, beavers, and river otters. Also found here are red and gray fox, mink, long-tailed weasel, southern bog lemming, eight species of bats, raccoon, muskrat, various squirrels, chipmunks, voles, and mice.

There are about 144 bird species, including the pine warbler and other songbirds. The Pine Barrens is an important migrating and wintering location for birds, and it provides a nesting, breeding, feeding, and resting area for waterfowl, such as ducks and geese. It is also home to raptors, such as the bald eagle, red-shouldered hawk, and osprey.

The Pine Barrens is the global stronghold for the Pine Barrens treefrog (*Hyla andersonii*), and other amphibians such as the carpenter frog (*Rana virgatipes*), green frog (*Hyla cinerea*), and the southern leopard frog (*Lithobates sphenocephalus*).

Reptiles of the Pine Barrens include the timber rattlesnake (*Crotalus horridus*), the only venomous species in the Pinelands. The most common snake of the Pinelands is the northern water snake (*Nerodia sipedon*). Eastern hognose (*Heterodon platirhinos*), also known as the puff adder because it often spreads its neck, cobra-like, when alarmed, is found here.

Native fish include the banded, blackbanded and mud sunfish (*Enneacanthus obesus, E. chaetodon, Acantharchus pomotis*), pirate perch (*Aphredoderus sayanus*), swamp darter (*Etheostoma fusiforme*), and yellow bullhead (*Ameiurus natalis*).

Fish species found in the Pine Barrens and elsewhere in New Jersey are American eel (*Anguilla rostrata*), bluespotted sunfish (*Enneacanthus gloriosus*), eastern mudminnow (*Umbra pygmaea*), redfin and chain pickerel (*Esox americanus, E. niger*), among others.

Environmental Threats

Wildfire is historically the dominant disturbance in the Pine Barrens. The pitch pine and

The bright green Pine Barrens treefrog (Hyla andersonii) *depends on specialized but threatened breeding habitats found in the Pine Barrens. (U.S. Fish and Wildlife Service)*

many understory plants have become adapted to resprout following naturally occurring wildfires, and the suppression or acceleration of this dynamic process is one of the most serious threats to the Pine Barrens. A conservation dialogue is just beginning among conservation organizations, aimed at updating government agency policy to include prescribed fires set during the growing season that closely mimic a natural-disturbance regime.

Over the past century, human activities have continued to threaten Pinelands habitats. Land use has significantly altered the fire regime relative to pre-settlement conditions. Some other anthropogenic disturbances in the region include habitat fragmentation; cutting forests for lumber, fuelwood, and charcoal; mining for iron ore and sand; impoundment of water; real estate development; and agriculture.

Despite pressures caused by these activities, the poor soils fortunately have limited the extent of agricultural development and urban sprawl in the NJPB. Therefore, it remains one of the best-conserved areas in the eastern United States, and serves as an unparalleled recreational and scientific resource.

Rising temperatures due to climate change threaten forest composition as species will migrate northward and upward, so that current oak and hickory forests will recede and possibly be replaced with oak, pine, and loblolly pine forests here. Also, rising temperatures and changes in precipitation could increase the risk of forest fires by 10–20 percent. If this happens, wildlife habitats would be threatened.

DANIELA SHEBITZ

Further Reading

Forman, R. *Pine Barrens: Ecosystem and Landscape.* New Brunswick, NJ: Rutgers University Press, 1998.

McPhee, J. *The Pine Barrens.* New York: Farrar-Straus-Giroux, 1981.

The Nature Conservancy. "Climate Change Impacts in New Jersey." http://www.pinelandsalliance.org/downloads/pinelandsalliance_55.pdf.

New Zealand Intertidal Zones

Category: Marine and Oceanic Biomes.
Geographic Location: New Zealand.
Summary: A contrast between upwelling and nonupwelling regimes exerts great biota variation in intertidal community structures of this biome.

Southern Hemisphere latitudes similar to New Zealand's are comparable to those located close to Oregon in the Pacific Northwest region of the Northern Hemisphere. For both biogeographic regions, the taxonomic composition of rocky intertidal flora and fauna are quite similar, despite differing almost completely at the generic level. The rocky shores of New Zealand harbor intertidal communities that are similar to those of other temperate rocky coasts. Barnacles dominate high zones; mussels dominate middle zones; and a mixture of algae, sessile invertebrates, and bare space rule the low zones. Processes such as predation and competition on New Zealand shores are comparable to those in Europe, North America, South America, and Australia.

The climate in New Zealand ranges from semiarid in central Otago, with rainfall of approximately 12 inches (300 millimeters), to very wet—with up to 314 inches (8,000 millimeters) in areas west of the southern Alps. In the summer, temperatures are generally 77 degrees F (25 degrees C) over most of the country, and may rise to 86 degrees F (30 degrees C) in the east. In the winter, temperatures are 50–59 degrees F (10–15 degrees C) in the North Island, and 41–50 degrees F (5–10 decrees C) in the South Island.

North Island Biodiversity

On the North Island, a narrow band of the brown barnacle (*Chamaesipho brunnea*), sometimes accompanied by a broader band of the black stubbly lichen (*Lichina confinis*), is found near the high-tide level. Between high- and mid-tide levels is the column barnacle (*Chamaesipho columna*), accompanied by a few Pacific oysters (*Crassostrea gigas*) and green seasonal sea lettuce (*Ulvalactuca*).

Frequent ornate limpets (*Cellana ornata*) graze microalgae on the rocks at all levels, and are joined between mid- and low-tide levels by golden limpets (*Cellana radians*) and the snakeskin chiton (*Sypharochiton pelliserpentis*).

Predation determines composition in the low zone and the distribution of the dominant sessile animals. Several studies have shown that whelk predation prevented the establishment of persistent populations of barnacles such as *C. Brunnea* and *Epopella Plicata*. Comparably, the distribution of the New Zealand green-lipped mussel (*Perna canaliculusis*) was determined by the predation of the keystone sea star (*Pisaster ochraceus*). The selective feeding of starfish (*Stichaster*) prevents this competitive filter feeder from monopolizing the lower intertidal region, thus allowing the coexistence of numerous inferior invertebrates and seaweeds.

South Island Biodiversity

On the South Island, rocky shores on the east coast have nearly solid covers of the *Chamaesipho columna* and *Epopella plicata* in the high zone, and the mussels *Mytilus galloprovincialis* and *Perna canaliculus*, and the red coralline alga *Corallina officinalis* in the mid and low zones. On the shore, a canopy of the brown algae *Durvillea willanar* and *D. Antarctica* are dominant.

Lepsiella Scobina is the most common whelk, occurring only in the middle and high zones. The whelks *Thais orbita* and *Haustrum haustrorum*, along with sea stars *Coscinasterias calamaria* and *Patiriella regularis*, are common in the mid and low zones.

Zonation for the rocky shores on the west coast have dense populations of the barnacle *Chamaesipho columna* in the high zone and dense populations of the mussels *Mytilus galloprovincialis* and *Xenostrobus pulex* in the mid zone. Mussels are scarce in the low zone, and primary space is dominated by algal turfs (upper low zone), crustose and foliose red algae (lower low zone, mostly *Gigartina decipiens* and *G. clavifera*), and bare space. Qualitative observations indicate that sea snails *Lepsiella* and *Thais* are rare on the west coast, but similar in size to the east coast animals. The sea star *Stichaster australis* is abundant.

On the east coast, whelks and oyster-catchers are top predators, with *Lepsiella* feeding exclusively on barnacles, mussels, oysters, limpets, snails, and tubeworms. On the west coast, the diet of *Stichaster australis* is very similar to the one observed on the North Island. *Stichaster* is considered a generalized predator, consuming mussels, barnacles, whelks, and other gastropods. Although similar biota occur on both sides of the island, the sea star *Stichaster australis* is far more abundant on the west coast.

Near-Shore Biodiversity

Near-shore oceanographic conditions vary from coast to coast and seem to offer contrasting environments. For the west coast, the eastward-flowing Tasman Current splits into the northeasterly flowing Westland and Southland Currents. The Westland Current flow is enhanced periodically by northeastward winds, which in turn are influenced by the orographic effect of the Southern Alps, creating upwelling-favorable conditions.

During strong northward winds, surface waters move westward offshore, drawing nutrient-rich water from the deep to the surface along the west coast, eventually leading to phytoplankton blooms. An evident consequence of this west coast upwelling is relatively high concentrations of chlorophyll a at shallow inshore depths.

To the contrary, with predominantly southwesterly winds, the east coast of the South Island generally is a downwelling ecosystem. Some upwelling can occur when northerly winds bring deep nutrient-rich waters into the intertidal zones.

Several studies have shown that bottom-up processes (recruitment, mussel growth, nutrients, and chlorophyll a concentration) and top-down processes (predation and grazing) appear to be greater on the west coast, creating the differences in community structure.

Environmental Issues

The intertidal zone is a carefully balanced environment of extremes, with creatures adapted to the harsh conditions. As such, the environment is sensitive to variations in water temperature, tides, winds, and currents. The rising temperatures because of climate change could lead to enhanced trematode infections, among other stressors, and possible local extinctions of intertidal animals.

MARIA JOSÉ GONZÁLEZ-BERNAT

Further Reading

Chiswell, Stephen M. and Davis R. Schiel. "Influence of Along-Shore Advection and Upwelling on Coastal Temperature at Kaikoura Peninsula, New Zealand." *New Zealand Journal of Marine and Freshwater Research* 35 (2001).

Grace, Roger V. "Zonation of Sublittoral Rocky Bottom Marine Life and its Changes From the Outer to the Inner Hauraki Gulf, North Eastern New Zealand." *Tane* 29 (1983).

Menge, Bruce A. "Top-Down and Bottom-Up Community Regulation in Marine Rocky Intertidal Habitats." *Journal of Experimental Marine Biology and Ecology* 250 (2000).

Morton, John E. and Michael C. Miller. *The New Zealand Sea Shore*. London: Collins, 1968 and 2009.

Wood, Spencer A., et al. "Organismal Traits are More Important Than Environment for Species Interactions in the Intertidal Zone." *Ecology Letters* 13 (2010).

Nicaragua, Lake

Category: Inland Aquatic Biomes.
Geographic Location: Central America.
Summary: The largest inland water body in Central America is home to unique freshwater and marine fauna.

Located in southern Nicaragua, north of the border of Costa Rica, Lake Nicaragua forms the center of a large tectonic basin shared by Lake Managua in the northwest and drained by the San Juan River to the Caribbean Sea in the southeast. It is the largest lake in Central America, with a surface area of more than 3,089 square miles (8,000 square kilometers), an average depth of 43 feet (13 meters), and an elevation of 112 feet (34 meters) above sea level.

Formed by volcanic activity, hundreds of beautiful volcanic islands and islets are found in Lake Nicaragua, the largest of which are Ometepe and Zapatera Islands, and the Solentiname Archipelago. Referred to as Cocibolca by native peoples, the lake historically has been an important source of fish, domestic water, irrigation, and transportation for the region's inhabitants. Several important towns and ports are located on the shores of Lake Nicaragua, the largest being the city of Granada.

In the past, the lake was occupied by pirates from the Caribbean, and once was considered for the construction of a trans-Nicaragua canal before the completion of the Panama Canal in 1914. Being so immense, the lake can assume oceanic conditions and generate strong winds, large waves, and powerful storms.

Nicaragua has a tropical climate and two seasons: dry and rainy. During the January-to-June dry season, there is virtually no rain and trees and plants start to dry out. During the June and July rainy season, everything starts growing again, turning green and flowering. The months of August and September usually experience brief, daily downpours.

There are three temperature zones in Nicaragua. In the lowlands found on the Pacific and Atlantic coasts, temperatures fluctuate from 72 degrees F (22 degrees C) at night to 86 degrees F (22–30 degrees C) during the day. In May, temperatures can reach 100 degrees F (38 degrees C). The central part of the country is about 9 degrees F (5 degrees C) cooler, and the mountain areas about 18 degrees F (10 degrees C) cooler.

Biodiversity

Although Lake Nicaragua is mostly surrounded by agriculture and second-growth forest on its perimeter, many of its islands still host native tropical dry forest, with cloud forest found at the top of Ometepe Island. This inland water body is home to hundreds of bird species, including egrets, herons, cormorants, hawks, and kites.

Many unique fish species are found in the lake, including the Caribbean bull shark (*Carcharhinus leucas*), whose high tolerance for freshwater has enabled it to travel up the San Juan River. Atlantic tarpon (*Megalops atlanticus*) also is found in the lake, along with more than 15 endemic cichlid fishes, and numerous other endemic (found nowhere else) gobies, silversides, and caracids.

The lake also is home to two sawfish species, *Pristis perotteti* and *P. pectinatus,* both of which are considered critically endangered. These long-lived species have little capacity to recover from decades of overfishing, and their populations in Lake Nicaragua collapsed following intensive fishing in the 1970s.

Environmental Threats

This vast lake is not immune to environmental problems. Discharge from urban-zone domestic and industrial wastewater and sewage enter Lake Nicaragua directly or indirectly; and agricultural runoff, including fertilizers and pesticides, flows into the lake from large-scale agriculture areas on its borders.

Deforestation for agriculture and livestock production has exposed significant areas to erosion, causing increased sedimentation and water turbidity in the lake and its tributaries. Navigation on the San Juan River and in Lake Nicaragua is affected by the progressive sedimentation of both water bodies. Navigation is an important source of pollution of the water resources, as boats are washed and serviced in both water bodies. Large amounts of introduced tilapia fish are being raised in the lake, also generating large amounts of waste and possibly introducing new diseases that threaten endemic fish.

Conservation Efforts

Currently, little effective legislation has been implemented locally to protect Lake Nicaragua and the valuable natural resources it supports. However, currently the area is being considered for a United Nations Educational, Scientific, and Cultural Organization (UNESCO) World Heritage Site designation.

The Nicaraguan government imposed a temporary moratorium on targeted fishing for sawfish in Lake Nicaragua in the early 1980s. As little recovery occurred, protection was bolstered in 2006 with a complete ban on fishing for sawfish in the lake.

The lake has large potential for tourism, as well as for the generation of renewable hydroelectric energy. However, the Lake Nicaragua watershed is shared by Costa Rica and Nicaragua, thereby complicating natural-resource management.

The lake region is characterized by extreme poverty, high population growth, low incomes, subsistence economies, poor sanitation conditions, and a relative imbalance in employment and income-generating opportunities. Uncontrolled migration exacerbates the situation by exceeding the capacity of existing institutions to meet the increasing sanitation, health, and educational needs. Resolving the environmental and ecological threats to Lake Nicaragua, therefore, is dependent on simultaneously addressing

View of the heavily wooded shoreline of a small island in the Solentiname Archipelago within Lake Nicaragua. At 3,089 square miles (8,000 kilometers) in area, the lake is the largest in Central America. (Wikimedia/Stoschmidt)

the socioeconomic needs of the biome's approximately 1 million inhabitants.

Climate change poses a devastating scenario for the Lake Nicaragua area, especially to the nation's indigenous communities who rely on the natural resources for their subsistence. Temperatures across Central America are expected to rise, and rainfall will decline by as much as 25 percent by 2070, increasing the likelihood of droughts, more frequent severe hurricanes, and unseasonal flooding. Such dramatic weather changes will present harsh consequences to the local human population, as well as to the environment and natural species of the lake.

BETH POLIDORO
SARAH WYATT

Further Reading

Belt, T. *The Naturalist in Nicaragua.* Chicago: University of Chicago Press, 1985.

Montenegro-Guillén, Salvador. *Lake Cocibolca-Nicaragua. Experience and Lessons Learned Brief.* Managua: Research Center for Inland Waters of Nicaragua, 2005.

Waid, R. M., R. L. Raesly, K. R. McKaye, and J. K. McCrary. "Zoogeografica Ictica de Lagunas Cratéricas de Nicaragua." *Encuentro* 51 (1999).

Nicoya, Gulf of

Category: Marine and Oceanic Biomes.
Geographic Location: Central America.
Summary: Costa Rica's largest gulf supports diverse habitats and the country's fishing industry, both of which are threatened by pollution and overfishing.

The Gulf of Nicoya represents an ecologically and economically invaluable asset of Costa Rica. Located on the Pacific coast of the country on the Nicoya Peninsula, it contains a remarkably diverse collection of landscapes, habitats, and organisms. It is vital to the fishing industry and is a popular destination for national and international tourists. Development, overfishing, and pollution threaten the wealth of resources offered by the Gulf of Nicoya.

The funnel-shaped gulf lies within the curve created by the Nicoya Peninsula and the western shore of Costa Rica. The geological history of this gulf lends to its unique shape and varied landscapes. Once a hilly terrain, it plunged below sea level thousands of years ago due to volcanic activity along a fault line. Only the hilltops remain above water and form small islands located throughout the gulf. At 50 miles (80 kilometers) in length and 31 miles (50 kilometers) across at its widest point, Nicoya is Costa Rica's largest gulf. At the gulf's narrowest point, a ferry serves as the primary form of transportation between the mainland port city of Puntarenas and the southern part of the Nicoya Peninsula.

The Nicoya Peninsula has two seasons: dry, which runs from the end of November until May, and the rainy season, which is September to November. The intervening months between the two seasons, known as "little summer," experiences periods of rain and showers. During the rainy season, the peninsula turns into a rainforest. Temperatures during this period—considered winter—may sometimes drop to 63 degrees F (17 degrees C). During the summer dry season, the temperature may reach 100 degrees F (38 degrees C). Rainfall during the rainy season exceeds 10 inches (250 millimeters) per month.

The Tempisque River drains fresh water into the northern, or upper, part of the gulf, creating a rich estuarine habitat. Organic material carried in by the river builds up on the floor of the relatively shallow upper gulf, which is 13–39 feet (4–12 meters) deep.

The lower section of the gulf has much deeper water than the upper regions, roughly 98–591 feet (30–180 meters) in depth, and lacks the other estuarine characteristics. However, this part of the gulf has two rivers, the Río Grande de Tárcoles and the Barranca, emptying into it. Rocky shores and steep bluffs border the gulf and connect the sparsely populated land with the water. Small fishing villages dot the shorelines. The deepwater here is rich with fish species, but supports relatively few mollusks or shrimp.

Biodiversity

Gulf waters have a high nutrient content, allowing algae to bloom profusely, which supports many fish and mollusks. Zooplankton supports large schools of sardines and anchovies, which then feed the larger fish. Present are sharks, dolphins, turtles, manta rays, tuna, snappers, groupers, dorado, roosterfish, jacks, trevallys, sea bass, mackerel, catfish, porgy, bonito, and sailfish. Between August and December, humpback whales visit the gulf.

The upper Nicoya gulf contains the gulf's largest island, Islá Chira. Wading birds such as roseate spoonbills (*Platalea ajaja*) and herons (*Ciconiiformes* spp.) feed and nest on Chira, which also attracts tourists and bird watchers to its banks. Other birds of Nicoya include grooved-billed ani (*Crotophaga sulcirostris*), scissor-tailed flycatcher (*Tyrannus forficatus*), hummingbird (*Trochilidae* spp.), brown pelican (*Pelecanus occidentalis*), magnificent frigatebird (*Fregata magnificens*), fiery-billed and collared aracari (*Pteroglossus frantzii* and *P. torquatus*), and bananaquit (*Coereba flaveola*).

This estuary also serves as prime habitat for mangrove trees. Their roots grow down into the water, creating a web of underwater structure that protects many species of fish, shellfish, shrimp, and other marine life. The biodiversity found in the mangroves is protected by the Costa Rican government. There are seven species of mangrove trees in Costa Rica including red mangroves (*Rhizophora mangle* and *R. harrisonii, Rhizophoraceae*), black mangroves (*Avicennia germinans* and *A. bicolor, Verbenaceae*), tea mangrove (*Pelliciera rhizophorae, Theaceae*), white mangrove (*Laguncularia racemosa, Combretaceae*), and the buttonwood mangrove (*Conocarpus erectus, Combretaceae*).

The gulf features many mammal, reptile, amphibian, and insect species. Monkeys are some of the most visible inhabitants of the rainforests surrounding the gulf, including white-faced or capuchin (*Cebus capucinus*), squirrel (*Saimiri*), howler (*Alouatta monotypic*), and spider monkeys (*Ateles*). Other local animals include mantled howler monkey (*Alouatta palliata*), fishing bulldog bat (*Noctilio leporinus*), black ctenosaur (*Ctenosaura similis*), house gecko (*Hemidactylus frenatus*), paca (*Agouti paca*), and milk frog (*Phrynohyas venulosa*).

Environmental Threats

There are many threats to the gulf, including overfishing, industrial runoff, red tide, conversion of mangroves for energy uses, and climate change. Overfishing has caused the Costa Rican government to implement periodic mandatory fishing bans, although with limited success.

The Nicoya has been compromised by runoff from industry, agriculture, and urban areas. In fact, the most polluted river in Costa Rica, the Río Grande de Tárcoles, empties water and toxins into the gulf. Although habitat destruction such as the removal of mangrove forests was once popular, many of the gulf's most important habitats are now under government protection. Boat traffic and development continue to take a toll on the marine life within the gulf and tidal zones.

Red tides—single-cell organisms called dinoflagellates—bloom in the early rainy-season months of April and May. Dinoflagellates use the sun to photosynthesize and grow, and can secrete toxins absorbed by shellfish such as mussels and oysters that can have deadly effects on humans who consume them. While this is not often the case, there have been isolated deaths. Red tide is naturally occurring but can be exacerbated by pollution, fed by agricultural runoff and raw sewage that provide the nutrients for the algae to grow.

The mangrove forests are being destroyed to convert their areas to fish pens, rice paddies, salt-drying ponds, cattle pasture, tourist developments, and human settlements. Mangrove wood makes good fuel and charcoal, and red mangrove is an important source of tannin, used in processing leather. Stripping of the bark to get the tannin kills the tree.

Climate change is likely to aggravate these human-induced disruptions of the waters here, forcing changes to the natural habitats on which the terrestrial and marine species depend.

KIMBERLY M. KELLETT

Further Reading

Almeyda, Angelica M., et al. "Ecotourism Impacts in the Nicoya Peninsula, Costa Rica." *International Journal of Tourism Research* 12, no. 6 (2010).

Voorhis, Arthur D., "The Estuarine Character of the Gulf of Nicoya, an Embayment on the Pacific Coast of Central America." *Hydrobiologia* 99, no. 3 (1983).

Whelan, Tensie. "Environmental Contamination in the Gulf of Nicoya, Costa Rica." *Ambio* 18, no. 5 (1989).

Niger Delta Flooded Savanna, Inner

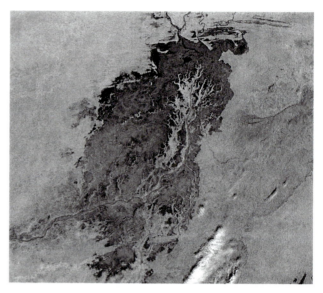

A satellite view of the Inner Niger Delta with extensive dark patches (areas of green vegetation) at the end of the rainy season in fall 2007. (NASA)

Category: Inland Aquatic Biomes.
Geographic Location: Africa.
Summary: This biome, located in the impoverished nation of Mali, is home to a wide range of animal and plant life, but threatened by poor local governance and climate change.

The Inner Niger Delta comprises an expanse of lakes and floodplains in the otherwise semiarid Sahel-Sahara Desert area of the west African country Mali. In recent years, the region has been of socioecological importance because of the discovery and exploitation of crude oil and the effect these activities have had on the region's plant and animal life.

The 54-mile- (87-kilometer-) wide lush delta extends for 264 miles (425 kilometers), tapering into a braided river near Timbouctou, where the Niger River curves to the east. The Niger is the main flood source of the savanna, and its main tributary is the Bani River in Côte d'Ivoire and Burkina Faso; smaller streams flow down from the Dogonland Plateau. The Niger originates on the Fouta Djalon highlands of Guinea and extends 2,548 miles (4,100 kilometers) before flowing into the Atlantic Ocean on the coast of Nigeria.

The surface level of the flooded savanna drops only 26 feet (8 meters) from beginning to end. It consists of a vast network of river channels with levees separated by low, clay-based floodplains that expand to cover 7,722 square miles (20,000 square kilometers) during the four-month rainy season of June to September; it then shrinks to 1,506 square miles (3,900 square kilometers) during the dry season.

The delta's climate is tropical; most of the year the region is hot and dry with temperatures rising to 104 degrees F (40 degrees C), with hot winds blowing in from the Sahara. Precipitation varies across the region. In the south, the rainy season of July to October records annual rainfall of 29 inches (750 millimeters), and in the north, the slightly shorter rainy season of July to September provides a mean annual precipitation of approximately 10 inches (250 millimeters).

Biodiversity
The southern portion of the delta is a low-lying floodplain filled with such grasses as *Acroceras amplectens*, *Echinochloa pyramidalis*, *E. stagnina*, and *Eragrostis atroviriens*. In the areas used for animal grazing, *Andropogon gayanus*, *Cynodon dactylon*, and *Hyparrhenia dissolute* dominate. Along the water areas, *Mimosa asperata* and *Salix chevalieri* grow above an understory of *Cyperus*

maculatus. In the south, the tree types include *Diospyros* spp. and *Kigelia africana,* while in the north the delta features emergent sand ridges and *Hyphaene thebaica* and *Borassus aethiopum* palms, plus *Acacia nilotica, Guarea senegalensis, Mimosa asperata,* and *Ziziphus mauritiana* in elevated areas.

The Inner Niger Delta Flooded Savanna biome provides essential habitat for numerous wetland birds. Wintering in the delta are approximately 500,000 garganey (*Anas querquedula*) and up to 200,000 pintail (*A. acuta*), plus ferruginous duck (*Aythya nyroca*), white-winged tern (*Chlidonias leucopterus*), ruff (*Philomachus pugnax*), black-tailed godwit (*Limosa limosa*), and more. The delta provides important breeding locations for African spoonbills (*Platalea alba*) and purple swamp-hens (*Porphyrio porphyrio*).

Other birds in the delta include cormorant, heron, spoonbill, ibis, and black crowned crane *(Balearica pavonina pavonina).*

Although most large mammals no longer use the delta—having been removed by humans—the African manatee (*Trichechus senegalensis*), also known as sea cow, lives in the rivers and feeds on underwater plants.

Approximately 110 different fish species are found in the inner delta, but few of them are endemic (found only here). Among those are the *Mochokidae* catfish (*Syndodontis gobroni*) and *Gobiocichla wonderi.* There are 20 endemic freshwater species from five families: *Denticipidae, Pantodontidae, Phractolaemidae, Hepsetidae,* and *Gymnarchidae.*

Aquatic mammals present in the delta include the hippopotamus (*Hippopotamus amphibius*) and the vulnerable pygmy hippopotamus (*Hexaprotodon liberiensis*), semi-aquatic sitatunga (*Tragelaphus spekei*), marsh mongoose (*Atilax Paludinosus*), and the spotted-necked otter (*Lutra maculicollis*).

Shore mammals include: Libyan striped weasel (*Ictonyx libyca*), African civet (*Civetticutus civetta*), caracal (*Felis caracal*), serval (*F. serval*), striped hyena (*Hyaena hyaena*), patas monkey (*Erythrocebus patas*), sand fox (*Vulpes pallida*), and African wild cat (*Felis silvestris*). Scien-

tists believe that small populations of roan antelope (*Hippotragus equinus*), red-fronted gazelle (*Gazella rufifrons*), dorcas gazelle (*G. dorcas*), and dama gazelle (*G. dama*) still exist.

Human encroachment has decimated many animal populations of the delta. Hunters and poachers have severely reduced the populations of the hippo, although small populations of gazelles still exist.

Human Activity and Climate Change

The deadly combination of limited rains, poor governance, increasing human population, and land-use changes has amplified pressure on the delta. Extreme grazing and severe droughts in the past 20 years have resulted in land degradation and pasture loss. Deforestation, erosion, and governmental negligence have contributed to the deterioration of the habitat.

Several attempts have been made to salvage the area, but the impact has been minimal. Since 1972, $100 million has been spent in the region, but conditions have not changed. The delta region is largely unprotected, although there are three Ramsar sites: Lac Horo, Lac Debo, and the Séri floodplain.

Climate change is expected to further stress the already-hot environment with rising temperatures that could increase the frequency of drought, accelerate biodiversity loss, and increase crop pest infestation. It also could diminish both water quality and quantity, and augment climate-related human health issues in an already impoverished region of the world.

PETER ELIAS

Further Reading

Dodman, T., H. Y. Béibro, E. Hubert, and E. Williams. *African Waterbird Census 1998.* Waginingen, Netherlands: Wetlands International, 1999.

Dugan, P. J, ed. *Wetland Conservation: A Review of Current Issues and Required Action.* Gland, Switzerland: International Union for Conservation of Nature, 1990.

Food and Agriculture Organization of the United Nations (FAO). "Irrigation Potential in Africa:

A Basin Approach." *FAO Land and Water Bulletin* 4 (1997).

Ticheler, H. *Fish Biodiversity in West African Wetlands.* Wageningen, Netherlands: Wetlands International, 2000.

Niger Delta Swamp Forests

Category: Forest Biomes.
Geographic Location: Africa.
Summary: This biodiverse forest ecosystem is threatened by oil exploration and deforestation.

The Niger Delta Swamp Forests biome is located in the largest wetland in Africa, laden with rich diversity in people, culture, fossil-fuel reserves, and biological resources, representing massive economic and ecological significance to the local people and all of Nigeria.

However, oil exploration, with frequent oil spills and gas flaring, is causing untold environmental disaster to the region's people. Worse, deforestation and the spread of an invasive forest plant are increasing at an alarming rate, threatening this remaining tract of primary forest in Nigeria. A small portion of the ecoregion is reserved and protected, but the most effective conservation method is animal worship—which in some of the communities includes snakes.

The Niger Delta region is a vast floodplain built up by the accumulation of silt washed down the Niger and Benue rivers. It is composed of four main ecological zones—coastal barrier islands, mangroves, freshwater swamp forests, and lowland rainforests—whose boundaries vary according to the patterns of seasonal flooding.

The Niger Delta swamp forests are categorized as both mangrove and freshwater swamp forests. These forests are rich in plant and animal diversity, and are part of the Guinean forests of West Africa, numbered among the 25 biological hot spots in the world.

The local climate is tropical, with fairly consistent temperatures averaging 75 degrees F (24 degrees C) during the coolest month to 79 degrees F (26 degrees C) during the warmest month. The ecoregion is extremely humid, with the wet season lasting 10–11 months; December is usually the only dry month. At Forcados, in the western part of the delta, rainfall averages 149 inches (3,800 millimeters) per year, but decreases eastward to Port Harcourt, where the annual rainfall is 97 inches (2,480 millimeters).

Biodiversity

The mangrove forest of Nigeria is the third-largest in the world and the largest in Africa, covering 2.5 million acres (997,700 hectares). Nigerian mangroves are dominated by red and white mangroves (*Rhizophora* and *Avicennia*), and a few other mangrove species, most of which are thickly clustered shrubs, making the area almost impenetrable.

A variety of birds, mammals, and reptiles inhabit the mangroves, including a few endemic (found nowhere else) species, such as Sclater's guenon and the Nile Delta red colobus monkey. About 200 fish species have been identified in the ecoregion; the mangroves are habitat and nursery grounds for many fish, crustacean, and mollusk species that are harvested locally and in offshore fisheries. Nigerian shrimp farming is especially strong, and shrimp are now being exported to many developed countries.

Local residents also use the mangroves for firewood and for dyeing their fishing nets, in addition to several other forest products collected from the understory. However, the mangroves are being destroyed for the construction of navigation canals and oil rigs.

Freshwater swamp forests cover 4.1 million acres (1.7 million hectares), which is more than half of the whole Niger Delta, and are the most extensive such forests in west and central Africa. The freshwater swamp forests are storehouses of biological diversity and appear to be the largest tract of remaining primary forests in Nigeria. They help in regulating coastal water flow and eliminating silt, sediment, and pollutants from moving water.

The region supports diverse animal populations, which include several species of monkeys, otters, civet cats, mongooses, leopards, warthogs, duikers, pythons and other snakes, antelopes, crocodiles, monitor lizards, snails, turtles and tortoises, and large numbers of bird species. Some of these species are endemic to the area, including the white-throated guenon and fawn-breasted waxbill; a number of birds here are highly endangered.

The fertile soil supports agriculture, which makes the area notable for production of cassava, palm oil, pineapple, cocoyam, and rubber, among many other crops.

Similarly, the freshwater swamp forests occurring around the creeks support fishing; gathering seafood and fuelwood; distilling gin from raffia palm trees; collecting wild fruits; and weaving mats and other items from screw pine, rattan palms, and bulrushes. Vast natural bamboo plantations in the region are being used locally for fuelwood, houses, furniture, and arts and crafts.

Environmental Threats

The freshwater swamp forests—initially protected from deforestation because of their relative inaccessibility arising from the swampy nature of the ecosystem—are now being destroyed at an alarming rate for expansion of agriculture, fuel-wood collection, timber harvesting, urbanization, and petroleum exploration.

The Niger Delta covers 12 percent of Nigeria's surface area and supports about 30 million people within 40 different ethnic groups, living in 13,329 communities, only 98 of which are urban centers. The main ethnic groups include Izon, Efik, Ijaw, Itsekiri, Urhobo, Ndoni, Ikwerre, Ibibio, and Ogoni. Local occupations involve fishing, arts and crafts artisans, alcoholic beverage production, and farming. Solid minerals such as limestone, marble, lignite, clay, gold, and granite are found in large quantities in the region.

The discovery of crude oil in the region in 1956 marked a turning point in Nigeria's economy. The delta's natural-gas reserves are ranked seventh in the world and the highest in Africa. A huge reserve of bitumen (as asphalt) was discovered in the late 1990s. Crude-oil exploration forms up to 80 percent of Nigeria's total income and 95 percent of its foreign-exchange earnings.

However, the oil has been a source of environmental disaster and poverty. Most communities do not have access to clean water; many sources have become polluted by oil, while lands have become too toxic for any meaningful farming. Fishing, which is the primary occupation of some of the ethnic groups, has been significantly reduced by surface water pollution, and the number of fishers working has also been reduced. Fishing is becoming an endangered occupation in the area.

Worse, oil spillage and gas flaring characterize oil exploration in the region. A United Nations Development Programme (UNDP) report stated that there were 6,817 oil spills between 1976 and 2001. In 2010, another source reported that 9 million to 13 million barrels have been spilled in the Niger Delta since 1958. The effects of petroleum spills on mangroves are known to acidify the soils, halt cellular respiration, and starve roots of vital oxygen.

As the soils supporting mangroves become too toxic, a nonnative invasive species of palm, nipa palm (*Nypa fruticans*), quickly colonizes the area. This invasive species has a shallower root system that destabilizes the banks along the waterways, impedes navigation, and reduces fishing potential.

The volume of gas flaring in Nigeria is the highest in Africa and second-highest in the world, after Russia. About 2.5 billion cubic feet (70 million cubic meters), or about 70 percent of the 3.5 billion cubic feet (100 million cubic meters) of associated gas produced annually, is wasted via flaring. This equals about 25 percent of the United Kingdom's total natural-gas consumption and is equivalent to 40 percent of the entire African continent's gas consumption in 2001.

Gas flaring releases large amounts of methane, which has high global-warming potential. Gas flares have potentially harmful effects on the health and livelihoods of the people in their vicinity, as they release a variety of poisonous chemicals that can cause cancer and severe respiratory problems. Despite legal actions in recent years, gas flaring has not stopped nor been significantly reduced in Nigeria.

Temitope Israel Borokini

Further Reading

Akani, Godfrey C., Luca Luiselli, and Edoardo Politano. "Ecological and Conservation Considerations on the Reptile Fauna of the Eastern Niger Delta (Nigeria)." *Herpetozoa* 11, nos. 3–4 (February 1999).

Blench, Roger. "Mammals of the Niger Delta, Nigeria." Cambridge, UK: Kay Williamson Education Foundation, 2007.

United Nations Environment Programme (UNEP). "Environmental Assessment of Ogoniland." http://postconflict.unep.ch/publications/OEA/UNEP_OEA.pdf.

Niger River

Category: Inland Aquatic Biomes.
Geographic Location: West Africa.
Summary: Largest river in western Africa, the Niger waters 10 nations and supports up to 100 million people—but its flow, habitats, and native species are at risk from a host of natural and human-made threats.

West Africa's Niger River is the third-largest river on the continent, at 2,600 miles (4,180 kilometers) long, and it has the ninth-largest drainage basin in the world, at approximately 808,000 square miles (nearly 2.1 million square kilometers). The Niger has a rambling water course that runs through seven countries, most of which is in Nigeria. The river originates just 150 miles (240 kilometers) from the Atlantic Ocean, and flows for 800 miles (1,287 kilometers) into the Sahara Desert before forming the largest delta of Africa in the Gulf of Guinea. More than 200 fish species are found in the river, 20 of which are found exclusively in this region. The Niger River has another delta at the midrange of its course, referred to as the Inner Niger Delta.

The Niger originates from the landward side of the Fouta D'jallon highlands in southern Guinea and emerges from a deep ravine about 2,800 feet (about 800 meters) below sea level to the Tembi River. Within a short distance of flow, the Tembi is joined by two rivulets, the Tamincono and Falico, which are joined by some tributaries to form the Niger. Instead of flowing to the Atlantic Ocean, the river flows in the opposite direction into the Sahara Desert, cutting through Guinea, Mali, Niger, and the Benin Republic until it finally empties into the Gulf of Guinea in Nigeria. All told, the river has about 14 tributaries

The Niger basin encompasses parts of 10 countries: Algeria, Benin Republic, Burkina Faso, Cameroon, Chad, Côte D'Ivoire, Guinea, Mali, Niger, and Nigeria. Geographically, the river is divided into four sub-basins: the upper Niger Basin, the Central Delta, the middle Niger Basin, and the Lower Niger Basin.

Temperatures in the region range from an average of 63 degrees F (17 degrees C) in January

Women doing laundry and carrying water away from the Niger River in Mali. The river currently supports as many as 100 million people, who depend on it for water, power, transportation, and agriculture. (Thinkstock)

to 94 degrees F (34 degrees C) in April and May, with lows dipping to 62 degrees F (16 degrees C) in winter, and highs rising to 106 degrees F (41 degrees C) in April. Mean annual precipitation decreases northward, from more than 160 inches (4,100 millimeters) in the delta area to less than 10 inches (250 millimeters) in Timbuktu. Rainfall declines from about 90 inches (2,300 millimeters) near the Niger's source to 10 inches (less than 250 millimeters) in the bend between Timbuktu and Bourem, then increases as the river flows southward to about 160 inches (4,000 millimeters) at its mouth.

Biodiversity
The river harbors 36 families and nearly 250 freshwater fish species, including catfish, African carp, Nile perch, tiger fish, barbel, lungfish, and tilapia. Twenty of these fishes are found nowhere else on Earth (they are endemic to this biome), and 11 of the 18 families of freshwater species are endemic to Africa and represented in the river. One of these is *Gobiocichla wonderi*, a small cichlid with an elongated body.

The inland wetlands and floodplains support a wide variety of animals such as manatees, crocodiles, hippopotamuses, mongoose, African otter, snakes, lizards, elephants, lions, and leopards. Various other species, such as antelopes and monkeys, live in many parts of the river basin, and buffaloes and jackals have more limited ranges.

The Inner Niger Delta serves as a major gathering spot for migratory birds from Europe. Some of these include black-crowned cranes, Egyptian plover, rock pratincole, goliath heron, and spur-winged goose.

The coastal delta of the river in Nigeria is the largest in Africa, and is rich in plant species, including mangroves that form the largest remaining tract in Africa. These mangroves, primarily the red and white types but with other mangrove species represented as well, cover an area of about 4,054 square miles (1 million hectares). The moist woodland savanna vegetation of the upper basin gives way to progressively drier savanna and semidesert conditions in Mali, where the river bends at the edge of the Sahel-Sahara Desert. After the river turns southward, it flows through increasingly moist and lush savanna, and through tropical forest near Onitsha. In the delta, the swamp forest contains many oil palms where the water is fresh, and mangroves where the water is brackish.

Human Influence
The River Niger is the source of water and livelihood to more than 100 million people. Fishing is an important activity, supplying food to regions along and beyond the river's course. The floodplains of the delta support agriculture and animal rearing. Dams built upstream provide water for agriculture and hydroelectric power for two countries. The discovery of crude oil and natural gas in the river's coastal delta in 1962 has led to the mixed economic fortunes of Nigeria. The river is threatened by climate change, increasing temperatures, siltation, repercussions from dam construction, pollution, and desertification.

Herders, mainly the Fulani, depend on the river for water and on its floodplains for dry season pastures for their cattle, sheep, and goats. Water from the river supports agriculture in all the countries the river runs through, enlivening such crops as rice, wheat, cotton, and sugarcane in Mali; millet and sorghum in northern Nigeria; and yams, cocoyams, and maize in the lower regions of Nigeria.

Several dams were built to foster irrigation on the Niger and along its tributaries. Other dams are proposed, built, or underway in Guinea, Mali, and Niger Republic. The river has been used as a source of hydroelectric power in Nigeria. Among the power-generating dams are Kainji, the largest on the river, providing Nigeria with about one-sixth of the nation's hydroelectric power.

The Niger also is used for navigation, with about 3,106 miles (5,000 kilometers) of the river and its tributaries navigable. The Niger is a center of energy resources, including petroleum and natural gas reserves. Including the delta, the Niger River system is the setting for most of the hydrocarbon extraction industry that forms some three-quarters of Nigeria's total income and 95 percent of its foreign exchange earnings.

Environmental Concerns

In spite of the river's significance, it faces serious environmental threats, the most significant of which is related to rising temperatures from climate change. In June 1985, the Niger completely dried up for the first time in history in Malanville, Benin Republic, and in Niamey, Niger Republic. Water volume and river flow has been very low since 1984, and was reported to lose about two-thirds of its flow at the Inner Delta between Segou and Timbuktu, Mali, due to seepage and evaporation.

Rising temperatures due to climate change will result in higher evaporation rates, shrinkage of natural wetlands, and reduction of fish diversity. Warming temperatures will increase both torrential rainfall and runoff, which will transport heavier loads of silts and sediments, leading to increased turbidity in the river and its tributaries, and degradation of water quality.

Other environmental threats include unsustainable agricultural practices, overgrazing, bush fires, and deforestation, all of which remove vegetation cover. Similarly, the Niger Basin Authority (NBA) has observed mining activities and deforestation around the Fouta D'jallon area, leading to pollution and sedimentation of the riverbed. Other threats include pollution from industrial wastes, and the recent proliferation of water hyacinth and water lettuce on the waterways, which steal oxygen from fish and other animals. Climate aridification and desertification will likely catalyze the rate of vegetation loss along the riverbanks, thereby increasing siltation still more.

In addition, the building of dams on the river for hydroelectric power and irrigation will also greatly reduce the flow of the river. The NBA and other authorities have their work cut out for them.

TEMITOPE ISRAEL BOROKINI

Further Reading

Gleick, Peter H. *The World's Water, 2000–2001: The Biennial Report on Freshwater.* Washington, DC: Island Press, 2000.

Stock, Robert. *Africa South of the Sahara: A Geographical Interpretation.* New York: Guilford Press, 2000.

Zwarts, Leo, Peter van Beukering, Bakary Kone, and Eddy Wymenga. *The Niger, A Lifeline: Effective Water Management in the Upper Niger Basin.* Dordrecht, Netherlands: Altenburg & Wymenga, 2005.

Nikumaroro Coral Atoll

Category: Marine and Oceanic Biomes.
Geographic Location: Pacific Ocean.
Summary: This small atoll is uninhabited, rich in fauna and flora, and was made famous as being a possible crash site of American aviator Amelia Earhart's final flight.

Marked by sharp coral, thick tropical foliage, and numerous coconut crabs, Nikumaroro Coral Atoll, formerly known as Gardner Island, is a small coral island, or atoll, located north of Western Samoa and the Tokelau Islands in the central South Pacific Ocean. Part of the Republic of Kiribati and the Phoenix Islands chain, Nikumaroro is a string of eight atolls and two submerged reefs that represents one of the last largely intact coral archipelago ecosystems in the world.

Approximately 4 miles (6 kilometers) long and 1 mile (2 kilometers) wide, Nikumaroro has three major sections: a central lagoon, a surrounding terrestrial rim, and an extensive ocean reef that has a steep drop. Nikumaroro is triangular and elongated in a northwest-southeast orientation.

The terrestrial rim, comprised of beach-rock ridges, rises through more than 16,000 feet (4,877 meters) of water. Nikumaroro forms part of a volcanic chain that includes a large seamount to the north and Carondelet Reef to the south. Other islands in the Phoenix group—including Kanton, McKean, Manra, and Orona—rest on parallel volcanic chains.

A low-lying atoll, Nikamaroro's highest point is about 20 feet (6 meters) above sea level. The rim forms an elongated barrier that surrounds a narrow and shallow lagoon with coral heads. On the northeastern edge, the rim is continuous. On the southwestern edge, it is broken by two passages: the seminavigable west-facing Tatiman Passage, and the non-navigable south-facing Bauareke Passage, which acts as an overflow channel for the lagoon.

Outside the rim is a reef flat that is 650–820 feet (198–250 meters) wide and descends steeply into the ocean. On the northeast shore is a sandy beach about 65–130 feet (20–40 meters) wide.

Biodiversity

Just south of the Equator and west of the 180th meridian, Nikumaroro has a hot, equatorial climate with temperatures averaging 81 degrees F (27 degrees C) year round. Daily temperatures range from warm to hot, 77–90 degrees F (25–32 degrees C). Rainfall averages 40 inches (102 centimeters) annually. In the extreme north, rainfall totals as much as 120 inches per year (305 centimeters).

The local Kiribati climate is tempered by easterly trade winds, with high humidity during the November-to-April rainy season. The atoll does experience occasional gales and tornadoes. Severe droughts are a periodic condition.

The equatorial climate helps support a variety of wildlife and teeming vegetation, including palm trees, flowering plants, and low-lying brush. The tree flora consists mainly of grand devil's claws (*Pisonia grandis*), also known as buka or mapou, and kou (*Cordia subcordata*), also called kanawa. Much of the kanawa has been cleared for its fine-grained wood, used to make furniture. These indigenous trees have been crowded out by plantations of coconut palm (*Cocos nucifera*), which have given rise to a population of feral coconuts.

The island's dense underbrush consists primarily of the fan flower (*Scaevola* spp.), which has tough interwoven stalks, while the shores feature the low-growing, littoral-zone octopus bush (*Tournefortia argentea*).

Other recorded plant species include beach mulberry (*Morinda citrifolia*), a small tree common to volcanic terrain that bears the edible "cheese fruit," named for its pungent odor; a Pacific hibiscus (*Sida fallax*) adapted to sandy soil; and purslane (*Portulaca* sp.), a tropical flowering plant also known as rose moss.

Though there have been reports of feral dogs and cats, the only mammal that assuredly lives on Nikumaroro is the rat. In the 1980s, dogs were systematically eradicated from the island. The other terrestrial vertebrates on the island are sea turtles (*Chelonioidea* and *Dermochelyidae*) and several seabirds, including boobies (*Sula*); gannets (*Morus*); and the rarest members of the *Phaethontidae* family, red-tailed tropicbirds (*Phaethon rubricauda*). Nikumaroro is part of the Phoenix Islands Protected Area and has been designated as an Important Bird Area by the Audubon Society.

Terrestrial invertebrate species are primarily crabs, such as hermit (*Coenobitidae* spp.) and fiddler (*Ocypodidae*), as well as a large population of coconut crabs (*Birgus latro*), the world's largest terrestrial arthropods. Other invertebrates include insects such as spiders, ants, and bees.

The waters around the atoll support a diverse collection of fish species, including the gray reef shark (*Carcharhinus amblyrhynchos*), blacktip shark (*Carcharhinus limbatus*), whitetip reef shark (*Triaenodon obesus*), grouper (*Serranidae*), trevally (*Carangidae*), and barracuda (*Sphyraenidae*). One marine mammal species that has been reported around the atoll is the bottlenose dolphin (*Tursiops*).

Climate change is a factor in the future of all living things here. Warming sea temperatures are causing water levels to rise, which threatens the atoll with flooding and habitat disruption. Seawater temperature, too, has proven to be critical to coral species, as warmer waters have been correlated with coral bleaching events. The

Unripe fruit of the beach mulberry (Morinda citrifolia) on the branch. The fruit and its seeds are considered "famine food" but are widely consumed in the Pacific islands. (Thinkstock)

Kiribati government is looking into ways to protect the coral island and its species.

Human Settlement Efforts

First sighted by Europeans in 1824 and originally named Gardner Island, Nikumaroro has been the site of several colonization attempts, which have failed because it is so difficult to land vessels along its shores, and there is a lack of freshwater.

After the United Kingdom laid claim to the island in 1856, a coconut plantation was established, along with a small settlement of people from Micronesia, but the project was closed within a year.

During World War II, the United States Coast Guard established a long-range navigation (LORAN) station on the island with 25 crewmen. In the mid-1950s, the population reached a high of about 100 people, but drought and lack of freshwater led to the island's evacuation in 1963. Two years later, Gardner Island was uninhabited and remains so today.

The name was officially changed to Nikumaroro when the Republic of Kiribati gained independence from the United Kingdom in 1979. The island is perhaps most famous for being the possible crash site of American aviation pioneer Amelia Earhart, who disappeared, along with her navigator Fred Noonan, somewhere in the Pacific on July 2, 1937, en route to Howland Island during her final flight to circumnavigate the globe.

REYNARD LOKI

Further Reading

Allen, Gerald and Steven Bailey. "Reef Fishes of the Phoenix Islands, Central Pacific Ocean." *Atoll Research Bulletin* 589 (2011).

Bryan, Edwin H., Jr. *American Polynesia and the Hawaiian Chain.* Honolulu, Hawaii: Tongg Publishing, 1942.

Maude, Henry Evans. *Of Islands and Men: Studies in Pacific History.* Melbourne, Australia: Oxford University Press, 1968.

Stackpole, Edouard A. *The Sea-Hunters, The New England Whalemen During Two Centuries: 1635–1835.* Philadelphia: Lippincott, 1953.

Nile Delta Flooded Savanna

Category: Marine and Oceanic Biomes.
Geographic Location: Africa.
Summary: A magnet for millions of migratory birds and once an amazing papyrus reed expanse, the Nile River Delta's flooded savanna biome is at risk of disappearing entirely under the rising waves of the Mediterranean Sea.

The flooded savanna of the Nile River has historically contained some of the richest soils in the world, that in turn support a rich, biodiverse ecosystem. In recent times, however, dams along the river and the impacts of global warming have been threatening the entire biome. Land continues to slip into the sea, and habitats are being threatened or eradicated by increased salinity, pollution, and haphazard water management.

The fertile soil surrounding the Nile River Delta lends itself to agriculture and draws people to the banks. In Egypt, about half of the nation's 83 million people have settled within the Nile River Delta region, the generally low-lying area toward the mouth of the Nile where it flows into the Mediterranean Sea. This area is approximately 109 miles (175 kilometers) long and 162 miles (260 kilometers) wide. The flooded savanna, a dominant habitat form here, is considered critically endangered.

In general, the grasslands and wooded shrublands of the delta are flooded during the annual summer rains. This was far more dramatic historically, but since major dams have gone up on the Nile River, most flooding is prevented from taking place—with the result that the characteristic *Cyperus papyrus* swamps of the wettest areas are mostly gone. The flooded areas that still exist tend to be along the lakes and lagoons that form alongside of the Nile in and around its branching delta. Today, the chief lakes located within the Nile Delta are El Mannah, El Qatta, Faraontya, Sinnéra, and Sanel Hagar. The main coastal lagoons are Manzala and Miheishar.

The Nile Delta has a mediterranean climate, with temperatures averaging 86 degrees F (30

The 72 fish species present in the Nile Delta in 1972 have declined to only 25. A fisher holds a nile tilapia, which is now dominant in the region, making up a full 75 percent of the total catch today. (Thinkstock)

degrees C) and scant rainfall of 4–8 inches (100–200 millimeters) annually. During the summer, temperatures can rise to 118 degrees F (48 degrees C), and in the winter they dip to 50–66 degrees F (10–19 degrees C).

Biodiversity

The Nile Delta Flooded Savanna biome is home to a variety of localized ecosystems, some of them attracting millions of birds that use the delta as a landing point on their migrations between Europe, Africa, and Asia. The wetland areas of Egypt are a major migration site for the white and black stork, European crane, and white pelican. Endangered species include the red-breasted goose, white-headed duck, and sociable lapwing.

Flooded savannas tend to attract birds of prey, and those that frequent the Nile Delta include various types of eagle, including the short-toed, booted, steppe, and lesser-spotted eagle; steppe and honey buzzard; and levant sparrowhawk. Hundreds of thousands of waterbirds spend the winter months in these flooded savanna areas, including what is thought to be the largest concentration of little gulls and whiskered terns in the world. Other waterbirds include the shoveler, teal, wigeon, garganey, grey heron, pochard, ferruginous duck, Kentish plover, and cormorant. Additionally, the largest breeding population of slender-billed gulls in the Mediterranean is found in the Nile delta.

In the area around Lake Manala is a large population of swamp cats, and elsewhere are otter mongoose, red fox, and the Nile monitor lizard. The delta has one endemic (found nowhere else) frog species, various aquatic reptiles, two types of marine turtles—the endangered loggerhead and green turtles—and the endangered Egyptian tortoise. In the water are sole and striped mullet.

Experts suggest that only 10 percent of waters from the Nile River now flow into the Mediterranean, resulting in many species of fish being prevented from reaching their traditional breeding grounds. In 1972, there were 72 fish species in the delta region, but now only 25 remain. Only tilapia, which survives in the current environment, has thrived, now making up three-fourths of the total catch. Invasive bird species that threaten the ecosystem by their feeding habits and ability to crowd out native avians include the cattle egret, rock pigeon, mute swan, purple swamp hen, and Eurasian collared-dove.

Like the water systems of Niger and Chad, which were connected to the Nile River in the ancient past, the delta has a rich ecosystem of plants. It is home to at least 552 plant species, eight of which are endemic. Once known for its large papyrus swamps that are now largely absent, the Nile Delta flora now consists predominantly of reeds (*Phragmites australis* and *Typha capensis*), and sea rush (*Juncus maritimus*), with some small sedges.

Hornwort (*Ceratophyllum demersum*) and pondweeds (*Potamogeton crispus* and *P. pectinatus*) grow along the southern shore areas, while *Najas pectinata*, *Eichhornia crassipes*, and *Cyperus* and *Juncus spp.* grow along lakes and small lagoons. The salt-tolerant *Halocnemum spp.* and *Nitraria retusa* grow in marshes along the Mediterranean coast, while *Phragmites* and *Typha* grow along riverbanks that were previously bare. Reed swamp, which attracts various waterfowl, grows on the islands along the river between Luxor and Kom Ombo.

Environmental Threats

Throughout the area, the soils of the delta have become poor. The use of chemical fertilizers is on the rise, and runoff from these pollutants, along with wastewater and sewage, is seeping into the sediments of the flooded savanna. Studies have revealed that one species of catfish has begun to accumulate mercury, iron, copper, and other metals within its muscles and liver. One lake has been tainted with the pesticide DDT and other toxins. Oil industries are adding their own pollutants to the delta.

The ecosystems of the flooded savanna also are being threatened by erosion as the sea-level rises. Away from the water, the outer rims of the delta are dry-eroding. High salinity levels are threatening ecosystems that have thrived in the area for centuries. Hunters are shooting and trapping within this fragile ecosystem. As elsewhere, global warming is a constant threat, as air and water temperatures rise toward habitat and species tipping points.

In the late 1980s, scientists around the world became deeply concerned about the ecological situation in the Nile Delta. Daniel J. Stanley, an oceanographer with the Smithsonian Institution, predicted that in 2100, major flooding would occur in the area. Stanley believes that the key flood will occur at a point between the Suez Canal (to the east) and somewhere just to the west of the Nile River, immersing a swath up to 20 miles (32 kilometers) long in the delta.

Although the gradual sinking (subsidence) of the delta in conjunction with rising sea levels has been occurring on the Nile for thousands of years, the process was accelerated by the building of the Aswan Dam, along with a vast network of drainage and irrigation canals. The deposits of sediment and silt from the Nile that used to offset the sinking of the delta have been interrupted by the Aswan Dam, which diverts those silts far upstream, mainly into Lake Nasser.

In 1990, a team of experts from the Smithsonian Institution and *National Geographic* learned that the northern area of the delta is sinking into the Mediterranean Sea at a rate of 0.2 inch (5 millimeters) each year, further threatening the ecosystem of the flooded savanna. By 2009, both independent agencies and the Egyptian environment minister had issued warnings that suggested that the entire fertile area of the northern delta could disappear if steps were not taken to halt the pace of global warming and climate change.

As early as 2020, some 15 percent of the Nile Delta, encompassing urban areas, farms, and great areas of the flooded savanna biome, could be under immediate threat by the rising Mediterranean Sea waters. The United Nations Environment Programme (UNEP) contends that as many as 6 million people could be displaced in the short term. Also, many of the ecosystems within the flooded savanna biome could be eradicated, and altered migration patterns of birds could create environmental problems of international proportions.

Conservation Efforts

The richest habitats remaining in the Nile Delta region associated with the flooded savanna biome are in the lake and coastal lagoon areas. Over the past several decades, much of these lands have been converted to agricultural fields—and the government generally has been reluctant to declare delta areas protected, as a consequence. Two such areas have gained reserve designation, however: the Ashtoun el Gamil-Tanee Island Natural Area, devoted chiefly to protecting gravid fish and their young; and Lake Burullu, a Ramsar Wetland of International Importance.

To halt erosion, walls of rocks have been placed along the banks of the Nile. Other attempts are focusing on building concrete walls around beaches. The most ambitious plan to mitigate the possible loss of the Nile Delta has involved building an entire new river in the desert, via a complex system that would pump water from the Nile and channel it into a fertile valley on somewhat higher ground.

ELIZABETH RHOLETTER PURDY

Further Reading

Collins, Robert O. *The Nile.* New Haven, CT: Yale University Press, 2002.

Godana, Bonaya Adhi. *Africa's Shared Water Resources: Legal and Institutional Aspects of the Nile, Niger, and Senegal River Systems.* London: F. Pinter, 1985.

Hoke, Franklin. "Nile Delta Losing Ground." *Environment* 32, no. 10 (1990).

Mayton, Joseph. "Egyptian Officials, Farmers Debate Effect of Climate Change on Fertile Nile Delta." *Washington Report on Middle East Affairs* 28, no. 1 (2009).

Wohl, Ellen E. *A World of Rivers: Environment Change on Ten of the World's Great Rivers.* Chicago: University of Chicago Press, 2011.

Nile River

Category: Inland Aquatic Biomes.
Geographic Location: Africa.
Summary: The world's longest river offers its ecoregion amazing gifts—but 11 countries contend over the disposition of these hydrological and ecological riches.

The Nile River is the longest river in the world, extending some 4,040 miles (6,500 kilometers) across the northeast sector of the African continent. The Nile River basin comprises 1.3 million square miles (3.4 million square kilometers), watering at least part of 11 countries: Burundi, Democratic Republic of the Congo, Egypt, Eritrea, Ethiopia, Kenya, Rwanda, Republic of Sudan, South Sudan, Tanzania and Uganda. Major lakes account for about 2.4 percent of the basin, while swamps cover another 2 percent.

Located in high-altitude swampland, the most upstream tributary to the Nile is the Kagera River, an important feeder to Lake Victoria, the largest of Africa's Great Lakes. North of the Great Lakes region, the river is referred to as the Upper White Nile where it flows through a hot, arid area with high evaporation losses in South Sudan. The White Nile stretches from the Sobat River to Khartoum, capital of the Republic of Sudan. Here, the river combines with the Blue Nile, which is comprised of flows from Lake Tana and other pools on the Ethiopian Plateau. Once the Blue and White Nile merge in Khartoum, the last remaining major Nile River tributary is the Atbara River, a seasonally dry stream flowing down from Ethiopia that meets the Nile in Egypt.

The Nile flows into the vast storage reservoir in Aswan, Egypt, known as Lake Nasser. From here, water is released into a narrow, sinuous stream, which ultimately passes through the Egyptian capital, Cairo. Approximately 14 miles (23 kilometers) north of Cairo, the Nile breaks into two branches, the Rosetta and the Damietta. These branches form the main bulk of the Nile Delta, where the river flows into the Mediterranean Sea.

The majority of the Nile River basin flows are highly seasonal and affected by rainfall variability. As much as 30 percent of the rainwater is lost to evapotranspiration. As the Blue Nile passes through eastern Sudan during the hot summer months, it can lose up to half its flow to evaporation. During the June-to-November rainy season in the Ethiopian highlands, however, surges in flow can cause major soil loss, streambed erosion, and flooding. The Blue Nile provides on the order of 20 percent of the total Nile volume, the White Nile roughly 80 percent, on average.

Eight major dams are located on the Nile River basin to reduce flooding, generate electric power, and store water. Major dam construction began in 1902 with the Old Aswan Dam. It was followed in 1925 by the Sennar Dam in Sudan on the Blue Nile, then in 1937 by Sudan's Jebel Aulia Dam on the White Nile, and then others.

Although damming the Nile has helped some geographic areas, it has caused detrimental effects in others. Since the construction of the Aswan High Dam in the mid-20th century, the lower Nile basin and Nile Delta have received very little of the fertile sediment previously accumulated during floods. Now, most of this sediment settles out upstream in Lake Nasser. The dam decreased the Nile's flow rate, depressed water oxygen levels, and increased the salinity of large parts of the ecosystem.

Further upstream, Lake Victoria's only outflow, the White Nile, was dammed in 1952 by British colonial engineers. The goal was to utilize the flow to provide power for the region. In 2002, Uganda constructed the Kiiva Power Station along the White Nile, and by 2006, Lake Victoria's water level was at an 80-year low.

Climate

The climate of the River Nile basin varies greatly throughout its long run. The temperature is generally around 100 degrees F (40 degrees C) during daylight hours in the summer's hot, dry season. The winters are generally mild. Temperatures are hottest in the northern reaches of the river (mainly in Sudan and Egypt), but also quite warm in the rainforest wetlands in central and eastern Africa.

The northern region of the biome is affected by a year-round high-pressure system; as the temperature increases, southern *khamsin* winds cause heat waves.

In contrast to the north region, the southern basin has significantly more rainfall, especially during summer, and is affected by sea surface temperatures and summer monsoon winds. Significantly less rainfall occurs in winter.

Biodiversity

The flora of the tropical forests found in the Great Lakes region of the upper Nile River basin comprise many diverse plant species including ebony, banana, rubber, bamboo, and coffee shrub. Further north, there is an area of unique dryland biodiversity. On the Sudanese plains, papyrus, reed mace, ambatch, turor and the South American water hyacinth can be found. On the Ethiopian Plateau, mixed woodlands and savannas dominate; in the Sahel and Sahara zones, deserts predominate and very little vegetation exists except along the river edges. There are more than 95 species of aquatic plants belonging to 33 families along the Nile.

The Nile River basin is home to reptiles, birds, and fish, including the Nile crocodile and hippopotamus, which are found mainly in the wetter southern region. Other reptiles, such as soft-shelled turtles, monitor lizards, and more than 30 types of snakes, also are found in the basin.

Currently, there some 800 fish species are in the Nile, the majority of them from the *Cichlidea, Cyprinidae, Mormyridae,* and *Mochokidae* families.

In the past 100 years, the Nile River system has experienced a significant decrease in fish population diversity. The native fish *Haplochromis* are small, bony fish that were not very profitable for the local fishing industry, but were very prevalent.

Water Hyacinth

Eichhornia crassipes, the water hyacinth, was introduced by Belgian colonists to Rwanda. Although pretty, the water hyacinth forms thick mats of vegetation that pose problems throughout the ecosystem. A nonnative species, the water hyacinth has grown to such levels that it has negatively impacted transportation, fishing, hydroelectric power generation, and drinking water supply. To combat the plant's growth, the water hyacinth weevil (*Neochetina eichhorniae*) was bred and released in Lake Victoria to limit the growth of the plant.

The purple blossoms and large leaves of the invasive water hyacinth (Eichhornia crassipes) floating on the surface of a lake. (Thinkstock)

In the mid-1900s, the Nile perch, which can weigh over 175 pounds (80 kilograms), was introduced into the river, and by the end of the century, Nile perch was so successful that it ranked as Uganda's second-biggest export (behind coffee). Although the fisheries increased up to four times their yield, the carnivorous fish preyed on smaller native fish. This has caused as many as 150 native Nile River fish to become extirpated.

In the Nile River Basin, there are both cosmopolitan and endemic—found only here—bird species. The most common of these include the osprey (*Pandion haliaetus*) and the moorhen (*Gallinula chloropus*). The blue-winged goose (*Cyanochen cyanopterus*) is among the most abundant of all the endemic species.

Environmental Issues

Besides the variety and always-shifting nature of its climate zones and hydrology, the Nile River basin faces many sociocultural challenges. By 2025, population growth is predicted to contribute to water scarcity in half of the river basin's countries. Major challenges face the 11 countries, chiefly finding a way to work together to better manage their shared water resources, complicated by the region's high rates of poverty and famine, and the frequency of drought. These nations also are challenged within the parameters of past Nile River treaties, specifically the 1959 Egypt-Sudan treaty that allocates the Nile's water annual flow and largely excludes other upstream countries.

The socioeconomic issues not only affect the allotment of water resources, but also the pollution along the river. Industrial waste is released directly into the Nile and associated irrigational canals. In Egypt, about 85 percent of the water available is used for agriculture, but the runoff carries pollutants such as pesticides, manure, salts, and wastewater. Agricultural drains and household waste also is directly discharged into the Nile. In some of the major drainage points, referred to as black zones, the water exceeds the European Community Standard of fecal contamination.

The lack of education, government planning, and enforcement continues to undermine Nile River cleanup efforts. Climate change poses other obstacles. Rising water temperatures could lead to even higher evaporation—and shore erosion from violent storm flooding. Unpredictable precipitation and river flow, land degradation, siltation, waterweed infestation, droughts, deforestation, species loss, and possible increased incidences of disease are all projected potential outcomes of global warming here. Improving education and better-coordinated regional water management schemes will be needed to form part of the answer to these daunting challenges.

JENNIFER STOUDT WOODSON

Further Reading

Beyene, Tazebe, et al. "Hydrological Impacts of Climate Change on the Nile River Basin: Implications of the 2007 IPCC Climate Scenarios." Seattle: University of Washington; and Wageningen, Netherlands: Wageningen University and Research Centre, 2008.

Dumont, Henri J., ed. *The Nile: Origin, Environments, Limnology and Human Use.* Dordrecht, Netherlands: Springer Science and Business Media, 2011.

Eaton, Lauren and Alex Couture. "Human Impacts on the Nile River." University of Michigan. http://www.sitemaker.umich.edu/sec004_gp5/foreign_species_and_biodiversity.

Osman, M. M. Ali. "Aquatic Plants of the Sudan." In Henri J. Dumont, ed. *The Nile: Origin, Environments, Limnology and Human Use.* Dordrecht, Netherlands: Springer Science and Business Media, 2011.

Ningaloo Coral Reef

Category: Marine and Oceanic Biomes.
Geographic Location: Australia.
Summary: Famous for mass spawning of coral organisms, Ningaloo is among the best-developed fringing coral reefs anywhere and the only major reef on the west coast of any continent.

Located offshore of the state of Western Australia, Ningaloo Reef is the only major coral reef on the west coast of any continent. It extends for more than 186 miles (300 kilometers) parallel to the coast in the Gascoyne region. Considered to be perhaps the best-developed fringing reef in the world, the Ningaloo is the site of one of nature's most spectacular mating rituals.

After the March full moon at neap tide, when tidal motion is at a minimum between 8:00 P.M. and 10:00 P.M., the corals spawn simultaneously in the still waters. Wind-driven waves later wash the larvae out to sea. Sometimes, however, the waves do not appear on schedule. The growing coral larvae deplete the oxygen in the lagoon, leading to asphyxiation of the larvae, fish, and even the parent corals themselves. Since its discovery in 1989 at Ningaloo, this accidental mass-suicide phenomenon has been observed at other reefs.

While the landscape onshore of the reef is arid, occasional summer tropical storms can cause flooding, as happened in December 2010, when some areas received more than 9 inches (22 centimeters) of rain. Coastal areas from Shark Bay southward receive additional moisture in the form of winter rains, which support a more savanna-like vegetation and spring wildflower displays.

Biota

South of the Ningaloo Reef, the vegetation is dominated by eucalyptus trees, while the northern, more desert-like areas are dominated by acacia shrubs. This area was not always a desert. Until 3 million years ago, northwestern Australia was covered with rainforest. At this time, the Indonesian archipelago rose from the ocean floor, changing ocean currents and blocking warm water traveling south along the Australian coast. This led to the decline of precipitation in Australia, and was the death blow to rainforest vegetation.

Ningaloo Reef supports 200 species of coral, 500 species of fish, 600 species of mollusks, and 90 species of echinoderms. There are large populations of dugongs, marine turtles, whales, and whale sharks. The dugong, a large aquatic marine mammal, is found feeding on seagrass beds around sandy lagoons. Among the fish, whale sharks are especially attracted to the coral spawning.

Two offshore currents, one south-flowing (Leeuwin Current) and one north-flowing (Ningaloo Current), interact in the area of the reef, enhancing biological growth and reproduction, and attracting whale sharks, manta rays, humpback whales, sea snakes, large predatory fish, and seabirds.

Cape Range National Park has a limestone karst landscape, supporting 500 caves and sinkholes with rare subterranean aquatic fauna. These animals have evolved in isolation as the Australian continent separated from Gondwana and became increasingly arid as it moved north. Part of the Ningaloo World Heritage Area, the Cape Range karst area formed when seawater dissolved limestone rocks. The aquatic community of the Bundera sinkhole here is unique in the Southern Hemisphere for its remipede crustaceans. Rock wallabies, kangaroos, emus, echidnas, and lizards are common animals.

Offshore, in deeper water, the Cape Range Canyons support upwellings that help maintain the biological diversity of the Ningaloo Reef system. Early peoples used the Cape Range area for exploitation of marine resources, with records dating back at least 32,000 years.

Protected Areas

Ningaloo Reef is protected by Ningaloo Marine Park. Together with adjacent properties at Cape Range National Park and Muiron Islands Marine Reserve, this area has been named a World Heritage Site.

Also part of the World Heritage Site is the Muiron Islands Marine Management Area. These islands are a northward extension of the Cape Range, and contain important turtle nesting areas used by loggerhead turtles, an endangered species. The limestone reef at Muiron Islands contains thousands of cardinalfish. In addition to Ningaloo Marine Park, Cape Range National Park, and Muiron Islands, the World Heritage Site includes Bundegi Coastal Park on Exmouth Peninsula, Jurabi Coastal Park on Exmouth Peninsula, and Learmonth Air Weapons Range; these are all in the Gascoyne region.

To the south of Ningaloo is Shark Bay, another World Heritage Site. Constituent units of the Shark Bay site include Dirk Hartog Island National Park, Francois Peron National Park, Hamelin Pool, and the Bernier and Dorre islands nature reserves. The bay itself is a marine park that supports vast assemblages of dugong, sharks, turtles, and whales. The world's largest seagrass beds are here. The entire Shark Bay-Ningaloo area is designated as an Australian National Landscape for conservation and tourism purposes.

North of Ningaloo Reef, Montebello Islands Marine Park, Lowendal Islands Nature Reserve, and Great Sandy Islands Nature Reserve were the site of British nuclear tests in 1952 and 1956.

Today, these parks protect about 300 islands with mangroves, corals, tropical fish, and seabirds. A whale migration path also passes by the islands. The Lowendal Islands are an Important Bird Area for the crested tern and bridled tern. The Montebello Islands are an Important Bird Area for the fairy tern, roseate tern, and sooty oystercatcher.

Environmental Threats

Environmental issues in the area include the management of commercial fishing, the potential for oil drilling around the Muiron Islands, climate change and its potential for warm-water-driven coral bleaching events, and the management of ecotourism effects on the Ningaloo Reef itself.

HAROLD DRAPER

Further Reading

Finkel, Elizabeth. "Bio-Inventory in Oz." *Science* 289 (2000).

Perkins, Sid. "When Islands Rose, Australian Rainforests Fell." http://news.sciencemag.org/sciencenow/2011/06/when-islands-rose-australian-rai.html

Richardson, Sarah. "A Slickness Unto Death." *Discover* 15, no. 4 (1994).

Western Australia Department of Environment and Conservation. "Management Plan for the Ningaloo Marine Park and Muiron Islands Marine Management Area, 2005–2015." http://sponsored.uwa.edu.au/wamsi/__data/page/3748/NMP_MPA.pdf.

Niue Island Coral

Category: Marine and Oceanic Biomes.
Geographic Location: Pacific Ocean.
Summary: This large coral atoll and island nation in the South Pacific faces challenges because of global warming.

Niue is a large tropical, uplifted coral atoll in the South Pacific Ocean, located 1,500 miles (2,414 kilometers) northeast of New Zealand in the middle of a triangle formed by the Samoas to the northwest, the Tonga Archipelago to the southwest, and the Cook Islands to the southeast.

Formed out of a volcano and comprised primarily of coral and limestone, the oval-shaped Niue is commonly known as the Rock of Polynesia. It has an area of 77 square miles (199 square kilometers) and a coastline of around 40 miles (64 kilometers). Its lowest point is at sea level, and it rises to a maximum height of about 230 feet (70 meters) above sea level at a point near the village of Mutalau on the northeastern coast.

The terrain of Niue is characterized by a rocky shore, isolated beaches in protected shoreline coves, a series of steep coastal limestone cliffs formed in the Pleistocene epoch, numerous coastal limestone caves; and a central plateau that rises to around 200 feet (61 meters) above sea level. The thin layer of volcanic ash over the exposed coral limestone terraces indicates that volcanic activity uplifted the island.

The island is almost encircled by a coral-reef platform that has a single substantial breach near the capital, Alofi, along the central western coast. These reef flats feature submerged caves below and rock pools on top. The western coast of Niue is marked by two large bays, Alofi Bay in the center and Avatele Bay in the south, which are separated by a promontory, Halagigie Point. The southwest coast is marked by Blowhole Point, a small peninsula near the village of Avatele. Niue also includes three outlying nonterrestrial coral reefs: Beveridge Reef, 150 miles (241 kilometers) southeast; Antiope Reef, 110 miles (177 kilometers) southeast; and Haran Reef, 182 miles (293 kilometers) southeast.

The Niue Island Coral ecosystem experiences a tropical climate that is affected by southeast trade winds, the South Pacific Convergence Zone, and the tropical cyclone belt. Cyclones hit the island quadrennially. In 2004, Cyclone Heta, the largest recorded tropical cyclone, caused widespread destruction to the island. Rain falls primarily during the wet season, from November or December to April. The average annual precipitation is 82 inches (208 centimeters). The mean temperature is 83 degrees F (28 degrees C).

Waves crashing on the steep platforms that extend out from the shore on Niue Island. These reef flats have submerged caves below and rock pools on top. (Thinkstock)

Vegetation

Niue is designated as part of the Tongan Tropical Moist Forests ecoregion by the World Wildlife Fund, with about one-third of the island covered by tropical forests. The rest of the vegetation is primarily low-lying, saline-resistant plants. The flora of Niue include orchid (*Orchidaceae*), hibiscus (*Malvaceae*), frangipani (*Plumeria*), bougainvillea (*Bougainvillea*), rhododendron (*Rhododendron*), poinsettia (*Euphorbia pulcherrima*), and stands of ancient ebony (*Diospyros*). The screwpine *Pandanus niueensis* is endemic to Niue, that is, found nowhere else.

Following Cyclone Heta, the destruction of much of the forest cover led to the growth of a host of opportunistic and invasive heliophilous (sun-loving) plants, such as rattlepod (*Crotalaria*), white leadtree (*Leucaena leucocephala*),

white shrimp plant (*Justicea betonica*), bitter vine (*Mikania micrantha*), blue porter weed (*Stachytarpheta cayennensis*), light-blue snakeweed (*Stachytarpheta jamaicensis*), and marigold tree (*Tithonia diversifolia*), as well as the indigenous morning glory vine (*Merremia peltata*).

Fauna

Niue supports several endemic birds, including the Polynesian triller (*Lalage maculosa whitmeei*), Polynesian starling (*Aplonis tabuensis brunnescens*), and purple-capped fruit-dove (*Ptilinopus porphyraceus whitmeei*). Two extinct bird species, the Niue night heron (*Nycticorax kalavikai*) and the Niue rail (*Gallirallus huiatua*), were also endemic to Niue. The combtooth blenny (*Ecsenius niue*) is one of Niue's endemic marine fish.

Other endemic animal species include a land snail (*Vatusila niueana*); a crab (*Orcovita gracilipes*); a seed shrimp (*Dantya ferox*); a sea snail (*Tectarius niuensis*); and several insects, including a leafhopper (*Empoasca clodia*), a weevil (*Elytrurus niuei*), and a scale (*Paracoccus niuensis*). The Niuean flat-tailed sea snake (*Laticauda schistorhynchus*) may be endemic.

Conservation and Threats

In 2001, the Niue government established the National Biodiversity Strategy and Action Plan to preserve the island's unique plant and animal life.

The beaches of Niue are relatively young and dynamic, eroding and growing in response to specific storm conditions, particularly cyclones. The high level of foraminifera, a protozoan, in its northwestern beaches indicates that coral growth can be reestablished relatively quickly following cyclone devastation, but it also indicates that anthropogenic climate change most likely affects the state of these sites.

Agriculture, and in particular subsistence agriculture, is important to Niue, which has some 80 square miles (207 square kilometers) of land available for agricultural use. Though the majority of Niue's volcanic-ash-based soil is rich in potassium and phosphorous, some areas lack essential plant nutrients like nitrogen and thus impede the growth of agriculture. Cows were introduced in the 1960s, and goats arrived in the late 1980s. There is a growing concern that the practice of slash-and-burn agriculture is affecting soil fertility. The effects of global warming pose a continuing threat.

Reynard Loki

Further Reading

Forbes, D. L. *Coastal Geology and Hazards of Niue.* Suva, Fiji: Pacific Islands Applied GeoScience Commission (SOPAC), 1996.

Gardner, Rhys Owen. *Trees and Shrubs of Niue: An Identification Guide to the Island's Indigenous and Naturalised Woody plants.* Waitakere, New Zealand: Katsura, 2010.

Hekau, Maihetoe, et al. *Niue: A History of the Island.* Suva, Fiji: Institute of Pacific Studies and Government of Niue, 1982.

Kreft, C. S. *The Climate and Weather of Niue.* Wellington: New Zealand Meteorological Service, 1986.

Nunn, P. D. "Myths and the Formation of Niue Island, Central South Pacific." *Journal of Pacific History* 39 (2004).

North Sea

Category: Marine and Oceanic Biomes.
Geographic Location: Europe.
Summary: This ecologically rich North Atlantic basin is the nautical corridor connecting northern Europe ecosystems with the wider seas.

The North Sea is a semi-closed, continental-fringe sea in the North Atlantic Ocean. Given its strategic position as a doorway to the Atlantic, it has had a major part in shaping European history—and this has heavily affected its ecology. Commercial productivity of the region lies mainly in its fisheries, hydrocarbon fuels, and renewable energy. Conservation is also of the utmost importance, with many of Europe's most valuable natural habitats occurring around its perimeter.

The North Sea lies between Great Britain and the coastline of continental Europe, with Norway at its northern extremity and France to the south. It is the only marine interface for Belgium and the Netherlands, and also a major component of the marine territory of Great Britain, Germany, Denmark, Sweden, and Norway. It spans approximately 289,577 square miles (750,000 square kilometers); and its gently sloping topology is shallowest in the south, typically around 164 feet (50 meters) deep, reaching 328 feet (100 meters) in the center of the basin, and more than 2,297 feet (700 meters) in the Norwegian Trench.

Scattered throughout the basin are several large sandbanks, including the Dogger Bank, the Broad Fourteens, and the Deep Forties. With an area of approximately 6,795 square miles (17,600 square kilometers), the Dogger Bank is one of the most productive fishing grounds in the North Sea.

Current flow tends to be counterclockwise, with the greatest inflow entering via the Norwegian Sea. Atlantic water also enters via the English Channel, but this flow is somewhat more restricted. Brackish water also infiltrates the North Sea from the Baltic Sea via the Kattegat and Skagerrak. The salinity and temperature of North Sea waters fluctuate with the North Atlantic Oscillation.

Species and Biomass

The North Sea has an extremely diverse coastal extent, including fjords, sandbanks, mudflats, large estuaries, and deltas, but the seabed itself is mostly sandy or muddy. There is a general pattern of increased species richness around the circumference of the North Sea. A variety of invertebrates are found on the North Sea benthos, including polychaete worms, sea urchins, brittle stars, shrimp, and lobsters.

Extensive beds of coldwater coral *Lophelia* have been mapped along the shores of Norway, particularly along the Norwegian margin of the Skagerrak, where a unique yellow variety has been recorded. *Sabellaria* reefs are also known to exist, although their distribution has yet to be fully studied.

The pelagic fish community is dominated by herring, with summer peak abundances in mackerel and horse mackerel. Cod, haddock, whiting, and saithe are also important, in addition to flatfish such as sole and dab. The relative abundance and distribution of fish varies from year to year in accordance with climatic conditions. In recent decades, highly migratory species such as tuna and halibut that once were abundant have now disappeared or become rare. Spiny dogfish, the region's major shark species, is now considered to be depleted. Catches of rays and skates have also decreased significantly over time.

Although overfishing is often blamed for the disappearance of species, scientists also suspect that climate change has had major impacts on the food chain in the North Sea, reducing the food supply for larger sea creatures. Evidence is also at hand that the average size of individual fishes is shrinking, also based on increasing difficulty of finding suitable foods, among related pressures.

The basin is a key ecosystem for seabirds, and around 5 million birds of 28 species breed here each year. These birds occupy niches from surface feeders to divers and waders. Foraging expeditions can range over hundreds of miles (kilometers) at a time. Many of these seabirds interact with fisheries in the area through competing for resources and consuming discards.

Marine mammals in the North Sea tend to be transient, with a few resident populations of harbor and gray seals, occurring particularly along the British coast. Seals compete for resources with fishing boats and commonly depredate passive fishing gear. Three main marine mammal species frequent this sea: minke whales, white-beaked dolphins, and harbor seals. Minke whales tend to remain in the northern reaches of the sea, while dolphins are more abundant in the south, and it is thought that the North Sea may be the most important habitat in the world for these animals.

Resources

The commercial and recreational value of the North Sea basin is considerable, especially in terms of fishing and boating. It is also a major conduit for freight ships, military vessels, and underwater cables. Considerable oil and gas deposits lie beneath the North Sea. For this reason, the area has attracted investment from the world's energy producers. Several European states now have well-established operations within the basin.

European nations have been harnessing wind power in the North Sea since the early 1990s; the world's largest offshore facility, the Thanet wind farm, was inaugurated in 2010 and is located close to the shore of southeast England. There are also test facilities established in the Scottish Orkney Isles to explore the potential of wave and tidal energy. Plans are under way to establish an electricity supergrid connecting renewable energy resources from across the North Sea area.

Conservation Efforts

European waters are conserved through several international policies, including the Habitats Directive and the Birds Directive. Numerous locations have been designated as Natura 2000 sites to meet the objectives of these two directives, while the Ramsar Convention designates wetlands of international importance, and other conventions aim to establish a network of marine protected areas.

Perhaps the most important region of conservation interest is the Wadden Sea, which follows 311 miles (500 kilometers) of coast from the northern reaches of the Netherlands, along the German shoreline, to the western seaboard of Denmark. Its extensive mudflats provide important habitat for seabirds and shorebirds, and rich feeding grounds for migratory species. It is an immensely valuable ecosystem because of its rich diversity, and part of the Wadden Sea is now listed as a World Heritage Site.

RUTH M. HIGGINS

Further Reading

European Environment Agency. "Natura 2000 Data—The European Network of Protected Sites." http://www.eea.europa.eu/data-and-maps/data/natura-2000.

Furness, Robert and Mark Tasker. "Seabird-Fishery Interactions: Quantifying the Sensitivity of Seabirds to Reductions in Sand Eel Abundance, and Identification of Key Areas for Sensitive Seabirds in the North Sea." *Marine Ecology Progress Series* 202 (2000).

International Council for the Exploration of the Sea. "ICES Advice Book 6: North Sea." http://www.ices.dk/committe/acom/comwork/report/2008/2008/6.1-6.2 North Sea ecosystem overview.pdf.

Pedersen, Søren Anker, et al. "Natura 2000 Sites and Fisheries in German Offshore Waters." *ICES Journal of Marine Science* 66 (2009).

Northwestern Mixed Grasslands

Category: Grassland, Tundra, and Human Biomes.
Geographic Location: North America.
Summary: These extensively altered semiarid grasslands are characterized by drought-resistant plants and animals; the vegetation dynamics are largely shaped by grazing and burrowing rodents.

The Northwestern Mixed Grasslands ecosystem is located in the central Great Plains of North America. It is part of the temperate grasslands biome, covering the southern portions of Alberta and Saskatchewan in the north, and extending southward into Montana, North and South Dakota, Wyoming, and Nebraska. These grasslands gradually increase in elevation from east to west, reaching their maximum height at the base of the majestic Rocky Mountains. In the few areas still in natural vegetation, the grasslands provide expansive views of mildly rolling topography that eventually changes into forested habitat at the base of the Rockies.

The northwestern mixed grasslands are part of the iconic history of the American West, where bison used to number in the millions and early settlers traveled in covered wagons to homestead Western lands. Today, little of the northwestern mixed grasslands remain untouched by human activity. While they are an important economic resource because of their utility for agriculture, the associated loss of native biodiversity makes them of conservation concern.

The northwestern mixed grasslands have a semiarid climate, with precipitation ranging from 12 to 18 inches (30 to 45 centimeters) annually. Precipitation decreases along an east-to-west gradient, with the westernmost portion of the grasslands located in the rain shadow of the Rocky Mountains. The climate is mid-continental, with pronounced seasons characterized by warm summers and extremely cold winters. Marked variations in temperature, coupled with low water availability, make the climate harsh. Heavy snow blankets the region in winter and provides much of the annual precipitation.

These grasslands are also windy and subject to severe summer storms. However, summer rains are often short-lived and may provide only enough precipitation to infiltrate the shallow topsoil layers. As a result, plants are often water-limited because they lose more water to the atmosphere through evaporation than enters the system through precipitation. For this reason, many of the plants as well as animals have physiological or behavioral adaptations to tolerate low water availability. Grasses have several adaptations for surviving drought, including dense, fibrous root systems that effectively capture moisture, and narrow leaf blades that lose less water to the atmosphere than broader leaves do. Small mammals, including some species of rodent, can extract water effectively from food items such as seeds.

Vegetation

The dominant natural vegetation of the northwestern mixed grasslands is composed of both short- and mid-height grasses. Representative species include blue grama (*Bouteloua gracilis*), needle-and-thread (*Stipa comata*), and western wheat grass (*Agropyron smithii*). Although grasses dominate in terms of their abundance, forbs (flowering plants that are not grasses) make the main contribution to plant-species diversity. It is common to see colorful explosions of the showy flowers of asters and legumes (members of the pea family) during the spring rains.

Several shrub species, including sage (*Artemisia tridentata*), contribute to the structural diversity of the grasslands by providing safe havens from predators and inclement weather to the rodents and rabbits.

Fauna

Historically, bison (*Bison bison americana*) fed on the grasses that dominate the northwestern mixed grasslands. As a result, some of the grasses, especially those that are of smaller stature, are tolerant to grazing by ungulates. Unfortunately, the bison were nearly exterminated by early European settlers, despite having numbered an estimated 50 million head in the early 1800s. They now exist predominantly in managed populations. In fact, pronghorns (*Antilocapra americana*) are the only large native ungulates that are still common here.

The black-tailed prairie dog (*Cynomys ludovicianus*) is also an important herbivore in the northwestern mixed grasslands system. It is considered to be a keystone species because it has marked effects on soil and vegetation characteristics, which influence the distribution and abundance of associated plant and animal species.

The endangered black-footed ferret (*Mustela nigripes*) is entirely dependent on prairie dogs, both as prey and because the ferret lives in the extensive burrow systems constructed by the prairie dogs. As prairie-dog populations declined because of extermination campaigns (they were viewed as pests by livestock owners), loss of habitat, and sylvatic plague, the ferret populations declined along with them. Today, captive breeding programs have brought the black-footed ferret back from the brink of extinction.

An abandoned farmstead in eastern Montana. Very little of the northwestern mixed grasslands remain untouched by human activity; because of past conversion to agricultural lands and other uses that have degraded the land, only 15 percent of the biome remains intact in its native state. (U.S. Geological Survey/Terry Sohl)

Effects of Human Activity and Threats

Grassland ecosystems worldwide have been important in shaping patterns of human settlement, because the fertile soils that underlie grasslands support many crop species, and the native grasses provide good forage for cattle and other livestock. Today, only 15 percent of the northwestern mixed grasslands remains in its native state, with much of it having been converted for agriculture or having been degraded due to various human activities. In Canada, the losses are particularly pronounced, with only 2 percent of the native grasslands remaining.

In areas used for livestock grazing, there is the potential to maintain a healthy, functioning ecosystem by using appropriate stocking rates. In fact, rangelands are considered to be conservation resources, because grazing, when managed properly, simply mimics the historic disturbance regime imposed by native grazers such as bison.

A conservation issue more difficult to address is the total conversion of these grasslands for other types of land use. In the Canadian mixed grasslands, disturbances such as road building are associated with increasing oil and gas development. These are fragmenting the remaining native habitat, while many rangelands are being converted for hay crops. Long-term effects from climate change are unknown, but the robust and rugged mixed grasslands may be resilient enough to withstand climate change. The biome has survived many drought periods, including super-droughts, in the past.

Christina Alba

Further Reading

Coupland, Robert T., ed. *Natural Grasslands: Introduction and Northwestern Hemisphere*. Amsterdam, Netherlands: Elsevier, 1992.

Ricketts, Taylor H., et al. *Terrestrial Ecoregions of North America: A Conservation Assessment*. Washington, DC: Island Press, 1999.

Thompson, et al. Grasslands, Savannas, Shrublands— Northern Mixed Grassland. U.S. Geological Survey Professional Papers, 2007. http://pubs .usgs.gov/pp/p1650-e/wwf/graphical_displays/ WWFV90.PDF.

Novel Ecosystems

Category: Grassland, Tundra, and Human Biomes.
Geographic Location: Global.
Summary: Human activities affect the distribution and abundance of species, often resulting in new combinations of species that did not occur previously in a given place. This change has unpredictable consequences for the biodiversity and ecology of a place.

Novel ecosystems are defined as ecosystems that have new assemblages or combinations of species resulting directly and indirectly from human action. They are called novel, or new, because species occur in combinations and in relative abundances unlike those that occurred previously within a given biome. Another term commonly used to describe novel ecosystems is *emerging ecosystems*.

Novel ecosystems are formed by changes in the physical and biological components of ecosystems; by species invasion of ecosystems; and/or by the abandonment of intensively managed ecosystems, such as agricultural or urban areas.

The distribution and abundance of species can fluctuate naturally. Changes in the physical environment through disturbances such as fires, landslides, floods, and volcanic eruptions cause the removal of some or all species. The opening of space and availability of resources allows for the colonization and reestablishment of species. In fact, disturbance can play an important role in maintaining species diversity in some ecosystems, by preventing the dominance of one species to the exclusion of others.

Ecological changes that occur after a disturbance are called *succession*. Primary succession occurs on a landscape that has been wiped clean of its original ecosystem, such as on a lava flow or a surface recently exposed by retreating glaciers. Secondary succession occurs when there are changes in species due to a disturbance that has affected but not removed the original ecosystem, as when a fire or a fallen tree in a forest opens space for plants to colonize.

On a global scale, changes in Earth's climate during the Pleistocene glacial and interglacial periods resulted in changing species assemblages. Species differ in their ability to adapt to changes in climate, and in their capacity to disperse or migrate in response to changes in their environment. By studying pollen records in lake cores and in sediments, scientists have found that past climatic changes resulted in new combinations of species through immigration of new species, emigration of species to a more suitable area, or extinction of those species that were unable to adapt to changing conditions. Paleoecologists call these communities no-analog communities because the resulting combinations of species have no analog in Earth's modern history.

Novel ecosystems differ from no-analog communities and from changes in species composition during natural succession in two ways: The changes in emerging ecosystems are driven by human activities which differ from natural past climatic changes and primary succession; the communities include new species that never coexisted there before, which differs from secondary succession. In addition, no-analog communities reconstructed from past environmental data are compared with modern systems, while novel ecosystems are compared with recent historical conditions.

While the emergence of novel ecosystems depends on some environmental change that is directly or indirectly caused by human activities, novel ecosystems—unlike agricultural ecosystems—do not require human management for their maintenance. Another important element of novel ecosystems is the unpredictability of the emerging species combinations, because they do not follow the well-studied patterns of species replacement during primary or secondary succession.

Effects of Human Activity

Researchers are realizing that the influence of human activities extends all over the globe, which has led some to rename our current geologic epoch *Anthropocene*, for the Age of Humans. Human activities that change the landscape and species distributions include deforestation; con-version of natural forests, grasslands, and wetlands to agriculture (crops and grazing lands); and fragmentation of the landscape by urbanization. Through the domestication and trade of species, humans have deliberately transported plant, animal, and microbial species around the world, with the result that many species now live in biomes where they never existed before and where they would not have been able to get to on their own.

Human activities have also caused the inadvertent movement of species around the world. The consequence is that many species are now finding themselves in new communities, interacting with species with which they have not co-evolved.

At the same time, humans have driven many species to extinction through overhunting, disease, habitat loss, and a breakdown of biological interactions required for the survival of a species. Through the fragmentation and degradation of landscapes, humans have created barriers to the natural dispersal or movement of some species from one place to another. All these activities result in communities becoming disassembled as individual species are affected in different ways. The reestablishment of the original species is prevented by changes in the physical environment—such as loss of soil nutrients or physical habitat—or by a reduction in the original species pool, such as through the loss of a seed bank or seed source.

Effects on Ecological Processes

Much recent research has focused on studying the role of novel ecosystems and their ability to restore forest and grassland cover to human-disturbed landscapes. Some novel forests harbor native species in the understory and can include a high biodiversity of plants and animals. However, because they are new ecosystems, they are less studied, and it is often unknown how changes in composition will affect important ecological processes.

Some ecological processes that are affected by which species inhabit an ecosystem are the amount of energy and food available for higher trophic levels through photosynthesis; plant-animal interactions, which can affect pollination, seed dispersal, and germination, as well as food-web dynamics; the breakdown of dead organic matter and recycling of important plant nutrients by bacteria, fungi, and invertebrates; and water availability for plant growth through species' effects on hydrology. The unpredictability of species combinations in novel ecosystems increases uncertainty about their continued ability to provide services that are important for humans and other organisms, such as habitat for biodiversity, watershed protection, regional climate stabilization, and carbon sequestration.

Examples of Novel Ecosystems

Examples of novel ecosystems can be found around the world. Some are relatively recent and others have been around for a long time, because humans have been modifying their environment and moving species around for thousands of years. South Africa's fynbos shrublands are being invaded by pine species that are especially adapted to germinating rapidly after fire and thus preventing the reestablishment of the native shrub species. In Brazil, New Zealand, and parts of the United States, the introduction by humans of recurring fire and of intensive livestock grazing have caused the replacement of native forests by grasslands and shrublands.

Environmental pollution by nitrogen and other elements is changing soil chemistry in forests and grasslands in Europe and North America, allowing invasive species to outcompete those that were already present and that were better adapted to the old conditions.

In Puerto Rico, extensive deforestation for sugarcane, coffee, and cattle pasture at the beginning of the 20th century has been reversed due to changes in economic policies. This Caribbean island has experienced a threefold increase in forest cover due to the large-scale abandonment of agriculture. The emerging forests on former agricultural lands have different combinations of tree species not found before on the island. Researchers are actively studying these subtropical forests to see how ecological processes are affected by the dominance of introduced species.

Examples also exist in the aquatic world. Human control of rivers for irrigation and flood prevention has altered water and sediment flows, changing the habitat of riverine species and facilitating the introduction of nonnative fish and aquatic plants. Many important estuaries, such as San Francisco Bay in California, are now completely dominated by new species combinations, due to the confluence of ships from all parts of the world that bring along hitchhikers.

Conservation Efforts

Novel ecosystems are important because they bring attention to the extent of human influence on the environment. Many conservation practices attempt to restore ecosystems to their original condition. These efforts often fail because enough changes have happened to the ecosystem that the original species are not able to come back without intensive and expensive human assistance. Furthermore, many of these species are not able to survive and replace themselves without costly management. Often, information on the original species composition is lacking.

Novel ecosystems provide opportunities for scientists to study how ecosystems respond to environmental change. While those who value the preservation of wild nature devalue novel ecosystems because they often are dominated by nonnative, or exotic, species, others argue that novel ecosystems are nature's way of dealing with change, by creating new combinations of species that are better adapted to the transformed environment. Novel ecosystems on soils where nutrients have been severely depleted through intensive agriculture, or on soils contaminated by industrial pollution or overfertilization, for example, harbor species that are able to tolerate extreme conditions. Over time, these species can actually improve soil conditions and facilitate the establishment of other species, some of which might have been there before the environmental change.

Controversy

There is a vigorous debate in the ecological community on the role of species that have been introduced by humans, and what place they should have in different types of ecosystems. Similar questions arise about novel ecosystems that comprise numbers of these "unwanted" species. Some argue that these ecosystems are ecological disasters, unworthy of ecological study because they are "unnatural," and suggest that energy should be spent on studying and preserving the last wild places on Earth. Those who study novel ecosystems respond that more and more of our landscape contains these new combinations of species, and that it is important to understand how they work.

The topic of novel ecosystems provides a philosophical context for discussing the extent to which one species—humans—is modifying the planet's surface and atmosphere, and altering the composition of ecosystems and biomes at a much faster rate than past climatic or environmental changes. Novel ecosystems suggest that, in some places, irreversible changes in physical and biological conditions may have already occurred. Emerging ecosystems provide opportunities for ecologists to study how terrestrial and aquatic communities respond to disturbances caused by human activities, changes in nutrients, and species invasions. Global warming will further drive the emergence of new combinations of species.

ERIKA MARÍN-SPIOTTA

Further Reading

Bridgewater, Peter, et al. "Engaging with Novel Ecosystems." *Frontiers in Ecology and Environment* 9 (2011).

Hobbs, Richard J., et al. "Novel Ecosystems: Theoretical and Management Aspects of the New Ecological World Order." *Global Ecology and Biogeography* 15 (2006).

Lugo, Ariel E. "The Emerging Era of Novel Tropical Forests." *Biotropica* 41 (2009).

Marris, Emma. "Ragamuffin Earth." *Nature* 460 (2009).

Seastedt, Timothy R., Richard J. Hobbs, and Katharine N. Sundig. "Management of Novel Ecosystems: Are Novel Approaches Required?" Ecological Society of America. http://www.esajournals.org/doi/abs/10.1890/070046.

Williams, John W. and Stephen T. Jackson. "Novel Climates, No-Analog Communities, and Ecological Surprises." *Frontiers in Ecology and the Environment* 5 (2007).

Nubian Desert

Category: Desert Biomes.
Geographic Location: Africa.
Summary: This large desert, once home to much wildlife, now has a small amount of flora and fauna striving to survive.

The Nubian Desert covers some 154,441 square miles (400,000 square kilometers) of the eastern Sahara Desert. It has no oases and almost no rainfall. In northeastern Sudan and southern Egypt, it is a very arid region, made up largely of sandstone plateaus and wadis, or seasonal streams, that flow toward the Nile River during some wet periods.

In spite of this seemingly very hostile environment today, it is clear that there were many animals and plants in the region in ancient times, and it seems probable that the environment must have been different, probably more akin to the savanna areas of central Sudan. As such, the transformation of the landscape and the ecosystems in the Nubian Desert may be an ominous portent of what can happen in an environment after the destruction of the flora and fauna of a region.

Ancient Settlement

Reliefs on temples at Thebes in Egypt show Nubians bringing tribute of ivory, skins, fruit, fish, and amulets. The Egyptians of the First Dynasty (2920 to 2770 B.C.E.) regarded Nubia as important; they were involved in fighting there for more than 1,000 years as they sought to control it and exploit its resources, which included ivory, ebony, gum resins from trees, cattle, dogs, and a range of exotic animals. A scene in the Temple of Beit el-Wali in Nubia, dating from the reign of Ramses II (1279 to 1213 B.C.E.), shows tribute including ebony logs, elephant tusks, ostrich eggs, feathers, and live animals: a giraffe, a gazelle, a leopard, and some monkeys. Some of the Nubians are shown wearing leopard skins as loincloths.

Nubia also was a source of amethyst and copper for ancient Egypt. Soon, it became vital when significant deposits of gold were found there; the gold from Nubia was shown on a wall painting in the tomb of Treasurer Sebekhotep at Thebes.

It has been suggested that the importance of copper and gold to ancient Egypt may have been the cause of the overexploitation of the region. To maintain their civilizations in ancient times, the Egyptians and Nubians would have needed access to wood for housing and for construction of mines. Nubia was a source of some precious woods that are not found there now. Coupled with the decline in the number of trees, the gradual degradation of agricultural land would have contributed to the destruction of the environment.

Ancient Biodiversity

One tree species that has survived since ancient times is the date palm *Medemia argun*, found only in Egypt and Sudan, and listed as critically endangered. Dried dates from these palm trees found in tombs in ancient Egypt suggest that the dates were delicacies brought from Nubia to Egypt in ancient times in a manner that is not possible now.

The loss of the trees in Nubia may have led to the gradual destruction of the ancient ecosystem and also its civilization there. Although giraffes were there as late as medieval times, there are none in the Nubian Desert today. Many giraffes, however, do live on acacia trees, which grow to the south and west of the Nubian Desert biome today.

Two other species that were known to be in Nubia are no longer there. One of these is the Nubian lion. The last wild member of this species, also known as the Barbary lion, was killed in the Atlas Mountains in 1922. Also originally found in the Nubian Desert was the Nubian bustard (*Neotis nuba*). Its habitat now is the Sahelian Acacia savanna, in the grasslands south of the desert and in countries to the west of Sudan. Even in its new environment, it has been affected by hunting and overgrazing of its habitat.

The nubian ibex (Capra nubiana) is one of four species of mammals living in the southern Sudan. (Thinkstock)

Current Biodiversity

Wildlife ekes out a precarious existence in the Nubian Desert. Animals have gradually moved south as the desert has become steadily more hostile. Four types of mammals currently live in the southern Sudan: the antelope, Barbary sheep, Nubian ibex (*Capra nubiana*), and some species of monkeys.

Also in the desert are a range of small lizards and the Nubian flapshell turtle (*Cyclanorbis elegans*), found across West Africa but now regarded as being near-threatened. There are also some insects, including a range of butterflies.

Effects of Human Activity

The transformation of the landscape by humankind continued in the 20th century, with the construction of the Aswan Low Dam in 1902 and the Aswan High Dam in the 1960s.

These projects resulted in the need to move many ancient monuments and sites, including the famous Great Temple at Abu Simbel and the temple of Beit el-Wali. The moves transformed the nearby landscape, allowing more land to be used for cultivation.

There were many other environmental changes, one of the first being the reduction in river sediment, which led to an increase in aquatic weeds and a decline in fish further down the Nile. These changes, however, affected only the area of the desert in the immediate vicinity of the river; the rest of the desert remained desperately parched. Climate change impacts on this harsh desert environment may reduce rainfall further, something which the remaining flora and fauna living in the Nubian desert can ill afford.

JUSTIN CORFIELD

Further Reading

Crowfoot, Grace M. *Flowering Plants of the Northern and Central Sudan.* Leominster, UK: Orphans Printing Press, 1928.

Friedman, Renée, ed. *Egypt and Nubia: Gifts of the Desert.* London: British Museum Press, 2002.

O'Connor, David B. *Ancient Nubia: Egypt's Rival in Africa.* Philadelphia: University Museum of Archaeology and Anthropology, University of Pennsylvania, 1993.

Shinnie, P. L. *Ancient Nubia.* London: Kegan Paul International, 1995.

Taylor, John H. *Egypt and Nubia.* London: British Museum, 1991.

Wilson, C. E. "Butterflies of the Northern and Central Sudan." *Sudan Notes and Records* 30, no. 2 (1949).

Ob River

Category: Inland Aquatic Biomes.
Geographic Location: Eurasia.
Summary: The Ob River watershed faces environmental issues including water diversion, nuclear waste, and oil and gas development.

The Ob River watershed is the fifth-largest drainage basin in the world, and the Ob is the westernmost of the three great Siberian Rivers, extending from northern Kazakhstan to the Arctic Ocean. The total river length is 2,300 miles (3,700 kilometers). Ob River drainage begins in the temperate grasslands and coniferous forests of the Altai Republic (near Lake Teletskoye) and flows northwestward, eventually entering the Kara Sea, part of the Arctic Ocean, through the Yamal-Nenets Autonomous Region of Russia in Arctic tundra. To the west of the catchment are the Ural Mountains, and to the east is the Central Siberian Plateau.

The largest tributary of the Ob is the Irtysh. Its drainage begins in the far western mountains of Mongolia and the Xinjiang Uyghur Autonomous Region of northwestern China, and flows northwest through the steppes of Kazakhstan to Omsk in Russia, where it enters the forests of western Siberia and eventually joins the Ob in the Khanty-Mantsi Autonomous Region.

Climate

The Ob crosses several climate zones. The lower reaches of the Ob are Arctic tundra, and the river is ice-bound 100 miles (160 kilometers) above its mouth from the end of October to the beginning of June each year. Mid-sections of the river flow through taiga, or boreal forest, while the upper reaches flow through the warmer temperate forest zones or steppe regions.

In general, the Ob basin has short, warm summers and long, cold winters. Average January temperatures range from minus 18 degrees F (minus 28 degrees C) on the shores of the Kara Sea to 3 degrees F (minus 16 degrees C) in the upper reaches of the Irtysh. July temperatures for the same locations, respectively, range from 40 degrees F (4 degrees C) to above 68 degrees F (20 degrees C).

Precipitation varies greatly in the Ob River watershed: less than 16 inches (40 centimeters) per year in the north, 20–24 inches (50–60 centimeters) in the taiga zone, and 12–16 inches (30–40 centimeters) on the steppes. The western slopes of the Altai Mountains receive as much as 62 inches (160 centimeters) per year.

Flora and Fauna

The Ob River estuary is a gulf that extends 500 miles (800 kilometers) north from where the river first enters the sea, making it the longest estuary in the world. The Taz River estuary here is a branch of the Ob River estuary, further extending its reach. The vast floodplain of the Ob River from Khanty-Mantsi north to the Gulf of Ob is one of the richest waterfowl habitats and nesting areas in the world. Three large wetlands of international importance, as designated under the Ramsar Convention, are present in this area. One of them, the Islands in Ob Estuary Ramsar Site (Nizhne-Obskiy Nature Reserve), is a vast delta consisting of numerous islands and temporary lakes used for duck migration.

Before reaching the sea, the river flows through a swampy taiga floodplain about 12–19 miles (20–30 kilometers) wide. A network of interconnected lakes and channels characterize the lower river within the extensive Western Siberian Lowland. This is a boggy area with belts of taiga, generally evergreen conifer, forests along rivers. Trees are widely spaced, with an understory of dwarf shrubs. Fish here include Arctic migratory species such as tugun, peled, muksun, and char; and boreal river species of pike, dace, and perch.

Of some 50 species of fish found in the river or in the gulf, the most valuable, economically speaking, are several varieties of sturgeon. Shallow floodplain water bodies called sors serve as major areas of feeding for migratory fish. Other, smaller river systems that are tributaries to the Kara Sea have a similar fish fauna. Lake Teletskoye, near the Ob River headwaters, has endemic (found nowhere else) species of whitefish.

The Turgay Plateau of Kazakhstan is the southern watershed boundary, separating north-flowing polar rivers from the temperate Volga or Caspian Sea drainage to the west. The steppe grassland habitats in this area contain patches of forests, and a narrow band of deciduous forest grows between the steppe and boreal forest. In the steppes in the southern part of the Ob River drainage, aspen, birch, and pine grow in small groves called koloks.

Within this steppe region are lakes used by migratory waterfowl; many of the lakes are saline and do not have connections with a river system. Thousands of lakes and wetlands dot the steppe and deciduous and conifer forests, providing vital habitat for waterfowl in this otherwise-dry region. There have been more than 170 bird species identified here.

Mammals of the Ob watershed include European and Siberian mole, Siberian and American mink, ermine, fox, wolf, elk, white hare, water rat, muskrat, otter, and beaver. Notably, the Siberian subsoil contains the remains of as many as 150 million mammoths, which have been frozen for thousands of years. Ivory tusks are found as more and more tundra melts away as the world warms. Common areas for exposure of these prehistoric fauna are along the eroding banks of rivers. Since the 19th century, about a dozen soft-tissue specimens of mammoths have been recovered. In May 2007, a complete, month-old baby mammoth corpse was found lying on a sandbar in the Yuribey River in Yamalia.

Human Impact

In the north, the river flows through the low-lying West Siberian Plain. It is the site of the world's second-largest gas field, the Urengoy field, in Yamal-Nenets; it is also the home of most of Russia's oil production, in Khanty-Mantsi. The Samotlor Oil Field is Russia's largest. The Urengoy and Nadym gas fields extend southward from the Gulf of Ob and the Gulf of Taz. To the north of the Taz River, the Yamburg Gas field is under development as the world's third-largest gas field. And the Bovanenkovo Production Zone on the Yamal Peninsula is one of the newest and northernmost gas fields being developed. A railroad to Bovanenkovo was completed in 2009. These oil and gas developments also include networks of pipelines and gravel roadbeds. Presently, environmental management, including the prevention of chemical spills and effects on indigenous peoples, are key issues.

In the southern parts of the Ob River watershed, water diversions have been discussed to provide water for drier southern areas. Currently, China is working on a project to divert the upper Irtysh River to provide water to drier parts of Xinjiang. During the 1960s and 1970s, a Soviet scheme to divert the Irtysh to southern Kazakhstan was discussed but never implemented.

In the watershed in Chelyabinsk and Sverdlovsk Oblasts, are areas affected by the legacy of nuclear weapons production. At Ozersk (Chelyabinsk-65), a 1957 explosion in a radioactive-waste storage area dispersed radioactivity into Sverdlovsk and Tyumen Oblasts. In the southern watershed along the Irtysh River in Kazakhstan, the Semipalatinsky Test Site was where much of the former Soviet Union's nuclear testing took place. Toxic remains from these activities remain a concern today.

Global climate change is now affecting the glacier melt in the Altai Mountains, and therefore water flow in the Ob River basin. Greenpeace reports that the melting permafrost soil not only affects the way of life of the indigenous nomadic Nenets people, but also adds burden to existing climate change because of the massive release of methane and carbon dioxide as decomposition is activated in the defrosting soil. More research is needed to conclude how much of an impact changing temperatures and precipitation patterns will have across the entire watershed of the Ob.

HAROLD M. DRAPER

Further Reading

Abell, Robin, et al. "Freshwater Ecoregions of the World: A New Map of Biogeographic Units for Freshwater Biodiversity Conservation." *Bioscience* 58 (2008).

Edwards, Mike. "Lethal Legacy: Pollution in the Former U.S.S.R." *National Geographic* 186, 2 (1994).

Gelfan, A. N. "Modeling Hydrological Consequences of Climate Change in the Permafrost Region and Assessment of Their Uncertainty." *Water Problems Institute of the Russian Academy of Sciences: Region Hydrology in a Changing Climate* (2011).

Kramer, Andrew E. "Ivory for the Taking, From Beasts Beyond Caring." *New York Times*, March 26, 2008.

Mueller, Tom. "Ice Baby." *National Geographic* 215, 5 (2009).

Surazakov, A. B., V. B. Aizen, E. M. Aizen, and S. A. Nikitin. "Glacier Changes in the Siberian Altai Mountains, Ob River Basin, (1952–2006) Estimated With High Resolution Imagery." *Environmental Research Letters* 2 (October-December, 2007).

Oder River

Category: Inland Aquatic Biomes.
Geographic Location: Europe.
Summary: The Oder River is an international waterway in east-central Europe that has received close monitoring of pollution to maintain and improve its ecosystems.

The Oder River is located in the east-central part of Europe and divides Poland from Germany. It originates in the Hruby Jesenik Mountains within the Czech Republic, and is fed by Warthe and Lausitzer Neibet River tributaries. Its width varies from 1 to 6 miles (1.5 to 10 kilometers). The endpoint of the Oder River is at the Baltic Sea, after it flows through the Szczecin Lagoon, close to Szczecin, the largest seaport in Poland.

Flooding in the Lower Oder Valley International Park, which is managed by both Germany and Poland. The Oder River frequently floods in spring. (Thinkstock)

The Oder flows for about 530 miles (850 kilometers); its watershed drains approximately 46,000 square miles (118,860 kilometers). Sometimes during the autumn and summer, the Oder River can become very low, yet during spring it can flood heavily, due to melting snow or heavy rains.

Called Odra in Poland, it is the second-longest river in that country. During the reign of the Roman Empire, it was part of the Amber Road; Germanic tribes used it as an important trade route. Water wheels and dams were already being constructed in the 13th century, diverting the flow for agricultural and power needs.

An international waterway, because it allows sea access several nations, the Oder links to the Vistula River in Poland via the Warta River. Many other canals off the Oder link as far away as parts of Russia and Ukraine. While designed for human transport, these channels also have permitted the migration of hosts of exotic flora and fauna.

The Oder River begins as a freshwater river and then turns into brackish water as it travels through the Szczecin Lagoon, until it is a fully saline habitat at the Pomeranian Bay on the Baltic Sea. Nutrient loads are found to be generally heavier in the upstream waters. Mountains-and-valley terrain is the rule in the upper and middle segments; lakes and lowlands are found in the lower section. Some side tributaries are used in mining alluvial sands.

Biodiversity

Under new European Union guidelines, the Oder River, until fairly recently a very heavily polluted waterway, could potentially begin to recover as a thriving ecosystem. The basin is a fairly good area for fish because of many different habitats. There are fertile, well-drained river meadows, at least 42 distinct wetland areas, and 15 varieties of forest community.

The lower Oder Valley has flora and fauna that are unique only to this area, or endemic. This area was originally settled for agricultural purposes, which did not work out well, due to the low level of land and extensive drainage problems. Eventually, it returned to a more natural state, and had the benefits of natural floods. There are now a

peat bog, reeds, thickets, and swaths of alder trees. These habitats support a great variety of fish life: Eel, bream, dace, catfish, carp, and trench are a few types found here.

Bird species of the Oder include black redstart, icterine warbler, golden oriole, red-backed shrike, hawfinch, crested tit, both nightingale and thrush nightingale, hooded crow, serin, crested lark, red kite, hobby, marsh warbler, penduline tit, and white spotted bluethroat. Near Szczecin, the breeding crane population thrives. The lesser spotted eagle and white-tailed eagle can also be found here.

Beavers and otters are among mammals that are increasing in numbers here. Others found in the region include: boar, roe deer, velvet shrew, bats, forest mice, migratory rats, and weasels. Amphibians include comb newt, grass frogs, water toad, and grey toad. There are also grass snakes.

Human Impact

Sewage, industrial waste, and farm runoff have polluted the Oder River for centuries. Early on in the 18th and 19th centuries, the Oder was shortened twice to improve it for the purpose of navigation. As a result of the building of 23 dams, fish suffered not only from pollution but also the lack of fish ladders built to allow them to return upstream to their spawning grounds.

After 1945, the need for the river for transportation was not as great, so some of the river was able to return to its natural ecological state, including wet meadows, meanders, and alluvial forests. Fauna and flora have increased in number and variety, helped by such developments as the founding of the Lower Oder Valley International Park, a reserve shared by Germany and Poland.

In 1999, the International Commission for the Protection of the Oder (ICPO) was established between Poland, Germany, and the Czech Republic. Some of the ICPO's goals were: to reduce pollution, and prevent further pollution, of the Oder and the Baltic Sea by contaminants; to achieve the most natural aquatic and littoral ecosystems possible with corresponding species diversity; to permit utilization of the Oder, in particular the production of drinking water from bank filtrate and the use of its water and sediments in agriculture; and to provide for precautions against the risk of flood damage.

In 2000, oxygen levels tested in the whole Oder River area were better than levels recorded in 1992. Organic pollution, total nitrogen, and phosphorous have become lower, but still need to be reduced. Bacterial pollution needs to continue to be lowered as well. As a result of programs implemented, scientists are now hoping to return sturgeon back to the Oder River, to encourage the ecosystem to flourish.

Climate change scenarios predict an increased risk of extreme weather events globally. Ongoing sea-level rise and a sinking coast, as well as changes in precipitation in the Oder River watershed, will increase the flooding risk in the river basin and at the coast.

Along the Baltic Sea coast, an increased risk of storms and storm surges will have immediate negative effects on coastal erosion, protection measures, and tourism infrastructure. Climate change will affect not only the coast, but also the river basin itself; integrated coastal and river basin flood protection will be needed.

WILLIAM FORBES
BLAKE BROADDUS

Further Reading

Bethge, Philip. "Bringing the Sturgeon Back to Germany." *Spiegel*, October 31, 2006.

Developing Policies & Adaptation Strategies to Climate Change in the Baltic Sea (ASTRA). "Germany: Oder/Odra Estuary." http://www .astra-project.org/02_germany.html.

Perkins, Sid. "River Stats Trickle In." *Science News* 164, no. 11 (September 2003).

Schnalitz, G. and A. Behrendt. *Heavy Metal and Contaminant Loading in the Oder River Floodplains and Consequences for Livestock.* Paulinenaue, Germany: Center for Agricultural Landscapes and Land Use Research, 1999.

Torsi, Agnieska and Arkadusz Nedzarek. "Oceanological and Hydrobiological Studies." *International Journal of Oceanography and Hydrobiology* 39, no. 3 (2010).

Ogilvie-Mackenzie Alpine Tundra

Category: Grassland, Tundra, and Human Biomes.
Geographic Location: North America.
Summary: This biome is home to the unique Ogilvie mountain lemming, and is a distinctive combination of ecozones with diverse terrains.

The Ogilvie-Mackenzie Alpine Tundra biome is located in a mountainous band from the Alaska-Yukon border southeastward to the border of the Yukon-Northwest Territories. This taiga-cordillera region corresponds in the north to the North Ogilvie Mountains, the Mackenzie Mountains running east to west through the Yukon, and southward to the Selwyn Mountains of the Northwest Territories (NWT). The northern part extends across a section of the Ogilvie and Wernecke mountains and the connected intermontane basins.

The middle section includes the westernmost parts of the Ogilvie and Wernecke mountains, the Backbone Ranges of the interior, and the Canyon Ranges of the east. The southern region lies in the rain shadow of the Selwyn Mountains. On the Yukon-NWT border is a mountainous area that includes the Selwyn and the southern portion of the Mackenzie Mountains. The Ogilvie-Mackenzie Alpine Tundra is predominately a wilderness area with no permanent settlements.

The biome is classified as tundra; it spans 80,500 square miles (208,400 square kilometers) and features a continental sub-Arctic climate. This type of climate has long, cold winters and short, cool summers. The average annual temperature of the Ogilvie-Mackenzie Alpine Tundra ecoregion is around 27 degrees F (minus 3 degrees C). Winter temperatures range from minus 3 to minus 7 degrees F (minus 19 to minus 22 degrees C), while the average summer temperature is 48 degrees F (9 degrees C). In the winter, frigid temperatures of minus 58 degrees F (minus 50 degrees C) are common. Annual precipitation ranges from 12 to 30 inches (30 to 75 centimeters), with the higher elevations receiving significantly more precipitation.

The Ogilvie-Mackenzie Alpine Tundra biome is composed of a diverse collection of terrains that range from rounded and flat-top hills, to extremely rugged and steep mountainous areas, to alpine and glacial valley areas. Portions of the Ogilvie and Wernecke mountains are unglaciated here. The valleys and river bottoms are V-shaped in the unglaciated areas, and U-shaped in areas carved out by glaciers.

The tundra is covered predominately by sub-Arctic and subalpine coniferous forests, followed by Arctic-alpine tundra and rocklands, with the smallest area being lakes and wetlands. Most of the terrain of northern part of the region lies from 3,000 to 4,400 feet (900 to 1,350 meters) elevation. Moving in a southeastern direction, several peaks reach 6,890 feet (2,100 meters), with the highest elevation being Mount Keele—9,678 feet (2,950 meters)—in the Selwyn Mountains. The North Ogilvie and Mackenzie Mountains have continuous permafrost, while the Selwyn Mountains region has irregular, patchy, low-ice-content permafrost.

Flora and Fauna

The Ogilvie-Mackenzie Alpine Tundra biome is characterized by Arctic-alpine tundra at higher elevations and subalpine-sub-Arctic coniferous forests at lower elevations. Tree species include paper birch, balsam fir, trembling aspen, balsam poplar, lodgepole pine, dwarf birch, and willows (blue-green and net-veined). Alpine fir, black and white spruce, and dwarf birch are commonly stunted, crooked, bent, and twisted (sometimes called krummholz). Much of the alpine and talus slopes are devoid of vegetation.

Existing vegetation consists of: lichens (fruticose, crustose, rock tripe, and reindeer), herbs (alpine blueberry, lupine, osha, mountain tobacco, horsetail, and northern Labrador tea), moss (feather and peat), forbs (entire-leaf mountain-avens, three-toothed saxifrage, and naked-stem wallflower), saxifrages (prickly, purple, and yellow mountain), ericaceous shrubs (Arctic bell-heather, white dryas, mountain bearberry, and bog cranberry—dwarfed in alpine regions), and sedge. Cottongrass is common in wetter areas and at higher elevations. All vegetation is vulnerable to frost at any time.

Permafrost Melting

Becuase of global warming trends, permafrost in the Ogilvie-MacKenzie Alpine Tundra region is starting to melt, and the pent-up carbon is already leaking into the air in the form of carbon dioxide and methane, powerful greenhouse gases. Scientists estimate there may be almost 100 billion tons of carbon in the first meter of soil alone, across all northern permafrost regions of the world. This is nearly twice as much as previous estimates, and is equivalent to about one-quarter of the amount currently in Earth's atmosphere. Though there are many theories, the exact direct implications of a thawing Arctic are not known; most scientists agree, however, that the general outcome will not be positive for habitats and native species here.

Most of the fauna found in this region are at their northern range limit. Amphibians here are the wood frog, western toad, and striped chorus frog. There are no reptiles. Common mammals include Dall's sheep, moose, mountain goats, and mule deer. Caribou herds are found in this region, such as the northern mountain woodland caribou and Yukon's only boreal caribou. Predators common to the area are wolverines, grizzly and black bears, red foxes, wolves, lynxes, and martens.

A unique mammal, that may be the only one exclusive to the Yukon, is the Ogilvie mountain lemming. Other small animals include Arctic ground and northern flying squirrels, beavers, hares (snowshoe and tundra), American and collared pikas, short-tailed weasels, hoary marmots, bushy-tailed wood rats, mice (meadow jumping and deer), least chipmunks, singing voles, and bats (little brown myotis).

An abundance of avian species makes the region home, such as sparrows (Lincoln, violet-green, cliff, chipping, and Savannah), dark-eyed juncoes, pine grosbeaks, white-winged crossbills, Alder flycatchers, blackpolls, snow buntings, wandering tattlers, horned larks, gray-crowned rosy finches, Townsend's solitaires, water and American pipits, warblers (tree and yellow), rusty blackbirds, flycatchers (olive-sided and yellowbellied), and thrushes (varied, northern water gray-cheeked, and Swainson).

Year-round residents include common ravens, ptarmigans (rock, willow, and the rare white-tailed), gray jays, and chickadees (black-capped and boreal). Common birds of prey are great horned and short-eared owls, bald and golden eagles, American kestrels, peregrine falcons, Northern harriers, and gyrfalcons.

Waterfowl can be found in the low elevations near lakes and rivers, and in the limited wetlands. These are summer breeding and nesting areas for waterfowl species such as Canada geese, northern pintails, trumpeter swans, loons (common, Pacific, and red-throated), mergansers (common and red-breasted), horned grebes, American widgeons, surf scoters, Northern shovelers, ducks (mallard, long-tailed, and harlequin), scaups (greater and lesser), green-winged teals, and Barrow's goldeneyes.

Human Impact

Land use in the Ogilvie-Mackenzie Alpine Tundra area consists mostly of big game, sport, and subsistence hunting and trapping; as well as tourism, recreational activities, and natural resource exploration. The area is considered to be 95 percent intact. Threats to the area are mainly in the form of increased road-building connected to natural resource development. Roads damage the fragile vegetation, as well as open the area to multiple human activities. Beyond the road-system areas, disturbance to the area is minimal.

ANDREW HUND

Further Reading

Benke, Arthur C. and Colbert E. Cushing, eds. *Rivers of North America.* Maryland Heights, MO: Academic Press, 2005

Reilly, Michael. "Arctic Tundra Holds Global Warming Time Bomb." *Discovery News,* August 25, 2008.

Ricketts, T., E. Dinerstein, D. Olson, C. Loucks, W. Eichbaum, D. DellaSalla, et al. *Terrestrial Ecoregions of North America: A Conservation Assessment.* Washington, DC: Island Press, 1999.

Ohio River

Category: Inland Aquatic Biomes.
Geographic Location: North America.
Summary: A major river in the United States and a key tributary of the Mississippi River, the Ohio and its basin are home to tens of millions of people in several different ecosystems.

A long, shallow, often wide, and heavily interconnected river in the eastern United States, the Ohio River is the main tributary of the Mississippi River, flowing through a region of varying climates and ecosystems. It has been a significant transportation route from before European settlements, for Native Americans, through today. The Ohio River flows southwest almost 1,000 miles (1,600 kilometers) through six states: Pennsylvania, Ohio, West Virginia, Kentucky, Indiana, and Illinois. Along the way, it serves as the characteristically wiggling border for several of these states, including the southern borders of Illinois, Indiana, and Ohio; and parts of the northern borders of Kentucky and West Virginia. The entire watershed of the Ohio River extends even further, including parts of New York, Maryland, Virginia, North Carolina, Georgia, Tennessee, Alabama, and Mississippi.

Hydrology and Climate

The source of the Ohio is the Allegheny River in western Pennsylvania. In Pittsburgh, Pennsylvania, the confluence of the Allegheny and Monongahela Rivers form the beginning of the Ohio, earning this area the nickname Three Rivers. Near Cairo, Illinois, the Ohio enters the Mississippi River. At this point, there is usually more water coming from the Ohio than from the northern section of the Mississippi, helping to make the Ohio the main tributary of the Mississippi.

The river is often very wide, stretching as far as one mile across near Louisville, Kentucky. It is not, however, naturally deep, but locks and dams have greatly increased depth, up to about 170 feet (52 meters) near Louisville.

Because it travels so far, helping to divide the Northeast, Midwest, and Southern regions of the United States, the Ohio River is a part of several different climates. In its northern regions, especially eastern Ohio, northern West Virginia, and Pennsylvania, the weather is typical of areas with a humid continental climate. Similar to that of much of the Northeast, it features warm summers and cold winters, including frequent snow. The Allegheny River, and sometimes the Ohio around Pittsburgh, have been known to freeze in the winter, although this is becoming less common due to climate change.

Below Pittsburgh, the Ohio rarely freezes, and by western Kentucky it is always free of ice. In these areas, the weather is more similar to the upper South and characteristic of a humid subtropical climate, with hot summers, moderate winters, and only occasional snow.

The two uniting features about the climate of the Ohio River are relatively frequent rainfall, which helps to keep the river's volume high, and relatively high humidity, especially in the summer. Because of this relatively frequent rainfall and low

Debris collecting at a dam in the Ohio River. There are 20 operating dams on the river that contribute to the build up of silt, harming bottom-dwelling fauna. (Thinkstock)

overall depth, flooding is common on the Ohio River. In particular, at least four severe floods occurred in just the last 125 years, with waters rising in some cases to over 60 feet (18 meters), and more than doubling local flooding heights.

Flora and Fauna

The Ohio is home to a variety of ecosystems and species. Two major factors influence this. The first is the large variation in climate, described above. The second is the mixture of glacial and nonglacial soils around the river. Because large glaciers covered the region to the north of the Ohio River tens of thousands of years ago, but did not extend to the areas south of the river, the soils contain different mixtures of rocks and minerals, and are suitable for different types of plants.

The northern portions of the river ecosystem are known as mixed mesophytic and western mesophytic forests, meaning they are temperate, diverse, largely deciduous forests. Common trees include several species of hickory and oak, river birch, basswood, poplar, some eastern cottonwood, and the famous yellow buckeye, from which Ohio get its nickname, The Buckeye State.

Toward Illinois, Indiana, and western Kentucky, the banks of the river once included large bottomland hardwood forests. Many have been cut down, but some still remain, and others are growing back on abandoned farmland. Throughout the system, but particularly in the southern portions, many trees are well adapted to the river's flooding. Species common here include bald cypress, tupelo gum, and more isolated patches of oaks and hickories than in the northern portions.

Especially in its southern reaches, the river is home to several species of large fish, including paddlefish and catfish. In total, over 150 types of fish call the river home. Although rare, bull sharks have even been found in the Ohio River as far north as Indiana. The river is also known for containing more than 50 different species of mussels. In part because of its fish, and because much of the river supports the dense hardwood and mesophytic forests nearby, many birds, including migrating birds, make use of the river. These include warblers (notably the prothonotary warbler and cerulean warbler), wrens, sparrows, and other small birds. Larger birds include the bald eagle, snow goose, and mute swan.

Many types of butterflies also take advantage of the flowers growing near the banks of the river, including hairstreaks, swallowtails, the tiny falcate orangetip, gemmed satyr, and monarch. Other typical insects here include praying mantis, solider beetles, diving beetles, tulip scale insects (which feed on tulip poplars), mayflies, carpenter ants, and cicadas.

Typical river amphibians in and near the Ohio and its tributaries include multiple types of salamanders, spring peepers, cricket frogs, Fowler's toad, fence lizard, box and snapping turtles, and several types of garter snakes and milksnakes. Water snakes, such as the copperbelly, may live in tributaries that are slower moving than the Ohio.

Common mammals in and around the river include the river otter, white-tailed deer, rats, raccoons, squirrels, muskrat, mink, beavers, and bats. Bear, bobcat, and mountain lion still inhabit some forested areas surrounding the river.

A number of the plants and animals that live in or near the river are endangered. Pollution and other human impacts have killed more than 30 types of mussels formerly found in the river. At least three are completely extinct. Fish are still plentiful, but significant restrictions have been placed on eating Ohio River fish, due to contamination. Catfish and carp consumption, for example, is generally prohibited. Other endangered species include turtles, snakes, dragonflies, and reedgrass. Many small islands exist throughout the river, and these have often served as havens for small mammals, birds, amphibians, insects, and reptiles, even when surrounding lands along the shores were degraded.

Several invasive, alien species have been accidentally introduced by people to the river ecosystems, and are competing with and sometimes leading to the death of native species. Zebra mussels are one such pest, causing the decline of other types of mussels, and the animals that depend on them. Several kinds of Asian carp are also present in the river, where they compete for the same food with other, native fish.

Human Impact

Close to 8 or 9 percent of the population of the United States, nearly 30 million people, now live in the Ohio River's basin, with the river itself the source of drinking water for more than 3 million people. Although the river is no longer the largest source of transportation in the region, it is still a significant source of recreation such as boating and fishing. Some types of river freight traffic still makes frequent use of the Ohio. The economic health of Ohio River cities depends, in part, on the value citizens place on having attractive views and easy access to the river.

Today, invasive species, pollution from mine and agriculture runoff, and flooding in certain areas threaten both the people and other species that depend on the river. Although much progress has been made, especially in the last few decades, to reduce pollution, heavy storms also cause city sewer systems to overflow into the river, poisoning it in some places and requiring expensive water treatment facilities in others. The 20 dams still in operation on the river have helped to control its flow, but also allow silt and mud to pile up, reducing water visibility, river appearance, and bottom-dwelling animals.

Climate change appears to have already had some effects on the river basin, but much more research needs to be carried out. Any future changes in the Ohio River water flow may impact human health, agriculture, water supplies, property and infrastructure, as well as tourism and recreation. Government and civic efforts continue today to focus on reducing these impacts, while maintaining desirable recreational and commercial use of the Ohio River.

KYLE GRACEY

Further Reading

Banta, R.E. *The Ohio (Ohio River Valley Series).* Louisville: University Press of Kentucky, 1998.

Jackson, Tom. *The Ohio River (Rivers of North America).* New York: Gareth Stevens Publishing, 2003.

Pearce, John Ed and Richard Nugent. *The Ohio River.* Louisville: University Press of Kentucky, 1989.

Schulte, Jerry G. "The Implications of Climate Change on the Ohio River and the Ohio River Valley Water Sanitation Commission." *American Water Resource Association* 1 (2009).

Okanagan Dry Forests

Category: Forest Biomes.
Geographic Location: Western North America.
Summary: This ecoregion includes alpine forests and grasslands of varying types; a major concern is the human interaction with fire regimes.

The Okanagan Dry Forests biome is found in inland western North America, with about 65 percent of the region in south-central British Columbia and 35 percent in north-central Washington State. The ecoregion is situated between the Rocky Mountains to the east, and the Coastal Ranges to the west. Impacted by the hot, dry air of the Canadian Basin and the colder, arctic air to the north, this region encompasses both the coldest climate in Washington, and some of the hottest and driest climate of British Columbia.

This biome is characterized by very warm to hot, dry summers and moderately cool winters with little snowfall. Average summer temperatures range from 59 to 62 degrees F (15 to 16.5 degrees C); and average winter temperatures are about 28 degrees F (minus 2.5 degrees C). Precipitation ranges from 10 inches (25 centimeters) to over 39 inches (100 centimeters) per year, depending on elevation.

Biodiversity

This area includes areas of grasslands and low elevation dry forests, as well as conifer forests. The conifer forests consist primarily of Douglas fir and ponderosa pine. Some areas sustain stands of lodgepole pine (*Pinus contorta*), quaking aspen (*Populus tremuloides*), white spruce (*Picea glauca*), grand fir (*Abies grandis*), and western larch (*Larix occidentalis*). These forests change to shrub-steppe in the valleys in the east-

ern part, and grasslands in the western portion of the region. The desert-like south Okanagan Valley is one of the most remarkable grasslands in British Columbia. It is not quite dry enough and too cold on average to be a true desert, and is more correctly termed a shrub-steppe. Nevertheless, its cactus, rattlesnakes, and scorpions hint at its relationship to the true deserts of the southwestern United States.

Dry desert sage brush and conifer trees provide abundant habitat for the many bird species found here. Approximately 190 species of birds breed in the south Okanagan. Animal habitat is diverse, and supports species such as the California bighorn sheep, mountain goat, mule deer, black bear, cougar, coyote, bobcat, grouse, and California quail.

This diverse ecosystem has several species classified as threatened or endangered, including the badger, tiger salamander, burrowing owl, pallid bat, Behr's hairstreak butterfly, night snake, Grand Coulee owl-clover, Great Basin spadefoot toad, western rattlesnake, lark sparrow, northern Pacific rattlesnake, Swainson's hawk, long-billed curlew, interior western screech-owl, yellow-breasted chat, brewer's sparrow, grasshopper sparrow, badger, and Great Basin gopher snake. Preserving this habitat is essential for their continued survival.

Human Impact

Lightning is the primary non-human cause of forest fires to the area, causing over 50 percent of fires in western Canada and the Northwest Territories. High temperatures, low humidity, and extended periods of drought added to high lightning probability can produce significant fires. Forest fires are not necessarily bad, and can in fact lead to the health of the forest. A fire within the forest will remove the tallest and most mature growth.

Also removed with this loss is the susceptibility of disease and insect infestation in mature trees. Leaves and debris found on the forest floor are removed with a surface fire, allowing the growth of new saplings. However, there is often public concern when wildfires burn unrestricted, that is, with no human intervention.

Research indicates that the future climate of these dry forests and conditions throughout the Canadian west will continue to become drier and warmer in the coming years. Fires that were once smaller, cooler, and more frequent have become larger and more intense, destroying resources and natural habitat. As a result, it is predicted that forest fires within Canada, including British Columbia, can be expected to increase over the next century.

It is estimated that only about 20 percent of the Okanagan Dry Forest ecoregion remains as intact habitat. Some areas have been heavily altered due to growing urban expansion, conversion of land into agricultural production, and uncontained wildfire. This is particularly true of the valleys and basins. Upper elevations have been impacted more by livestock grazing, logging, open pit mines, agriculture, and transmission and pipeline corridors.

Some grasslands have been seriously overgrazed by livestock. Climate change is already impacting the forests here: warmer temperatures affect tree growth rates, mortality rates, disturbance patterns (such as storms, drought, fire), and the distribution of tree species after disturbances. Impacts are cumulative and interconnected. For example, drought can stress trees, making them more susceptible to attack by insects and disease.

SANDY COSTANZA

Further Reading

Hessburg, Paul F. and James K. Agee. "An Environmental Narrative of Inland Northwest United States Forests, 1800–2000." *Forest Ecology and Management* 178 (2003).

Pacific Northwest Research Station. *Dry Forests of the Northeastern Cascades Fire and Fire Surrogate Project Site, Mission Creek, Okanogan-Wenatchee National Forest.* Portland, OR: United States Department of Agriculture—Forest Service, 2009.

The Partnership for Water Sustainability in British Columbia. "Geographical Analysis of Cumulative Threats to Prairie Water Resources: Mapping Water Availability, Water Quality, and Water Use Stresses." http://waterbucket.ca/okw/.

Okavango Delta

Category: Inland Aquatic Biomes.
Geographic Location: Southern Africa.
Summary: One of the world's largest inland deltas is surrounded by desert and is an indispensable seasonal home to savanna wildlife.

The Okavango Delta is located in the northwest of the Kalahari Desert, in the Republic of Botswana (district Ngamiland). It is one of the largest sites in the Ramsar list of Wetlands of International Importance, although size estimates vary greatly among authors. Gumbricht and collaborators estimate the total area, including permanently dry islands, to be 11,000 square miles (28,000 square kilometers). The delta is situated in an endorheic basin, meaning the drainage of the river is closed and landlocked, rather than flowing into a lake or ocean.

The two main tributaries of the Okavango River, the Quito and Cubango, both rise in the central part of the neighboring state Angola, draining catchments of 25,000 and 44,000 square miles (65,000 and 115,000 square kilometers), respectively. They flow through the Caprivi strip (of neighboring Namibia) into the so-called Panhandle and subsequently into the delta, which spreads out among three major fault lines: the Gumare fault in the northwest, and the Kunyere and Thamalakane faults in the southeast. These lines of weakness in the basement rock represent an extension of the East African rift system. Between them, the crust of the Earth has collapsed, and by subsequent deposition of Okavango river sediments into this depression, a gently sloping, slightly conical alluvial fan has developed. McCarthy and collaborators determined a gradient of 1:5,570 in the Panhandle region, and a gradient of 1:3,400 on the delta.

Seasonality and Volume

The dry season in Botswana lasts usually from May until October. The climate is semiarid, with potential evapotranspiration exceeding rainfall in all months of the year. Both in the Angolan catchment and the delta, rainfall usually occurs from December to February and peaks in January. Inflow into the Panhandle at Mohembo peaks four months later, in April, and peak outflow from the delta into the Thamalakane River at Maun occurs in August.

The distance between Mohembo and Maun is only about 100 miles (250 kilometers), but the flood wave spreads outward as it leaves the confinement of the Panhandle, and generally proceeds slowly due to the low topographic gradient and relief. Authors also disagree on the amount of surface inflow, and indeed rainfall and water discharge vary greatly among years, decades, and in space. Ellery and collaborators give a mean figure of 14 billion cubic yards (11 billion cubic meters) per year for total inflow via the Okavango, and 6.5 billion cubic yards (5 billion cubic meters) per year of summer rainfall over the Okavango Delta.

Most of the river system drains terrain covered by Kalahari sand. Although the concentration of dissolved solids—mostly silica and calcium and magnesium bicarbonates—is very low, the sheer volume of water introduced ensures that they are found in considerable quantities. Considering that more than 96 percent of inflow plus rainfall is lost to the atmosphere by evapotranspiration, while only 2 percent leaves the delta as surface outflow and presumably less than 2 percent exists as subsurface outflow, it is remarkable that the system maintains drinking water quality without human intervention.

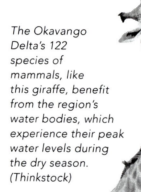

The Okavango Delta's 122 species of mammals, like this giraffe, benefit from the region's water bodies, which experience their peak water levels during the dry season. (Thinkstock)

Two ecological processes probably account for this phenomenon: Continuous spatial and temporal shifts in flooding regimes assist in the regeneration of soils and water bodies, and a multitude of islands develop continuously. Initially, any elevation above the level of flooding, such as termite mounds, enables growth of woody plants. Dissolved salts are either taken up by these woody plants via transpiration, or else precipitate as silica and carbonates (calcite) in the soil—resulting in vertical and lateral expansion of islands—or concentrate in the groundwater beneath the island center. Because the water table is generally beneath that of the surrounding floodplain, no lateral movement of salt water occurs away from the island.

Three Zones

Ecologists broadly subdivide the delta into three zones: The panhandle comprises a meandering river system flanked by permanent swamps, dominated by up to 13-foot-high (four-meter-high) stands of papyrus sedge (*Cyperus papyrus*). Further vegetation types which receive and require a very steady water supply predominate in the panhandle, the delta apex, and northeast of Chief's Island. Frequently flooded communities are characterised by the sedge *Schoenoplectus corymbosus* and wireleaf daba grass (*Miscanthus junceus*). Open water and permanent swamp communities contain, besides papyrus, fully or partly submerged plants such as common hornwort (*Ceratophyllum demersum*) and hippo grass (*Vossia cuspidata*).

Seasonal or temporary floodplains are characterised by a multitude of islands. The largest of these, several thousand square feet (several hundred square meters) in extent, are believed to be of tectonic origin, whereas smaller islands develop as described above. Shrubs, trees, and the Makalani fan palm (*Hyphaene petersiana*) grow on the highest elevations in older islands, often on a fringe around the islands, while the center of the islands turns barren, or supports only few salt-tolerant plant species. Between these islands and the main floodplain lies a grass-dominated zone, characterized by grazing-tolerant species such as Bermuda grass (*Cynodon dactylon*).

Wildlife and Human Needs

With 1,300 species of plants, 71 of fish, 33 of amphibians, 64 of reptiles, 444 of birds, and 122 of mammals identified so far, Ramberg and collaborators consider the species diversity of the Okavango Delta "normal for the southern African region." Its great importance for wildlife lies in the extent of its water bodies, and the fact that these have their peak water level during the dry season. Unfortunately in this regard, Ngamiland was furnished with a cross-pattern of veterinary fences from the late 1950s onward in order to avoid contact and possible spread of diseases between wildlife and domestic herds, beef production being one of the mainstays of the economy of Botswana. Some of these fences have posed a severe threat to wildlife, obstructing migratory routes and access to the water resources of the delta.

There is also a constant demand for the water resources of the delta by humans. Various plans for extensive water extraction not only from the Botswana section of the Okavango system, but also from the tributary rivers in Angola and Namibia were proposed and, partly as a consequence of national and international impact assessments, shelved. The Permanent Okavango River Basin Commission (OKACOM), a joint commission of the three countries Angola, Namibia, and Botswana, recently approved of a Transboundary Diagnostic Analysis, which will serve as a basis for a joint management plan for the whole river basin.

Julia-Maria Hermann

Further Reading

Alonso, Leeanne E. and Lee-Ann Nordin, eds. "A Rapid Biological Assessment of the Aquatic Ecosystems of the Okavango Delta, Botswana: High Water Survey." *RAP Bulletin of Biological Assessment* 27 (2003).

Ellery, William N., Terence S. McCarthy, and J. M. Dangerfield. "Biotic Factors in Mima Mound Development: Evidence From the Floodplains of the Okavango Delta, Botswana." *International Journal of Ecology and Environmental Sciences* 24, nos. 2–3 (1998).

Gumbricht, Thomas, Jenny McCarthy, and Terence S. McCarthy. "Channels, Wetlands and Islands in the Okavango Delta, Botswana, and Their Relation to Hydrological and Sedimentological Processes." *Earth Surface Processes and Landforms* 29, no. 1 (2004).

McCarthy, Terence S., M. Barry, A. Bloem, and William N. Ellery. "The Gradient of the Okavango Fan, Botswana, and its Sedimentological and Tectonic Implications." *Journal of African Earth Sciences* 24, nos. 1–2 (1997).

Ramberg, Lars, Peter Hancock, Markus Lindholm, Thoralf Meyer, Susan Ringrose, Jan Sliva, Jo Van As, and Cornelis VanderPost. "Species Diversity of the Okavango Delta, Botswana." *Aquatic Sciences* 68, no. 3 (2006).

Okeechobee, Lake

Category: Inland Aquatic Biomes.
Geographic Location: North America.
Summary: Lake Okeechobee plays a pivotal role in the long-term ecological and economic health of southern Florida. Significant restoration is needed to prevent ecological collapse.

At 700 square miles (1,800 square kilometers), Lake Okeechobee is the largest freshwater lake within the southeastern United States. The lake sits in the center of southern Florida; due to this location and its role as a major water source, it has been referred to as the Liquid Heart of South Florida. The lake has suffered major changes over the past century, and may require significant restoration projects to prevent an ecological collapse.

Lake Okeechobee is located about 40 miles (65 kilometers) northwest of West Palm Beach. The lake is shallow, with an average depth less than 9 feet (3 meters), and is about 35 miles (55 kilometers) long. Its surface area is approximately 730 square miles (1,900 square kilometers), within a total watershed exceeding 4,000 square miles (10,400 square kilometers). The rainwater-fed Kissimmee River is its chief source, flowing southward into Lake Okeechobee via a chain of lakes on the north side. The Caloosahatchee River is its main outlet, flowing southward into the Everglades.

Lake Okeechobee forms the main source of water for the Everglades ecosystem, as well as for a great part of southern Florida's agriculture. The tension between watering the natural habitats of the Everglades, on the one hand, and providing agricultural water use, on the other, is the basis of a passionate debate over the past few decades, in light of the decline of the Everglades.

The climate is considered to be subtropical, and can be divided into a dry season (November through May) and a wet season (June through October). Temperatures range from 59 degrees F (15 degrees C) in the winter to 86 degrees F (30 degrees C) in the summer. Average annual rainfall is 40 to 65 inches (102 to 165 centimeters), with more than half of the rainfall occurring in the wet season. Historically, rainfall from the Kissimmee Valley filled the lake during the summer rainy season, and excess water spilled out on the southern side.

Flora and Fauna

Aquatic and wetland plants of Lake Okeechobee provide habitat for birds and fish, and help improve water quality. However, they can also pose some ecological problems, especially those that are invasive. Because the lake is heavily used for fishing, hunting, and birding, plant growth must be managed. Presently, there is concern over larger and more frequent phytoplankton blooms in the lake because they are composed of blue-green algae (cyanobacteria), which can produce chemicals in the water that are toxic to fish. It is believed these blue-green algae blooms are mainly stemming from runoff from urban areas that drain into the Kissimmee River, and from local agricultural settings that use large amounts of fertilizers.

There are four types of vascular plants sustained in the lake. Some of these emersed (or emergent) plants include bulrushes and cattails, which are important to nesting birds such as flycatchers and the endangered snail kite. There are also arrowhead varieties and maiden cane (an aquatic grass), spike rush, southern amaranth, swamp rose mal-

low, common reed, wild taro, torpedo grass, primrose willow, and smartweed.

Submersed aquatic plants include: pondweed (peppergrass), eelgrass, southern naiad, coontail, and hydrilla. Floating-leaved plants in Okeechobee include: American lotus, floating hearts, spatterdock, and water lily. And, finally, floating plants include: water hyacinth, water lettuce, tropical water grass, alligator weed, pennywort, floating bladderwort, frog's-bit, duckweed, and mosquito fern.

Lake Okeechobee is also home to some unique trees, vines, fruits, and mosses. Forests of pond apple trees once dominated the southern rim of the lake, but by 1930, about 95 percent of them had been removed to make way for farmland and water-level management. The trees are now the focus of restoration efforts away from the dike or canal banks. Moonvine, or evening glory, can be seen hanging from the pond apple trees or along the shoreline of lagoons. Spanish moss also thrives here.

Lake Okeechobee forms a pivotal part of the ecological and economic health of the state of Florida. The lake provides foraging and nesting habitats for a wide range of bird species. Some that are seen here include: fulvous whistling-duck and the mottled duck; shorebirds such as black-necked stilt; waders such as roseate spoonbill, wood stork and least bittern; and specialty raptors including crested caracara, snail kite, and barred owl.

There are more than 250 fish species in the lake. Some of the most popular (as far as game fish go) are: the largemouth bass (*Micropterus salmoides floridanus*), speckled perch or black crappie (*Pomoxis nigromaculatus*), bluegill (also known as bream, blue bream, and sun perch), red-ear sunfish, blue tilapia, and spotted tilapia. Other common fish here are: oscars, snook, cichlids, catfish, gar, shad, and bowfins. Lake Okeechobee maintains a fisheries industry valued at $100 million annually.

Human Impact

The earliest known name for this great lake was *Mayaimi*, meaning *big water*, as it was called by the dominant indigenous tribe in the 16th century, the Calusa. Its current name, Okeechobee, also means big water (*oki* for *water* and *chubi* for *big*) from the Hitichi tribe, who were Muskogean-speaking Seminoles. Due to the uninhabitable character of the prairies, swamps, and marshlands of southern Florida, the lake received little attention from European migrants until the early 1800s, with the start of the Seminole Wars (also known as the Florida Wars).

The lake originated just 6,000 years ago during the oceanic recession; it at first had vast marshes to the west and the south. The southern marsh was contiguous with the Everglades, and thought to be slightly eutrophic (nutrient-overloaded).

Modern-day Lake Okeechobee differs in many ways. The morphology of the lake has changed dramatically. Originally, the lake was not encircled by a dike, as it is now; instead, a marsh surrounded the lake. With its discovery becoming known throughout the United States in the 19th century, interest escalated in using this region for agriculture.

The taming of Lake Okeechobee became a reality in the early 20th century, when it was decided that the lake could still be navigable with 6 feet (2 meters) less of water depth. This resulted in the construction of levees and canals, draining 6 million acres (2.4 million hectares) of land to the west and south of the lake. This event set the framework for the current system of levees and canals across southern Florida.

The hurricanes of 1926 and 1928 caused massive floods and thousands of deaths, prompting the federal government to instruct the U.S. Army Corps of Engineers to build a comprehensive system of levees, gates, and channels. This system protected the surrounding area from floods, allowing the use of the lake as a reservoir, and the Okeechobee Waterway allowed boats to move between the east and west coasts of Florida. The major levee was later named the Herbert Hoover Dike, after the president who was pivotal in its development.

In addition, taming the flow of water enabled the conversion of 50 percent of the Everglades ecosystem to agricultural land, mainly for sugarcane production. As a result, the direct link between Lake Okeechobee and the vast Everglades was cut off, and the reduction in water supply has significantly affected the Everglades ecosystem ever since.

The diking of Lake Okeechobee resulted in significant changes to this landscape, resulting in numerous changes to its ecology. Diking caused significant reductions in average water levels and a 30 percent reduction in the size of the pelagic zone, which consisted of the majority of the clear and clean water in the lake. The dike itself formed an artificial shore, allowing for the formation of a previously nonexistent littoral zone. The unnatural connection directly to the sea also had consequences for the estuaries into which the lake drains. During high-rain events, a large influx of freshwater to the estuaries changes the salinity dramatically and threatens the health of these ecosystems.

From the 1950s to 1970s, changes in land use caused further degradation of the lake ecosystem in the form of eutrophication. The high influx of nutrients causing this oxygen depletion in the water was the result of increased use of fertilizer in agriculture, and the explosion in population growth, which resulted in large-scale sewage discharges. The increased regulation of regional land-use and water management have significantly improved the water quality of the lake, but with growing human demand for water and projected extremes in rain events with global climate change, continued effort is required to restore and to maintain this great lake.

The future health of Lake Okeechobee is unclear. The Comprehensive Everglades Restoration Plan (CERP) contains the projected plans for Lake Okeechobee which the Congress directed the Corps of Engineers to implement. The goal of CERP for Lake Okeechobee is to maintain sufficient water-storage capacity, and to maintain water quality and ecological functioning. Instead of overexploiting Lake Okeechobee for intensive water use, focus has shifted to recharging aquifers for below-ground water storage. If employed, the strategies may improve the health of Lake Okeechobee and allow the lake to continue as a thriving ecosystem for years to come.

Peter Baas
Megan Machmuller

Further Reading

Hanna, Alfred J. and Kathryn A. Hanna. *Lake Okeechobee: Wellspring of the Everglades.* Indianapolis and New York: Bobbs-Merrill Co. Publishers, 1948.

Havens, K. E., et al. "Rapid Ecological Changes in a Large Subtropical Lake Undergoing Cultural Eutrophication." *Ambio* 25 (1997).

Janosky, Jim. *Okeechobee: A Modern Frontier.* Gainesville: University Press of Florida, 1997.

Steinman, Alan D., et al. "The Past, Present, and Future Hydrology and Ecology of Lake Okeechobee and Its Watershed." In James W. Porter and Karen G. Porter, eds., *The Everglades, Florida Bay, and Coral Reefs of the Florida Keys: An Ecosystem Sourcebook.* Boca Raton, FL: CRC Press, 2002.

Okefenokee Swamp

Category: Inland Aquatic Biomes.
Geographic Location: North America.
Summary: Debates continue over how to best protect this richly inhabited major freshwater swamp from naturally occurring events such as droughts, lightning strikes, and forest fires.

The Okefenokee Swamp is the largest freshwater swamp in the United States, and is located in southern Georgia. The low pH balance, or high acidity, of the swamp leads to a limited flora, but supports numerous, diverse fauna.

The climate in South Georgia is subtropical; typically, summers are hot and rainy. In the drier winters, occasionally the temperature drops below freezing. The Okefenokee Swamp receives the majority of its water from precipitation throughout the 1,400-square-mile (3,600-square-kilometer) watershed. Surface springs also provide a minor contribution. The swamp itself covers an area of roughly 680 square miles (1,770 square kilometers).

Annual precipitation, which occurs mainly in the hot summer, averages 50 inches (125 centimeters); however, droughts are common throughout the region. The swamp is sometimes called the largest blackwater swamp in North America, the name referring to its tea-colored water. This color is produced by tannic acid, generated by decaying vegetation. Tannins cause the water pH to drop from a neutral 7.0 pH to an acidic 3.7 pH.

Water leaves the swamp either by evaporation or directly via two major rivers, the Suwannee and the St. Marys. The east side of the swamp has a natural barrier, called the Trail Ridge, that funnels a majority of the water into the Suwannee River, which then cuts across north-central Florida, draining into the Gulf of Mexico. The St. Marys River discharges east into the Atlantic Ocean. Despite the low pH of the swamp water, the rivers flow over calcareous limestone deposits, which eventually increase the pH and neutralize the river chemistry.

Flora and Fauna

The low pH and sandy soil limit the flora which can survive in Okefenokee Swamp. Naturally occurring floods and wildfire maintain the sensitive ecosystem. The most abundant tree in the swamp is the cypress (*Taxodium ascendens*). Although most of the cypress was logged in the early 1990s, some giant pond cypresses still exist. The pond cypress prefers lakes and ponds, whereas the bald cypress prefers to grow in moving waters. Less widespread trees such as the slash pine (*Pinua elliotii*), black gum (*Nyssa sylvatica*), water ash (*Fraxinus caroliniana*), Ogechelme (*Nyssa ogeche*), loblolly-bay (*Gordonia lasianthus*), swamp-bay (*Peresa plaustris*), and sweet-bay (*Magnolia virginiana*) are also found here.

The lack of phosphate- and nitrogen-enriched soils have attracted many carnivorous plants. Carnivorous plants trap and ingest insects to survive. Three varieties of pitcher plants (*Sarracenia flaua, S. minor, S. psittacina*) survive in the swamp by first attracting flying insects into their pools of bacteria, then drowning them. Sundews (*Drosera rotunditolia*) attract insects by use of their sticky hairs. Insects such as gnats become stuck and later ingested. Butterworts and species of bladderworts (family *Lentibulariaceae*) use similar strategies for attracting creatures living on the water surface.

The water lily (*Nymphaea odorata*) is one of the most common flowers found in the swamp.

Canoeing in the Okefenokee Swamp in 2010. In 2011, a massive fire burned an estimated 75 percent of the Okefenokee Swamp National Wildlife Refuge. The drought conditions that helped the fire spread throughout much of the refuge are thought to have been related to global climate change. (U.S. Fish and Wildlife Service)

Vines such as the greenbriar (*Smilax walteri*) and the climbing heath (*Pieris phillyreitolia*) are also present here. Spanish moss is a notable "air plant" often seen hanging from tree branches.

Peat deposits up to 15 feet (5 meters) thick cover much of the Okefenokee floor. These deposits are so unstable in spots that trees and surrounding bushes may tremble if the adjacent surface is heavily trodden. It is this trembling ground that gave reason for the Choctaw Indians to refer to the swamp as *okafenoke*, meaning "quivering Earth." Floating peat batteries—mainly composed of decaying water lilies and cypress—form small islands. Up to 60 of these floating peat batteries can be found in the Okefenokee Swamp, and some of them will support the weight of a human.

It is estimated that over 400 species of vertebrate animals live in the Okefenokee Swamp, half of which are birds. This number reflects 39 fish, 37 amphibian, 64 reptile, 50 mammal, and 235 bird species. Currently, seven species are threat-ened or endangered, including the sand hill crane, wood stork, and great blue heron. The ivory-billed woodpecker (now considered extinct), Carolina parakeet, and passenger pigeon are no longer present in the swamp. The 235 bird species present in the swamp include both year-round and seasonal residents. Common birds include the white ibis (*Eudocimus albus*), anhinga (*Anhinga anhinga*), great egret (*Casmerodius albus*), various vultures, and warblers.

Amphibians require wetland depth fluctuation for proper reproduction, and have found such cycles suitable here. There are over 20 species of frogs and toads throughout the Okefenokee Swamp, including 10 members of the tree frog family. Salamanders are also plentiful in the swamp, with 16 types present. Additionally, two types of skink can be found.

Within the swamp, there are also 14 families of fish, featuring 36 species of freshwater fish. Some of the most common are the Florida gar,

American eel, the Okefenokee pygmy, sunfish, bowfins (*Amia calva*), five species of catfish, and the chain pickerel (*Esox niger*).

Of the 64 species of reptiles present in the biome, the swamp alligator is the most common one specifically associated with the Okefenokee Swamp. Due to hunting in the middle of the 20th century, the alligator almost became extinct; it was listed as endangered in 1973. By 1984, the alligator population had recovered and it was removed from the Georgia Protected Species List. The 36 snake species (five of which are venomous) also contribute to the swamp's fauna. The snake population includes the venomous pit viper and coral snake. Also found here are the rare rainbow snake and federally protected threatened species, the indigo snake. There are also 11 species of lizards and 14 species of turtles.

The Florida black bear, *Ursus americanus floridianus*, is the Okefenokee Swamp's largest mammal. The majority of these bears live in the eastern region of the swamp, along the Trail Ridge Section. The bobcat (*Felis rufus*) and raccoon (*Procyon lotor*) are also quite common during the nighttime hours. Although there is an abundance and variety of animals living in the Okefenokee Swamp, no plant or animal is endemic, that is, unique to the region.

Human Impacts

Attempts were made in the late 1800s to drain parts of the swamp. Though the drainage projects were not successful, much of the swamp area was logged during the early 1900s. In 1936, the United States Biological Survey paid the Hebard Lumber Company for the approximate 470 square miles (1,200 square kilometers) which initially formed the basis for the Okefenokee Swamp reserve. In 1937, President Franklin D. Roosevelt signed an executive order establishing the Okefenokee National Wildlife Refuge. In 1960, the Suwannee River Sill was built to increase the water level in the swamp. By 1974, part of the swamp was designated as a part of the National Wilderness Preservation System. The swamp is now managed mainly by the Department of the Interior, Fish and Wildlife Services.

Fire is a dominant influence in the Okefenokee Swamp ecology. The Okefenokee Swamp has been classified as a high lightning-frequency area by the Forest Service. The naturally occurring swamp peat is about 85 percent combustible. An area with a high probability of lightning strikes that contains a large and combustible fuel source will have naturally occurring forest fires. Without natural fires, peat can accumulate and will eventually lead to invading shrubs and hardwoods.

As cypress and gum trees recover rapidly from fire damage, the present vegetative types cannot be sustained without fire. Human influence has not impacted the swamp in recent times as much as the naturally occurring wildfires. Although there is usually public demand to manage and quickly extinguish the wildfires, the naturally occurring swamp fires help with vegetative stability within the swamp.

The care of the swamp, in regards to unpredictable natural fires, remains a critical issue. Most neighbors and land owners in the area would prefer the government to control and prevent the natural spread of the fire. A wildfire which began with a lightning strike near the center of the refuge in May 2007 eventually merged with another wildfire that had begun earlier near Waycross, Georgia, due to a tree falling on a power line. By the end of the month, more than 935 square miles (2,400 square kilometers) had burned in the region, much of it in the refuge.

Again, in 2011, fire ravaged the area; some estimates are that 75 percent of the refuge was burned. The drought conditions that led to the spreading fires are believed to be related to global climate change. If so, this will be another factor to consider in terms of fire-control plans. However, many environmental groups would still prefer the National Wildlife Refuge to remain a naturally sustained ecosystem.

The Okefenokee Swamp Alliance is a conservation group that works for continued preservation of the swamp, which was named a Wetland of International Importance under the Ramsar Convention, and is on the United Nations Environmental, Scientific, and Cultural Organization (UNESCO) Tentative List for World Heritage Site status.

JENNIFER STOUDT WOODSON

Further Reading

Lenz, Richard J. *The Longstreet Highroad Guide to the Georgia Coast and Okefenokee.* Marietta, GA: Longstreet Press, 1999.

McHugh, Paul, ed. *Wild Places: 20 Journeys into the North American Outdoors.* Emeryville, CA: Avalon Travel Publishers, 1996.

Wilson, James. "The Okefenokee Swamp." *Georgia Wildlife Press: Natural Georgia Series* 6, no. 1 (1997).

Okhotsk, Sea of

Category: Marine and Oceanic Biomes.
Geographic Location: Russia.
Summary: A wild and unique seascape, this biome is now threatened by climate change and ongoing development schemes, fueled by national and international interest.

The Sea of Okhotsk is located in the Russian Far East region, adjacent to the western Pacific Ocean. It is partially enclosed by the Russian Kamchatka Peninsula to its east, Japan and the island of Sakhalin to its southwest, and the Kuril Islands to the southeast. This ecoregion now shows global warming trends, including changed patterns of sea ice. The area of the sea stretches across some 611,000 square miles (1,583,000 square kilometers).

The Sea of Okhotsk is the coldest sea in east Asia. Its temperatures during winter compare to the Arctic. Its western and northern regions experience severe winter weather due to influences from the continent. From October through April, temperatures are bitter and the area is ice-covered and dry, with little precipitation. A milder maritime climate occurs to the south and southeast, due to oceanic influences. February's average monthly air temperature in the northeastern region is minus 4 degrees F (minus 20 degrees C); north and west of the sea it is minus 11 degrees F (minus 24 degrees C); and in the southern and southeastern parts it averages 19 degrees F (minus 7 degrees C). The average August temperature is 54 degrees F (12 degrees C) in the northeast; 57 degrees F (14 degrees C) in the north

and west; and 64 degrees F (18 degrees C) in the south and southeast.

Annual precipitation varies from 16 inches (40 centimeters) in the north, to 28 inches (70 centimeters) in the west, and about 41 inches (104 centimeters) in the south and southeast.

During the winter, navigation on the Sea of Okhotsk becomes difficult, or even impossible, due to the formation of large ice floes. Large amounts of freshwater from the Amur River, which reaches the sea at the Straight of Tartary at the southwestern end of the Sea of Okhotsk, lowers the salinity in this area—and results in lowering the freezing point of the sea here. The thickness and distribution of ice floes depends on many factors: the location, the time of year, water currents and wind systems, and the sea temperatures.

Wildlife

The Sea of Okhotsk presents a vast shoreline mostly made up of gravel and cliffs, interspersed with large bays and tidal mudflats. The region is relatively poorly populated, apart from various fishing villages. Only Magadan in the north, and Sakhalin Island in the southwest, harbor larger population centers. Other cities and ports are Korsakov, Ayan and Okhotsk. The volcanic Kuril Islands chain is sparsely populated, mostly in the south. (Tensions between Russia and Japan exist surrounding some of the southern islands and associated marine regions of the Kurils).

The sea is recognized globally for its fisheries, namely pollock, herring, sardine, and salmon (all Pacific salmon species occur here); species such as crab, squid, and sea urchin are also heavily pursued in southern regions, destined for Asian markets. Caviar (*ikra* in Russian) is highly sought after, and local sturgeon species in the Amur and Sakhalin region are of global conservation concern. Overfishing, illegal fisheries, and poor straddling stock management are also concerns.

Sea of Okhotsk is also home to many marine mammals, including the endangered species of Kuril harbor seals and gray whales. Other marine mammals that inhabit the region are: North Pacific right whales, bowhead whale, Baird's beaked whale, Arnoux's beaked whale, beluga, Dall's porpoise, harbor porpoise, sea otter, Stellar's sea lion, Pribilof fur seal, Okhotsk ringed seal, Pacific bearded seal, ribbon seal, and spotted seal.

Some of these species show a connectivity with Bering Sea populations, including with Alaska and British Columbia. Walrus now appears to be extirpated in this region. Other wildlife in the surrounding area includes many sea- and shore-birds. Large migrations of waterbirds occur here, including endangered species. The Kuril Islands are home to many millions of seabirds, including northern fulmars, tufted puffins, murres, kittiwakes, guillemots, auklets, petrels, and cormorants.

Human Impact

Indigenous populations inhabited the Sea of Okhotsk region for millennia. Today, there are still several indigenous communities living along the long shoreline (most of the Japanese-oriented Ainu have withdrawn from the Kuril Islands, however). The native spiritual and material cultures with their subsistence practices still show a high degree of interdependency with the sea. Therefore, they are some of the first to experience and suffer from changes in the environment, such as decline of maritime resources due to human expansion and climatic changes.

The Sea of Okhotsk shelf, which runs along the coast, has been identified as having significant zones of potentially recoverable oil and gas accumulation. Pressures are growing for offshore oil and gas exploitation and extraction. These demands are not only fueled by Russian interests, but also by Japan, China, Korea, and India. The construction of one of the largest liquefied natural gas plants in the world, and its associated pipeline, on Sakhalin is an example. International financing plays a major role for such endeavors. Subsequently, with consumption and the amount of shipping required on the rise, tanker traffic and extreme weather conditions in the region have increased the risk of oil spills.

Because of climate change, rising average air temperatures, changing amounts and duration of sea ice cover, changing sea surface temperatures, changing air circulation patterns, and circulation cycles of carbon in this ecoregion have all been affected. Migration routes of salmon and other

> ### *Whaling*
>
> The Sea of Okhotsk was a hotbed for whaling in the mid-1800s. American whaling ships began hunting right whales in the southeastern part of the Sea of Okhotsk near the Kuril Islands, and from 1849 bowhead whales dominated the catch; but later, the majority of the fleet focused on bowheads in the Bering Strait region. Following poor catches there, whaling would shift again to the Okhotsk Sea. By 1860, the number of whaling vessels in the region had decreased significantly, and whaling there stopped altogether by the early 1900s.

fish have already been impacted. Sea ice is used by marine mammals to facilitate their migration routes, and this has already been compromised.

The Sea of Okhotsk comes mainly under the governance of Russia. But the issue of who has sovereignty over the Kuril Islands involves Japan. Sakhalin Environment Watch and the California-based Pacific Environment and Resources Center are a few of the environmental organizations that have led independent experts to the Kurils to review local monitoring methods, spill prevention, and response measures.

The North Pacific Marine Science Organization (known as PICES) has provided a forum for the exchange of marine data in response to previous spills. However, there is still major international concern over this issue. There also remains much research to be done on the continuing changes in climate factors, and how these will impact species throughout the ecoregion.

FALK HUETTMANN

Further Reading

Gerasimov, Yuri and Falk Huettmann. "Shorebirds of the Sea of Okhotsk: Status and Overview." *Stilt* 50, no. 15 (2006).

Huettmann, Falk. "Marine Conservation and Sustainability of the Sea of Okhotsk in the Russian

Far East: An Overview of Cumulative Impacts, Compiled Public Data, and a Proposal for a UNESCO World Heritage Site." In M. Nijhoff, ed., *Ocean Year Book*. Halifax, Nova Scotia, 2008.

Newell, Josh. *The Russian Far East: A Reference Guide for Conservation and Development, 2nd ed.* Seattle, WA: Daniel and Daniel Publishers, 2004.

North Pacific Marine Scientific Organization. "Third Workshop on Okhotsk Sea and Adjacent Areas." http://www.pices.int/meetings/regional_NPESR/Okhotsk/default.aspx.

Olango Island Group Coral Reef

Category: Marine and Oceanic Biomes.
Geographic Location: Asia.
Summary: These coral reefs are highly diverse but threatened by overexploitation—although government programs and education have slowly helped increase the fish population.

The Olango Island Group is a cluster of seven islands in the central Visayas region of the Philippines, just off the coast of Cebu. This seven-island group has a total land cover of 3.9 square miles (10.3 square kilometers). The islands are raised coral reefs; in the center of the group is a tidal flat that supports a 3.5-square-mile (9.2-square-kilometer) area of the Olango Island Wildlife Sanctuary.

The Olango Island Group has a tropical climate, with temperatures ranging from 73 to 86 degrees F (23 to 30 degrees C) and average rainfall of 57–62 inches (144–160 centimeters). Approximately 20,000 people live in this island group, and 75 percent engage in fishing or related activities for subsistence.

More than half of Olango is composed of diverse coastal and marine habitats. This island series, which is no more than 33 feet (10 meters) above sea level, is especially known for its low-lying, limestone intertidal mudflats, mangroves, wide fringing coral reefs, seagrass beds, and hard reef-building coral. This cluster of islands is accompanied by six satellite islets that are bounded by a steep coral reef wall to the west and a sloping reef to the east of Olango.

The Philippines has a coral cover of approximately 10,425 square miles (27,000 square kilometers), of which only 5 percent is in excellent condition. Of this area, Olango Island Group's coastal habitat, including seagrasses and coral reefs, occupies only 15 square miles (40 square kilometers). Seagrass represents an average of 44 percent of live cover, while coral represents only 22 percent.

Biodiversity

The complex structure of coral reefs offers shelter to diverse fauna. There are 97 bird species that have been identified in the Olango Island Group, of which 48 are migratory. Among the migratory species are: Chinese egret, Asiatic dowitcher, tern, eastern curlew, plover, sandpiper, black-tailed godwit, and red knot. At the southern end of the islands is the large intertidal bay that is now the Olango Island Wildlife Sanctuary, which in 1994 was declared a Wetland of International Importance under the Ramsar Convention. The area has been identified as an East Asian-Australian Flyway.

There are more than 2,000 coral-reef-associated fish species in the Philippines, including those of the Olango Island series. Species richness and diversity in fish have been shown to correlate with certain characteristics of coral diversity, such as architectural complexity, biological diversity, species richness, abundance, colony size, coverage of living coral, and coverage of massive and encrusting coral. Overall, the Olango Island reefs have an annual fish yield of 5.5 tons (5 metric tons) per 0.4 square mile (1 square kilometer), which is quite low compared with the 17–22 tons (15–20 metric tons) per 0.4 square mile (1 square kilometer) estimated for other Philippines reefs.

Human Impact

Although illegal fishing has decreased considerably in the region due to government intervention, fish yields are still below their true potential, due to years of damaging harvesting methods. The Olango coral reefs, although diverse in nature, are considered to be in generally poor condition. Coral reefs in the area have fallen victim to what is known as the "tragedy of the commons." As oceans and coral reefs are considered to be a shared pool of resources, they are frequently exploited.

Over 75 percent of households in the island cluster are engaged in fishing or related livelihoods. The Olango coral reefs are greatly depleted by local fishers, due to incorrect fishing methods and siltation from island logging activities. As coral cover decreases, the diversity and abundance of reef fish also decrease. Studies have shown that there is a 2.4 percent decrease in reef-fish biomass for every 1 percent annual decrease in coral cover.

Government programs protecting natural fish habitats, including mangroves, seagrasses, and mudflats, as well as education for local fishers about the consequences of incorrect fishing methods, have resulted in slow increases in fish yields. The government of the Philippines is currently using a conservation program called Community-Based Coastal Resource Management, in which the daily users of a resource set the rules and regulations that govern the resource. It is also using another method of conservation called co-management, a system in which responsibility is distributed equally between the government and resource users.

As a result of these methods of conservation, marine sanctuaries have been built to protect and nurture the region's coral reefs. Gilutongan Island Marine Sanctuary, a section of the Olango coral-reef complex, protects the designated marine habitat from fishers and other threats. The sanctuary has an average coral cover of 40 percent. Although coral recovery from damaging methods may be slow, it is essential that the government of the Philippines continue to protect this important biodiversity hot spot. In the meantime, research is being conducted on the future effects of global warming trends on the islands and surrounding reefs. With predicted rises in sea level, and increases in frequency and strength of storm surges, there is likely to occur more damage and threat of damage to these reefs and their hosted marine species.

YASMIN MANNAN
APRIL KAREN BAPTISTE

Further Reading

Finlayson, C. M. and Rick Van Dam, et al. *Vulnerability Assessment of Major Wetlands in the Asia-Pacific Region—Summary Report on the Project Workshops for Olango Island and Yellow River Delta.* Canberra, Australia: Environmental Research Institute of the Supervising Scientist (ERISS), 1998.

Sotto, Filipina B., Joey L. Gatus, Michael A. Ross, Ma. Fe L. Portigo, and Francis M. Freire. *Coastal Environmental Profile of Olango Island, Cebu, Philippines.* Cebu City, Philippines: Coastal Resource Management Project, 2001.

White, Alan, Michael Ross, and Monette Flores. "Benefits and Costs of Coral Reef and Wetland Management, Olango Island, Philippines." Asian Development Bank. http://www.poverty environment.net/node/190.

Oman, Gulf of, Desert and Semi-Desert

Category: Desert Biomes.
Geographic Location: Middle East.
Summary: This unique desert environment has important mangroves and is also a habitat for migratory birds.

The Gulf of Oman desert and semi-desert is the coastal ecoregion located in the Sultanate of Oman and the neighboring United Arab Emirates (UAE). Its western reaches, the coastal areas along the Persian Gulf, cover the lands around Dubai and former Sharjah to the Musandam Peninsula, an enclave belonging to Oman. Along the Gulf of Oman, the region includes the rest of the Musandam Peninsula—famous for its position controlling the Straits of Hormuz—and south through Fujairah, and then along the Al Batinah region to the capital, Muscat.

Although the hinterland of this part of Oman is hilly, indeed mountainous in parts, and although the equivalent area in the UAE is desert, the coastal region has lagoons, mudflats, and mangrove swamps. The parts bordering the Gulf of Oman have more rain than those of the Persian Gulf, which are dry; temperatures in both places often rise to 120 degrees F (49 degrees C).

Exposed tidal flats at low tide in a harbor on the Gulf of Oman. The coastal region of the Gulf of Oman desert and semi-desert coastal ecoregion is lined with mudflats, lagoons, and mangrove swamps. (Thinkstock)

Flora

Large mangrove swamps once existed along the coast, but some of them have eroded or been destroyed as building development has taken place in recent years. These swamps were largely of gray or white mangrove (*Avicennia marina*), which is also found elsewhere in the region, especially along the Red Sea and on parts of the east coast of Africa. The Christ's thorn jujube (*Ziziphus spina-christi*) is also found in the region, with its flowers being used by bees for the manufacture of honey. The umbrella thorn acacia (*Acacia tortilis*) is common in northern Africa, the Arabian Peninsula, and the east coast of Africa. It survives in this region so well because its roots can stretch up to 98 feet (30 meters) into the soil.

Other trees, such as the neem (which is related to mahogany) and date palms can still be found along the coast. Also along the coast of Oman are many examples of the pea tree (*Prosopis cineraria*), in addition to the umbrella thorn acacia. Both areas also have desert hyacinths, which often are really evident only after rain, when the flowers open.

Fauna

This ecosystem sustains a small variety of fauna. Mammals include some caracals (*Caracal caracal*) and occasionally packs of striped hyenas in more remote areas. The Arabian tahr (*Arabitragus jayakari*), a type of goat, can be found in the hilly areas, with the Arabian leopard (*Panthera pardus nimr*) living in mountains, especially in Oman.

The coastal regions were home to loggerhead sea turtles (*Caretta caretta*), but the building of coastal housing and the main highways along the coast have caused many problems for them and for olive ridley turtles (*Lepydochelys olivacea*), green turtles (*Chelonia mydas*), and especially the now-very-rare hawksbill turtles (*Eretmochelys imbricata*). Fortunately, the loggerhead sea turtles have found a refuge in southern Oman, especially on Masirah Island.

Reptile life includes the desert monitor lizard (*Varanus griseus*), which can grow up to 3 feet (1 meter) long; the spiny-tailed agama; and the sand skink (*Neoseps*). These species have managed to survive more easily than others. At least 37 reptile species have been identified in the area.

The biggest proliferation of fauna are birds, which have adapted well to the change in the environment. Many of them use the region when migrating between Africa and Asia, with as many as 320 migratory species passing through in the spring and autumn, and some spending the winter in the region. Those native to the region include the Socotra cormorant (*Phalacrocorax nigrogularis*); purple sunbird (*Cinnyris asiaticus*); crab plover (*Dromas ardeola*); and black-crowned sparrow lark (*Eremopterix nigriceps*), which, although it tends to nest on the ground, is abundant.

Human Impact

There are many threats to wildlife, the most important being the degradation of the land in the region. Although much of this damage has been done by humans through development and intensive farming, camels and goats have also contributed extensively to erosion in both the UAE and Dubai. Certainly, the massive building projects that have taken place in and around Dubai, as well as construction on the road north, have affected many animals, cutting off some from access to the sea.

Human habitation has also had a major effect in terms of waste, pollution, and the sheer presence of so many people, especially in areas such as Dubai Creek, where the vast majority of the mangroves have been removed.

The oil-industry wealth of the area has led to major land-reclamation projects, especially some of the well-known schemes off the coast of Dubai. These projects and marine pollution have affected turtle and fish populations, reducing their numbers and in turn affecting the birds. Off-road driving (mainly four-wheel-drive vehicles used for recreational purposes) causes much damage to the vegetation, which is slow to recover due to the limited annual rainfall. Another big worry in the region is oil spills, which have been a problem in the past and which continue to pose a major threat to the fragile ecosystem.

In the Al Bainah region of Oman, where there is less development, it has been possible to preserve much more of the traditional way of life, which has helped the ecosystems be maintained. Runoff from fertilizer used in intensive agriculture has been a problem, but the government of Oman has attempted to save many of the mangroves with the creation of the Khor Kalba Nature Reserve. Also, the government of the UAE has created the Ras Al Khor Wildlife Sanctuary (also called the Dubai Wildlife and Waterbird Sanctuary), located around the southern end of Dubai Creek, to help preserve the ecosystem and allow migratory birds to use the area, which has led to some 500 greater flamingoes (*Phoenicopterus roseus*) living there. This sanctuary alone has led to an interest in wildlife conservation in Dubai.

JUSTIN CORFIELD

Further Reading

Bailey, Roger. "The Sea Birds of the Southwest Coast of Arabia." *Ibis* 108, no. 2 (1986).

Browne, P. W. P. "Notes on Birds Observed in South Arabia." *Ibis* 92, no. 1 (1950).

Delahy, M. J. "The Zoogeography of the Mammal Fauna of Southern Arabia." *Mammal* 19, no. 4 (1989).

Emara, H. S. "Oil Pollution in the Southern Arabian Gulf and the Gulf of Oman." *Marine Pollution Bulletin* 21, no. 8 (1990).

Salm, R. V. "Coastal Zone Management Planning and Marine Protected Areas." *Parks* 12, no. 1 (1987).

Onega, Lake

Category: Inland Aquatic Biomes.
Geographic Location: Northern Europe.
Summary: The second-largest lake in Europe, surrounded by some of the region's oldest coniferous forests, has relatively good water quality and fairly rich animal diversity.

Lake Onega is an inland lake situated in the Karelian region of northwestern Russia, close to the border of Finland and eastern Europe. The second-largest lake in Europe, its surface area is approximately 3,700 square miles (9,700 square kilometers). The lake drains through the River Svir, which eventually empties into the Baltic Sea. Lake Onega is relatively deep, with an average depth of 98 feet (30 meters), with its deepest point being 417 feet (127 meters). There are more than 1,600 islands on the lake.

The lake is fed by about 50 tributaries; the three most important rivers flowing into Lake Onega are: River Vodla, River Shuja, and River Suna; combined, they account for 60 percent of the riverine inflow. Lake Onega is connected to both the White Sea and Arctic Ocean, as well as the Baltic Sea through a network of canals—and to the Caspian Sea through the River Volga and its canal links. The human population around Lake Onega is currently about 400,000. The majority live in the vicinity of Petrozadovsk, capital of Karelia.

Lake Onega is in the Atlantic continental climate zone. The average temperature in January is 18 degrees F (minus 8 degrees C), and the average temperature in July is 62 degrees F (16 degrees C). Average annual precipitation is 20–28 inches (50–70 centimeters).

Flora and Fauna

Lake Onega lies amidst the global coniferous forest belt, or boreal forest, that stretches from Canada to Siberia. In Russia, the forest is called the northern taiga. In these forests, northern and southern species meet, making them biologically very diverse. Human impact in the Karelian forests has remained minor. Consequently, these northwestern Russian forests are among the last remaining fairly pristine forests in Europe.

Lake Onega has interaction with numerous rivers where wild freshwater salmon spawn. Baltic salmon (*Salmo salar*) became landlocked in the Karelian lakes as a result of land uplift; they have been living in these lakes for about 130,000 years. Today, pollution and poaching threaten salmon populations; earlier, construction of dams and timber rafting were their main threats.

Lake Onega features a large variety of fish and aquatic invertebrates, including relics of the glacial period, such as lamprey. There are about 47 fish species from 13 families; they include: stugeon, brown trout, European smelt, grayling, whitefish, char, pike, common dace, silver bream, carp, perch, spined loach, wels catfish, European eel, rudd, gudgeon, ruffe, and burbot.

Near Kizhi Island, one of the largest in the lake, about 180 bird species from 15 families have been identified; perhaps one-quarter of them were observed on the island itself. Most of these species are migratory and stop on the island either for rest or nesting, such as swans, geese, ducks, lake seagulls, and sterna—but there are also more stationary birds such as sparrow, siskin, chaffinch, sklyark, jackdaw, and crow.

The banks of Lake Onega are low and often flooded; they are swampy and rich in reeds and other wetland vegetation, hosting ducks, geese and swans. Mammals include elk, brown bear, wolf, fox, hare, squirrel, lynx, marten, and European badger, as well as American muskrat and mink, which were introduced to the area in the early 20th century.

Human Impact

The beauty of lake and its islands attracts an increasing number of tourists, which bring with them some degree of ecological wear-and-tear. Kizhi Island, a United Nations Environmental, Scientific, and Cultural Organization (UNESCO) World Heritage site, is renowned both for its natural beauty and bird species, and for the 89 wooden churches—in particular Kizhi Pogost, a settlement that features a summer 22-dome church, a winter nine-dome church, and a striking belfry. The oldest church on Kizhi dates from the 1300s.

On the whole, the water quality of Lake Onega is good, but anthropogenic impact is intense around Petrozadovsk and the city of Kondopoga; both are situated in bays on the northwestern side of the lake. Biologically treated domestic sewage and raw industrial wastewater from these urban sites are discharged into Lake Onega. This has led to strong blooms of blue-green algae in Petrozadovsk Bay during the summer. The bay receives 1.6 times more total phosphorus and five times more total nitrogen than Kondopoga Bay, which is a deeper and larger bay.

The biggest polluter in Kondopoga Bay is a major paper pulp mill, one of the most important paper producers in Russia, and still in action after more than 80 years of service. After having discharged untreated sewage into Kondopoga Bay for more than 40 years, its sewage is now being treated.

Lake Onega's water column layering helps combat nutrient pollution. Strong stratification in the water column prevents much mixing of water, and helps isolate the deeper, central part of the lake from waters of the polluted bays. At any rate, since the 1990s, a diminishing nutrient and waste load in the catchment area has resulted in a decrease in levels of nitrogen and phosphorus in both Petrozadovsk and Kondopoga bays.

Climate change in the region has shown trends of milder winters with an increased number of thaw days, and wetter springs with significant amounts of precipitation. Changes in air temperature, precipitation, and other meteorological qualities can cause changes in water balance, lake level, thermal characteristics, and ice events, as well as hydrochemical and hydrobiological regimes across the lake ecosystem. What this means in terms of the response of the Onega Lake environment will depend on the future intensity of regional climate change.

JESSICA HAAPKYLÄ

Further Reading

Bilaletdin, Ä., T. Frisk, V. Podeschin, H. Kaipainen, and N. Filatov. "A General Water Protection Plan of Lake Onega in Russia." *Water Resource Management* 25 (2011).

Ozerov, M. Y., A. J. Veselov, J. Lumme, and C. R. Primmer. "Genetic Structure of Freshwater Atlantic Salmon (*Salmo Salar* L.) Populations From the Lakes Onega and Ladoga of Northwest Russia and Implications for Conservation." *Conservation Genetics* 11 (2010).

Podsechin, Victor, Heikki Kaipainen, Nikolai Filatov, Ämer Bilaletdin, Tom Frisk, Arto Paananen, Arkady Terzhevik, and Heidi Vuoristo. "Development of Water Protection of Lake Onega." *The Finnish Environment* 36 (2009).

Sabylina, A. V., P. A. Lozovik, and M. B. Zobkov. "Water Chemistry of Lake Onega and its Tributaries. *Water Resources* 37 (2010).

Ontario, Lake

Category: Inland Aquatic Biomes.
Geographic Location: North America.

Summary: Smallest of the North American Great Lakes in surface area, and third-largest in total volume of water, Lake Ontario has suffered from human activity since the arrival of the earliest European settlers.

Étienne Brûlé, the young scout of Samuel de Champlain, was the first European to see Lake Ontario, in 1615, and a few months later, he guided Champlain to the lake. *Ontario* stems from a Huron word meaning *sparkling water.*

Lake Ontario is, by surface area, the smallest of the North American Great Lakes, and if the shrinking Aral Sea of central Asia is excluded, Lake Ontario is the 13th-largest lake in the world. The lake forms an international boundary between the United States and Canada. It is bounded on the north and southwest by the Canadian province of Ontario, and on the south, New York state forms the boundary with the United States. Lake Ontario is the only Great Lake that is not in part bordered by the state of Michigan.

Lake Ontario is the last in the Great Lakes chain; it serves as the outlet to the Atlantic Ocean for the waters of all the lakes. The lake lies 243 feet (74 meters) above sea level, and is 325 feet (99 meters) below Lake Erie. Water that begins its voyage in Lake Superior flows through Lake Erie, over Niagara Falls, and down into Lake Ontario, where it joins the Saint Lawrence River and finally mixes with the Atlantic Ocean.

Lake Ontario is 190 miles (310 kilometers) long and 53 miles across (85 kilometers) at its widest point. Its average depth is 283 feet (86 meters), and it reaches a maximum depth of 802 feet (244 meters). Lake Ontario has a water-retention time of six years, nearly three times that of Lake Erie. The shoreline of Lake Ontario is the shortest of all the Great Lakes, extending for only 730 miles (1,170 kilometers), including its islands. Lake Ontario holds a large proportion of cold bottom waters and is mesotrophic, meaning that its nutrient load is less than that of eutrophic Lake Erie, but greater than that of the mainly oligotrophic Lakes Superior, Michigan, and Huron.

Its watershed includes parts of Ontario, New York, and a small portion of Pennsylvania, as well

Releasing young Atlantic salmon in October 2011 as part of a joint effort by the U.S. Geological Service and the Mohawk Tribe to restore the salmon population in Lake Ontario. (U.S. Geological Service/Tony David)

as the incoming Niagara River, Niagara Falls, and numerous other lakes and streams that empty into the lake before all the waters discharge into the St. Lawrence River.

Major urban industrial centers such as Toronto and Hamilton, Ontario, are located on its western shore. The Canadian shoreline of Lake Ontario is densely populated and heavily urbanized, while its United States shoreline is, for the most part, sparsely populated and used mainly for farmland and resorts.

Lake Ontario has suffered the effects of exploitation longer than any of the other Great Lakes, first by lumber mills and commercial fisheries, and later by agriculture and industries. Industry

and agriculture have added nutrients to the shallower regions of the lake, resulting in making Lake Ontario a mesotrophic lake with more oligotrophic deep basins.

Because of its great depth, the lake as a whole never freezes in winter, but an ice sheet covering between 10 and 90 percent of the lake area typically develops, depending on the severity of the winter. When cold winds pass over the warmer water of the lake, they pick up moisture and precipitate it as lake-effect snow. Because most of the winds are from the northwest, the southern and southeastern shoreline of the lake is referred to as the snowbelt, and may receive up to 20 feet (600 centimeters) or more of snow.

Lake Ontario is mainly in the humid continental climate zone. This region has warm and sometimes hot summers with colder, longer winters, and ample snowfall. Average daily maximum and minimum temperatures in Toronto in July are, respectively: 80 and 64 degrees F (26 and 18 degrees C); and for January: 30 and 19 degrees F (minus 1 and minus 7 degrees C).

Wildlife

Lake Ontario has an extensive dune system stretching for 17 miles (27 kilometers) along the New York coastline. Dune systems here support more unique species of plants and animals than any other ecosystems in the Great Lakes region. These dunes are composed of sand that was deposited in the basin by the melting glaciers 10,000 years ago. This sand is transported by wind and water to the lake shoreline, where plants such as marram grass (*Ammophila breviligulata*) helps trap the sand and begin dune formation.

The dunes of the Great Lakes basin support more than 315 species of vascular plants, 32 species of mammals, 68 species of shorebirds and passerines, and numerous insects and fungi. Of note, naturalist Henry Cowles's studies of succession at the Indiana Dunes, along the eastern coast of Lake Michigan, provided the first truly scientific look at how succession works and how the environment affects plant-community structure.

Lake Ontario and its shores provide a stopover for many migratory bird species. Millions of songbirds will travel here from the tropics each spring and nest, then return to southern regions of the United States, Mexico, Central America, the Caribbean, and South America. Some of the bird species found here include: warbler, oriole, tanager, flycatcher, thrush, loon, grebe, goose, merganser, gull, tern, and petrel. Endangered or threatened species include: black tern, pied-billed grebe, least bittern, American bittern, and northern harrier. Large flocks of swallows, including bank swallows, are also known to migrate through the area.

The lake provides critical habitat and migration area for shorebirds, songbirds, and waterfowl in the form of inland dunes and wetlands with extensive barrier beaches backed by shrub and forested lands. There are some rare or exemplary ecological communities: silver maple-ash swamp, rich shrub fen, medium fen, red maple-hardwood swamp, red maple-tamarack peat swamp, maple-basswood forest, and deep emergent marsh.

Lake Ontario was once home to the land-locked Atlantic salmon (*Salmo salar*), but in less than a century after European settlement, a resource that seemed limitless had all but vanished from the lake. The species was revered by the indigenous peoples living near the lake, and was an important part of their diet. European settlers were already familiar with this fish, and not long after settlements were established, commercial and recreational fisheries for Atlantic salmon developed on the lake. Average sizes of 18 inches (46 centimeters) and 2–4 pounds (0.9–1.8 kilograms) were probably typical in the historical population of Lake Ontario, and old records suggest that fish up to almost 45 pounds (20 kilograms) were occasionally caught. Habitat loss due to lumber mills along the streams, overfishing, and pollution led to the extirpation of this important native species by 1898.

After a series of failed recovery efforts beginning in the early 20th century, a reintroduction program was initiated in 1985 by the New York State Department of Environmental Conservation. Current recovery efforts appear to be working, and now, more than a century after the last harvest of these fish, the Atlantic salmon appears to have returned to Lake Ontario. A 35-inch, 24.3-pound (89 centimeter, 11-kilogram) fish caught in 1989 is the modern Ontario record for the species.

Other fish that made up the original Lake Ontario food web were lake trout; lake whitefish; and bloater chub, a species once abundant in the deeper water of the lake, along with the lake sturgeon, which fed on the bottom with a specially adapted suctionlike mouth. This feeding niche is now occupied by a nonnative fish, the carp, which can tolerate more pollution and is more prolific. The kiyi, a subspecies of cisco, is now extinct.

As species were lost, others immigrated and took their places. Alewives and rainbow smelt from the Atlantic have replaced the similarly planktivorous lake herring. Smallmouth, pumpkinseed, and rock bass live around the islands and shallow shoals in the eastern portion of the lake, feeding on minnows, frogs, and insects. Even these natives have been replaced by carp and other invaders as pollution begins to change the environment.

Freshwater eels are important scavengers in Lake Ontario, and as a result of their feeding habits, they have picked up large amounts of toxins from point sources along the shorelines. These fish have a life cycle that is the reverse of that of the Atlantic salmon. They spawn in the Sargasso Sea in the Atlantic. After hatching, they migrate up the Gulf Stream to the St. Lawrence River and on into Lake Ontario. Some even make it into other Great Lakes. When it is time to spawn, they make their way back to the St Lawrence and out to the Atlantic.

Once harvested for food, these eels are now too contaminated to eat. In fact, about 500 beluga whales feeding on eels in the St. Lawrence estuary have become contaminated with polychlorinated biphenyls (PCBs) and Mirex, an insecticide banned since the 1970s. These toxins are still lingering in the lake sediments, which the eels disturb while feeding.

Human Impact

The Erie Canal was opened in 1825, and served as a major pathway for species from the Atlantic Ocean and Hudson River to the Great Lakes. The opening of the Welland Canal in 1829 provided a passage around Niagara Falls for other invaders. With the opening of the St. Lawrence Seaway in

1959, there was an increase in ship travel into the Great Lakes. Ships emptied ballast water containing animals such as the zebra and quagga mussels and the spiny waterflea into the lakes, resulting in serious damage to the Great Lakes ecosystems.

In 1988, zebra mussels were inadvertently introduced to Lake St. Clair, and they quickly spread throughout the Great Lakes region following flooding in the Mississippi River drainage in 1993. Since then, they have caused severe problems at power plants and municipal water supplies, clogging intake screens, pipes, and cooling systems. These filter feeders compete with native clams and mussels, and have nearly eliminated the native clam population in the Great Lakes ecosystem. Mussel tissues are heavily contaminated with PCBs, and these toxins, among others, are excreted in their fecal pellets, which form part of the lake-bottom sediment. Bottom-feeding animals then help pass the toxins up the food chain, eventually to the salmon and other top predators, including humans.

Other invaders in Lake Ontario include the spiny waterflea (a competitor and predator of native zooplankton), the fish hook waterflea, the round goby (which eats the eggs of native fishes), and the sea lamprey (a parasitic fish that preys on large fish such as salmon). Recently, aquatic ecologists have determined that the numbers of a tiny crustacean called *Diporeia* are decreasing. *Diporeia* formerly represented up to 70 percent of the bottom-dwelling invertebrates and was food for bottom-feeding fish such as smelt and slimy sculpins, which in turn were eaten by salmon and trout.

Climate change has already been documented in the Great Lakes region, in the form of lower water levels, warmer air and water temperatures, less winter ice cover, and more extreme storms. With a shorter ice cover season, there is potentially more evaporation. Lake Ontario's water helps to moderate the local climate, so there may be other consequences in terms of local weather.

With all the combined changes to the entire ecosystem, economies and communities will ultimately be affected. The health of the biome and its components will be further compromised: the plants, the birds, the fish, and the intangible benefits derived from the ecoregion, in general. There are limits to adaptation, and when wildlife cannot adapt, there is a loss of diversity and shifting of species.

LIANE COCHRAN-STAFIRA

Further Reading

Ashworth, William. *The Late Great Lakes: An Environmental History.* Detroit, MI: Wayne State University Press, 1987.

Dempsey, Dave. *On the Brink: The Great Lakes in the 21st Century.* East Lansing: Michigan State University Press, 2004.

Dennis, Jerry. *The Living Great Lakes: Searching for the Heart of the Inland Seas.* New York: Thomas Dunne Books, St. Martin's Press, 2003.

Grady, Wayne. *The Great Lakes: The Natural History of a Changing Region.* Vancouver, BC: Greystone Books, 2007.

Spring, Barbara. *The Dynamic Great Lakes.* Baltimore, MD: Independence Books, 2001.

Sproule-Jones, Mark. *Restoration of the Great Lakes: Promises, Practices, Performances.* East Lansing: Michigan State University Press, 2002.

Orange River

Category: Inland Aquatic Biomes.
Geographic Location: Southern Africa.
Summary: The most vital river in southern Africa, the Orange supports agriculture, tourism, and a wide variety of animal life.

The Orange River, the fourth-longest river in Africa, flows some 1,400 miles (2200 kilometers) across southern Africa in a westerly direction, through four countries—Botswana, Lesotho, Namibia, and South Africa. The Orange River starts in the Maloti Mountains in northern Lesotho, where it is known as the Senqu, then flows southwest through Lesotho. It meanders northwest and west through central South Africa, forming the southwestern boundary of Free State and part of the South Africa-Namibia border, then through the southern part of the Kalahari and Namib Deserts, before entering the South Atlantic Ocean at Oranjemund by the Alexander Bay.

In Lesotho, at the source, the river receives up to 79 inches (200 centimeters) of precipitation per annum; as evaporation rates increase, and rainfall decreases in a westerly direction, precipitation can be as little as 2 inches (5 centimeters) per year at the river's mouth. In dry years, water does not even reach the mouth. The total catchment area of the Orange River system is approximately 376,000 square miles (973,000 square kilometers), equal to three-quarters of the entire land area in South Africa.

Wildlife

The ecology of the Orange River does not have true estuarine communities, as the sea hardly enters the river and there are few salt tolerant plants. The Ramsar designation, indicating a wetland of international importance, is comprised of sand banks and channel bars covered with pioneer vegetation, a tidal basin, a narrow floodplain, pans, the river mouth, and a salt-marsh on the south bank of the river mouth.

The wetland vegetation includes wetland marshes, saltmarsh, island, and bank vegetation; these consist mainly of freshwater species. The predominant presence of freshwater species in the island and bank vegetation is the result of the present regulated flow through the Orange River mouth system. The saltmarsh on the southern bank of the Orange River mouth system is cut off from the rest of the system by the embankment of an access road to the mouth.

Invertebrates in the Orange River include the sand prawns or bivalves that would normally be expected to dwell in the mud flats of estuaries. There are predominately freshwater fish species in the river. As far as fish life is concerned, this system cannot be regarded as an ecologically important system. One endangered fish species, the largemouth yellowfish, is found only in the lower reaches of the Orange River.

The Orange River mouth is regarded as the sixth-most-important coastal wetland in southern Africa, in terms of the number of waterfowl it supports. The river mouth, mudflats, intrafluvial marshlands, islets near the mouth, and adjacent pans provide a sizable area of sheltered shallow water suitable for concentrations of wetland birds, which use these habitats for breeding purposes or as a stopover on migration routes. The bird population can be as high as 20,000 to 26,000 individuals.

Of the 57 wetland species recorded, 14 are listed as either rare or endangered in one or both of the South African and Namibian Red Data Books. At times, the area supports more than 1 percent of the world population of three species endemic (not found elsewhere) to southwestern Africa: the Cape cormorant, Damara tern and Hartlaub's gull. The site also supports 33 mammal species, which include the Cape clawless otter, as well as one of the tourist attractions, the manatee.

Human Impact

The river was named by Robert Jacob Jordan after the Dutch Royal House. Although the river does not pass through any major cities, it plays an important role in the South African economy by providing water for irrigation, as well as hydroelectric power. Shoals, falls, irregular flow, and a sandbar at its mouth limit navigation, but the river is used extensively for irrigation. It is also used for rafting, fishing, and other recreational and tourism activities.

The South African Orange River Project, which includes the Gariep and Vanderkloof dams, provides water for irrigation, hydroelectric power generation, and municipal water supplies. The Orange feeds the Gariep dam, which at 318 billion cubic feet (6 billion cubic meters), creates the country's largest reservoir. This dam, near Colesburg, is the main storage structure on the Orange River. Considering the generally dry climatic conditions characterizing the sub-continent, it is essential for this freshwater resource to be utilized to the greatest benefit of the region and its peoples.

The Orange River Project increased the value of South African agricultural production: it has provided for the establishment of a large number of irrigated farms, and stimulated the production of meat, wool, milk, citrus, cotton, wheat, raisins, beans, and peas. The project has also promoted economic activity and development in the areas directly involved, counteracted the migration of the rural population to the cities by creating stable farming communities, created recreational facilities in the center of the interior and promoted tourism, leveled off moderate flood peaks in the course of the river, and, in the process, safeguarded riparian communities and irrigation schemes downstream.

South Africa is particularly vulnerable to climate change because, among other things, a large proportion of the population has low resilience to extreme climate events (poverty; high disease burden; inadequate housing infrastructure and location). Large parts of South Africa already have low and variable rainfall, and a significant proportion of surface water resources are already fully allocated.

Some effects of climate change may already be occurring, manifested as changes in rainfall (droughts and floods), temperature extremes, and cholera outbreaks. These have each been associated with extreme weather events, especially in poor, high density settlements. Agriculture and fisheries are important for food security and local livelihoods; those in poverty will be the most impacted.

JESSE KILLINGSWORTH

Further Reading:

Jacobs, Nancy J. *Environment, Power, and Injustice: A South African History (Studies in Environment and History)*. Cambridge, MA: Cambridge University Press, 2003.

Orange-Senqu River Commission (ORASECOM). "Four Nations, One River." http://www.orange senqurak.org/.

SA Places. "Orange River—South Africa." http://www.places.co.za/html/orangeproject.html.

Wetlands in South Africa. "Wetland: Orange River Mouth." http://www.ewisa.co.za/misc/wetlands/defaultorangemouth.htm.

Orinoco River

Category: Inland Aquatic Biomes.
Geographic Location: South America.
Summary: The third-longest river in South America, the Orinoco has been damaged by industrialization and exploitation of natural resources.

The Orinoco River is the third-longest river in South America, extending about 1,370 miles (2,200 kilometers) from its headwaters at the Cerro Delgado Chalbaud at the Venezuela border with Brazil, to its mouth on the North Atlantic Ocean. Throughout most of its course, the river flows through Venezuela and through a wide variety of landscapes, including the impenetrable rainforest, the vast savanna also known as the llanos, marshlands, and floodplains. The navigable capac-

Canoes by the side of the Orinoco River in Venezuela. The river has traditionally been so useful for transportation that its name means "a place to paddle." (Thinkstock)

ity of the Orinoco River makes it a transportation pathway that connects the different regions of the basin. In fact, the name *Orinoco* comes from the Guarauno dialect and means "a place to paddle," referring to its navigable features.

The Orinoco River basin, also known as Orinoquia, is a diverse ecosystem with a total area of 416,990 square miles (1.1 million square kilometers), covering about one-fourth of Colombia and four-fifths of Venezuela. The Orinoquia boasts rich ecological and cultural diversity, being one of the most vital reservoirs of South America, and home of 26 different Amerindian groups. The basin can be divided into four regions: the Andean mountains, the plains and savannas also known as the llanos, the highlands of Guayana, and the Delta region. The Orinoco's main tributaries flow mainly from the south and west—from the Guyana highlands in the southeast and from the Andes Mountains in the northwest. These tributaries include: Rios Caura, Paraqua, Caroni, and Guaviare from the south; Rios Arauca, Apure, and Meta from the west; and Rio Cojedes from the northwest.

Climate and Vegetation

The climate of the Orinoco River Basin is tropical, with small variations depending on the geographical features of each region. According to the precipitation range, there are two distinct seasons: a dry season from November to April, and a wet season from April to November. Rainfall, air temperature, and altitudinal zonation vary throughout the basin, thus defining diverse forms of landscape and vegetation.

The Andean Orinoquia is formed by irregular and uneven slopes, with moraines and terraces in the mountain front and canyons in the piedmont. The annual temperature varies from 35 to 54 degrees F (1.5 to 12 degrees C), and the annual precipitation ranges from 59 to 79 inches (150 to 200 centimeters). Three vegetational layers characterize the region: the paramo community, with herbaceous vegetation thriving above 13,100 feet (4,000 meters); the Andean bush, found at 9,840–13,100 feet (3,000–4,000 meters), dominated by shrubs; and Andean forest—below 9,840 feet (3,000 meters), with arborescent vegetation.

The Orinoco plains comprise a heavy, clay sediment layer from the Quaternary. This region suffers from poor drainage and therefore is susceptible to flooding during the wet season. The climate is continental, with mean temperature of 75–82 degrees F (24–28 degrees C). Every four to five years, this region experiences major floods.

According to the altitudinal zonation, the part of the biome is characterized by two well-defined landscapes: the high llanos, extending from the piedmont to 330 feet (100 meters); and the low llanos, below 330 feet (100 meters). There are four vegetation types: dry savannas, wet savannas, forests, and swamps. The boundary between forests and savannas is dynamic, due to the effect of fires, fluctuations of the hydrological regime, and human activities.

The Guayana region lies on the Precambrian Guayana shield, and most of the soils are acidic. The two well-defined physiographic units are the plains and peneplains, found below 1,640 feet (500 meters), with an annual mean temperature above 77 degrees F (25 degrees C) and a peak during the dry season of 93–104 degrees F (34–40 degrees C); and the hills, piedmont, and highlands, set at 1,640–9,840 feet (500–3,000 meters), with a mean temperature of 64–75 degrees F (18–24 degrees C) and mean annual precipitation of 59–79 inches (150–200 centimeters). The vegetation varies from dense evergreen and deciduous forests, to savannas with gramineous and wide-leaved herbaceous plants.

The Delta region is a vast area formed by Quaternary sediments. Marshlands, both marine and fluvial, dominate the landscape, with a slope less than 2 percent. The mean annual precipitation is 75 inches (190 centimeters), and the annual mean relative humidity is 80 percent. The prelittoral, intermediate, and littoral zones are the three geographical zones of the Delta region. The prelittoral zone lies inland and is the highest portion of the delta. At this point, the Orinoco River merges with the Delta. The river diversifies in multiple channels, with brown water bordered by sand banks 10–13 feet (3–4 meters) high.

Vegetation was once gallery forests and palm trees, but it has been completely altered by humans.

The intermediate zone is subject to seasonal flooding and twice-daily tides. The littoral zone is characterized by mangrove trees and swamp forests, and is permanently flooded due to heavy rainfall and tidal fluctuation. The intermediate and littoral zones comprise the Delta's wetlands.

Biodiversity

The Orinoco River basin has very high biodiversity, and is one of the most prolific regions in the world in terms of birds. The basin boasts more than 1,000 avian species. In the Andes area surrounding the river are: golden-headed and crested quetzal, emerald toucanet, the endemic (not found elsewhere) rose-headed parakeet, Andean condor, and more than 35 species of hummingbirds. There are several dozen varieties of tanagers, several species of hawks, fruiteaters, woodcreepers, and woodpeckers.

In the llanos or plains area there are: scarlet macaw, yellow-crowned parrot, three species of storks including jabiru, seven species of ibis including scarlet, more than 12 species of heron and egret types, more than 12 different shorebirds, wattled jacana, gray-necked wood-rail, three species of whistling-ducks that can be seen sometimes in flocks of several thousands, plus muscovy and comb duck, orinoco goose, horned screamer, great black and white-tailed hawks, ornate hawk-eagle, snail kite, king vulture, aplomado falcon, owls, the rare and primitive hoatzin, the near-endemic yellow-knobbed curassow, the pale-headed jacamar, nightjars, nighthawks, more than 40 flycatchers, manakins, antbirds, puffbirds and many more.

In the Guyana highlands area, there have been more than 500 bird species identified, including: red-billed toucan, black-necked and green aracari, Guianan toucanet, red and green macaw, caica, red-fan parrot, tepui parrotlet, hawks, eagles, falcons, hummingbirds, the guiana cock-of-the-rock, pompadour cotinga, spangled cotinga, the amazing capuchin bird, antbirds, woodpeckers, puffbirds, jacamars, tanagers, and flutist wrens.

Among the large diversity of fish are the piranha; the electric eel; and the laulao, a catfish that can weigh more than 200 pounds (91 kilograms).

Other fish species commonly found in various part of the river include: rubi tetra, cardinal tetra, checkerboard cichlid, twig catfish, leaf fish, peacock bass, pike cichlids, red hook, oscar, spotted leporinus, and basketmouth cichlid. These are just a few of the more than 1,000 fish species in the Orinoco. Also in the river are found the Orinoco crocodile (endangered), pink and grey river dolphins (threatened), giant river otters, the giant anaconda, and the arrau turtle.

Most mammals of the biome dwell in the gallery forests of the llanos region. Common dwellers of the forests are deer, rabbits, tapirs (which are an endangered species), jaguars (the third-largest feline species), armadillos, capybaras (the largest living rodent species), shrews, bats, and opossums. The spectacled bear, also known as the Andean bear, is the last remaining short-faced bear species. Its status is threatened.

Human Impact

Most of the indigenous groups of Venezuela dwell within the Orinoco River basin. Some of the major aboriginal settlements include the Maquiritare in the rainforest area; the Warao in the delta region; and the Guaica, Guahibo, Yaruro, and Yanomami in the llanos. Before the 1950s, the settlements were limited to a few villages, but after oil and gas strikes at el Tigre and Barinas, slow but steadfast industrialization and urbanization of the region began. Several towns grew into sizable cities of 10,000 or more inhabitants. Intensive agricultural practices were adopted along the river valleys and Andean piedmont.

The Orinoco River basin is rich in natural resources. The vast plains of the llanos region are used to raised livestock, in particular cattle. Plantations of cotton, rice, and sugarcane are also harvested on a large scale. The Guayana region is rich in mineral deposits, including iron, nickel, manganese, bauxite, vanadium, diamonds, chromium, and gold. The largest exploitations, however, are petroleum and natural gas in the Orinoco Llanos and Orinoco Delta regions, followed by industrial development around Guayana City, where steel, aluminum, and paper factories have been established.

Industrial success and hydrocarbon exploitation are possible due to the construction of the Macagua and Guri dams, which provide the energy needed. Industrialization also brought major road constructions to connect the newly formed cities with the rest of the country. A bridge across the Orinoco at Bolivar City connected the llanos with the Guayana region in 1967. Around that time, a bridge was built across the Caroni River to connect the industrial city of Puerto Ordaz with the San Felix port, creating the new city of Guayana. The latest one was later connected to Caracas by a major highway.

The predictions for impact of climate change for Venezuela in general, and the Orinoco basin in particular, are increases in temperatures, decreases in precipitation, and increases in dry land masses, prone to desertification. Floods may also severely affect agriculture, much as they already have in the past. In Venezuela, nearly 95 percent of agricultural land is rain-fed, making this sector vulnerable to any changes in climate.

Fortunately, much of the Orinoco River basin is still forested. Misión Árbol is an initiative the government created in 2006, aimed at increasing reforestation activities in the country to help increase carbon-capturing capacity of the land.

Rocio R. Duchesne

Further Reading

Echezuria H., J. Cordova, M. Gonzalez, V. Gonzalez, J. Mendez, and C. Yanes. "Assessment of Environmental Changes in the Orinoco River Delta." *Regional Environmental Change* 3 (2002).

Gasson, R. "Orinoquia: The Archaeology of the Orinoco River Basin." *Journal of World Prehistory* 6, no. 13 (2002).

Hilty, Stephen L. *Birds of Venezuela.* Princeton, NJ: Princeton University Press, 2002.

Lewis, W. Jr., F. Weibezahn, J. Sounders III, and S. Hamilton. "The Orinoco River as an Ecological System." *Interciencia* 15, no. 6 (1990).

World Bank. "Venezuela: Country Note on Climate Change Aspects in Agriculture." http://siteresources.worldbank.org/INTLAC/Resources/Climate_VenezuelaWeb.pdf.

Orinoco Wetlands

Category: Inland Aquatic Biomes.
Geographic Location: South America.
Summary: This still-pristine ecoregion provides habitats for several endangered animal and plant species, but industrial and agricultural development plans threaten the ecological balance.

The Orinoco delta region in northeastern Venezuela extends from near the Araya and Paria Peninsula southward to the Amacuro Delta, the world's seventh-largest river delta. There, vast floodplains were created by fluvial-marine sediments. The delta's distributaries, called *caños*, spread like a fan over an area of about 2,300 square miles (6,000 square kilometers), branching out from the main distributary, called the Rio Grande, and the second major distributary, Caño Manamo. The delta region of the Orinoco River has been declared an internationally significant wetland.

In this ecoregion, the climate is tropical. Precipitation varies throughout the area from 39 to 79 inches (100 to 200 centimeters) annually, with a wet season from April or May until December. The United Nations Educational, Scientific, and Cultural Organization (UNESCO) added this ecoregion in 2009 to the World Network of Biosphere Reserves (WNBR). A reference map categorizes the forest subregions of the Orinoco delta as inundated woodland savannas (with palms) of the Upper Delta, wetlands of the Middle Delta, and swamp wetlands of the Lower Delta.

Large permanent wetlands such as swamps, marshlands, tuberas (peatlands), swampy plains, and mangroves occur in patterns with seasonally flooded freshwater swamp forests. While the peatlands are dominated by water extensions with rich organic soils, supporting a herbaceous vegetation, moriche palms, and forests, the wetlands are subdivided by networks of fluvial and tidal channels. Westward, these flooded grasslands become drier, forming the llanos, and eventually retain stands of evergreen broadleaf trees. Common species of the riparian forests include bloodwood, peramán, Guiana chestnut, chaguaramo palm, coral trees, and jobo. Salt marshes are formed close to the coastal streams in the east. The mangroves in this region represent an important sink for atmospheric carbon dioxide gas.

Biodiversity and Conservation

The Orinoco basin, together with the Gulf of Paria, comprise the world's largest manatee refuge. Nonetheless, the Venezuelan government had to order a resolution to ban manatee hunting to prevent extinction of these marine mammals. Further threatened species are the Brazilian tapir, giant anteater, Orinoco crocodile, Amazon River dolphins (grey and pink), jaguar, bush dog, Orinoco goose, and harpy eagle.

Other wildlife inhabiting the delta include: the river otter, crab-eating fox, howler monkey, capuchin monkey, parrots, toucans, scarlet macaws, rufous crab-awk and the muscovy duck. There are reptiles such as the caiman, iguana, turtles and sea turtles, lizards, and many snake species. Fish include the lau-lau, catfish, piranha, tarpon, morocoto, coporo, sardine, rays, and a variety of coral reef fish. There is also an abundance of crab, shrimp, and other crustaceans, and mollusks.

There are many species of birds in the delta region, mainly shorebirds, many of these migratory. The delta provides habitat for nesting, breeding, and stopovers. The endemic scarlet ibis is found here, as is the black skimmer, gull-billed tern, short-billed dowitcher, lesser and greater yellowlegs, black-bellied whistling duck, and tricolored egret.

Conservation and Threats

At the beginning of the 1990s, several protected nature and indigenous areas were established, including the United Nations Environment Programme (UNEP) Delta del Orinoco Biosphere Reserve, Delta del Orinoco National Park, Turuïpano National Park, and Mariusa National Park, protecting a combined area of approximately 5,965 square miles (15,500 square kilometers). Endangered species (as listed in the Convention on International Trade in Endangered Species of Wild Fauna and Flora—CITES—Appendix I) include manatees and giant otters.

A brown caiman in the Orinoco River. The Orinoco delta region provides habitat for such threatened species as the Orinoco crocodile and Amazon River dolphins, and is the largest manatee refuge in the world. (Thinkstock)

Although some nature-reserve actions are already enforced and the Orinoco Delta is still largely untouched and intact, human activities already influence the ecosystems in this region. Current threats to the area include water construction projects such as water diversion and damming, and increasing demands of a growing human population. Upstream, in the northwestern delta, the constructions of the Volcán dam and the Raul Leoni dam (on Caroní River) are two examples. Dams provide water reservoirs and increase areas for farming and cattle ranching, but interrupt essential seasonal flooding.

Reduced water levels in the upper delta caused a chain of ecological consequences. The region became tidal, resulting in water levels that rise and fall by 3–7 feet (1–2 meters) daily. As a result, the dramatically increased salinity affects the flora and fauna.

The potentially greatest threat, however, is oil extraction—mainly by the Venezuela Oil Corporation (Corporación Venezolana de Petróleo), but also by transnational petroleum companies that explore and exploit abundant resources in the Gulf of Paria. Major sources of hydrocarbons include Lago de Asfalto de Guanaco and Caño La Brea in the delta of the San Juan River.

The floodplains of large rivers are not only among the most productive ecosystems, but also are most vulnerable to changes in water quality. If economic development—including agriculture, mining, oil exploration, and navigation of the rivers—accelerates, environmental risks also accumulate, such as saltwater intrusion in the delta, degradation of aquatic and floodplain habitat, reduction of ecosystem capacity, and interference with groundwater systems. Another consideration for this area is future impact to the coast due to climate change. Studies are already indicating rising sea levels, increased frequency and severity of storm surges, and changes in precipitation patterns.

Venezuela's government plans the construction of large navigation facilities. The Orinoco River system is still relatively pristine, so it is possible to avoid fatal disruption of the ecology of the river system and, thus, long-term costs of economic development. One successful application of new approaches is the model of a privately owned conservation network for the Orinoco river system. This is adapted from the llanos area. Currently, about 10 percent of the llanos ecosystem is conserved by families and corporations that mix ecotourism with cattle ranching in sustainable management schemes. Thus, gallery forests, rivers, and streams will stay protected as biological corridors, and the land can be maintained in a natural, balanced state even while facing economic growth.

Manja Leyk

Further Reading

Dinerstein, E., et al. *A Conservation Assessment of the Terrestrial Ecoregions of Latin America and the Caribbean.* Washington, DC: World Wildlife Fund and The World Bank, 1995.

Lasso, Carlos A., Leeanne E. Alonso, Ana Liz Flores, and Greg Love, eds. "Rapid Assessment of the Biodiversity and Social Aspects of the Aquatic Ecosystems of the Orinoco Delta and the Gulf of Paria, Venezuela." *The Rapid Assessment Program (RAP) Bulletin of Biological Assessment* 37 (2004).

Seeliger, Ulrich and Björn Kjerfve, eds. *Ecological Studies 144: Coastal Marine Ecosystems of Latin America.* New York: Springer, 2001.

Westlake, D. F., J. Kvet, and A. Szczepanski, eds. *The Production Ecology of Wetlands: The IBP Synthesis.* Cambridge, United Kingdom: Cambridge University Press, 1999.

Ozark Mountain Forests

Category: Forest Biomes.
Geographic Location: North America.
Summary: This distinct forest in northeastern Arkansas, dominated by oak and hickory, has recently been threatened by an oak-decline event.

The Ozark Mountain Forests occur in one of the oldest mountain ranges in the United States, and exist mainly in the Ouachita and Boston Mountains ranges. This is a temperate and broadleaf mixed forests ecoregion. The forest is dominated by oak-hickory stands, and has a rich history of European timber harvesting and regrowth. Historic forest structure and species composition have been altered, and some recent oak-decline events have been linked to forest management practices. This hardwood forest stretches from the very southern part of Missouri through northern Arkansas and just past the border into Oklahoma, in the central United States.

The current Ozark Mountain Forests biome composition is dominated by mixed oak-hickory-shortleaf pine. Dominant species are red oak (*Quercus rubra*), white oak (*Q. alba*), and hickory (*Carya* spp., especially *Carya texana*). Disturbed sites, shallow soils, and south- and west-facing slopes are dominated by shortleaf pine (*Pinus echinata*) and eastern red cedar (*Juniperus virginiana*).

During the Pleistocene era, spanning 2.6 million to 11,700 years B.C.E., the Ozark Mountain forests acted as refugia for lowland species. These forests represent some of the oldest in the country: There are thousands of acres of stunted, old growth oaks on sites in western Arkansas's Ouachita National Forest that have never been cut. The topography here is rugged valley and ridge; peak elevations rise to 2,600 feet (780 meters), with valley descents ranging from 40 to 1,480 feet (150 to 450 meters).

Mean monthly temperatures range from 37 to 81 degrees F (3 to 27 degrees C), with an annual mean of 61 degrees F (16 degrees C). Most precipitation occurs during spring and fall, totaling approximately 49 inches (124 centimeters). The growing season consists of 180 to 200 days. The karst landscape consists primarily of limestone, shale, and sandstone, with many caves and valleys, and the mountainous topography of the Ozark Plateau results in highly variable soils. Soils are typically acidic and rocky, with low organic-matter content.

Wildlife

Only about 3 percent of the original forests remain as intact habitat. However, there is still a fair amount of biodiversity in the mountain forests; there are more than 200 endemic (not found elsewhere) species of plants and animals found here. Mammals that inhabit the ecoregion include: white-tailed deer, raccoon, red and gray fox, coyote, opossum, otter, mink, eastern chipmunk, beaver, bobcat, short-tailed shrew, squirrel, rabbit, and other small mammals. Black bear and mountain lion can still be found here, but their numbers are limited. Elk and plains bison are now gone from the region.

Wild turkey, quail, pheasant, the great horned owl, and the bald eagle inhabit these forests. Other common bird species here are: black capped chickadee, blue jay, northern cardinal, Carolina wren, cedar waxwing, downy woodpecker, American goldfinch, house wren, northern flicker, red-tailed hawk, northern mockingbird, white-breasted nuthatch, red-eyed vireo, red-bellied woodpecker, indigo bunting, junco, tufted titmouse, and blue grosbeak.

Reptiles and amphibians are abundant in the Ozark forests. Some of these include: ringed salamander, long-tailed salamander, dark-sided salamander, four-toed salamander, Ozark zigzag salamander, Ozark hellbender, wood frog, and various snakes and turtles.

In the streams, creeks, and rivers that flow through the region, there are also many freshwater invertebrates and fish species, including: chert pebblesnail, black sandshell, Curtis's pearly mussel, Neosho mucket, salamander mussel, scales hell, freshwater shrimp, belted crayfish, big creek crayfish, coldwater crayfish, blue-stripe darter, blunt-face shiner, channel darter, eastern slim minnow, flathead chub, least darter, Neosho mad tom, Niangua darter, Ozark chub, paddlefish, pallid sturgeon, and brook lamprey.

Human Impact

Historic evidence shows that open oak savanna dominated the Ozark Mountain Forests biome before European settlement. This was partly as a result of an indigenous fire regime that annually burned hundreds of acres (hectares) and decreased undergrowth and tree densities by decreasing tree recruitment and establishment. Forest species composition thus historically favored shade-intolerant, fire-tolerant canopy species. Fire was an important feature of these landscapes before increased fire-prevention efforts around 1920. Increased European settlement, beginning around 1830, resulted in increased clearing for agriculture and grazing.

The Ozark National Forest was established in 1907 by President Theodore Roosevelt and the head of the U.S. Forest Service, Gifford Pinchot. This first national forest encompassed 55,000 square miles (142,500 square kilometers), but the new status of the forest did not much change the management until mandates were enacted in 1920 to combat large forest fires. Decreased forest fires meant establishment of closed-canopy forest stands, in contrast to historic open savanna ecosystems.

Further altering the ecosystem was the practice of industrial clear-cut timber harvesting, which

A white-tailed deer in the Ozark Mountains. While as little as 3 percent of the region's original forests are intact, over 200 endemic (found nowhere else) species of plants and animals live there. (Thinkstock)

boomed in northern Arkansas in the 1920s and 1930s, and resulted in even-aged stands of white, red, and black oak (*Quercus*); hickory (*Carya*); and shortleaf pine (*Pinus echinata*) in the Ozark Mountains range. Oak and shortleaf pines were the most important species in this landscape before European settlement, but changes in species and forest structure as a result of fire suppression beginning in the 1920s can be noted in changes in recent forest assessments in understory composition.

The Ozark chinquapin, recognized as a species in 1923, is an indigenous tree species in the Ozark Mountains and is currently being considered for endangered-species status. Ozark chinquapin (*Castanea ozarkensis*) and black oak (*Quercus velutina*) decreased in the forest composition from 1934 to 2002. Chestnut blight (*Cryphonectria parasitica*) is likely responsible for the decrease in chinquapin oak, but the decrease in black oak is less well understood. Also noted in the Ozark Mountain Forests was an oak-decline event in the early 21st century in the Boston Mountains range.

Ecosystem benefits are in part viewed as the economic, recreational, and social values of nature; thus, the ecosystem services provided by the Ozark Mountain Forests biome include water quality maintenance, timber-important species (predominantly red oaks), and potential carbon sequestration. Many of the communities of southern and middle Arkansas rely on the water resources located in the this region, which the forests in part help to purify for public use. Recreation opportunities in the Ozark Mountains include the Ozark Highlands Trail, which was constructed and maintained by more than 3,000 volunteers and is the longest hiking trail in the forest. The trail extends 168 miles (270 kilometers) from Buffalo National River to Lake Fort Smith State Park to the south.

The fragmentation of forest lands due to changing forest ownership and physical degradation are current threats to ecosystem health and functionality. The individual importance of tree species importance in the Ozark Mountains has related heavily to timber and, to a lesser degree, recreation. Oak-pine-hickory dynamics formerly represented the dominant forest species, with some decrease in the value of timber lands due to poor silvicultural management. The even-aged characteristics of the forest relate to the re-establishment of forests following agriculture, clear-cutting, and fire suppression. The U.S. Forest Service found that 72 percent of the hardwood forests existing in the Ozark Mountains regions currently are between 40 and 90 years old, with only 21 percent over 90 years old.

ALEXIS S. REED

Further Reading

Chapman, R. A., E. Heitzman, and M. G. Shelton. "Long-Term Changes in Forest Structure and Species Composition of an Upland Oak Forest in Arkansas." *Forest Ecology and Management* 236, no. 1 (2006).

Hunter, C. G. *Trees, Shrubs, and Vines of Arkansas.* Little Rock, AR: Ozark Society Foundation, 1995.

Ricketts, T. H., E. Dinerstein, D. M. Olson, et al. *Terrestrial Ecoregions of North America: A Conservation Assessment.* Washington, DC: Island Press, 1999.

Starkey, Dale A., et al. *Forest Health Evaluation of Oak Mortality and Decline on the Ozark National Forest, 1999.* Pineville, LA: United States Forest Service, 2000.

Pacific Coastal Forests, Northern

Category: Forest Biomes.
Geographic Location: Western region of North America.

Summary: Over the past 100 years, one of the world's largest temperate rainforests and its old growth have been widely lost because of human economic activities.

The Pacific Northern Coastal Forest biome forms a band of vegetation from Alaska to northern California. The far northern parts include the Tongass National Forest in southeastern Alaska and also cover south-central Alaska as well as British Columbia in Canada, and areas of coastal Washington and Oregon. Despite wide over-logging, all these regions still make for major conservation hot spots.

The existence of this biome is linked with the coastal mountain range topography of the Pacific Coast Mountains—which include, running northwest to southeast, the Kenai Mountains, the Chugach Range, the St. Elias Mountains, the Coast Mountains, and the Cascade Range—as well as with the climate regime of the northwestern Pacific Ocean. Likely, the glaciation period and its succession and climate shaped most of this unique forest biome. The vegetation is defined by its huge rainfall, and it classifies broadly as a temperate rainforest. Because of the high amount of rain and subsequent lack of frequent fires, trees can get very old and form unique old-growth vegetation.

Sitka spruce, yellow and western red cedar, mountain and western hemlock feature as the key conifers here—the dominant flora community. Broadleaf trees appear in the form of birch, alders, and cottonwood; these deciduous types are mainly restricted to riverine zones and damp bottomlands.

Salmon also make for a distinct habitat feature, with the biome's thousands of streams connecting to the Pacific making an exceptionally good base for their anadromous life cycle. Salmon have run here for thousands, and likely millions, of years, largely undisturbed. This formerly super-abundant food source allowed human habitations to take root and grow in the region, but salmon now is widely endangered throughout. Large predators, including a variety of bears and the bald eagle, are at least partly dependent upon salmon in their diet.

View of part of Oregon's old-growth Pacific Coastal Forest, which is a breeding ground for the endangered marbled murrelet seabird and habitat for the spotted owl. (U.S. Fish and Wildlife Service/David Patte)

Forestry Impacts

Forestry in this region became a dominant industry during the past 150 years, with practices such as clearcutting becoming a threat to many species of reptiles, amphibians, mammals, and birds. For example, while the now-endangered marbled murrelet seabird species still breeds in this habitat, its sharp decline has been linked to the loss of more than 80 percent of old-growth forest in the region due to commercial logging operations. Lawsuits about the listing and protection of marbled murrelets through the U. S. Endangered Species Act and various civil disobedience campaigns over several decades speak to the recognized unique value of this habitat.

In addition to the terrestrial impacts, this ecosystem is also inherently linked to the coastal marine environment. Many nutrients from fertilizer and pesticide use have saturated some streams and complicated their outflow absorption into the aquatic biomes. Rampant construction of dams has altered sedimentation patterns and decimated salmon runs, increasing the human impacts on native species. Aquaculture, or farming of salmon and other fishes, is yet another factor in an environmental scene that has changed permanently, and thus challenges those who would conserve it to devise more adaptive stewardship techniques,

as well as cooperation with fisheries, the timber industry, and other constituent groups.

Real estate development, increasing tourism activities, new invasive species, and altered marine chemistry each deliver new inputs and potentially increasing stress to this rich habitat area. Various entities—state and national governments, citizens, native population, and scientific—continue to address the need for protected areas, species support, and changes in public policy that will help sustain this biome.

Global warming effects on the Northern Pacific Coastal Forests biome are difficult to predict. Some research has shown, for example, that past periods of lower temperatures and lower rainfall have actually led to the establishment and expansion of various tree types on some coastal mountain ranges in this region. The future effect of a higher temperature/higher rainfall regime is unknown. Also unclear are how pervasive will be the inland influence of altered patterns and interaction among such offshore factors as the North Pacific West Wind Drift current, the California Current, coastal upwellings, and many impacts from Asia.

Falk Huettmann

Further Reading

Durbin, Kathie. *Tree Huggers: Victory, Defeat & Renewal in the Northwest Ancient Forest Campaign.* Vancouver, Canada: Mountaineers Books, 1998.

Huettmann, Falk. "From Europe to North America Into the World and Atmosphere: A Short Review of Global Footprints and Their Impacts and Predictions." *Environmentalist* 10, no. 7 (2011).

Lackey, Robert T., Denise H. Lach, and Sally L. Duncan, eds. *Salmon 2100: The Future of Wild Pacific Salmon.* Bethesda, MD: American Fisheries Society, 2006.

Ruth, Maria Mudd. *Rare Bird: Pursuing the Mystery of the Marbled Murrelet.* Emmaus, PA: Rodale Books, 2005.

Wolf, Edward C., Andrew P. Mitchell, and Peter K. Schoonmaker. *The Rain Forests of Home: An Atlas of People and Place.* Portland, OR: Ecotrust Publications, 1995.

Pacific Ocean, North

Category: Marine and Oceanic Biomes.
Geographic Location: Northern Hemisphere.
Summary: The North Pacific Ocean provides an assemblage of some of the most productive and diverse marine ecosystems in the world.

The Pacific Ocean is the world's largest, coldest, and deepest ocean, covering approximately one-third of the Earth's total surface. The Pacific Ocean is commonly divided into North and South Pacific segments. The North Pacific Ocean is a major world source of marine organisms. Harvest of these organisms plays an important role in the economies and international relations of bordering countries—and of the planetary food web. The region itself extends from the near-tropical waters off southern Japan to the Arctic waters of the Bering Sea.

The North Pacific Ocean is characterized by strong latitudinal gradients in surface temperature and salinity (high in the south and low in the north), multiple water masses, major oceanic gyres, mesoscale eddies of up to 124 miles (200 kilometers) across, and a complex bathymetry of deep trenches and remote seamounts. Climate-system variability around the world is intimately linked to the natural variability in this oceanic region.

High biological production is typical in the area, and it is responsible for major commercial fisheries of tuna and Pacific salmon. The North Pacific Ocean biome is also an important sink region for atmospheric carbon dioxide (CO_2), and therefore plays an important role in the ultimate fate of CO_2 on Earth.

Seven physically complex subregions are identified in the North Pacific Ocean: Yellow Sea and East China Sea, Okhotsk Sea, Oyashio Current System, Kuroshio Current System, Bering Sea, Alaska Current System, and California Current System. These seven subregions are part of the main gyre, the North Pacific Gyre. This gyre occupies approximately 7.7 million square miles (20 million square kilometers), with a clockwise circular pattern. The Aleutian Island chain constricts the main gyre, causing recirculation with two subgyres: the Western Subarctic and Alaskan gyres. These gyres and seas have distinct characteristics, often supporting different species and patterns of production.

Yellow Sea and East China Sea

The Yellow Sea is a shallow marginal sea with a surface area of 146,719 square miles (380,000 square kilometers) and an average depth of 144 feet (44 meters). It is broadly connected with the East China Sea to the south, and contains a semienclosed gulf in the north, the Bohai Sea. The People's Republic of China, the Democratic People's Republic of Korea, and the Republic of Korea are populations located on the Yellow Sea, making it one of the most densely populated regions in the world. Three main currents are found in the Yellow Sea: Kuroshio, Tsushima, and the Yellow Sea Warm Current (YSWC).

Due to its jagged coastline and many islands scattered around the shallow waters, the Yellow Sea has diverse marine habitats. Intertidal flats are the most significant type of coastal habitat, but mudflats, salt marshes, sandflats with gravel beaches, sand dunes, eelgrass beds, and mixed flats are also found. These habitats provide feeding, wintering, and summering grounds for migratory birds, and support important food resources and ecological niches for many species. Approximately 1,600 species are reported from marine and coastal habitats in the Korean part of the Yellow Sea, including 400 phytoplankton, 300 marine macroalgae, 50 halophytes (salt-tolerant species), 500 marine invertebrates, and some 389 vertebrate species. Among them, 166 zooplankton and 276 fish have been reported as resident species in the Yellow Sea. Additionally, 100 commercial species have been identified in the region, comprising demersal fish (66 percent), pelagic fish (18 percent), cephalopods (7 percent), and crustaceans (7 percent).

Sea of Okhotsk

The Sea of Okhotsk is a semienclosed marginal sea bounded by Russia to the north and west and Japan to the south, with an area of 590,003 square miles (1,528,100 square kilometers), similar in magnitude to the Bering Sea. The bottom topography is rugged, featuring the deep Kuril Basin. The shelf zone occupies almost 40 percent of the total area, and is somewhat isolated from direct water exchange with the open ocean by the Kamchatka Peninsula. Seasonal changes are distinctive, and ice can be found from November (rarely, in October) to June. In March, the Sea of Okhotsk is mostly covered by ice, except near Kamchatka and the Kuril Islands.

Oyashio and Kuroshio

The Oyashio is the western boundary current of the sub-Arctic North Pacific, and it is characterized by low temperature, low salinity, and high nutrient concentrations. The eastward-flowing Oyashio forms the Subarctic Front (Oyashio Front), which is a distinctive temperature front. The Oyashio Front mixes with the Kuroshio Extension Front, where cold and warm waters mix and many mesoscale features are formed. This region is called the Kuroshio-Oyashio-Transition Zone (KOTZ), or mixed water region. The Oyashio brings nutrient-rich water into the western North Pacific, resulting in high productivity in the confluence zone of the two currents.

Along the south side of the Japanese Archipelago flows the Kuroshio Current, a western-boundary warm current. The width of this current is about 62 miles (100 kilometers). The high temperature and high salinity qualities of Kuroshio water bring a high diversity of tropical marine life northward to the coastal area of the Japanese archipelago and off eastern Japan.

Although many species of invertebrates (such as corals) and vertebrates (such as butterfly fishes) can be found near the south coast of Japan, most of them cannot survive and reproduce in the coastal area because of the cold winter water temperatures and the low nutrient concentrations of the Kuroshio. Nutrient concentrations and biological production in the Kuroshio are much lower than in the cold western boundary Oyashio current. Nevertheless, the Kuroshio is a major spawning area for many species of pelagic fishes, such as Japanese sardine (*Sardinops melanostictus*) and Pacific saury (*Cololabis saira*), that have their nursery and feeding grounds in the Oyashio area.

Bering Sea

The North Pacific and Arctic Oceans are connected by a semienclosed sub-Arctic sea, the Bering Sea. Bounded by the Bering Strait to the north and the Aleutian archipelago to the south, the Bering Sea consists of a deep central basin, a northwestern shelf in the Gulf of Anadyr that reaches south along the Kamchatka Peninsula, and a broad eastern shelf that stretches from the Alaska Peninsula to Russia and the Bering Strait.

High biological productivity in the Bering Sea supports more than 400 species of fish and at least 15 species of squid. Of these, at least 40 species are of some commercial importance, and catches are dominated by walleye pollock (*Theragra chalcogramma*), flatfish (Pleuronectidae), Pacific cod (*Gadus macrocephalus*), crab (*Paralithodes* spp. and *Chioenocetes* spp.), rockfish (*Sebastes* spp.), and five species of Pacific salmon (*Oncorhynchus* spp.). Rich benthic communities with a large biomass and production of flatfish, Pacific cod, crab, and cephalopods are found in the broad shelves along the eastern and western margins of the Bering Sea.

Forage fishes such as capelin (*Mallotus villosus*), eulachon (*Thalichthys acificus*), Pacific sand lance (*Ammodytes hexapterus*), and juvenile walleye pollock and cephalopods can be locally abundant and provide important food sources to upper-trophic-level species. The surface waters of the central basin of the Bering Sea comprise important feeding areas for abundant Pacific salmon. The midwater community in the basin is poorly known, but it is likely dominated by lantern fishes (Myctophidae, in particular *Stenobrachius leucopsarus*), and deep-sea smelts (Bathylagidae).

Alaska Currents

The Alaska Current System is comprised by the shoreward Alaska Coastal Current (ACC); the offshore Alaska Current (AC); and the Alaskan Stream, eddies, and meanders. Currents extend along the continental shelf and shelf break, reaching as far south as the mouth of the Columbia River and as far west as the Aleutian Islands. The eastern Gulf of Alaska reaches abyssal depths of 9,843 feet (3,000 meters) and plummets to 22,966 feet (7,000 meters) in the Aleutian Trench. The continental shelf has a total area of approximately 142,858 square miles (370,000 square kilometers), and ranges from 3 to 124 miles (5 to 200 kilometers) in width. Islands, banks, ridges, and numerous troughs and gullies cut across the shelf, resulting in a complex bathymetry that promotes exchange between shelf water and deeper waters.

The ACC distributes sub-Arctic plankton communities around the region and into protected inside waters. During the summer months, the ACC has local reversals and small eddies that concentrate plankton and small fishes in convergence zones for foraging fish, birds, and marine mammals. Species including the southeast Alaska Pacific herring (*Clupea pallasii*), pink salmon (*Oncorhynchus gorbuscha*), sablefish (*Anoplopoma fimbria*), Alaska rockfish (*Sebastes* spp.), Pacific Ocean perch (*Sebastes alutus*), northern rockfish (*Sebastes polyspinus*), tanner crab (*Chionoecetes bairdi*), and Dungeness crab (*Metacarcinus magister*) are of commercial and ecological importance.

Marine mammals such as the stellar sea lion (*Eumetopias jubatus*), beluga whale (*Delhinapterus leucas*), killer whale (*Orcinus orca*), humpback whale (*Megaptera noraengeliae*), and sea otter (*Enhydra lutris*) are also found in the gulf.

California Current

The California Current is a year-round Equator-directed flow that extends 1,864 miles (3,000 kilometers) from the northern tip of Vancouver Island to Baja California Sur. Cool, fresh, and nutrient-rich water is carried through the California Current System. Fishery resources include invertebrate populations, especially in near-shore waters; groundfish populations along the continental shelf; and migratory pelagic species such as salmon (*Oncorhynchus* spp.), Pacific sardine (*Sardinops caeruleus*), Pacific hake (*Merluccius productus*), and Pacific herring (*Clupea harengus*).

At the southern end, the northern anchovy (*Engraulis mordax*) and market squid (*Loligo opalescens*) are important. This system also supports large and diverse seabird and marine-mammal populations.

Oceanic Biome Status

Throughout the North Pacific, annual rates of primary and secondary productivity are generally similar among regions. In general, biomass has been decreasing in the deepwater areas of the eastern and western North Pacific over the past 20 years. Phytoplankton and zooplankton biomass has decreased in the northern part of the Oyashio Current and in the southern part of California Current over the past 30 years, but increased in the Yellow Sea and the coastal parts of the Gulf of Alaska. In the Bering Sea, zooplankton biomass has remained similar to that observed throughout the 1990s.

Fish fauna of North Pacific Ocean coastal systems have declined around 50–75 percent. More than 50 species of marine mammals in all are

A killer whale (Orcinus orca) leaps out of the water near the Kamchatka Peninsula between the Sea of Okhotsk and the North Pacific Ocean. (Thinkstock)

found in the North Pacific. Although not all species are considered to be of commercial importance, the stellar sea cow (*Hydroanalus gigas*) is known to have been exploited to extinction before 1800, and the Japanese fur seal (*Zalophus californianus japonicus*) is thought to be extinct.

The North Pacific Ocean is known to be one of the most productive and diverse marine ecosystems in the world. This region has witnessed a rapid increase in human population, with important technological, economic, and social transformations during the past 50 years. Due to human pressures and the chemical deterioration of many regions in the North Pacific Ocean, significant losses of marine biodiversity have already been caused.

Observing an area as large as the North Pacific Ocean is a significant task, resulting in a lack of information for many coastal regions. Scientific research on many fronts has addressed global warming and its potential effects on Pacific Ocean water-layer modalities, salinity, acidity, current structure and speed, atmospheric changes, seasonal "marching," and island and shoreline effects—to say nothing of observations and theories on the climate-change impacts sustained by many affected marine species. Long-term monitoring programs in this ecosystem that sample physical and biological conditions are necessary to understand the long-term effects of global environmental changes and human activity.

Maria Jose González-Bernat

Further Reading

Bakun, A. "The California Current, Benguela Current, and Southwestern Atlantic Shelf Ecosystems: A Comparative Approach to Identifying Factors Regulating Biomass Yields." In "Characterising Meso-Marine Ecosystems of the North Pacific," S. D. Batten, et al. *Deep-Sea Research II* 53 (2006).

Barange, M. et al. *Marine Ecosystems and Global Change.* New York: Oxford University Press, 2010.

Beamish, Richard J. *Impacts of Climate and Climate Change on the Key Species in the Fisheries in the North Pacific.* Sidney, Canada: North Pacific Marine Science Organization, 2008.

Mackenzie, Fred T. and Judith A. Mackenzie. *Our Changing Planet: An Introduction to Earth System Science and Global Environmental Change.* Upper Saddle River, NJ: Prentice Hall, 1995.

McKinnell, S. M. and M. J. Dagg, eds. *Marine Ecosystems of the North Pacific Ocean 2003–2008.* Sidney, British Columbia: PICES, 2010.

Polovina J. J., et al. "The Ocean's Least Productive Waters Are Expanding." *Geophysical Research Letters* 35 (2008).

Segar, D. A. *Introduction to Ocean Sciences, 2nd Ed.* New York: W. W. Norton & Company, 2007.

Sherman, K. *Large Marine Ecosystem: Assessment and Management in the Large Marine Ecosystems of the Pacific Rim.* Oxford, UK: Blackwell Science, 1999.

Pacific Ocean, South

Category: Marine and Oceanic Biomes.
Geographic Location: Southern Hemisphere.
Summary: Amid this vast oceanic biome, island ecosystems are fragmented by separation, and threatened by climate change.

The Pacific Ocean is the largest geographic feature on the Earth's surface, an area greater than all the continents combined. The South Pacific is dotted with islands because of its geological structure; the divergent plate boundary of the East Pacific Rise is the formative source of the Pacific tectonic plate. In its westward movement, the plate passes over a number of volcanic hot spots; these have formed the eastern Pacific volcanic island chains of Pitcairn, French Polynesia, the Cook Islands, and the Samoa group. The Pacific plate subducts beneath the Australian plate to the east of the main islands of Tonga, Fiji, and Vanuatu, and north of the Solomons and Papua New Guinea, causing islands to the west to be larger and less isolated.

The South Pacific region includes 16 countries and territories comprising over 8.6 million square miles (22.3 million square kilometers) of ocean area, and less than 232,000 square miles (600,000 square kilometers) of land, with 84 percent of this land in Papua New Guinea. Most of the remaining land area is in the other larger Melanesian islands of the Solomons, Vanuatu, Fiji, and New Caledonia, again with the islands overall declining in size from west to east. There are approximately 1,735 islands in total. Land is a primary control on ecosystem diversity, and physical conditions of geomorphology and tropical climates exert controls on habitats and ecosystems.

South Pacific ecosystems are biogeographically fragmented by physical separation of habitats across these islands, and controlled by island type: whether the island groups are characterized by high topography, by limestone formation, or by coral atoll structure. Of these major island groups, the high type is found mainly in Papua New Guinea, Solomon Islands, New Caledonia, Vanuatu, Fiji, Wallis and Futuna, Samoa and American Samoa, and French Polynesia. The atoll types include Tuvalu, Kiribati, and Tokelau. Nauru and Niue are limestone types; Tonga is mixed high and limestone type; and the Cook Islands are mixed high and atoll type.

Ecosystem diversity is at a maximum in the west at Papua New Guinea, and decreases toward the east as island sizes become smaller, and distances between islands become larger. Endemic species (those found nowhere else) occur particularly in terrestrial ecosystems, such as freshwater habitats, where the greatest threats are from introduced weed species. In coastal ecosystems, the greatest threats are also from direct human impacts, but increasingly from climate change and sea-level rise.

Papua New Guinea, Solomons, New Caledonia, Vanuatu, Fiji, Wallis and Futuna, and Tonga are all west of the tectonic margin line, which influences the biogeography of island ecosystems, as larger islands to the west are older continental crust. The three geologically old, large islands of New Guinea, New Caledonia, and Viti Levu (largest of the Fiji Islands) form three diverse biogeographic sub-regions in the Pacific Islands, from which species have migrated to other islands. Southern New Guinea and New Caledonia are Gondwanan fragments, while the Fijian large islands were formed from uplifted crust over 40 million years ago. Other islands here were uplifted coral or formed volcanically in association with the subduction margin.

Island groups to the east include Nauru, Tuvalu, Kiribati, Tokelau, the Samoas, Niue, Cook Islands, and French Polynesia; these are "true islands," mainly atoll types. This has a great influence on ecosystem biodiversity, with these being low in native species, having only secured those adapted to long-distance oceanic migration.

The ecosystems of this region are biogeographically fragmented by physical separation of habitats within the coastal biomes of coral reefs, mangroves, seagrasses, and even more so within the freshwater wetland types. Ecosystems of the South Pacific have many analogous environmental settings, but due to the filter effect of oceanic separation, the species present in ecosystems all differ between countries (see Table 1). This results from two factors: first, the center for species diversity within all groups is on the margins of the western edge of the region; and second, island size tends to decrease and island isolation increase toward the east and north.

Ecosystem habitats on islands are dependent on island types. While volcanic islands can host crater lakes and river catchments, over time, as they start to erode and submerge, coastal plains, mangroves, and freshwater swamps are likely to develop. Fringing reefs become barrier reefs with further island submergence, increasing area and habitat variety in reef environments and introducing lagoons and seagrass habitats. Further subsidence leads to atolls, losing all freshwater wetlands—but increasing reefs and lagoons.

Four Biological Groups

The biological diversity of corals, reef fish, mangroves, and seagrasses of the Pacific region result from their biogeographic history. The Indo-Malayan archipelago to the west of the region is the center of greatest species diversity for all four groups. It is where they first appeared in the ancient Tethys Sea at equatorial latitudes. Species richness declines from this center toward the east.

Coral reefs are the most extensive ecosystem type in the Pacific island region, occurring offshore of nearly all coastlines that lack turbid river discharge. New Caledonia has coral reefs of around 9,300 square miles (24,000 square kilometers), including the second-largest barrier reef in the world surrounding the main island and enclosing a lagoon of 6,200 square miles (16,000 square kilometers). Papua New Guinea has 15,500 square miles (40,000 square kilometers) of coral reef, and some countries in Micronesia and Polynesia are almost entirely composed of coral atolls.

There are 5,000 square miles (13,000 square kilometers) of coral reefs in just the countries east of Samoa, compared with 2,300 square miles (6,000 square kilometers) of land. Coral reefs and freshwater mollusks are the best-studied wetland invertebrate groups in the Pacific island region; research on other groups has pointed to the need for baseline surveys and inventories. Some of the better distributional records of groups are summarized in Table 1.

Fiji's marine environment is better studied than most other Pacific Island countries, facilitated by a marine research and training capability at the University of the South Pacific. In 2000, Fiji's reefs suffered an extensive mass bleaching event, causing loss of 40–80 percent of stony corals. As a result,

Table 1: Distribution of biota across the South Pacific region

	Coral Genera	Shorefish Families	Strombus Species (Conch)	Seagrass Species	Mangrove Species	Bird Species (Endemics)	Freshwater Fish	Amphibians and Reptiles
Papua New Guinea	70	149	22	12	31	168 (48)		472
Solomon Islands	60		20	9	17	159 (59)	52	
New Caledonia	60	125	20	11	15	75 (20)	85	
Vanuatu	60			12	14	56 (7)	52	49
Fiji	60	116	14	6	7	57 (22)	80	29
Wallis & Futuna					2	11		
Tonga	40			4	7	21 (2)		20
Nauru					2	3 (1)		
Tuvalu				1	2	2		
Kiribati				1	4	2		
Tokelau						1		
Samoa	50	102	9	4	3	33 (10)	49	15
American Samoa	50	102	9	4	3	33 (10)	49	15
Niue					1	11		
Cook Islands	30	83		0	0	12 (6)	10	
French Polynesia	30	47	7	1	1	18 (11)		

reef monitoring commenced to allow ongoing surveillance of coral reef biodiversity and condition, and provide better understanding of natural and human impacts. In some coastal areas, poor inshore water quality with nutrients, suspended sediment, and trace metals have been found to be affecting inshore reefs. American Samoan reefs are well studied and a detailed inventory is available, thanks to United States government research. The reefs of Tonga and Vanuatu are little studied, compared with New Caledonia or Fiji, and other islands, such as Niue, even less.

The diversity of reef-building corals in the equatorial western Pacific at Papua New Guinea is close to the highest in the world. Diversity declines to the north and the south with cooler temperatures, and from west to east across the Pacific with increasing distance.

Similar regional patterns occur in the shallow water finfish, and nearshore shallow water conches, as seen in Table 1. The finfish families, which include 461 genera and approximately 1,312 species, are important sources of protein for the traditional village peoples of the South Pacific region. Reef and shore mollusks also decline in diversity from west to east across the region; species present are however dependent on habitat variety. Different molluscan assemblages live on reef slopes, outer lagoons, and inner lagoons, for example. Coastal sponges, polychaetes, and crustaceans also decline in diversity from west to east; however, the records of all these groups are patchy and more research is needed.

Six of the seven marine turtles are found in the South Pacific region; most common are the green turtle, hawksbill, and leatherback. Of these, the hawksbill is, however, critically endangered; the green and leatherback are endangered. Less common are the loggerhead, flatback, and olive ridley, of which the olive ridley is endangered, and the loggerhead and flatback turtle vulnerable.

Marine turtles are slow to reach maturity, and migrate long distances between nesting and feeding areas. Individuals tracked from a nest at Rose Atoll, American Samoa, moved to the rich seagrass and algae beds of Fiji, spending 90 percent of their time there. Impacts to turtles include bycatch from long-line fisheries, over-exploitation by local communities, and human hunting when nesting on beaches.

Mangroves occur on intertidal habitats of sheltered shorelines, which are most extensive at sedimentary estuaries. In the Pacific Islands region, the total mangrove area is about 2,196 square miles (5,687 square kilometers), with the largest areas in Papua New Guinea, Solomon Islands, Fiji, and New Caledonia. The species mixture is unique in each country or territory (see Table 1), and the mangroves provide useful ecosystem services. There are 31 species of mangroves of the Indo-Malayan assemblage, with highest global biodiversity in southern Papua New Guinea; diversity declines from west to east across the Pacific, reaching a present limit at Samoa, with a possibly introduced outlier in French Polynesia.

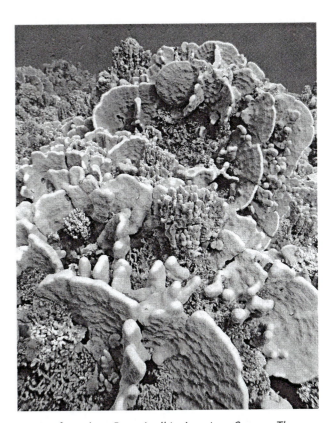

A mix of corals at Rose Atoll in American Samoa. The atoll is a nesting ground for marine turtles. American Samoa is home to 15 types of amphibians and reptiles and 50 genera of coral. (NOAA/Jean Kenyon)

Seagrasses occur mostly in shallow lagoons between coral reefs and shoreline mangroves in the South Pacific, with a close proximity and connectivity of these wetland types typical of island shorelines. Seagrasses evolved from terrestrial grass in the Cretaceous Tethys Sea, and spread into the Pacific. The center of biodiversity for seagrass is New Guinea, with up to 13 species; two families and seven genera extend across the South Pacific, declining in diversity with distance (as seen in Table 1), except *H. ovalis sp. bullosa*, which is endemic to Fiji, Tonga, and Samoa. In the eastern Pacific, seagrass is absent from the Cook Islands, and only one species is recorded from French Polynesia. The fruits of most seagrass species are not buoyant, hence reducing their ability to disperse over long distances.

Land-Based Fauna

The tropical Pacific terrestrial mammal fauna is immensely depauperate compared with elsewhere in the world, due to island isolation and lack of landbridges, and is dominated by bats. No mammals at all are present in French Polynesia, Tuvalu, Kiribati, the Marshalls, or Tokelau, and one bat species is the only native mammal species present in the Cook Islands and Niue. This most widespread species is the Pacific flying fox *Pteropus tonganus*, which roosts in tall coastal trees.

Pacific island bird distributions are shown in Table 1, which indicates that endemism is enhanced by isolation from other land masses and speciation encouraged. These totals are of bird species resident on islands recorded by Adler in 1992, although many flightless or ground nesting species became extinct as a result of human impact before that. Losses have included about 2,000 species of flightless rails, as well as petrels, shearwaters, terns, and kingfishers.

Unusual species remain, however, on remote islands; for instance, on the volcanic island of Niuafo'ou in Tonga is found the endangered endemic Niuafo'ou megapode, which incubates its eggs by burying them in warm sands near volcanic ducts. Recently chicks and eggs have been released on the predator-free Tongan islands of Late and Fonualei.

The estuarine crocodile occurs in the South Pacific, being fairly common in Papua New Guinea and extending to the Solomon Islands and Vanuatu, which is the limit of its Pacific range. The Solomon Islands once had substantial populations, but these are now decimated by human impact and cyclone damage. The crocodile is currently vulnerable in the South Pacific.

Numbers of amphibian and reptile species decrease rapidly with distance to the east of New Guinea, also due to decrease in island size, proximity of other islands, and island age. Amphibians are intolerant of salt water and therefore have difficulty migrating in this region, a factor also influencing the prawns, shrimps, and freshwater fish (Table 1). Recent studies of freshwater fish species on the large islands in Fiji found a total of 161 fish from 45 families.

Vulnerability

Pacific island ecosystems are vulnerable because of their relatively small size, poor state of protection or available knowledge, and their proximity to rapidly increasing human populations that put heavy demands on these natural resources to support their economic development. Weeds and introduced species are a great threat to these special biodiversity groups, particularly freshwater ecosystems. Climate change, sea-level rise, and cyclone damage are increasing vulnerability, particularly of coastal ecosystems. Gaps in ecosystem management in the region include weak or vague legislation for their protection in many countries, and the need for more knowledge and capacity to manage.

JOANNA C. ELLISON

Further Reading

Ellison, Joanna. "Wetlands of the Pacific Island Region." *Wetlands Ecology and Management* 17 (2009).

Ellison, Joanna and Monifa Fiu. *Vulnerability of Fiji's Mangroves and Associated Coral Reefs to Climate Change.* Suva, Fiji: World Wildlife Fund South Pacific Programme, 2010.

Mueller-Dombois, Dieter and Ray Fosberg. *Vegetation of the Tropical Pacific Islands.* New York: Springer-Verlag, 1998.

Steadman, David. *Extinction and Biogeography of Tropical Pacific Birds*. Chicago: University of Chicago Press, 2006.

Veron, Charlie. *Corals in Space and Time: The Biogeography and Evolution of the Scleractinia*. Sydney, Australia: UNSW Press, 1995.

Wilkinson, Clive. *Status of Coral Reefs of the World: 2004: Volume 2*. Townsville: Australian Institute of Marine Science, 2004.

Pallikaranai Wetland

Category: Inland Aquatic Biomes.
Geographic Location: Asia.
Summary: Government action is needed to stop the degradation of this wetland into a wasteland.

The Pallikaranai wetland is a freshwater swamp near the Bay of Bengal, about 12 miles (20 kilometers) south of Chennai, the capital city of the state of Tamil Nadu in southeastern India. It is the last remaining wetland ecosystem of the city of Chennai, and one of the few remaining in southern India. At present, the city is left with only 27 bodies of water. As one of the few remaining freshwater sources, the Pallikaranai wetland is crucial for supporting life and mitigating flooding in the connected water bodies in and around Chennai. The smaller wetlands that surround Pallikaranai serve as the only source of irrigation for the area. Together, these wetlands are an important resource for bird migration and nesting, but human activities, such as dumping of chemical and other toxic wastes, has severely impacted the area.

Wetland Essentials

Wetlands cover approximately 6 percent of Earth's surface, and are defined as lands transitional between terrestrial and aquatic systems where the water table is usually at or near the surface, or the land is covered by standing water that does not exceed 20 feet (6 meters). Human civilization originated in wetlands habitats such as the floodplains of the Indus, the Nile delta, and the Fertile

Crescent of the Tigris and Euphrates rivers. This is in large part because wetlands are home to diverse plant and animal life. Biologically, the wetlands are among the most fertile ecosystems.

India has 27,403 documented wetlands, of which 23,444 are inland and 3,959 are coastal wetlands. The coastal type occupy 2,606 square miles (6,750 square kilometers). The share of wetlands in India overall amounts to 17 million acres (7 million hectares). One-third of these, however, have been wiped out, altered, or degraded.

Among ecological services, wetlands help check floods and mitigate the effects of natural disasters, including cyclones and tidal waves. Wetlands can store water for a longer period than other habitat types. They assist in erosion control, as they are located between water bodies and high ground; the roots of the vegetation in the wetlands protect soil from high-impact events such as wave action, wind, and heavy rainfall. Given that one-third of all wetlands in India are gone or heavily damaged, preserving areas such as the Pallikaranai becomes even more important than ever.

Pallikaranai Biota

The flora of the Pallikaranai wetlands consists primarily of aquatic plants, reeds, liana, and sedges. Acanthus, water lettuce, water hyacinth, joyweed, swamp morning glory, snowbush, and paspalum are featured across marsh areas, as well as transition areas. Among the sedges are umbrella papyrus, purple nutsedge, and slender cyperus. Large tracts of marsh grasses and pastures adjoin the wetlands areas.

The Pallikaranai wetland is home to several endangered species, and it acts as a breeding, feeding, and sheltering ground for many species of migratory birds. At least 125 avian species, both transient and permanent, have been recorded here. Among them are greater flamingo, glossy and white ibis, purple and Indian pond heron, little grebe, bushlark, various bitterns, ducks, pipers, yellow-wattled lapwing, and spot-billed pelican. The birds here are joined by 21 reptile species, 10 mammalians, nine amphibian types, a suite of crustaceans such as the mud crab, mollusks including windowpane oyster, and seven identified butterfly species.

At least 45 species of fish inhabit or visit the Pallikaranai Wetland biome. The more frequently seen types include striped snake head, long-fin eel, and stinging catfish; other common species here are mullet, glass fish, razor belly, peninsular olive barb, and long-whiskered catfish.

Threats

Water for irrigation in Tamil Nadu and drinking-water availability in Chennai and environs have been shrinking at an alarming rate. During the monsoons, the connected water bodies bring along sewage, thereby making Pallikaranai a dump site. The topography of the swamp retains the excess water from the connected water bodies during their floods, but also releases the accumulated toxins from it back into the ecosystem—sometimes in greater concentration—during heavy weather events. The deterioration of the wetland proper may also result in damage to neighboring water bodies. Outrage at the amount of garbage and chemicals dumped in and around the area has led to many calls for clean-up, regulation, and enforcement to preserve the wetlands.

The external manipulation of this wetland ecosystem began in 1806, with the construction of the 262-mile (422-kilometer) Buckingham Canal. The anthropogenic changes in and around this wetland cut much of the natural drainage channel network, thereby aggravating the frequency and intensity of floods. Flooding has also increased due to rapid urbanization; the growing number of inhabitants in the city of Chennai has in recent decades largely outpaced the capacity of the 31-square-mile (80-square-kilometer) Pallikaranai wetland to process water and suspended and dissolved materials. Further, global warming is increasing the damage already sustained—and to come—from the threats of rising sea levels, saltwater intrusion, soil erosion, and invasive species penetration.

The government of Tamil Nadu, to protect the flora and fauna of the wetlands, at last in 2007 declared a major portion to be a reserve forest, thus setting its monitoring and preservation under the domain of the Forest Department.

RHAMA PARTHASARATHY

Further Reading

Ali, Salim. *The Book of Indian Birds.* Mumbai: Bombay Natural History Society, 1996.

Nikhil, Raj P. P., J. Ranjini, R. Dhanya, J. Subramanian, P. A. Azeez, and S. Bhupathy. "Consolidated Checklist of Birds in the Pallikaranai Wetlands, Chennai, India." *Journal of Threatened Taxa* 2, no. 8 (2010).

Oppili, P. "Pallikaranai Vulnerable to Devastating Floods: Study." *The Hindu,* August 25, 2007.

Ramesh, R., P. Nammalwar, and V. S. Gowri. *Database on Coastal Information of Tamil Nadu.* Chennai, India: Institute for Ocean Management, Anna University, 2008.

Pamir Alpine Desert and Tundra

Category: Grassland, Tundra, and Human Biomes.
Geographic Location: Asia.
Summary: This cold, dry alpine desert is rich in biodiversity, but some of its hosted animal species are endangered.

The Pamir is a high plateau located at the intersection of the world's tallest mountains of the Himalayas, Karakoram, Hindu Kush, Kunlun, and Tian Shan regions. These mountains are sometimes referred to as the Roof of the World. The Pamir is shared by Afghanistan, China, Kyrgyzstan, Pakistan, and Tajikistan. Because of its geopolitical importance, it is sometimes referred to as the Pamir Knot. The biodiversity in the Pamir is high thanks to the intersection of several climates among the various mountain ranges.

With an approximate area of 97–106 square miles (250–275 square kilometers), the Pamir extends up to the Kashghar and Tarim Basin in the east, the Trans-Alai Range in the north, the Hindu Kush to the south, and Tajikistan in the west. In this way, it is a biogeographic barrier between south Asia and central Asia. Average altitude is 13,780 feet (4,200 meters), but the highest peak in the Pamir is more than 24,278 feet (7,400 meters). Topography here is defined by steep, rolling hills and vast valleys. The landscape in most areas forms a desert with crumbling mountains and somewhat barren valleys.

While it is a dry ecoregion, it is the source of several rivers in the region, including the Amu Darya and Oxus. Well-known lakes in the area include Karakul and Zorkul. Ibn Sina Peak (also known as Lenin Peak), Ismoil Somoni Peak, and Peak Korzhenevskaya are key mountains. The region also features glaciers, including the 48-mile (77-kilometer) Fedchenko Glacier, known as the longest glacier outside the polar region.

Climate is sunny and dry, with mean annual precipitation of 2–6 inches (40–150 millimeters), ranging from cold desert to semi-desert. The area shows extreme temperature variations; it can vary from well below freezing to 50 degrees F (10 degrees C). Strong and steady winds are the norm here.

A herder's tent in the Pamir plateau in Tajikistan. Environmental concerns stemming from human activity in the area include overgrazing of livestock, fuel-wood collection, and hunting. (Thinkstock)

Flora

The Pamir has more than 620 floral species; these can be divided into multiple distinct ecological zones. At the lower elevations, the regime resembles a mediterranean climate. Between the lowest and highest altitudes, a lower steppe and an upper Eurasian steppe zone occur, with prickly cushion plants such as wormwoods, needles, and fescue grass. At the highest elevations are found alpine sedge meadows.

The Pamir includes characteristic central Asian plant varieties, such as *Eurotia ceratoides* and *Acantholimon diapensioides,* which are also found on the Tibetan Plateau, which resembles the Pamir ecologically but is a bit lower and warmer. The Pamir has more floral varieties than the Tibetan Plateau because of its more rugged topography. *Salix pycnostachia, Comarum salesvovianum,* and *Dasiphora dryadanthoides* are endemic, or unique to this region. The Pamir Alpine Desert and Tundra biome also shares many alpine varieties with Iran and Afghanistan, such as *Arabis kokanica* and *Saponaria griffithiana*. Its eastern extent, in China, has grassy steppe flora. Throughout its average altitude range, forb species abound; some of them have thickened stems, or caudex, as well as woody taproots, for protection against pests, temperature extremes, and arid conditions.

Fauna

Despite the Pamir's cold and dry climate, some bird species are also found here, especially around the lakes. Most notable are brown-headed gulls, bar-headed geese, and snowcocks. The western Pamir is richer in species because of its proximity to central Asia.

Several world-famous, high-altitude species are hallmarks of the Pamir. These include the snow leopard, Marco Polo sheep, blue sheep, and markhor. However, they are under increasing threat due to overgrazing and poaching.

The Siberian ibex is relatively abundant in the Pamir—as are the wolves that prey upon ibex herds. The Pamir also has some endangered species of brown bear. Most of these animals are protected in the Pamir National Park in Tajikistan, as well as the Taxkorgan Nature Reserve in China. Markhor have protected status in Pakistan, as the national animal.

Threats

Overgrazing, fuel-wood collection, hunting for meat, and killing the predators of livestock are some of the major concerns in the Pamir Alpine Desert and Tundra biome. Use of pesticides for agriculture is a more recent threat developing to jeopardize animals and the native natural environment. Climate change may impact this area through reduction of its glaciers, advanced seasonal flooding patterns, and changes to the mean temperature that will alter seed germination timing. The impacts of such threats to plants in the high alpine areas are being particularly closely monitored.

In other conservation efforts, Pamir National Park occupies more than 6 million acres (2.6 million hectares), where many species take shelter and thrive in its network of intact natural habitat areas, including Lake Karakul and the Zorkul, Muzkol, and Sanglyar Refuges.

MUHAMMAD AURANG ZEB MUGHAL

Further Reading

Bliss, Frank. *Social and Economic Change in the Pamirs*. London: Routledge, 2006.

Knystautas, Algirdas. *The Natural History of the USSR*. New York: McGraw-Hill, 1987.

P'Yankov, V. I. and A. V. Kondrachuk. "Basic Types of Structural Changes in the Leaf Mesophyll During Adaptation of Eastern Pamir Plants to Mountain Conditions." *Russian Journal of Plant Physiology* 50, no. 1 (2003).

Schaller, G. B. *Mountain Monarchs: Wild Sheep and Goats of the Himalaya*. Chicago: University of Chicago Press, 1977.

Strong, Anna Louise. *The Road to the Grey Pamir*. New York: Robert M. McBride & Co., 1930.

Toynbee, Arnold. *Between Oxus and Jumna*. London: Oxford University Press, 1961.

Pantanal Wetlands

Category: Inland Aquatic Biomes.
Geographic Location: South America.
Summary: One of the world's largest wetlands, located in the heart of South America, is a seasonally flooded area hosting an amazing diversity of wildlife that is the subject of expanding ecotourism and protection efforts.

Known as South America's Serengeti, the Pantanal Wetlands biome is one of the largest wetland areas in the world, extending some 66,000 square miles (170,000 square kilometers). The Pantanal is mostly located along the Paraguay River in Brazil, in the states of Mato Grosso and Mato Grosso do Sul; it also reaches into neighboring Bolivia and Paraguay. The southern edge of the Pantanal is near Fort Olimpo, Paraguay, where hills close in on both banks of the Paraguay River. The wetlands are flooded in the wet season but emerge as a vast grassland during the dry season.

Climate and Biodiversity

A summer rainy period from November to March deposits most of the annual rainfall of 39–59 inches (1,000–1,500 millimeters). The name *Pantanal* is from the Portuguese word *pantano*, for *swamp*, certainly an adequate descriptive term here. Winters are the dry season, and occasionally the temperature drops to near freezing. The area is mainly flat, with minor changes in elevation resulting in big changes in the water coverage. In the low areas are thousands of lakes, ponds, and river oxbows; these support great spreads of floating aquatic plants.

At slightly higher elevations, sedges and seasonally-flooded grasslands, known as *campos*, may be found. Fish enter flooded areas during the wet season to breed; birds such as roseate spoonbill and wood stork are attracted to the flooded areas to feed on the fish.

To the north and on hills adjacent to the flooded grasslands are cerrado vegetation, a woodland savanna that bridges the area between Amazon rainforest and more temperate forests. Because the cerrado is less frequently flooded, much of the vegetation is fire-adapted, with grasses, shrubs, and short trees. Along rivers are found riparian forests.

Large animals in the Pantanal include caimans (reptiles in the alligator family), anacondas, jaguars, giant otters, giant armadillos, monkeys, bats, anteaters, marsh deer, and pampas deer. The jaguars feed on large rodents, the capybara, that are slightly larger than beaver, as well as other mammals, including livestock owned by local ranchers.

Notable bird life includes more than 650 species, such as includes macaws, roseate spoonbills, wood storks, rheas, and seriemas. Storks, herons, and ibises nest in large rookeries here. The jabiru, a type of stork, is the symbol of the Pantanal. There are an estimated 400 species of fish inhabiting the Pantanal waters. The richness of wildlife in this biome makes the Pantanal one of the world's most critical wildlife areas, rivaling even the African savannas.

Threats

The surrounding plateaus include areas of extensive soybean, wheat, and rice plantations. These agricultural lands use large amounts of pesticides and herbicides, which wash downstream into the Pantanal and have caused fish kills. In the past, the Paraguay River was scheduled for large-scale channel constructions, which would have changed the hydrology of the area. However, this proposal, known as the Paraguay-Parana Hidrovia, is on hold in the upper Paraguay basin. The proposal is of concern because the Pantanal stores water and maintains the flow of the Paraguay River during the dry season, and reduces flood peaks during the wet season, thus playing a major role in the hydrological regime of southern South America.

Elsewhere in the region, smaller dredging projects have occurred in the Brazilian and Bolivian portions of the Pantanal. Overfishing is also a threat to the Pantanal Wetlands biome, as are hunting and poaching. Climate change threats faced by this biome include accelerated evaporation due to higher air temperatures, a dangerous trend that could severely unbalance an ecoregion evolved around very distinct seasonal moisture

regimes. In carbon-rich systems like the Pantanal, any significant long-term drying could release a great quantity of greenhouse gas, which would act to advance the pace of global warming in Earth's atmosphere.

Conservation

The Pantanal Matogrossense National Park, established near the center of the biome and along the east bank of the Paraguay River, protects a flooded habitat area of about 520 square miles (1,350 square kilometers), but little upland habitat. This area is listed as a Ramsar Wetland of International Importance and a World Heritage Site. The Pantanal Conservation Complex World Heritage site includes this park and Doroche Reserve, both in the Brazilian state of Mato Grosso, as well as the Acurizal and Penha Reserves, sited in Mato Grosso do Sul. The reserves were private ranches that now have become part of a system of voluntary Private Reserves of Natural Heritage. Landowners participating in the private reserve system are offered tax breaks and management assistance.

The Brazilian extent of the Pantanal as a whole is also a Man and the Biosphere Reserve. This is a designation of the United Nations Educational, Scientific, and Cultural Organization (UNESCO), in which protected areas and adjacent buffer zones for human use are managed to maintain the cultural heritage of indigenous communities while also protecting ecological sustainability. The Pantanal Biosphere Reserve includes such features as the national park, Taiama Ecological Station, and other areas in the region. Ecotourism and recreational fishing are examples of sustainable natural resource use activities that are encouraged by the Biosphere Reserve designation. Across the Paraguay River in Bolivia is the San Matias protected area. Bolivia also protects an area along the Otuquis River, at the common border with Paraguay and Brazil.

Various parts of the Pantanal are being developed as ecotourism destinations for bird watching and wildlife observation. A number of outfitters offer jaguar-viewing tours, for example. In another, the Estrada Transpantaneira is a dead-end road in southern Mato Grosso that is used for viewing caimans and wading birds. Boat rides and tourist activities also embark from Corumba and Miranda in Mato Grosso do Sul.

HAROLD DRAPER

Further Reading

Alho, Cleber J.R., Thomas E. Lacher, Jr., and Humberto C. Goncalves. "Environmental Degradation in the Pantanal Ecosystem." *BioScience* 38 (1988).

Gottgens, Johan F., James E. Perry, Ronald H. Fortney, Jill E. Meyer, Michael Benedict, and Brian E. Rood. "The Paraguay-Parana Hidrovia: Protecting the Pantanal with Lessons from the Past." *BioScience* 51 (2001).

Lowen, James. *Pantanal Wildlife: A Visitor's Guide to Brazil's Great Wetland.* Buckinghamshire, UK: Bradt Travel Guides, 2010.

Swarts, Frederick A., ed. *The Pantanal: Understanding and Preserving the World's Largest Wetland.* St. Paul, MN: Paragon House Publishers, 2000.

Papahanaumokuakea Coral Reefs

Category: Marine and Oceanic Biomes.
Geographical Location: Pacific Ocean.
Summary: This World Heritage Site consists of a thriving coral reef community and is significant both ecologically and culturally.

Papahanaumokuakea, located in the far northwestern Hawaiian Islands, has been a United Nations Educational, Scientific, and Cultural Organization (UNESCO) World Heritage Site since 2008 and a United States national monument since 2006. The area comprises 138,997 square miles (360,000 square kilometers) of ocean set around 10 islands and atolls, in an elongated 1,000-mile (1,600-kilometer)-long stretch of the North Pacific Ocean that is rich in coral reef habitats.

Because the Hawaiian archipelago is the most remote island archipelago in the world, histori-cally few people have established communities in the northwestern Hawaiian Islands. Today, the monument is managed by a partnership among the U.S. Department of Commerce, the state of Hawaii, and the U.S. Department of the Interior.

This protected area has been recognized for both ecological and cultural values. It not only contains many species that are endemic (found nowhere else) or endangered, but also has spiritual significance to the traditional native Hawaiian culture as an example of human-environment kinship. The name *Papahanaumokuakea* refers to the joining of the sky father and earth mother, the union that was to have given birth to the Hawaiian Islands and all the people there. It was bestowed upon this marine monument (considered part of the national marine sanctuary system) to honor that tradition.

Species Diversity and Interaction

The waters of Papahanaumokuakea harbor more than 7,000 species, many of which are found only in the Hawaiian Islands archipelago. Rare species such as the threatened green sea turtle (*Chelonia mydas*) and the endangered Hawaiian monk seal (*Monachus schauinslandi*) call these waters home. The coral reefs of this remote area have largely been spared from human stressors such as pollution and habitat destruction, which plague many other tropical reefs around the world. However, just like other reefs around the world, the Papahanaumokuakea reefs are subject to climate-change effects from rising ocean temperatures and increasing water acidity. Both warming and acidification cause physiological, biological, and ecological harm to coral reefs over both short and long periods.

Scientists around the world are striving to understand whether some reefs will be stronger in resisting such harm in the future, and to identify factors in their environment that could help strengthen their resistance to damaging impacts. In Papahanaumokuakea, continuous coral-reef health monitoring programs are carried out to measure changes and to understand the vulnerability and responses of these reefs, as well as their habitat neighbor species.

Aspects of the Papahanaumokuakea coral reefs being monitored include the movements and habits of top predators, such as sharks; the spread of disease among coral colonies; the symbiotic relationship between the coral animal and the photosynthesizing algae dwelling inside the coral; and genetic structures and dynamics of various fish populations. Sharks are important for any coral reef ecosystem because of the extent to which they control and shape the ecological community and help maintain various fish populations at levels that seem to contribute to sustaining healthier reefs.

The massive scale of shark fishing for their prized fins has reduced shark populations in the ocean dramatically in recent decades. Because many shark species live on or near reefs, it is important, both for reef communities and the overall balance of the marine food web, to maintain protected areas that can stand as strongholds for reef and shark conservation. Papahanaumokuakea is one such stronghold.

Genetic studies of fish populations found here are revealing some of the biological and biogeographical connections of this biome and its inhabitants to other marine ecosystems. This is yet another avenue for scientists to learn more about the ecological services reefs provide, as well as their vulnerabilities and in the precise ways intact reefs can act, as oceanographer Sylvia Earle has said, as "hope spots."

Above the sea, there are other natural riches here. On and around the small islands of Laysan and Nihoa, for example, bird populations of Laysan and Nihoa finches, Nihoa millerbirds, Laysan ducks—perhaps the world's most endan-

Divers freeing an endangered Hawaiian Monk Seal (Monachus schauinslandi) from a derelict fishnet at French Frigate Shoals in the northwestern Hawaiian Islands. Before the 138,997-square-mile (360,000-square-kilometer) protected area and national monument was established, these waters were open to fishing of all kinds. (NOAA/Ray Boland)

gered ducks—and seabirds such as the Laysan albatross feed, breed, and add their inputs to the ecosystem. There are 21 seabird species in residence around the seemingly cramped 6 square miles (16 square kilometers) of land within the Papahanaumokuakea biome. It turns out to be more than enough terrestrial base: up to 14 million birds breed and nest on the constricted land area here, including those belonging to several endemic species.

Human Presence

The land areas of the Papahanaumokuakea biome are of archaeological significance to Native Hawaiian culture; the highest number of sacred sites on any Hawaiian-archipelago island is found on Mokumanamana here. More recent history has also made a mark, with 19th-century commercial whaling in the region and a military presence on some atolls. Midway Atoll, near the northwesternmost extent of Papahanaumokuakea, figured large in the World War II struggle between Japan and the United States.

Before the establishment of the national monument, these waters were entirely open to fishing, and after decades of overfishing, many fish populations in the area have declined. Currently, the only uses allowed in the monument's waters are management activities carried out by jurisdictional agencies, research, education, and Native Hawaiian practices. Midway Atoll also accommodates a small number of recreational trips every year.

The underwater realm of the monument contains fascinating maritime history. Maritime archaeologists have discovered remains of ancient Polynesian sailing canoes, modern submersible and research vessels, U.S. Navy aircraft, British whaling ships, and other craft, all of which embody layers of the region's seafaring history. Such windows into the past add to the uniqueness of the area and underscore the transience of man's impact on the natural environment, as well as the importance of wise management of resources.

LIDA TENEVA

Further Reading

Dale, Jonathan J., Carl G. Meyer, and Christian E. Clark. "The Ecology of Coral Reef Top Predators in the Papahanaumokuakea Marine National Monument." *Journal of Marine Biology* (2011).

Kikiloi, S. *Reconnecting with Ancestral Islands: Examining Historical Relationships Between Kānaka Maoli and the Northwestern Hawaiian Islands.* Manoa, HI: Center for Hawaiian Studies, 2006.

Parrish, F. A., K. Abernathy, G. J. Marshall, and B. M. Buhleiert. "Hawaiian Monk Seals (*Monachus schauinslandi*) Foraging in Deep-Water Coral Beds." *Marine Mammal Science* 18, no. 1 (2002).

Papuan Rainforests, Southeastern

Category: Forest Biomes.
Geographic Location: Melanesia.
Summary: These rainforests in Papua New Guinea contain pristine areas of great biodiversity that are threatened by encroaching human activity.

New Guinea is the second-largest island in the world. It occupies a peculiar location in Melanesia, between southeast Asia and Australia, surrounded by tropical ocean. Its evolutionary history is well known in the scientific literature, and provides a unique laboratory for biology and humankind alike. New Guinea has diverse and steep topographic variations, with highlands that form particularly unique habitat areas.

The island is divided between two countries: the western half is a possession of Indonesia, while the eastern half forms the nation of Papua New Guinea (PNG). Within PNG, in southeastern New Guinea, the Owen Stanley highlands feature tropical forests with very high endemism (species that are found nowhere else). These rainforests have been sustained for millennia, and while they have long supported native human populations and still encompass pristine swaths of forest, the biome faces a range of threats today, from more industrialized development to climate change.

Biodiversity

The Southeastern Papuan Rainforests biome is a landscape of tropical and subtropical moist broadleaf forests, tropical wet evergreen forests, and tropical montane forests. As a result, it offers a tremendous diversity of both floral and fauna. Climax trees of the canopy include tava (*Pometia pinnata*), canarium nut (*Canarium schweinfurthii*), ficus, and various palms. Klinki pine (*Araucaria hunsteinii*) is among the tallest tropical trees found anywhere in the world, reaching heights of 280 feet (85 meters) here.

Oaks are key deciduous taxa in the montane forests; a range of laurel species features in the understory, as do climbing vines, ferns, and epiphytes (species that live upon the surfaces of other organisms). Many shrubs and herbs spread across the forest floor and in riparian habitats.

The range of fauna in these rainforests boasts such colorful species as the painted tiger parrot (*Psittacella picta*) and the sooty honeyeater (*Melidectes fuscus*), another avian. Many species here are endemic or near-endemic, including dozens of mammals; these include Van Deusen's rat (*Rattus vandeuseni*), the small-toothed long-eared bat (*Nyctophilus microdon*), and the long-footed hydromine (*Leptomys elegans*). There is also an Australasian marsupial, the tree kangaroo (*Dendrolagus spadix*).

Threats

While this ecosystem has remained relatively intact over centuries of shifting human settlement, the pressures of contemporary civilization are finally catching up with it. Recent and current economic development schemes are designated for most parts of PNG, exposing the southeastern Papuan rainforests to many threats in the form of road construction, mining, industrial-scale logging, clear-

A type of diving beetle (Hyphydrus elegans) photographed in Papua New Guinea in 2010. The southeastern rainforests of Papua New Guinea have a very high concentration of species found nowhere else. (Wikimedia/ Michael Munich)

ing for forest plantations, hydrocarbon extraction, and other activities.

Swelling populations are a challenge in their own right, as urban infrastructure has not kept pace with sprawl that threatens to upset biological balances in inland and coastal habitats alike. Port Moresby, the capital of PNG and located within the southeastern Papuan rainforests, is a prominent example: It now has high violent crime rates among its expanding population of more than 300,000, creating a growing footprint of social and ecological damage and decay.

Elsewhere in PNG, the human presence offers aspects that are fascinating to anthropologists and social scientists of every stripe. The rugged topography of New Guinea has spawned human language and cultural groupings that are as extremely diverse as the floral and faunal communities—and indeed, the people have co-evolved along with these other species. Small parts of this evolutionary heritage are, for example, protected in Variata National Park.

Impacts of climate change here are projected to include higher likelihood of periodic drought and subsequent deforestation. In the meantime, air, water, and soil pollution linked to "development remain the greatest human-induced threats in this biome.

FALK HUETTMANN

Further Reading

Beehler, B. M. "Biodiversity and Conservation of the Warm-Blooded Vertebrates of Papua New Guinea." *Papua New Guinea Conservation Needs Assessment, Vol. 2.* Washington, DC: Biodiversity Support Program, 1993.

Diamond, Jared. *Guns, Germs, and Steel: The Fates of Human Societies.* New York: Norton and Co., 1999.

Flannery, Tim. *Throwim Way Leg: Tree-Kangaroos, Possums, and Penis Gourds.* Sydney, Australia: Grove Press, 2000.

West, Paige. *Conservation Is Our Government Now: The Politics of Ecology in Papua New Guinea.* Durham, NC: Duke University Press, 2006.

Paraguay (Eastern) Plantation Forests

Category: Forest Biomes.
Geographic Location: South America.
Summary: These constantly changing forests currently feature eucalyptus as the most prevalent tree type; many native species are under duress here.

Paraguay, a landlocked country in the center of South America, is a well-wooded nation, but one that has not developed its forestry industry in a sustainable manner. Forests have been exploited over the years, and the small percentage of land that is being used for plantation forests is given over mainly to fast-growing exotic trees. Despite the high-valued timber of Paraguay's native trees, its ability to compete in the global market is suppressed by inconsistent enforcement of forestry laws, and its remoteness from ports and markets.

Before trees were planted for profit here, the Guanani Indians cultivated plants such as the native citrus and the yerba mate (*Ilex paraguaiensis*) tea plant. The native tribes practiced slash-and-burn clearance on parts of the land, to open space for cultivation. Their small-scale crop farming did not, however, permanently alter the forest because they allowed the land to rest and regrow.

The Guarani eventually held larger yerba plantations and also made a living off working on major cattle ranches. With the introduction of cattle from the Spanish colonists, the workers cleared greater amounts of land. This technique was quick and easy, but led to erosion and the acceleration of forest fragmentation. Laws to protect the forest were not firmly in place until the 1980s. Today, agriculture, plantation forests, and other industrial development have put enormous pressures on the land in Paraguay.

Biodiversity

In comparison to the original tree planting of the Guarani Indians, bare patches of soil no longer sprout forth with the seeds of nearby shelter trees of the Atlantic forest. Instead, plantations are surrounded by open agricultural fields. The diversity of plants and animals in the region has dramatically decreased.

Rather than native trees, plantations of fast-growing eucalyptus and various pine species have been planted. This is partly because the markets for fine, high-valued timber are more difficult to find, unless they are to distant countries, such as the United States.

To do business with closer countries that are not members of the Mercosur Trade Group, tariffs must be paid to export. Meanwhile, the neighboring countries that are Mercosur members—Argentina, Brazil, Uruguay, and Venezuela—have cheaper trading costs, but do not necessarily have high demands for the high-valued timber. Paraguay's internal market demand is much higher for fuelwood and charcoal; these are lower-value commodities and do not require the trees to be large. Thus, the economic value of old-growth forests that do remain here tends to be diminished.

Animal species that must be considered in any planned changes to plantation forest operations include more than 400 bird species here. Several are endangered and quite unusual, such as blue-winged macaw, black-fronted piping guan, and harpy eagle. The national bird of Paraguay, the bare-throated bell-bird, is also vulnerable.

Mammals in and around the biome include red-brocket deer, lowland tapir, and predators such as jaguar and wolf.

Threats

In addition to the industrial commercialization of trees, smaller family-owned plantations have also been encouraged throughout past decades. Funds from the government and international agencies have been encouraging reforestation from a grass roots level since the 1980s. Technical assistance for rural farmers has become stronger over the years, yet is still a limited. Rural farmers have found ways in which to incorporate trees into their fields of cassava and sugarcane among, other crops. The trees help to buffer them financially when the agricultural market plummets. Likewise, there are opportunities for cattle ranchers to employ silvopastoral practices—the art of managing trees within a pasture-type system.

Many organizations in Paraguay join efforts to sustainably manage the land with the government and landowners alike. Parks and reserves have taken shape, and environmental education continues to be a major subject of high school curricula.

The choice of more ecologically sound mixes of native and exotic tree types for plantations in the eastern forests of Paraguay is only one aspect of the equation for sustainable futures. Although rural farmers may outnumber the wealthy, the wealthy own larger tracts of land. Additionally, Paraguay has not yet developed its forest industry to manufacture and support larger markets. Forests have been exploited over the years, leaving few large-diameter trees and cheaper, low grade pulpwood.

Furthermore, it is difficult for businesses and the government to justify leaving 16.8 million acres (6.8 million hectares) of deforested land to regrow, when it could be given over to rapidly-growing plantation species that will turn in a predictable, if low, profit. Undetermined variables including the market, socioeconomic stability, and lack of infrastructure make it difficult to predict how the timber industry can be bolstered in a sustainable manner through plantation forests.

Calculation of global-warming effects upon the plantation forest habitats here is difficult. Factors to evaluate include projections for increased incidence of drought, which it is generally agreed, the climate trends support. Coupled with heavier storm events, drought would work to weaken soil structure and increase erosion problems. Additionally, it will lead to severe pressure on various native flora, which in turn will motivate those

fauna species capable of migrating permanently to do so. Therefore, some habitats will change entirely, with invasive species likely to colonize at faster rates.

Among protected areas that support native forest stands in eastern Paraguay are the Guyra Reta Reserve, an intact zone of some 8,000 acres (3,240 hectares). Another important site, albeit not as rigorously protected as Guyra Reta, is the San Rafael National Park.

MICHELLE CISZ

Further Reading

Galindo-Leal, Carlos and Ibsen de Gusmao Carara, eds. *The Atlantic Forest of South America.* Washington, DC: Island Press, 2003.

Global Forest Coalition. "Tree Plantations in Paraguay, and the Role of False Solutions to Climate Change." http://globalforestcoalition.org/wp-content/uploads/2012/03/Tree-plantations-in-Paraguay.pdf.

Kernan, Bruce S., William Cordero, and Ana Maria Macedo Sienra. *Report on Biodiversity and Tropical Forests in Paraguay.* Washington, DC: United States Agency for International Development, 2010.

World Land Trust (WLT). "Guyra Reta Reserve." http://www.worldlandtrust.org/projects/paraguay/guyra-reta-reserve.

Parana River

Category: Inland Aquatic Biomes.
Geographic Location: South America.
Summary: The second-largest river in South America, the Parana runs through Brazil, Paraguay, and Argentina; it has a rich ichthyofauna and high species endemism.

The Parana River is the 10th-longest river in the world, and the second-longest river in South America, second only to the Amazon River. With a length of 3,030 miles (4,880 kilometers), the Parana River runs through the countries of Brazil,

Paraguay, and Argentina. The name Parana comes from the word *para rehe onava*, which means *like the sea* in the Tupi aboriginal language.

The confluence of the rivers Paranaiba and Grande give life to the Parana River in southern Brazil. Then it runs southward until it reaches Paraguay at the city of Saltos del Guira and becomes the natural boundary between Brazil and Paraguay for the next 124 miles (200 kilometers).

The Guaira Falls, a series of majestic cascades, was the former border between Paraguay and Brazil. Unfortunately, in 1982 the construction of the Itaipu Dam flooded the falls. This binational hydroelectric plant started operations in 1984 and is run by both governments. It provides 19 percent of the electricity consumed by Brazil, and 90 percent of what is consumed by Paraguay. It is considered the second-largest hydroelectric station in the world after the Three Gorges Dam in China.

Further downstream, the Parana River merges with the Iguazu River and becomes the natural border between Argentina and Paraguay. The river continues for about 800 miles (1,288 kilometers), making a turn to the west until it encounters the Paraguay River, the largest tributary of the Parana. After merging with the Paraguay River, the Parana turns to the south, and continues through Argentina for the next 509 miles (820 kilometers).

At the city of Santa Fe, the Salado River merges with the Parana River, and at the city of Rosario, it turns east. During the last 310 miles (500 kilometers) of its journey, the Uruguay River merges with the Parana River to form the Rio de la Plata; it then empties into the South Atlantic Ocean. The Parana Delta has numerous arms; it is 37 miles (60 kilometers) wide.

The Parana River has three main segments. The Upper Parana River in Brazil runs from Porto Primavera Dam until the upper portion of the Itaipu Reservoir. The High Parana starts at the Itaipu Dam and ends at the tri-national border of Argentina, Brazil, and Paraguay. The Middle and Lower Parana start where the Parana River meets the Paraguay River, and ends where it meets Rio de la Plata.

The Parana River and its tributaries form a vast watershed that serves numerous cities and supports the regional economy. Fish species like the sabalo and suburi are of particular economic significance. The Parana River is navigable up to the dams, but the economic benefit from the hydroelectric outputs outweighs the potential drawbacks to commercial shipping. Nevertheless, much of its length serves as a corridor that connects many inland cities in Paraguay and Argentina with the Atlantic Ocean. The waters are deep enough to allow large ships to easily navigate its waters. The area of the upper Parana River Basin extends to about 553,000 square miles (891,000 square kilometers) in Brazil.

The climate is tropical, with an annual mean temperature of 59 F (15 C). Yearly rainfall is more than 59 inches (150 centimeters).

The 1982 Itaipu Dam on the Parana River, shown here, supplies 19 percent of the electricity consumed in Brazil, and 90 percent of what is consumed in Paraguay. It is considered the second-largest hydroelectric station in the world. (Thinkstock)

Biodiversity

A common understory plant found in the riparian areas of the Parana is the *Palicourea crocea* (*Rubiaceae*), a nectar-producing plant that feeds three species of hummingbirds. It blooms during the October-to-December rainy season, and is especially abundant in the Upper Parana. Bromeliads, or pineapple-family plants, are also common in this area.

The Parana waters host a high diversity of tiny rotifers, benthos, phytoplankton, zooplankton, periphyton, and aquatic macrophytes. The biological richness of the river sustains more than 300 fish species grouped in nine families. More than 50 percent of the fish species belong to the *Characiformes* and *Siluriformes* families. Endemism (species found nowhere else) is common in tributaries of the Parana River, in part because of isolation by waterfalls.

During springtime floods, the migratory fish species golden dorado (*Salminus maxillosus*), Atlantic sabertooth anchovy (*Lycengraulis grossidens*), and sabalo (*Prochilodus lineatus*) swim upstream to spawn. The sabalo is a key species in the Parana River and the base of the food chain, while the golden dorado is an endangered species in Paraguay. The South America lungfish inhabits the lower Parana River basin, and is the only species in the family *Lepidosirenidae*. Two species of piranha, the pirambeba (*Serrasalmus marginatus*) and the pirambeba dorada (*S. spilopleura*), live in these waters as well.

Three types of fishing are practiced here: artisanal, subsistence, and recreational. Fishermen who live along the river banks participate in artisanal fishing. Small farmers carry out subsistence fishing, while residents of major cities practice recreational fishing. Downstream, the Rio de la Plata delta is of enormous regional importance as a migratory and nesting bird habitat.

Threats and Conservation

The diversion of water for agricultural purposes and the construction of dams threatens the Parana River ecoregion. The massive generating capacity of the hydroelectric power plants have had a positive economic impact in the region, yet the ecological cost is underestimated; it may be incalculable. For instance, constructing dams blocks fish migration, which results in high endemism in the area, and endemism makes the species vulnerable to environmental changes. Furthermore, the water quality diminishes due to agricultural pesticides, fertilizer runoff, and untreated domestic sewage.

As a mitigation measure to the environmental damage of the dams, the local governments have enacted laws under which the dam operators had to implement either hatcheries or fish ladders to protect the ichthyofauna. The failure of the fish ladders mechanism led to new enforcement mandates in the 1960; in 1971, regulators again compelled each dam to build and maintain at least one hatchery in every sub-basin. Such struggles are ongoing.

Climate change impacts in the region include intensified El Nino activity, which increases the frequency of flooding in the Parana River Basin. This flooding, and the heavy rainfalls that accompany it, disturb the wildlife in and around the river, erode and bury various habitats, and periodically increase the amount of pesticide and fertilizer runoff, leading variously to toxic situations and eutrophication (nutrient overload).

The World Wildlife Fund (WWF) carries out preservation projects in the area, including the Upper Parana Atlantic Forest Project; its goal is to guard the forest alongside the Parana River. Some of the WWF aims include creating a green corridor to protect key species such as the jaguar, promoting biodiversity, and restoring various ecosystem components. The fact that the Upper Parana Atlantic Forest area is one of the most endangered rainforests in the world motivated the development of this project. About half of the plants and 90 percent of all amphibians found in this forest are endemic.

ROCIO R. DUCHESNE

Further Reading

Agostinho, Angelo A. and Luiz C. Gomes. *Biodiversity and Fisheries Management in the Parana River Basin: Successes and Failures.* Maringa, Brazil: Universidade Estadual de Maringa, 2002.

Barros, Vincente, Moira Doyle, and Ines Camilloni. "Potential Impacts of Climate Change in the la Plata Basin." *Regional Hydrological Impacts of Climatic Variability and Change* 1 (April 2005).

World Widlife Fund. "The Upper Parana Atlantic Forest Area—The 'Selva Parananese,' and What Undermines It." http://wwf.panda.org/who_we_are/wwf_offices/argentina/wwf_argentina_conservation/parana/area.

Parramatta River Estuary

Category: Marine and Oceanic Biomes.
Geographic Location: Australia.
Summary: This urbanized and heavily modified but biodiverse estuary on the east coast of Australia is threatened by urban development and contamination.

The Parramatta River forms one of Australia's most iconic estuaries; it is the main tributary to Sydney Harbor and Port Jackson. In geological terms, Port Jackson is a specific type of estuary called a *ria*. The word *ria* refers to an unglaciated river valley partly flooded by rising sea levels.

The river begins at the confluence of two smaller rivers just west of the city of Parramatta. An Aboriginal name, *Parramatta* stems from the word *baramada* or *burramatta*, and is commonly translated as *the place where the eels lie down.* Parramatta is the second-oldest settlement in Australia, established in 1788; today, it is an important city within greater Sydney. The estuary begins at the tidal limit, a weir in central Parramatta, and extends in an easterly direction for about 12 miles (19 kilometers) to an imaginary line between headlands at Yurulbin and Manns Point. Beyond this point, the estuary merges into Sydney Harbour. The total catchment of the river is approximately 50 square miles (130 square kilometers).

The Parramatta River estuary and Sydney Harbor form a large waterway that literally cuts Sydney in half. As a result, the many crossings and the regular ferries, called river cats, that stop at public wharves along the river are extremely important for transport and navigation in and around the area.

Biodiversity

The catchment supports a broad range of wildlife in diverse habitats including bushland, rivers, creeks, and wetlands. The estuary itself supports diverse assemblages of aquatic plants and animals, including large areas of mangroves that form important spawning ground, shelter, and feeding zones for fish and crustaceans. They also provide large amounts of organic matter for the base of the food web. Mangroves also help maintain water quality by filtering silt from runoff and recycling nutrients.

At the time of British colonization, mangroves were far fewer, but urban development has since allowed soil to be washed into the river, causing siltation in many bays and providing suitable conditions for mangroves. Between the 1980s and 2000s, the extent of mangroves increased by 25 percent. By contrast, large areas of seagrass were lost.

Eel, carp, long-necked turtles, and small mammals such as the ring-tailed and brush-tailed possum are common sights. Birds found in the Parramatta River Estuary biome include the yellow-tufted honeyeater and the white-napped honeyeater, both often seen near the upper portions of the river.

Effects of Human Activity

Over the past 200 years, extensive urban development along the river has meant that much of the natural vegetation has been removed. The original courses of creeks have also been altered, with many now canalized or straightened and lined with concrete. Large areas of agricultural land have been transferred to residential or commercial use. Land claim has occurred along many areas of the river, particularly at the heads of bays. The areas were often used as rubbish dumps before being converted to playing fields or large parks. Large amounts of shoreline are now protected by artificial seawalls.

Much of the foreshore still belongs to businesses or private residences, but an increasing amount of waterfront land is becoming available to the public as foreshore reserve. The estuary is important for recreational activities, and especially vital in its role of hosting nature reserves.

Approximately 1 million people live within the catchment area, and consequently, the area is under intense pressure from urban development. Researchers at the University of Sydney have shown that estuarine sediments in the Parramatta are heavily contaminated; this condition is worse on the southern side of the river, because that is where industrial development was greatest. Many of the contaminants, such as dioxins, lead, DDT, and polycyclic aromatic hydrocarbons, originated in chemical and paint factories here. Some areas have been remediated to remove contaminated sediment, or have been covered with concrete to prevent contaminants from being disturbed and incorporated into the food web.

In 2006, commercial fishing was banned throughout Sydney Harbor, including the Parramatta River, because of contaminated sediments in the river. Water quality in some areas of the estuary is monitored by the state Department of Environment and Climate Change for bacterial contamination. The water quality is generally acceptable except after heavy rains, when large volumes of water cause sewers to overflow, contaminating the estuary with waste, bacteria, and nutrients. Runoff from the land also transports litter, nutrients, and toxic chemicals into the estuarine waters and sediments, hindering wildlife health and undermining habitats.

The Parramatta estuary has several areas of wetlands, such as Bicentennial Park and Newington Wetlands, that are of national importance for birds, particularly migratory waders. Agreements exist with China and Japan that require each country to take measures for the management and protection of shared migratory bird species and their environments.

ANGUS C. JACKSON

Further Reading

Blaxell, Gregory. *The River: Sydney Cove to Parramatta.* Sydney, Australia: Halstead Press, 2010.

Cardno Lawson Treloar Pty. Ltd. "Parramatta River Estuary Data Compilation and Review Study." http://www.parramattariver.org.au/wp-content/uploads/Parramatta-River-Estuary-Data-Compilation-and-Review-Study.pdf.

Cole, Joyce. *Parramatta River Notebook.* Sydney, Australia: Kangaroo Press, 1983.

Patagonia Desert

Category: Desert Biomes.
Geographic Location: South America.
Summary: One of the world's most outstanding examples of a rain-shadow desert lies at the narrow southern tip of South America, where geography seems to rule out any but a cool and moist climate.

The Patagonia Desert lies in the southern conical tip of South America, also known as Tierra del Fuego, or *land of the fire.* The desert is considered to be the largest desert in the Americas, and the seventh-largest desert in the world. The Andes Mountains form a barrier immediately to the west, effectively shutting off the moist westerly winds blowing in from the South Pacific; this creates the rain-shadow effect. Half or more of this desert lies within only 200 miles (322 kilometers) of either the Atlantic or Pacific Oceans. Because of improper land use, more than 90 percent of Patagonian soils are degraded to some degree. In addition, severe desertification affects much of the fringe areas of the core desert region here.

Flora

Patagonia is largely covered by treeless shrub and grass steppes that give way to dwarf-shrub semi-deserts in the drier areas of the central plateaus here. Vegetation is characterized by the dominance of xerophytes, plants that have evolved remarkable adaptations to cope with severe water deficits. Shrubs, for example, have either very small, sclerophyllous (hard and waxy) leaves with abundant glandular hairs or leaves with thick cuticles. Many

Grasses in the Patagonia Desert. The largest desert in the Americas suffers from a high level of soil degradation, mainly because of poor land management practices and overgrazing by livestock. Over 90 percent of all Patagonian soils have been affected. This threatens surrounding areas, especially steppe regions, with increased desertification. (Thinkstock)

species have dwarf cushion growth, another means of conserving moisture.

Grasses commonly have leaves with thick cuticles, convoluted laminae, and bunch growth habits, and with fairly large accumulations of dead biomass that help shield the grass from transpiration.

Blended in the more arid steppe landscapes are small areas associated with rivers or permanent water sources, with more mesic (moisture-adapted) plant communities comprising mostly grasses, sedges, and rushes. These are referred to as riparian meadows. Although they account for a small proportion of the total area, they play a key role in livestock production, and in many cases suffer most from the effects of bad land management, specifically overgrazing.

Shrub steppes occur in transition areas between the grass steppes and the semi-deserts. Semi-deserts and shrub steppes exhibit similar latitudinal extents. The steppe grassland usually is found in areas that are less prone to moisture in the form of rain or groundwater. Steppe vegetation is well suited to this drier climate, and the grass generally is shorter than that on prairie grasslands. Steppes are virtually semiarid deserts in the making—they are highly threatened by overgrazing, and highly sensitive to global warming, with desertification advancing as temperatures and evaporation rates increase.

Together, both variants of semi-desert cover 22 percent of the region, with plant communities of low diversity, including on average 19 plant species. Dwarf shrubs with cushion habits are typical of this type of vegetation. *Nassauvia glomerulosa, N. ulicina,* and *Chuquiraga aurea* are dominant, accompanied by *Chuquiraga kingii, Brachyclados caespitosus,* and *Perezia lanigera.* Grasses, such as *Stipa humilis, S. ibarii,* and *S. ameghinoi;* and shrubs, such as *Chuquiraga avellanedae, Schinus polygamus,* and *Lycium chilense,* are secondary types in semi-desert plant communities.

Dwarf shrubs, such as *Azorella caespitosa, Mullinum microphyllum,* and *Frankenia* species; grasses, such as *Poa dusenii, P. ligularis,* and (less frequently) *Stipa neaei;* and shrubs, such as *Junellia tridens,* occur in clumps in ancient depressions and along natural drainage networks.

Patagonia's deserts and semi-deserts have a quite different non-gramineous (non-grass-family) flora from those in the Northern Hemisphere. *Stipa* and *Festuca* are common to the semi-deserts of all three continents, but the shrubs and forbs of Patagonia are mostly cushion plants belonging to the *Asteraceae, Verbenaceae,* and *Rubiaceae* families. The fact that these types of plants are favored here may be related to the fact that the strong, nearly constant winds create a large amount of transpiration stress. These plants have adapted to this harsh weather by having thicker cuticles and smaller stomatas that help prevent loss of water, which means less transpiration; they also have adapted photosynthesis cycles that allow them to flourish in this environment. The taxonomic uniqueness, however, probably is related to the distance of this tip of the South American continent from the rest of the world's land masses.

Fauna

During most of the Tertiary Period, there was an absence of land connections between South America and other continents. Thus, its fauna evolved in isolation, making it much more diverse than at present. Animals evolved that were unique to the continent, including giant sloths, tank-like glyptodonts, and saber-tooth marsupials. Because the isolation of this area was incomplete, however, caviomorphs and marsupials arrived during the Late Eocene. The greatest exchange seems to have taken place around 5 million years ago with the emergence of the land now known as Central America, which allowed for the fauna in North and South America to mingle. The Panamanian land bridge was fully in place 3 million years ago, and the interchange of fauna has continued to the present.

Animal life on the Patagonian steppe is comprised of grazing mammals and a wide variety of burrowing mammals such as ground squirrels and ferrets. Patagonia is left today mainly with ungulates, guanacos, and some rodents, dogs, and cats.

Birds may fill part of the niches that mammals have in similar habitats on other continents. The large flightless bird, the rhea, is a browser; ground-dwelling parrots are major carnivores. The heteromyids—long-tailed, nocturnal mice species—that are so abundant in the temperate deserts of the Northern Hemisphere are lacking in Patagonia, but are replaced by bird species and a few caviomorphs, a specialized clade of rodents. The avifauna of the driest portions of Patagonia include the large, ground-feeding bird family *Tinamus.*

Threats

The area remains sparsely populated, but mining for gas, coal, oil, and other natural resources may change this in coming years. Overgrazing is the long-standing threat to plants, soils, and animal habitats in some areas of the Patagonia Desert biome. The effects of climate change in this region of South America are difficult to model, due to countervailing variables such as potential changes in oceanic currents, Antarctic glacial melt rates, air temperature and pressure regimes, and uncertainty over long-term wind patterns. In sum, it is thought that the desert area may shrink or expand, depending on the complex interplay of these factors.

ALEXANDRA M. AVILA

Further Reading

Paruelo, J. M., E. G. Jobbágy, and O. E. Sala. "Biozones of Patagonia (Argentina)." *Ecologia Austral* 8 (1998).

Soriano, A. "Deserts and Semi-Deserts of Patagonia. Climate of Patagonia." In *Ecosystems of the World 5. Temperate Deserts and Semi-Deserts,* edited by N. E. West. Amsterdam, Netherlands: Elsevier, 1983.

Villamil, C. B. "Patagonia—Argentina and Chile." Smithsonian Institution. http://www.nmnh.si.edu/botany/projects/cpd/sa/sa46.htm.

Patagonian Grasslands

Category: Grassland, Tundra, and
Human Biomes.
Geographic Location: South America.
Summary: The Patagonian grasslands are
experiencing land degradation because of a
variety of stresses, mainly from overgrazing
and climate change.

Patagonia is a mostly cold, semi-desert region
in Argentina and Chile that covers a variety of
settings, mostly shrub-covered steppes, with
sub-Andean grasslands in the southern portions.
Almost all of Patagonia's grazing lands are on
the cool, semiarid steppes of the extra-Andean
territory of southern Argentina, and extending
into Chile around the Straits of Magellan. These
mainly grass-covered steppes comprise approximately 289,577 square miles (750,000 square
kilometers).

A description of Patagonia's climate is hampered by the low density and uneven distribution of weather stations, one station for every
15,444 square miles (40,000 square kilometers).
However, it is known that the climate is influenced mostly by warmer-than-land Pacific Ocean
air masses that are forced inland by prevailing
westerly winds. The Andes Mountains range
stands between this moist air and the Patagonian steppes, creating an extensive rain shadow
that controls regional climatic patterns. There is
a very steep gradient of mean annual precipitation, decreasing toward the east from 157 inches
(4,000 millimeters) in the eastern foothills of the
Andes to 6 inches (150 millimeters) on the central
plateau, some 112 miles (180 kilometers) east of
the mountains.

Annual precipitation in different areas of the
Patagonian grasslands reaches variations greater
than 45 percent from the more moist to the drier
end of the gradient. The eastern coastal strip is
influenced by moist air from the Atlantic, leading
to somewhat higher annual precipitation rates of
8–9 inches (200–220 millimeters) that is evenly
distributed seasonally, as opposed to the generally
heavier winter rainfall in most of Patagonia.

Some of the variation found in the precipitation rates can be associated with the El Niño-La
Niña cycles of temperature and humidity influence, but scientists report a longer-term cycle
for southern Patagonia: a significant decrease in
precipitation during 1930 to 1960, followed by a
reversal of this trend, with significant increases
over the subsequent 30 years.

Mean annual temperatures range from 61
degrees F (16 degrees C) in the north to 41
degrees F (5 degrees C) in the far south of Tierra
del Fuego. Mean temperatures of the coldest
month, July, are above the frost mark, although
absolute minimum temperatures can be below
minus 4 degrees F (minus 20 degrees C).

Scientists reported a significant increase in
mean annual temperatures over the last 60 years
of the 20th century for Río Gallegos, a provincial
capital located on the steppes surrounding the
Straits of Magellan, which has one of the longest
regimes of weather record-keeping. This trend
is also seen as consistent with predictions of climate change from global circulation models that
simulate enhanced atmospheric carbon-dioxide
concentration—but conclusions from weather-
station data analysis are still preliminary.

Strong, persistent westerly winds are an outstanding characteristic of Patagonia's climate.
Because there is relatively little land in the Southern Hemisphere to act as a barrier, westerlies can
gain impressive momentum, with annual intensities of 9–14 miles per hour (15–22 kilometers
per hour) and frequent gusts of more than 62
miles per hour (100 kilometers per hour), mostly
in spring and fall. Strong winds increase evaporation, and can have a considerable influence on
the productivity of sheep husbandry, through the
chill effect.

Soils and Biota

Soil textures can explain a large portion of the
variation in dominant plant life form, such as
grasses versus shrubs, across the region. Small-
scale, uneven distribution of various of soils tends
to increase with aridity, and important differences
in leaching and salinity occur over short distances,
possibly causing soils within a taxonomic group

Patagonian Ranching

There are currently more than 12,000 family- or company-owned sheep farms in Patagonia, with flocks ranging from fewer than 1,000 to more than 90,000 head of sheep. Patagonia's grasslands have been grazed by sheep for just over a century. Sheep numbers peaked in 1952, at more than 21 million, and since then numbers have been shrinking gradually, to about 8.5 million in 1999.

Ranchers raise un-herded Merino or Corriedale flocks in continuously grazed large pastures, usually for wool. Wool production is fairly insensitive to the forage scarcity associated with a large population of animals grazing on the land, and is somewhat insensitive to drought. On the other hand, some research blames present-day land degradation on these intensive and widespread wool-oriented grazing operations.

to function differently. More than 90 percent of Patagonian soils are degraded to some degree, mostly because of improper land use. More than 70 percent of the topsoil is coarse-textured, ranging from sand to sandy-loam. The sandy loam is composed of sand, silt, and clay.

In addition, severe desertification affects 19–30 percent of the region. Some of the most dramatic erosion processes occur in the form of sand accumulations that, in the early 1970s, covered approximately 32,819 square miles (85,000 square kilometers). Both aerial photography and satellite imagery indicate that many of these accumulations are about 100 years old, suggesting that the rate of wind-driven erosion has been accelerated by the introduction of domestic livestock, due to the fact that the livestock trample the fragile vegetation of the grasslands, resulting in native plant die-offs. Fewer plants means that fewer roots hold the Earth together, and nothing retains moisture in the ground, which results in the desertification of the grasslands. The Patagonian grasslands, by capturing and storing carbon, are essential to helping manage climate change effects in the area; their degradation undermines this vital ecoservice.

Although the Patagonian Grasslands biome was generally considered to have evolved under light grazing pressure, pre-European numbers of guanacos (*Lama guanicoe*), the only native ungulates, may have been higher than previously thought. Recent counts show that guanaco populations here are fairly stable now, at approximately 500,000.

The lesser rhea (*Pterocnemia pennata pennata*) and the upland goose (*Cloephaga picta*) are the most conspicuous birds. The Patagonian hare (*Dolichotis patagonum*), skunk (*Conepatus humboldtii*), and the small armadillo (*Zaedyus pichyi*), together with the lesser rheas, are important zoogeographical indicators. There are significant numbers of predators, such as red fox (*Dusicyon culpaeus*), gray fox (*Ducisyon griseus*), and puma (*Felis concolor*). Red foxes and pumas are responsible for most predation; lamb losses due to red-fox predation can be as high as 75–80 percent.

ALEXANDRA M. AVILA

Further Reading

Elbers, Joerg and Javier Beltran. *Patagonian Steppe—Developing a Transboundary Strategy for Conservation and Sustainable Management.* Quito, Ecuador, and San Carlos de Bariloche, Argentina: Temperate Grasslands Conservation Initiative, 2010.

Herzog, Sebastian K., Rodney Martinez, Peter M. Jorgensen, and Holm Tiessen, eds. *Climate Change and Biodiversity in the Tropical Andes.* São Jose dos Campos, Brazil: Inter-American Institute for Global Change Research (IAI), and Scientific Committee on Problems of the Environment (SCOPE), 2011.

Nature Conservancy, The. "Argentina—Sustainable Sheep Grazing." http://www.nature.org/ourinitiatives/regions/southamerica/argentina/placesweprotect/sustainable-sheep-argentina.xml.

Suttie, J. M., S. G. Reynolds, and C. Batello. *Grasslands of the World. Plant Production and Protection Series No. 34.* Rome, Italy: Food and Agriculture Organization (FAO) of the United Nations, 2005.

Peninsular Malaysian Rainforests

Category: Forest Biomes.
Geographic Location: Southeast Asia.
Summary: These biologically diverse rainforests feature continental Asiatic and Australasian floristic elements.

Peninsular Malaysia in southeast Asia was once fully covered by tropical rainforest, before the arrival of humans, and even now, almost all the tropical rainforest formations are still present here. The rich biological diversity in this biome may be attributed to geographical history and to the complex terrain and soil conditions present on the peninsula.

Geographically, peninsular Malaysia is the southernmost extreme of the Asian continent. Recent geological evidence suggests that a broad continuous mountain range connected eastern Asia to the Indonesia islands via peninsular Malaysia in the early Tertiary period, about 60 million years ago. This land connection permitted many of the mainland Asiatic flora to migrate to the southern latitude during the global cooling event by the mid-Tertiary.

In the Middle Miocene, about 10 million years ago, after the Australian Plate fully collided with the Asian Plate, some of the original Australian taxa crossed the shallow sea that was present at that time, and entered mainland Asia via peninsular Malaysia. Thus, peninsular Malaysia was an important pathway for the exchange of biological elements from two continents, and acquired some of its rich biological resources through those biogeographical events.

Biodiversity

The ever-wet tropical climate, with annual rainfall exceeding 98 inches (2,500 millimeters), and rather uniform diurnal temperature throughout the year of approximately 79–91 degrees F (26–33 degrees C), permits the establishment and growth of many tropical plant species. Except for the extreme northwestern region of peninsular Malaysia, which experiences a few dry months every year, the dry spells in a large part of the peninsula are usually short. Hence, in such humid tropical conditions, the changes in land relief, soil type, and soil water level are among the key environmental factors that regulate the distribution and pattern of the vegetation in peninsular Malaysia.

Lowland evergreen rainforest is the main forest formation found in lowland peninsular Malaysia, and it sustains the highest biological diversity compared to any other vegetation type. The forest here developed on well-drained soils, and is dominated by members of a single tree family, *Dipterocarpaceae*; hence, the forest is often referred to as a Dipterocarp forest. The forest is well known for its intense plant diversity, with numerous plant species growing together but usually occurring in low numbers.

The main mountain range runs along the length of the peninsula, with its maximum height of about 6,890 feet (2,100 meters) above sea level. Most of the mountains in the peninsula are granitic; in fact, two-fifths of the peninsula is comprised of granitic rocks. Change in land relief has a significant influence on the local temperature and rainfall distribution, which consequently shapes the vegetation found there. On this mountainous region, lowland evergreen forest is replaced by lower mountain forest, with upper mountain forest occupying the summit region.

Beach forests are found along the east coast of Peninsular Malaysia, with mangroves more common along the western coasts. Among the mangrove swamps, brackish water vegetation is common, especially in areas where fresh and saltwater mixes during high tides. Peat swamp forests develop in areas that are permanently waterlogged, and the soil there is highly acidic and mineral deficient. The tualang tree is one of the tallest trees found in the Peninsular Malaysian Rainforest biome. It can reach heights of over 250 feet (75 meters). Over 6,000 species of trees have been identified here, making the area quite rich in flora.

Animal species found throughout the rainforest include about 450 species of birds, such as the crestless fireback pheasant and other pheasants, woodpeckers, hornbills, and pigeons. Asian elephants, rhinos, and other large mammals lumber through the forests, while the endangered Sunda otter-civet preys upon local amphibians. The Malaysian tapir and brush-tailed porcupine are also found in this rainforest. This area is also one of the few remaining natural habitats for the rare Malayan tiger.

Threats

In peninsular Malaysia, forest loss is traditionally associated with commercial agricultural development, timber extraction, and mining. It has been reported that the forest area declined by almost 50 percent between 1971 and 1989, during the peak of agricultural expansion. The rate of deforestation has only been slowing since the late 1980s, when accessible tin deposits were exhausted and the country had shifted its emphasis from agriculture to industrialization.

However, timber and forest-related products are still important export commodities that bring significant revenue to Malaysia. Of the 13.24 million acres (5.36 million hectares) of forested area (40.7 percent of Peninsular Malaysia's total land area) still present in 2008, nearly half of it—6.9 million acres (2.8 million hectares)—is categorized as "production forest," meaning it is still subjected to timber extraction, albeit with appropriate forest management schemes to be followed and monitored.

Individual Malaysian states have complete jurisdiction over forestry matters; forest commodities have been critical in providing revenue for state budgets. Although a central body, the National Forestry Council was set up by the federal government to serve as a forum for discussing and coordinating the forestry policies among the 11 states in peninsular Malaysia. There has been considerable variation in actual implementation and enforcement by different state governments.

However, the fact remains that the high biodiversity found in the Peninsular Malaysia Rainforest biome is largely attributed to the terrain and soil complexities that exist here. A proper federal management plan that takes into account

the protection of various natural ecosystems found in different states is essential in order to safeguard the extreme range of habitats. Climate change impacts must also be worked into the equation; these are likely to include altered rainfall timing, and perhaps changes in patterns of seed germination, food availability, and breeding seasons, and thus new stresses on plant and animal habitats here.

KIEN-THAI YONG

Further Reading
Holttum, Richard Eric. *Plant Life in Malaya.* London: Longmans, 1953.
Jomo, Kwame Sundaram, et al. *Deforesting Malaysia.* London: Zed Books, 2004.
Manokaran, N. "An Overview of Biodiversity in Malaysia." *Journal of Tropical Forest Science* 5, no. 2 (1992).
Morley, Robert J. *Origin and Evolution of Tropical Rain Forests.* Chichester, UK: John Wiley & Sons, 2000.
Whitmore, Tim C. *Tropical Rain Forests of the Far East.* Oxford, UK: Oxford University Press, 1984.
Wong, Khoon Meng. "Patterns of Plant Endemism and Rarity in Borneo and the Malay Peninsula." *Academia Sinica Monograph Series* 16 (1998).

Persian Gulf

Category: Marine and Oceanic Biomes.
Geographic Location: Middle East.
Summary: The Persian Gulf offers rich biodiversity, but oil and gas extraction threatens the ecosystems here.

The Persian Gulf is a shallow marginal sea located in the northwestern Indian Ocean between southwestern Iran to the east; the Arabian Peninsula to the west and south; and Iran, Iraq and Kuwait to the north. The number of countries in total that surround the Persian Gulf are eight, and include:

Iran, Iraq, Kuwait, Saudi Arabia, Bahrain, Qatar, United Arab Emirates, and the Sultanate of Oman. It is sometimes also referred to as the Arabian Gulf by people of the Arab nations here. The Persian Gulf has a maximum length of 614 miles (989 kilometers), a maximum width of 224 miles (362 kilometers), a surface area of 155,964 square miles (251,000 square kilometers), average depth of 164 feet (50 meters), and maximum depth of 295 feet (90 meters).

The Indian Ocean connects to the southeast of the Persian Gulf through the Gulf of Oman via the Straits of Hormuz. The dominant river discharge into the Persian Gulf is from the north, by way of the Shatt al Arab estuary that delineates the boarder between Iran and Iraq. The Shatt al Arab receives freshwater from the Tigris and Euphrates River system that flow through Iraq, as well as the Karun River that flows from Iran. Other rivers that discharge freshwater into the gulf are exclusively on the Iranian side; these flow down from the Zagros Mountains and include the Mand, Hilleh, and Hendijan Rivers.

Due to the arid climate, the Persian Gulf experiences evaporation rates of some 5–7 feet (1.5–2 meters) per year off of the total surface area, an extreme situation that causes sharp water density and salinity variations in the Gulf. Surface salinity varies between 36 and 41 practical salinity units (PSU) in the summer, with a range of 38–43 PSU in the winter. Surface temperatures here are in the range of 75–93 degrees F (24–34 degrees C) in the summer, and a range of 59–68 degrees F (15–20 degrees C) in the winter. Circulation in the gulf is cyclonic, or counter-clockwise, and deflects freshly entering water from the Indian Ocean initially toward the Iranian coast.

Biodiversity

The arid environment of the Persian Gulf region creates a unique biome. Mangroves here are an ecologically vital marine coastal habitat, one that has been under threat because of human activities and development. The dominant mangrove species of the Persian Gulf is *Avicennia marina* or grey mangrove, which is distributed along both Iranian and Arabian Peninsula coasts. The grey mangroves inhabit the salt flats, or *sabkha,* adjacent to the coast. The grey mangrove typically is not submerged underwater; the root system consists of pneumatophores or aerial roots.

The less-common Asiatic mangrove, *Rhizophora muronata,* is more restricted in its distribution along both coasts; it occasionally fringes embayments on either side of the gulf. The Asiatic mangrove has a similar habitat to other species of the genus *Rhizophora,* which grow submerged roots in seawater. There have been recent efforts of restoring the Asiatic mangrove around the Persian Gulf, such as the UAE Agency for the Environment-sponsored program in 2008.

The Persian Gulf biome is home to several charismatic megafauna that are threatened by extinction due to region-wide conflicts, unregulated coastal development and expansion, and accelerated climate change. One of the most widely known is the dugong (*Dugong dugon*). The dugong is an herbivorous marine mammal belonging to the order Sirenia; its closest relative of the same order is the Atlantic Ocean manatee (*Trichechus* spp.), whose distribution includes both sides of the Atlantic.

The geographic distribution of the dugong is throughout the Indo-Pacific region spanning from the Persian Gulf to the coasts of Australia. In the Persian Gulf, the dugong can be found off the coasts of Saudi Arabia, Iran, Bahrain, Qatar, and the United Arab Emirates. All these countries recognize that the International Union for Conservation of Nature (IUCN) Red List has the dugong listed as vulnerable, and discourage hunting of the species.

The dugong lifespan is similar to humans, and they can reach an age of up to 70 years. The adults can reach a length of 9 feet (3 meters) and approximately 881 pounds (400 kilograms). The young reach sexual maturity between the ages of 8 and 18; the female dugong typically bears a single calf once every three years, and rears the calf for 18 months.

The dugong inhabits shallow seagrass beds, which are also their main source of food. When seagrasses are scarce, dugongs may feed on deepwater algae and small invertebrates. Dugongs can hold their breath underwater for about eight minutes, similar to dolphins, but short compared to other marine mammals such as whales (at least 20 minutes and up to 2 hours), seals (70 minutes), or walruses (30 minutes).

There have been five species of whales, three species of dolphins, and one species of porpoise recorded in the Persian Gulf. The whale species include Bryde's whale (*Balaenoptera edeni*), blue whale (*Balaenoptera musculus*), humpback whale (*Magaptera indica*), fin whale (*Balaenoptera physalus*), and false killer whale (*Pseudorca crassidens*). Whales generally have a global distribution, and may have traversed the gulf but gone largely unnoticed due to the lack of a whaling industry here. The same is true in identifying dolphin and porpoise species in the Persian Gulf, which are also a globally cosmopolitan group of mammals.

Three species of dolphins and one species of porpoise have been documented in the Persian Gulf; these include speckled dolphin (*Sotalia lentiginosa*), Indo-Pacific bottle nosed dolphin

Over 70 species of migratory birds visit the Persian Gulf. The photo shows a black socotra cormorant (Phalacrocorax nigrogularis), right, and gray western reef heron (Egretta gularis) in the United Arab Emirates. (Wikimedia/Nepenthes)

(*Tursiops aduncus*), pantropical spotted dolphin (*Stenella attenuate*), and Asiatic black finless porpoise (*Neophocaena phocaenoides*). The warm and highly productive waters of the Persian Gulf are likely to be appealing as a feeding ground for cetaceans of all sizes. The great abundance of fishes and shrimp in the Persian Gulf appears to be able to support and sustain large populations of whales, dolphins and porpoises. However, the activities associated with the oil industry and decades-long armed conflicts may have rendered these species endangered in these waters.

The other charismatic megafauna group is the sea turtle. There are five species of sea turtles identified in the Persian Gulf: loggerhead turtle (*Caretta caretta*), green sea turtle (*Chelonia mydas*), hawksbill (*Eretmochelys imbricata*), olive ridley (*Lepidochelys olivacea*), and the leatherback sea turtle (*Dermochelys coriacea*). The IUCN Red List has the loggerhead and green turtles listed as endangered, the hawksbill and leatherback turtles listed as critically endangered, and the olive ridley turtle as vulnerable. Females of all these species are known to lay their eggs on the shores of islands around the Persian Gulf; many of the known egg-laying sites are protected by the respective governments.

The Persian Gulf hosts over 70 species of migratory birds that use the Gulf as a flight path, rest stop, and feeding grounds as they fly south to winter in Africa, or fly north to summer in northern Eurasia. These include various species of pelicans, cranes, flamingos, and warblers. Most habitats for migratory and endemic birds are in a grave state of degradation due to development. One rare species of kingfisher that is in danger of extinction is the collared kingfisher (*Todirhamphus chloris kalbaensis*).

Coral reefs are the hot spots of diversity in the Persian Gulf biome. Coral reefs and coral species are distributed throughout the shallow waters here, except for the northernmost reaches that are dominated by alluvial mud flats associated with the Shatt al Arab outflow. In general, a Persian Gulf coral reef consists of several coral species and can go up to 34 species. The dominant hard (scleractinian) coral genus in the Gulf is

Acropora spp., which is endemic (found nowhere else); the most common endemic species in the gulf is *Acropora arabensis*. Other genera include *Porites*, *Siderastrea*, *Pseudosiderastrea*, *Favia*, and *Platygyra*. Anthropogenic activities and climatic factors have led to serious degradation of the coral reef habitat—jeopardizing many species of marine reef fishes and invertebrates.

Fishes of the Persian Gulf number over 700 species, and over 500 of these are associated with coral reefs. Among important commercial crustacean species of the Persian Gulf are the green tiger prawn (*Penaeus semisulcatus*) and the penaeid shrimp (*Metapenaeus affinis*), which uses the southern Iraqi marshes as a nursery ground by traveling up the Shatt al Arab as juveniles and returning to the Persian Gulf upon maturity. There are other brackish water shrimp that are part of the northern Persian Gulf biome that have no commercial value; however, they are important players in the aquatic ecosystem of this extreme environment.

Threats

Threats to the flora and fauna in this region come from the potential for oil spills and other human-made disasters. As was seen from the BP oil disaster in the Gulf of Mexico, such spills can harm the local environment and wildlife, causing untold damage. Climate change threats to the Persian Gulf region include the potential for food shortages, as acidification of the waters may decrease fish populations. Many people in the region rely upon fish as both a food and income source, and any decreases in the available amount and types of fish can create economic hardship and habitat disruption.

NASSEER IDRISI

Further Reading

Alosairi, Y., J. Imberger, and R.A. Falconer. "Mixing and Flushing in the Persian Gulf (Arabian Gulf)." *Journal of Geophysical Research* 116 (2011)

Idrisi, N., and S. D. Salman.. "Distribution, Development, and Metabolism of Larval Stages of the Warm Water Shrimp, Caridina Babaulti Basrensis (Decapoda, Atyidae)." *Marine and Freshwater Behavior Physiology* 38 (2005).

Riegl, B. "Corals in a Non-Reef Setting in the Southern Arabian Gulf (Dubai, UAE): Fauna and Community Structure in Response to Recurring Mass Mortality." *Coral Reefs* 18 (1991).

Riegl, B. and Sam Purkis. *Coral Reefs of the Gulf: Adaptation to Climatic Extremes.* New York: Springer, 2012.

Persian Gulf Desert and Semi-Desert

Category: Desert Biomes.
Geographic Location: Asia.
Summary: This important habitat for Palearctic waders is under threat from oil spills and other anthropogenic activities.

The low desert plains of the Persian Gulf are located on the northeastern coast of the Arabian Peninsula. This ecosystem includes a stretch of the United Arab Emirates (UAE) coastline at Abu Dhabi, the island of Bahrain, the coastline of Qatar, parts of Kuwait, and parts of the Eastern Province of Saudi Arabia extending up to the Rub'al-Khali Desert. The region is marked by red-brown sandy mounds and sparse vegetation. It is an important stopover for the birds traveling to Asia and Africa. Overgrazing, poaching, and oil spills are some of the major threats to the bird population.

This ecosystem comprises low desert plains, usually dry, west of the Arabian Gulf into the Al-Dahna Desert, which connects the an-Nafud Desert with the Rub'al-Khali Desert in the southeast; all are divisions of the Arabian Desert. Due to the limestone and sandstone formations containing microscopic shell material, the coastal sands appear to be whiter than the red sands of the Al-Dahna and the Rub'al-Khali deserts. The entire coastline is about 746 miles (1,200 kilometers). The Rub'al-Khali Desert is mostly uninhabited, hence the name, meaning *the Empty Quarter*. It is one of the largest continuous bodies of sand on Earth, and holds one of the largest oil reserves in the world.

Dry lakebeds in the desert indicate the presence of water here thousands of years ago. The terrain is covered with sand dunes, and sprinkled with gravel and gypsum plains. The presence of feldspar gives the sand a reddish-orange color. The windblown sand accumulates at the base of plants, forming sand hummocks.

Long, hot, dry summers with northerly winds sometimes cause sandstorms in the region. This ecosystem is sometimes classified as hyperarid. Annual rainfall ranges from 1 to 6 inches (35 to 150 millimeters). It has an average temperature of 73 degrees F (23 degrees C) in winter. In summer, the temperature is usually 95 degrees F (35 degrees C), but may rise to 122 degrees F (50 degrees C). Along the coast, the humidity may reach 90 percent in summer.

Several freshwater artesian aquifers, along with wells, form oases in the inland, which help with wheat and alfalfa cultivation in this climate. UAE, Kuwait, and Qatar are increasingly re-landscaping most of the area to make it cultivable where an adequate supply of water is possible.

Biodiversity

The vegetation in the area is sparse, with limited varieties. Although date palm groves are in abundance, reeds like phragmites, mace, *Pamarix aphylla,* and *Prosopis juliflora* are also found. Various small shrubs, such as *Rhanterium epapposum* and *Hammada elegans,* as well as grass tussocks such as *Panicum turgidum* and *Stipa capensi,* are also seen in this region. *Avicennia marina,* or grey mangrove, is found in the muddy channels of the seashore.

The desert is an important habitat for migratory seabirds, as it connects Africa with Eurasia. More than 250 species of birds are estimated to be found here. The coastline is especially vital for the black-necked and great crested grebe, and for Saunders's little tern. The Socotra cormorant's breeding population is estimated to be above 95 percent of the world population. Tarut Bay on Saudi Arabia's gulf coast, a nursery for fish and shrimp, hosts a huge number of wintering and migrating waterbirds.

The Eastern Province of Saudi Arabia is the only site for the breeding of quail, spotted sandgrouse, and great grey shrike. Bahrain and Qatar have relatively fewer species and varieties, but waders are common in Bahrain on the coast. Many species are rare or occur accidentally in Bahrain, a haven for globally threatened species such as the houbara bustard. Although reptiles and mammals find it difficult to survive in the desert conditions, some animals are very well adapted the environment. Red foxes, Cape hares, and Ethiopian hedgehogs are typical in this ecosystem, while plenty of marsh frogs and Caspian pond turtles are also found.

Human Activity, Threats, and Conservation

The populations of seabirds and migratory birds face threats from oil spills along the coastline and pollution caused by other anthropogenic factors. In the 1990s, the damage of Kuwaiti oil facilities caused oil spills that released toxins into the atmosphere, causing the deaths of thousands of seabirds and rendering parts of the area uninhabitable. The spoiled wells released a huge amount of oil into the desert and seasonal lakes, contaminating the soil. During the Gulf War, armed forces applied depleted-uranium-tipped projectiles in Kuwait and Iraq, particles of which are thought to pollute the surrounding areas, including the Arabian Desert and its divisions.

The tern population is reported to be threatened on Saudi Arabia's gulf coral islands. Most of the people living in this region are nomads or seminomads; overgrazing by camels and goats poses a serious threat to the area's vegetation. Animals such as the jackal and honey badger are extirpated here, while others are endangered. Hunting and poaching also threaten species.

Re-landscaping the arid desert for cultivation is also critical. On one hand, there is a possibility of introduction of new species from surrounding areas, affecting the local desert bird species in particular; on other hand, the increased salinity of water and soil can cause disequilibrium of prevailing conditions. Coastal degradation due to oil spills, and wastes from oil refineries is another threat to habitat. Climate change may impact the desert areas, causing some areas to increase while others shrink; the likely effects are not clearly known.

In recent years, there has been an increasing need to form protected areas in this region. Realizing the threats to wildlife and the natural environment, governments in the region are beginning to do so.

MUHAMMAD AURANG ZEB MUGHAL

Further Reading

Al-Khalili, A. D. "New Records and Review of the Mammalian Fauna of the State of Bahrain, Arabian Gulf Region." In *Wildlife of Bahrain,* edited by T. J. Hallam. Manama, Bahrain: Bahrain Natural History Society, 1999.

Child, G. and J. Grainger. *A System Plan for Protected Areas for Wildlife Conservation and Sustainable Rural Development in Saudi Arabia.* Riyadh, Saudi Arabia: National Commission for Wildlife Conservation and Development, 1990.

Edgell, H. Stewart. *Arabian Deserts: Nature, Origin and Evolution.* Dordrecht, Netherlands: Springer, 2006.

Halwagy, R., A. F. Moustafa, and S. M. Kamel. "On the Ecology of the Desert Vegetation in Kuwait." *Journal of Arid Environments* 5 (1982).

Hirschfeld, Erik. "Migration Patterns of Some Regularly Occurring Waders in Bahrain 1990–1992." *Wader Study Group Bulletin* 74 (1994).

Ticehurst, D. C. B. and M. R. E. Cheesman. "The Birds of Jabrin, Jafura, and Hasa in Central and Eastern Arabia and of Bahrain Island, Persian Gulf." *Ibis* 67, no. 1. (1925).

Uttley, J. D., C. J. Thomas, M. G. Green, D. Suddaby, and J. B. Platt. "The Autumn Migration of Waders and Other Waterbirds Through the Northern United Arab Emirates." *Sandgrouse* 10 (1988).

Pescadero Marsh

Category: Marine and Oceanic Biomes.
Geographic Location: North America.
Summary: This marsh in northern California is an important stopover and nesting ground for migrating waterbirds, and a critical breeding and nursery habitat for many endangered species of fish, amphibians, and reptiles.

Located just 50 miles (80 kilometers) south of San Francisco, California, the Pescadero Marsh biome is an estuary that connects the redwood forests of the Santa Cruz Mountains to the kelp forests of the Pacific Ocean. The upper Pescadero watershed is extensively wooded, including willow-alder riparian corridors above the estuary. The marsh itself is not large, at only 340 acres (138 hectares), but it is home to at least six threatened or endangered species, including steelhead trout, coho salmon, the tidewater goby, San Francisco garter snake, and the California red-legged frog. During the winter, more than 200 species of birds use the marsh to nest and feed during their annual migrations.

Pescadero Marsh is located at the confluence of two streams, Pescadero Creek and Butano Creek, which together drain more than 80 square miles (200 square kilometers) of the Santa Cruz Mountains. As the largest watershed between the Golden Gate Bridge and the coastal town of Santa Cruz, the Pescadero-Butano watershed and Pescadero Marsh represent an oasis of critical spawning and nursery habitat for fish, amphibians, reptiles, and shellfish along a well-developed coastline. It supports one of the largest remaining runs of steelhead within the northern California region; it also supported a large coho salmon run as recently as the late 1960s, although few if any coho have returned to spawn in recent years.

The Pescadero watershed and marsh typically has cool, wet winters and mild, mostly dry summers. Fog and overcast are common, particularly during the summer months, because of the cold, adjacent Pacific Ocean. January is the coolest month, with an average maximum of 60 degrees F (16 degrees C) and an average minimum of 40 degrees F (4 degrees C). September is the warmest month, with an average maximum of 72 degrees F (22 degrees C) and an average minimum of 49 degrees F (9 degrees C). Winter temperatures seldom drop below freezing, so there is rarely snow. Average annual precipitation is approximately 30 inches (75 centimeters).

Fifty miles south of the city of San Francisco, the Pescadero Marsh's 340 acres (138 hectares) provide habitat for six threatened or endangered species, including steelhead trout, coho salmon, the tidewater goby, San Francisco garter snake, and the California red-legged frog. The marsh, shown above, is also visited by 200 species of migrating birds. (Thinkstock)

Human Impact

Humans have been modifying and exploiting the marsh since before Europeans arrived in North America. The Portola Expedition, the first Spanish exploration of the California coast, found the area inhabited by the Ohlone people, who burned the grassy meadows of the marshes to stimulate growth of the grasses that they depended on for food. Throughout the 1800s, the foothills above the marsh were used for cattle ranching and farming.

Logging of the massive redwood trees found up in the mountains began in the 1850s and continued for the next 100 years. Occasionally, the sawdust released from sawmills into the streams would cause the waters downstream to become toxic to people and to fish. In the 1870s, lawsuits were brought against the sawmills to stop the practice, and in 1877, fish stocking began in Pescadero and Butano Creeks as an effort to rebuild the trout population.

During the 1920s and 1930s, farmers began to construct extensive levees along the creeks and upper marsh to divert water for crops. Also during this time, a bridge was built across the marsh as part of the construction of California Highway 1. The combined effects of hundreds of years of human activity on the marsh from fire, ranching, farming, logging upstream, and channelization of the creeks downstream has been to increase sedimentation in the marsh and degrade the suitability of the habitat for native species.

A common feature of many coastal estuaries is the formation of a seasonal sandbar across the mouth of the estuary. Behind the sandbar, the marsh gradually transforms into a freshwater lagoon. The formation of the lagoon here is critical to the growth of juvenile salmon and trout that live in the marsh. However, because of the increased sediment load in the marsh and diversions of water for crops, the lagoon may form too late in the season or be too shallow to function as nursery habitat for the fish of the marsh.

Runoff (mainly nitrates and phosphates) in the marsh waters from fertilizers used by farmers upstream eventually filter down into this area, causing it to become eutrophic, or overloaded with nutrients; the high levels cause algal blooms that reduce the oxygen in the water and increase decomposition in the sediments. Sometime in the fall, the sandbar barrier is breached, and the marsh transforms again from a freshwater lagoon back into a tidal estuary. Since 1995, the breaching of the sandbar has coincided with large fish kills in the marsh. Water low in dissolved oxygen and high in hydrogen sulfide is released during the flushing of the lagoon; this makes the marsh waters too toxic for fish to live. Native species populations in the wetland, for this and related reasons, are at critically low levels.

Conservation

Efforts have been made to alleviate some of the stressors placed on the watershed and the marsh. The cessation of sawdust dumping by the sawmills and the eventual protection of the old-growth redwood forest in the mountains of Butano and Big Basin Redwoods State Parks were vital to protecting the watershed downstream. In the late 1930s, California golden beavers were reintroduced to the upper watershed to reduce sedimentation downstream. Beavers can dramatically modify watersheds by holding back water and debris with their dams. Their reintroduction, following extirpation by fur traders, has been credited with reducing sediment flow downstream and represents a natural method of watershed protection.

In 2004, an extensive assessment of the Pescadero-Butano watershed, including the Pescadero Marsh, was conducted. This assessment was performed to evaluate the health of the entire watershed and to determine how best to manage the watershed and marsh to best support the critically endangered runs of steelhead trout and coho salmon into the future. The problems facing Pescadero Marsh are the same faced by coastal estuaries around the world.

Logging of upstream forests increases sediment flow downstream, and farming and ranching in the watershed increase the amount of nutrients in the estuary waters. Encroaching development encourages channelization and reduces seasonal flooding events, which actually may alleviate some effects of sedimentation and eutrophication. Butano Creek is presently identified as a California Critical Coastal Area, as it flows into a Marine Protected Area (Pescadero Marsh Natural Preserve). However, flooding and erosion already pose a threat to human and wildlife communities along the California coast, and there is strong evidence that these risks will increase in the future due to global climate change. California coastal areas are already vulnerable to storms, extreme high tides, and rising sea levels. Even with protection, the future of bay and wetland regions remains uncertain.

ROBERT D. ELLIS

Further Reading

Bay Area Open Space Council. *The Conservation Lands Network: San Francisco Bay Area Upland Habitat Goals Project Report.* Berkeley, CA: Bay Area Open Space Council, 2011.

Heberger, Matthew, Heather Cooley, Pablo Herrera, Peter H. Gleick, and Eli Moore, eds. *The Impacts of Sea-Level Rise on the California Coast.* Oakland, CA: Climate Change Center—Pacific Institute, May, 2009.

Monterey Bay National Marine Sanctuary Foundation. "Pescadero-Butano Watershed Assessment." Environmental Science Associates. http://montereybay.noaa.gov/resourcepro/reports/sedrep/pescadero.pdf.

Phang Nga Bay

Category: Marine and Oceanic Biomes.
Geographic Location: Asia.
Summary: This ecoregion, which was damaged by the Indian Ocean tsunami in 2004, faces continuing threats from increasing tourism and development.

Phang Nga Bay is a 154-square-mile (400-square-kilometer) area of water that lies between the Malay peninsula of southern Thailand and the island of Phuket, and is part of the Andaman Sea. The bay is shallow in nature and contains 42 islets, together with intertidal forest areas, mangroves, and coral reefs. Historically, the Phang Nga area represented a means of subsistence for the comparatively few fishing communities located in the region. Recently, tourist development on Phuket Island in particular has placed considerable pressure on the bay, in which coral diving, water sports, economic development, and attendant pollution have become prominent.

In 2004, the Indian Ocean earthquake and tsunami, one of the deadliest natural disasters in recent history and the third-largest ever recorded on a seismograph, forced water into the bay and led to mud displacement and sediment deposits. More than 5,000 people were confirmed to have been killed in Thailand by this event, 259 of them in Phuket and more than 4,000 in the province of Phang Nga as a whole. Investigations have indicated that perhaps three previous tsunamis have affected the region in the preceding 3,000 years.

The climate of Phang Nga Bay is tropical marine, with characteristic high annual rainfall and year-round high temperatures. Rain is abundant in the southwest monsoon season from May to October. The average annual rainfall is 140 inches (356 centimeters). The temperature fluctuates between 73 degrees F (23 degrees C) and 90 degrees F (32 degrees C).

Diverse Species

In 2002, an extensive portion of the bay was designated as a Ramsar site known as Ao Phang Nga Bay National Park. This was an effort to protect vulnerable species, including the dugong (*Dugong dugon*), Malaysian plover (*Charadrius peronii*), and black finless porpoise (*Neophocaena phocaenoides*). More than 80 bird and fish species are represented in the park, with three amphibian, 26 reptile, and 17 mammal species living in the associated coastal area. Other animal species within the park include: pantropical spotted dolphin, Indonesian white dolphin, wild boar, smooth-coated otter, long-tailed macaque, and the Sunda pangolin.

Fish found in the bay include: moray eel, puffer fish, and a variety of fish that live among the coral. Other marine species include 14 species of shrimp, 15 species of crabs, and 16 species of rays and sharks, including squat-headed hammer head shark (*Sphyrna tudes*) and freshwater stingray (*Dasyatis bleekeri*).

There are blue crabs, swimming crabs, mud-skippers, humpback shrimp, mud-lobsters, pomfrets, goliath grouper, butterfly fish, sole, anchovies, scad, and rock cod, as well as rainbow cuttlefish, soft cuttlefish, musk crab, mackerel, spinefoot, grouper, black sea cucumber, brain coral, staghorn coral and flowerlike, soft coral. The park also contains well-known karst formations, limestone caves, and a lengthy history of occupation dating back to 10,000 years before present, when the whole area was above sea level.

Human Impact

The damage inflicted by the tsunami on the coastal mangroves in Phang Nga and elsewhere affected many trees and plants that had been recently planted or replanted. The processes of economic development and globalization in Thailand had taken a severe toll on the environment before the tsunami, particularly in the case of mangrove forests. The total amount of mangrove forest had declined to a low of about 413,000 acres (167,000 hectares) in the mid-1990s, which represented only about 30 percent of the historical extent. Much of this was due to intensive farming of brackish-water shrimp.

Some of this loss had been reversed before the tsunami, when, by 2004, about 605,000 acres (245,000 hectares) of mangroves existed, much of which were located in the south of Thailand, particularly on and around the muddy river mouths similar to Phang Nga Bay.

Recuperation of the mangroves and other forest areas, which has received support and consciousness-raising around the nation, has taken place in conjunction with maintenance of the important fishing industry in the bay and, particularly, prawn fishing. Because of environmental degradation and overfishing by trawlers in the Gulf of Thailand, fishing in Phang Nga and elsewhere has been placed on a community basis, in which fishers are entrusted with responsibility for sustainability issues in their area of operation.

The majority of the population in the Phang Nga Bay area recognize the importance of biodiversity, not just for its own sake, but also because of the attraction it represents for tourists. The

The eroded limestone islet of Ko Tapu in Phang Nga Bay near Phuket is included in the Ao Phang Nga Marine National Park, established in 1981. (Thinkstock)

are often ignored, and inadequate enforcement permits industrial fishing to take place unhindered. Global warming, too, is exerting a cost, as sea levels are rising, delivering more stress on the intertidal zone, pushing saltwater further inland where vegetation is less salt-tolerant, and leading to greater erosion problems in the aftermath of storm events.

JOHN WALSH

Further Reading

Choowong, Montri, Naomi Murakoshi, Ken-ichiro Hisada, Punya Charusiri, Thasinee Charoentitirat, and Vichai Chutakositkanon, et al. "2004 Indian Ocean Tsunami Inflow and Outflow at Phuket, Thailand." *Marine Geology* 248, nos. 3–4 (2008).

Kashio, M. "Tsunami Impact Assessment in Mangroves and Other Coastal Forests in Southern Thailand." United Nations Food and Agriculture Organization—Regional Office for Asia and the Pacific. http://www.fao.org/forestry/8553-0964ce43 fc83d711b589172adceaedf1e.pdf.

Seenprachawong, Udomsak. "An Economic Valuation of Coastal Ecosystems in Phang Nga Bay, Thailand." Thailand National Institute of Development Administration. http://web.idrc.ca/en/ev-29434 -201-1-DO_TOPIC.html.

variety of natural resources available on land and in the water has contributed to the presence of a network of diverse villages specializing in different forms of resource use, although with some activities held in common. Inevitably, large-scale tourism development around the bay, including hotel construction and the creation of water-based leisure activities, is placing increased pressure on the environment as a whole—as well as rising efforts to ease such pressure.

Some high-profile projects in this regard have received popular support, but results have been mixed. Damage to the coral reefs in the bay and the declining sea-turtle numbers are easy to understand, and have resonance on emotional and spiritual levels. Nevertheless, official exclusion zones on dynamite and poison fishing in the bay

Philippine Sea

Category: Marine and Oceanic Biomes.
Geographic Location: Asia.
Summary: The Philippine Sea has been identified as a world center of marine biodiversity, but mismanagement of resources, lack of environmental law enforcement, and a high poverty rate have meant that ecology often takes a back seat to economic pressures.

Lying to the east and north of the Philippine Islands, the Philippine Sea stretches across some 2 million square miles (5 million square kilometers) of the North Pacific Ocean. North of the sea is Japan; to the northwest is Taiwan and the Sea of China; due west and extending southward is the Philippine Archipelago. To the south lies Papua New Guinea; Guam and the Northern Mariana Islands are to the southeast. Several seas—including the Celebes Sea, Sulu-Sulawesi Sea, South China Sea, and the East China Sea—are adjacent to the Philippine Sea but separated by various islands.

Because of geological formations that have resulted from plate tectonic activity, the floor of the sea is marked by faults and fractures that lie within the Philippine Sea Plate. One of the most significant features is the deep-sea trench, the deepest point on the entire Earth, that has been labeled the Philippine or Mariana Trench.

Because the Philippine Sea is located in the tropical climate zone, it is hot and humid all year, averaging around 80 degrees F (27 degrees C). It also has some of the warmest sea surface temperatures in the world. This part of the Earth is an area of intense tropical convection, a dominant heat source for global atmospheric circulation. It contains complex ocean current systems that link the Pacific and Indian Oceans, and is intermediate between the Asian monsoon regime to the north and the Australian monsoon to the south.

Biodiversity

The Philippine Sea is home to a vast aquatic ecosystem. More than 500 species of hard and soft coral lie within its waters. The Philippine Sea includes part of the Coral Triangle, which consists of waters off Indonesia, Malaysia, Papua New Guinea, the Philippines, the Solomon Islands, and East Timor. Near the apex of the triangle lies the Sulu-Sulawesi Sea, containing an ecosystem that is home to some 2,500 species of fish.

About one-fifth of all species of shellfish in the world live in the Philippine Sea, which is also home to sharks, moray eels, octopi, and sea snakes. The Japanese eel, tuna, and various species of whales regularly spawn in the sea waters.

While much is still unknown about the marine diversity here, it is estimated that at least 5,000 species of clams, snails, and mollusks live in these waters, along with more than 900 species of bot-

tom-living algae and thousands of other organisms of various kinds. A 2004 survey revealed the presence of 1,200 species of crabs and shrimp, 600 mollusk species, and hundreds of species that have yet to be identified.

Of the seven known species of sea turtles, six are found in the Philippine Sea. At least 2,824 marine fish species have been recorded in the Philippine Sea, 33 of which are endemic to the Philippine Sea, that is, found nowhere else. More than 1,700 of these are reef-associated fish species, with 169 pelagic and 336 deep-water species.

In addition to live species, there is fossilized evidence of extinct snails. The area around the Philippine Sea is also home to 86 species of birds and 895 species of butterflies. Some 352 of the latter are endemic to the Philippines.

Human Impact

The Philippine Sea is recognized as a center of marine biodiversity because of its wealth of marine ecosystems. However, environmentalists around the world have also identified the Philippines as a region of environmental concern, and in 2000, the International Union for the Conservation of Nature and Natural Resources (IUCN) labeled the area as a biodiversity hot spot, noting that, at a minimum, 418 species here are threatened. Current levels and methods of harvesting fish and other marine life are not sustainable, among other problems.

The Asian Development Bank has estimated that in some areas of the Philippine Sea, the quantity of marine organisms had declined by as much as 90 percent. Major concerns have been raised at both the national and international levels about the lack of responsible management of marine resources here. For much of its history, the Philippines was exploited by colonial powers, and the local population was unable to make decisions about protecting the environment. Even in the 21st century, the situation is complicated by the fact that China controls the adjacent Western Philippine Sea (South China Sea), and Chinese interests do not always run in tandem with those of the Philippines. There is little regulation of either commercial or local fishing in the area, and Chinese fishermen are among those who have been known to use dynamite, which wipes out fragile ecosystems.

The waters and the coast of the Philippine Sea are vulnerable to a host of environmental practices that combine to damage ecosystems in the area. Only about one-tenth of sewage in the entire country of the Philippines is adequately treated, for example. This means that 90 percent is deposited elsewhere, and most of that waste goes directly into the sea. Furthermore, the coastal mangrove swamps that are important fish breeding grounds are being stripped to provide wood for development. Abandoned mines, meanwhile, have been responsible for highly toxic mercury seeping into surrounding waters.

Habitat loss and degradation, increasing pollution, destructive fishing on both commercial and local scales, and over-exploitation as a result of expanding markets for marine products are all damaging this complex ecosystem. Compared with several other Asian nations, including Vietnam, Thailand, and Indonesia, the Philippines was ranked the highest in reef degradation and the lowest on taking action to protect its fragile reefs. Environmentalists accuse the government of being swayed by the fact that the fishing industry employs more than 1 million people and comprises a major portion of the Philippine economy. Another environmental concern for marine life here is the red tide algal blooms that first came to the attention of scientists in 1983. Believed to occur naturally, but also influenced by environmental changes including rising sea-surface temperatures, red tide is a high concentration of algae that depletes oxygen levels in water, and releases toxins that are poisonous to some fish.

Climate change is already having a big impact on marine and coastal ecosystems in the Philippine Sea by warming, acidifying, and raising sea levels. It is believed that rising global temperatures severely challenge, and could even eliminate, many coral-dominated reef systems. Global warming also means there will likely be more variability of monsoon events, and more intense cyclones and typhoons. There is an urgent need for improved management and cooperation between countries that surround this sea in conserving and protecting their resources against this swarm of threats.

ELIZABETH RHOLETTER PURDY

Further Reading

Adraneda, Katherine. "Scientists: The Philippines Is the World's 'Center of Marine Biodiversity.' Announce Alarm Over Threats." Underwater Times, June 8, 2006. http://www.underwatertimes.com/news.php?article_id=64017239105.

Batongbacal, Jay L. and Porfirio M. Aliño. "More Than Just Oil in the West Philippine Sea." *Philippine Daily Inquirer,* June 25, 2011.

Castilla, Juan Carlos and Tim R. McClanahan, eds. *Fisheries Management: Progress Toward Sustainability.* Ames, Iowa: Blackwell, 2007.

Goldoftas, Barbara. *The Green Tiger: The Costs of Ecological Decline in the Philippines.* New York: Oxford University Press, 2006.

One Ocean. "The Philippine Marine Biodiversity: A Unique World Treasure." United States Agency for International Development. http://www.oneocean.org/flash/philippine_biodiversity.html.

Pindus Mountains Mixed Forests

Category: Forest Biomes.
Geographic Location: Europe.
Summary: The mixed forests of the Pindus Mountains offer a glimpse into the human activities of ancient and classical Greece, and the former biodiversity of the lowland Mediterranean basin.

Altered by millennia of human activity, the remaining flora and fauna of the Mediterranean Sea basin reflect only a fraction of prehistoric biodiversity. Due to geographical remoteness and the ancient Greek respect for high-altitude vegetation, the remnants of a once-flourishing mixed forest of pine, oak, beech, and fir has survived on a distinct landform, the Pindus Mountains, also known as the "backbone of continental Greece." The Pindus Mountains mixed forests provide a significant contrast to one of the globe's highest concentrations of historical deforestation, which has occurred across the expanse of lowland Greece.

The Pindus Mountain range is located in northern Greece and southern Albania. The maximum elevation is found at Mount Smolikas at 8,650 feet (2,637 meters).

The climate here is predominately mediterranean, with cooler, alpine mediterranean temperatures at higher elevations. The area is characterized by average annual rainfall of 47 inches (120 centimeters), but at high altitudes this can be more than 79 inches (200 centimeters). Snow typically falls in the mountains during the winter months, and the minimum average temperatures are below freezing. Pindus is sometimes referred to as the southern part of the greater Balkan range. It extends across some 15,300 square miles (40,000 square kilometers).

Flora and Fauna

From Albania to the Gulf of Corinth, the limestone ridges of the Pindus Mountains provide sanctuary for the majority of Greece's remaining timber. Black pine (*Pinus nigra*) dominates other species from 1,640 to 6,230 feet (500 to 1,900 meters); Balkan pine (*Pinus heldrecheii*) commonly coexists and grows up to 7,870 feet (2,400 meters). Several species of fir cover the northern Pindus, including the Cephalonian fir (*Abies cephalonica*) and the silver fir (*Abies pectinata*), and both thrive at elevations up to 6,560 feet (2,000 meters). An economically significant tree, the Valonea oak (*Quercus aegilops*), has served as a source of vegetable dye and accompanies a multitude of tall oak varieties.

At altitudes above 4,920 feet (1,500 meters), perhaps the most common arboreal species in the continent—the European beech (*Fagus sylvatica*)—grows in sparse numbers due to the calcareous soils of the Pindus. The European horse chestnut (*Aesculus hippocastanum*), found among the mixed conifer and broadleaf forest, is native only to Greece and Albania. Additionally, juniper,

sycamore, poplar, and cypress complete the extensive forests of the Pindus range.

As a result of the extensive duration of human-driven forest depletion here, faunal diversity remains low except for the extremely variegated bird population. Pelicans, herons, spoonbills, and egrets inhabit the montane lakes of the Pindus. Other woodlands species include woodpeckers, thrush, blackbirds, and nightingales.

Large wild mammals are scarce; however, brown bears (*Ursus arctos*), wolves (*Canis lupus*), and jackals (*Canis aureus*) persist in limited numbers.

Human Impact

In ancient Greece, people worshipped woodland gods and spirits, demonstrating a familiarity with forests and arguably illustrating an abundance of woodlands. In particular, Greeks revered the oak tree and associated the species with Zeus, the god of sky, rain, and thunder. The correlation derived from the common understanding of Zeus residing atop the Greek mountains where the oak trees grow. Furthermore, the classical poet Homer's works detail a forested region with abundant olive, oak, cedar, and pine trees.

However, human challenges to Pindus Mountain vegetation, including deforestation, initiated as early as 6,000 B.C.E. The combination of agriculture and livestock grazing threatened forest proliferation; woodlands were cleared to accommodate both methods of human subsistence, fueling population growth in city-states. Moreover, the forests served as a source of fuel, and approximately 90 percent of timber consumption satisfied fuel needs during the classical period, circa 500–300 B.C.E.

Deforestation for the purpose of shipbuilding—a common practice of all classical Western civilizations jockeying for control of Mediterranean resources—further exacerbated timber scarcity and subsequent soil erosion. Consequently, historical forest-clearing and loss of soil have contributed to a cycle of frequent, often prolonged periods of aridity that have prevented natural regeneration of the mixed forests, which are now found only at higher altitudes. Scrub vegetation replaced the former stands of evergreens and deciduous trees that once covered most of the Greek landscape at the middle and lower elevations.

Although the Pindus Mountains have escaped modern-era urbanization, human encroachment continues to threaten the fragile ecoregion. The largest urban center in the area, Ioannina, contains a population of 50,000, but most settlements are agricultural villages of fewer than 200 residents. Mining, road construction, and illegal logging continue the tradition of deforestation. Likewise, mountain tourism intensifies forest degradation.

From antiquity to the 21st century, humans have consistently transformed the Mediterranean basin landscape and consequently affected flora, fauna, and climate. The old-growth mixed forests of the Pindus Mountains remain in jeopardy despite the global awareness of Greece's historical failures in forest management. The Pindus Mountains mixed forests and the adjacent scrub vegetation of lower-elevation terrain have, however, become a model in forestry and in the natural-resource dialogue for catastrophic outcomes resulting from ecosystem exploitation. Hence, the conifer and broadleaf evergreens of the Pindus maintain significance as both endangered relics and reminders of Greek civilization's ecological errors.

As global warming intensifies, these forests may yet suffer even more loss of habitat, depending on how successful efforts might be to combat higher temperatures, increased invasive species prevalence, and potentially worse erosion from heavier storms.

MATTHEW ALEXANDER

Further Reading

Mather, Alexander S. *Global Forest Resources*. Portland, OR: Timber Press, 1990.

McNeill, J. R. *The Mountains of the Mediterranean World: An Environmental History*. New York: Cambridge University Press, 1992.

Williams, Michael. *Deforesting the Earth: From Prehistory to Global Crisis*. Chicago: University of Chicago Press, 2003.

Woodward, Susan L. *Biomes of Earth: Terrestrial, Aquatic, and Human-Dominated*. Westport, CT: Greenwood Press, 2003.

Piney Woods Forests

Category: Forest Biomes.
Geographic Location: North America.
Summary: The humid eastern end of Texas contains pine forest more typical of the southeastern United States—and distinct from the remainder of the mostly arid and semiarid state.

The forests of eastern Texas are known within the state as the Piney Woods, or Pineywoods. The term evolved to distinguish pine forest in this area from drier post-oak savanna immediately to the west. This temperate coniferous forest occurs primarily due to higher, year-round rainfall—35 to 50 inches (89 to 127 centimeters) per year—which is derived mainly from moist air rising from the warm Gulf of Mexico. Humidity and temperatures are typically high.

Average temperatures in August range from 72 to 95 degrees F (22 to 35 degrees C), and average temperatures in January range from 36 to 58 degrees F (2 to 14 degrees C). The Piney Woods region extends roughly from Texarkana southwest to Tomball, Texas; east to Orange, Texas; and north back to Texarkana. It is part of a larger pine-hardwood ecosystem extending into the states of Louisiana, Arkansas, and Oklahoma. The biome covers approximately 54,000 square miles (141,000 square kilometers).

Flora and Fauna

Geologic rifting and uplift after Pangaea led to several features, notably the Coastal Plain (southeastern Texas), the East Texas Basin (east-central Texas), and the Sabine Uplift (northeastern Texas). Sediment washing onto receding coastline created belts of sand underlain by clays that influence current vegetation distribution, including longleaf pine concentrated on sandy ridges in the southern Coastal Plain portion of the biome. Sandy ridges also harbor seeps, springs, and bogs, which greatly influence endemic (found only here) botanical diversity, evident in such manifestations as carnivorous pitcher plants and other species.

The endangered red-cockaded woodpecker (Picoides borealis), shown here, depends on longleaf pine ecosystems, which have been greatly reduced in the Piney Woods Forests biome. (U.S. Fish and Wildlife Service)

The rolling East Texas Basin further inland, with elevations of 200 to 500 feet (60 to 150 meters), hosts predominantly loblolly pine, with post-oak savanna at its western end. The Sabine Uplift to the northeast harbors mostly mixed shortleaf pine and hardwood forest. Cutting down through these landscapes over time are (east to west, respectively) the Sabine, Neches, and Trinity River courses and basins. Along river bottoms are depositional sediments with fertile soils, and hardwood forests of water oak, sweet gum, magnolia, elm, tupelo, and ash that harbor the highest biodiversity of any Texas habitat, providing food, water, and cover for wildlife.

Tree species composition in the Piney Woods Forests biome can change with a small variation in average water depth along riverbanks. Bottomland systems support as many as 189 species of trees and shrubs, 42 woody vines, 75 grasses, 802 herbaceous plants, 116 species of fish, 31 species of amphibians, 54 species of reptiles, 273 species of birds, and 45 species of mammals. Some of the mammals found here include: armadillo, white-tailed deer, Virginia opossum, gray fox, bobcat, ring-tailed cat, rabbits, and many other small mammals.

Conservation

Several national forests were established in the 1930s and 1940s on cut-over and abandoned eastern Texas timberlands. The southernmost area of the region is called the Big Thicket for the forests that once thrived there. The Big Thicket is ecologically different from the rest of the region; the land today is mostly low-lying wetlands and swamps. An interesting and diverse mix of habitats has also developed in this area.

Big Thicket is a combination of bottomland forest, sandy upland habitat, and a mixture of western and eastern plant species in the southern part of the biome. Cacti and roadrunners live near orchids and cypress trees. It is so renowned for its high biological diversity that it was set up as a National Preserve (part of the National Parks system) in 1974. The last Texas sighting of the ivory-billed woodpecker occurred in the Big Thicket. Another species seriously reduced through logging and hunting is the endangered red-cockaded woodpecker, largely dependent on open, fire-adapted longleaf pine ecosystems that have been replaced with denser, faster-growing loblolly pine stands. Longleaf pine has been reduced to approximately 5 percent of its former distribution in the southeastern United States.

There is a trend toward restoration of longleaf pine ecosystems within the Big Thicket National Preserve, East Texas National Forests, and smaller nongovernmental nature reserves in the biome. Prescribed burning and planting of longleaf pine is occurring. The Louisiana black bear is also slowly making a comeback in the Piney Woods region. Private timberlands have recently changed ownership from long-term forest management firms such as Temple-Inland to shorter-term timber management organizations (TMOs), which can be more inclined to sell land after timber harvest.

This trend could lead to increased fragmentation of some Piney Woods forest into smaller ownership parcels, some of which may change to residential suburban or exurban land use.

In 2009, The Conservation Fund joined more than 20 nonprofits and government agencies in the America's Longleaf Initiative to rebuild the shrinking stands of longleaf pines across the south and southeast. It is hoped that the regeneration of some of these forest lands will mitigate the effects of global climate change in some capacity. This could include: increasing carbon sequestration; the creation of new, much-needed bird stopovers (due to rising temperatures); and the building of more buffering zones, such as wetlands, to absorb the impact of more severe hurricanes and flooding events.

WILLIAM FORBES

Further Reading

Crocket, G. L. *Two Centuries in East Texas: A History of San Augustine County and Surrounding Territory.* Dallas, TX: Southwest Press, 1962.

Frye, Roy G. and D. A. Curtis. *Texas Water and Wildlife: An Assessment of Direct Impacts to Wildlife Habitat from Future Water Development Projects.* Austin: Texas Department of Parks and Wildlife, 1990.

Gunter, Peter Y. A. *The Big Thicket: An Ecological Reevaluation.* Denton: University of North Texas Press, 1993.

Jackson, J. A. *In Search of the Ivory-Billed Woodpecker.* Washington, DC: Smithsonian, 2004.

Maxwell, R. S. and R. D. Baker. *Sawdust Empire, The Texas Lumber Industry 1830–1940.* College Station: Texas A & M University Press, 1983.

Po River

Category: Inland Aquatic Biomes.
Geographic Location: Europe.
Summary: The longest river in Italy, the Po supplies water for irrigation to Italy's primary agricultural region, but is threatened by pollution and overuse.

The Po is the longest river in Italy, at about 410 miles (650 kilometers), and also has the largest river basin, encompassing more than 27,000 square miles (70,000 square kilometers), about half of which drains montane regions, and the rest draws flow through plains. The Po basin in general constitutes a broad plains area between two mountain ranges, the Alps and the Apennines; it is the largest region of fertile agricultural land in Italy. Almost one-quarter of the population of Italy, or about 16 million people, live in this region.

The Po rises in the Cottian Alps in northwest Italy and flows through the regions of Lombardy, Liguria, Piemonte, Emilia Romagna, Veneto, Valle d'Aosta, and the province of Trento; it has 141 tributaries including the Adda, Dora Riparia, Dorea Baltea, Mincio, Oglio, Tanaro, and Ticino. The Po originates as a mountain stream on Monte Viso in the Alps; most of its vertical drop occurs between its source and Piacenza, after which it becomes a marshy and heavily diked river, with many meanders. In the upper Po plain, numerous lakes are created by the natural dams formed by low moraines; two of the largest are Lake Garda and Lake Como. The Po enters the Adriatic Sea in a wide cuspate delta southwest of Venice, created in part by the considerable sediment transported by the river.

The Po River Valley has a mild continental climate to a humid subtropical climate, depending on the section of valley. Summers and winters are both more extreme in the lower elevations along the river. Average temperatures here range between 50 and 59 degrees F (10 and 15 degrees C). The precipitation range is 25–45 inches (63–115 centimeters) per year. Generally, the precipitation amount increases in the region moving from south to north and from east to west. The Po's waters are highest around the month of May, because of snowmelt from the Alpine region, and November, due to rainfall.

Wildlife

The Po River Delta has been protected by the institution of two regional parks in the region of these wetlands: the Veneto and the Emilio-Romagna. The Po Delta Regional Park in Emilia-Romagna is the larger of the two, and in 1999 was designated a World Heritage Site by the United Nations Environmental, Scientific, and Cultural Organization (UNESCO). The park contains wetlands, forest, dunes, and salt pans. It has a high biodiversity, with 1,000 to 1,100 plant species and 374 vertebrate species, of which 300 are birds.

The ecoregion's wetlands are very important breeding and resting grounds for many bird species. Commonly seen here are egret, tern, heron, cormorant, and globally threatened ferruginous duck (*Aythya nyroca*). Several endemic (found nowhere else) species of fish are present, and the upland valleys of the Po River function as spawning grounds for many species, including some that are rare and/or threatened.

Human Impact

Although humans have been making modifications to the Po since ancient times, the most significant changes in the Po began in the 1950s. From this time forward, human activities altered the Po in several ways, including dredging to deepen the riverbed, forcing the river into channels, and erecting embankments (*argini*); the latter now exist along more than half the length of the river. Another major source of alteration in the river is sediment mining for construction purposes. In 2002, it was estimated that 33 million cubic yards (25 million cubic meters) of gravel and sand were removed annually from the river.

The Po River is used primarily for irrigation rather than for transport, and its floodplain provides a good growing environment for crops such as rice, maize, and sugar beets. Agricultural development in the area preceded the Romans. Today, about half the land in the Po Valley is irrigated, and some 12,000 square miles (31,000 square kilometers) are cultivated. As of 2003, about 70 percent of water withdrawn from the Po was used for agriculture, amounting to some 26 billion cubic yards (about 20 billion cubic meters) annually.

As there are no effective plans to promote efficiency or reduce water use, inefficient methods of irrigation continue to be practiced. In addition,

about 4.2 million head of cattle and 5.2 million hogs are raised in the area. The natural vegetation of the Po has also been altered; along about one-quarter of its length, natural vegetation has been replaced by poplar trees grown to be harvested for cellulose.

The Po basin is hugely important in the nation's economy, as 38 percent of the production of Italy is generated in this area, and 42 percent of the national workforce is employed here. Major industrial cities within the Po River biome include Cremona, Ferrara, Pavia, Piacenza, and Turin—and the metropolis of Milan is connected to the Po through a series of channels. The demands of this economic productivity have placed high burdens on the environment, while cultural and organizational factors have made it difficult or impossible to engage in effective resource management.

The combination of dense human population, agriculture, and animal husbandry produces a high organic load, and has created eutrophication (nutrient overload) in segments of the river with low flow rates. Although there is a permit system for point-source pollutants, it is not effective in controlling the problem. Pollutants from the many industrial installations in the area also contribute to a degradation of water quality in the Po. Although the Po Basin Authority is responsible for planning in the region, implementation is the responsibility of many other institutions (e.g., city councils, provinces) that often favor local interests, complicating the execution of an integrated management plan. One result is that many regulations that are in place are not enforced.

Major changes in water use will be necessary in northern Italy's economically strategic Po River basin because of climate change. Scientists are predicting that rising sea levels, reduced rainfall, and lack of snow in the Alps will combine over the next few decades to render the last 60 miles (100 kilometers) of the river useless in many cases, mainly because of saltwater intrusion. Aspects of the change are clearly well underway. Meteorological records indicate that the total number of rainy days in Italy decreased by 14 percent from 1951 to 1996.

Because so much of the water from the Po is used for agriculture, farming techniques, especially those dealing with irrigation, will have to be altered to become more efficient. At the same time, mitigation efforts, barriers, and other contingency plans will have to be developed to preserve the ecological function and habitat support capacity of the delta and estuary segments of the Po River.

SARAH BOSLAUGH

Further Reading
Grove, A. T. and Oliver Rackham. *The Nature of Mediterranean Europe: An Ecological History.* New Haven, CT: Yale University Press, 2003.
Marchi, Enrico, Giorgio Roth, and Franco Siccardi. *The November 1994 Flood Event on the Po River: Structural and Non-Structural Measures Against Inundations.* Genoa, Italy: Instituto di Idraulica, 1995.
Zwingle, Erla and W.A. Allard. "Italy's Po River. Punished for Centuries by Destructive Floods, Northern Italians Stubbornly Embrace Their Nation's Longest River, Which Nurtures Rice Fields, Vineyards, Fisheries—and Legends." *National Geographic* 201, no. 5 (2002).

Potomac River

Category: Inland Aquatic Biomes.
Geographic Location: North America.
Summary: The Potomac River is the second-largest tributary to Chesapeake Bay; its condition reflects the various land uses in its watershed.

The Potomac River is the second-largest freshwater input to Chesapeake Bay. The name *Potomac* is a European spelling of an Algonquian name for the Powhatan tribe, who inhabited the upper reaches of the Northern Neck Peninsula in the vicinity of modern-day Fredericksburg, Virginia. The spelling of the name has been simplified over the years from *Patawomeke*, which may mean *place of tribute* or *place of trade*. The Potomac begins at Fairfax Stone, West Virginia, and meanders

380 miles (620 kilometers) from the Appalachian plateau through the Ridge and Valley, Blue Ridge, and Piedmont provinces, finally tumbling 76 feet (23 meters) over the fall line at Great Falls to the Coastal Plain, where it empties into the Chesapeake Bay at Point Lookout, Maryland. The river is tidally influenced for the 100 miles (170 kilometers) below Great Falls, and is brackish for the final 50 miles (80 kilometers).

The river has many major tributaries. The Potomac proper forms at the confluence of its North and South branches. From there, it receives water from the Shenandoah and Cacapon rivers to the south; Conococheague and Antietam Creeks and the Monocacy River from the north; and the Anacostia and Occoquan rivers below the fall line.

The 14,700-square-mile (38,073-square-kilometer) watershed encompasses parts of four states—Pennsylvania, Maryland, West Virginia, and Virginia—and the District of Columbia. It is the fourth-largest watershed on the East Coast of the United States. A total 57 percent of the basin today is forested, 32 percent is in agricultural use, and 5 percent is suburban or urban development. The remainder comprises wetlands and open water. The Potomac is home to more than 6 million people, and thousands of species of plants and animals.

Biodiversity
The Potomac watershed contains several unique aquatic species and community assemblages. It is home to more than 15 species of freshwater mussels, many of which are the targets of conservation efforts due to their rarity and limited ranges. These bivalves live partially buried in the sediments and feed by filtering particles from the water flowing over them. Young are brooded on the gill surfaces of the parent mussel and then released into the water, where they attach to the gills of fish to complete their development.

The eastern elliptio (*Elliptio complanta*) is the most common mussel in the watershed, in the rivers and streams of the Piedmont and Coastal Plain. It appears to use the American eel exclusively as a host fish for its offspring. Other mussels exploit several host species during their reproduction. Mussels in the Potomac are negatively affected by land-use activities that alter hydrological regimes, water temperatures, or sediment and nutrient inputs.

Shortnose sturgeon (*Acipenser brevirostrum*), which were placed on the Endangered Species List in 1967, are also notable inhabitants of the Potomac River. Shortnose sturgeon are anadromous fish, spawning in rivers such as the Potomac but spending much of their time in the brackish waters of the Chesapeake Bay and coastal Atlantic Ocean marine habitats. They have a complex life history and are very long-lived, with the oldest recorded female reaching 67 years of age. The fecundity of the species is very low; females are slow to mature and thereafter spawn only once every three years. Their numbers in the Potomac were decimated in the early 1900s, due to by-catch from fishing pressure on a related species, the Atlantic sturgeon.

Several species of shad and river herring also migrate into the Potomac to spawn. American shad (*Alosa sapidissima*) spend much of their life in the Atlantic and return to natal rivers such as the Potomac to spawn in the spring. Juvenile shad spend their first summer in the river, then migrate to the open ocean by autumn. Shad have a relatively short maturation period of three to six years. Upon reaching sexual maturity, they return to the Potomac to spawn. However, numbers of migrating shad have declined sharply over the past 150 years.

As an economically important seasonal fish, American shad were subjected to extreme fishing pressure and were over-harvested through much of the 1800s and 1900s. The annual harvest for shad went from 17.5 million pounds (7.9 million kilograms) at the turn of the century to less than 2 million pounds (907,000 kilograms) in the mid-1970s. The fishery was closed in 1982. While harvests played a part in the drastic decline of the shad population, poor water quality, the loss of spawning habitat, and the construction of dams that hampered migration also contributed to the fish's decline.

Since the mid-1990s, shad restoration efforts have included stocking more than 20 million shad fry and the building of a fish ladder at Little Falls, a dam used as a drinking-water supply for Washington, D.C. After more than two decades of restoration efforts, the shad population in the Potomac is still well below its historic levels, but is increasing slowly. In 2003, a very small commercial fishery was re-established in the river, but recreational fishing is still limited to catch-and-release at the present time.

Many exotic plants and animals have been introduced into the Potomac watershed. Some of them have become invasive species that are having significant effects on the ecosystem. The northern snakehead (*Channa argus*) is an invasive fish that was first discovered in the river in 2004. It is now firmly established, and can be found from Great Falls to the Chesapeake Bay.

Native to northern China and eastern Russia, the snakehead's introduction was likely related

A brook floater (Alasmidonta varicosa), at left, and a northern lance (Elliptio fisheriana), at right. These two native mollusks were taken from the Potomac River at Shepherd's Island in Maryland. (U.S. Fish and Wildlife Service/Phillip Westcott)

to its popularity as a food fish among Asian communities and its availability in live-food markets. Juvenile snakeheads feed on a variety of benthic organisms (bottom-dwellers); adults are mainly piscivorous, or fish-eaters. While the impacts of the northern snakehead in the Potomac River have not been quantified, it is thought that these invaders may outcompete some native fish species for food and habitat.

The Asiatic clam (*Corbicula fluminea*), a bivalve mollusk native to Asia, the Middle East, and parts of Africa, was introduced along the west coast of the United States during the 1930s, and arrived in the Potomac River by 1977. Since that time, the clam population has varied widely from year to year, sometimes becoming very large in the tidal portions of the river. However, scientists have yet to discern negative effects due to this invasion. While they may be competing with native mussels for food and habitat, Asiatic clams are also helping filter the water in the Potomac and acting as a food source for aquatic birds and muskrats.

Likewise, a 1980s invasion of the aquatic plant hydrilla (*Hydrilla venticillata*) has appeared to have had beneficial effects on the tidal freshwater Potomac. A 17-year study indicated that while hydrilla became a dominant macrophyte in the river, it did not completely displace any native species of aquatic vegetation. In fact, the populations of native aquatic grasses expanded. Similarly, waterfowl populations increased with an increase in the growth of hydrilla and other nonnative plant species.

Human Impact

In addition to the introduction of nonnative species, land-based human activities have directly altered water quality in the Potomac watershed. Before 1994, the North Branch of the Potomac River was badly polluted by acidic coal mine drainage, rendering the environment desolate and devoid of all life for many decades. When some of the surface coal mines were abandoned in the 19th and 20th centuries, water flooded the mines. The influx of water and oxygen caused the oxidation of surrounding rock, which produced an array of chemicals (mainly sulfuric acid and iron precipi-

Endocrine Disrupters

In 2008, a study performed by the U.S. Geological Survey found astonishing occurrences of fish that exhibited a mixture of male and female characteristics in the same individual. This condition is known as intersex. Of smallmouth bass collected from the Shenandoah and the South Branch of the Potomac, 80–100 percent were found to be intersex. The study found that the rate of intersex within smallmouth bass was correlated with both human population density and farming intensity.

It is thought that intersex may be due to a new class of emerging contaminants, known as endocrine disrupters, that can alter the signaling ability of hormones in the fish. The use of chicken litter to fertilize farmland within the watershed has scientists concerned that the leachates from poultry manure, which ultimately enter the Potomac, are responsible for eutrophication of the waterways and possibly the development of intersex. This hypothesis is supported by recent research indicating that intersex occurs in adult fish when fish larvae are exposed to environmentally relevant concentrations of chicken-litter leachates. There are an abundance of industry-supported chicken farms in the Virginia, West Virginia, and Maryland headwaters of the Potomac, most established during the last two decades.

tates) that ultimately decreased the pH level of the water, increasing its acidity. This polluted water flows through the water table and drains into the headwaters of the Potomac.

Remediation of these effects continues to be a huge challenge. The pollution caused by acid mine drainage in the North Branch has not been eliminated, but through a series of lime-dosing treatments (to neutralize the water pH value) that began in the early 1990s, water quality has improved enough for some aquatic life to survive. A trout-stocking program has allowed this portion of the river to become fishable again, but due to current pollution levels above Jennings Randolph Lake, natural populations of trout are not supported in that portion of the river.

By far the major point sources of nutrient pollution into the Potomac are sewage-treatment plants. The largest such facility in the Potomac River watershed is Blue Plains Sewage Treatment Plant, just south of downtown Washington, which processes an average 370 million gallons (140 million liters) of wastewater per day from the city and its Maryland and Virginia suburbs, as well as run-off from city streets.

Conservation Efforts

Recently, Blue Plains made dramatic improvements in the way it treats wastewater by adding a biological nutrient-removal step to its processing system. This step has resulted in a dramatic reduction in nitrogen-loading to the tidal Potomac.

Despite these improvements, more work needs to be done to reduce treated effluent entering the Potomac from Blue Plains. Areas within the watershed on combined sewer systems, where storm water and wastewater flow down the same pipe, are faced with the threat of overflow during periods of heavy rain. Some pipes are not big enough to deliver all the water to Blue Plains for treatment during heavy rains. In these cases, overflow from combined pipes is directed to local waterways. Blue Plains has made a pledge that by 2025, it will reduce the number of combined sewage overflows in the Potomac and Anacostia Rivers and Rock Creek by 96 percent.

It is thought that an additional threat to the health of the Potomac will be from climate change. There will likely be more intense and frequent flooding, and with rising sea levels, existing tidal marshes and coastal habitats may shift. This will impact lower portions of the river and the wildlife there.

In spite of the problems currently facing the Potomac River biome, there is hope that local and federal efforts to restore the river from what President Lyndon Johnson famously labeled "a national disgrace" are slowly working. From the reduction of acid-mine drainage effects in the headwaters, to increases in water clarity and native submerged aquatic vegetation in the tidal freshwater por-

tion of the river, the Potomac is on its way back to improved health and sustainability.

M. Drew Ferrier
Claire L. Hudson

Further Reading

Johnson, D. and K. Hallberg. "Acid Mine Drainage Remediation Options: A Review." *Science of the Total Environment* 338 (2005).

Lookingbill, Todd R., et al. "Altered Ecological Flows Blur Boundaries in Urbanizing Watersheds." *Ecology and Society* 14, no. 2 (2009).

Odenkirk, J. and S. Owens. "Expansion of a Northern Snakehead Population in the Potomac River System." *Transactions of the American Fisheries Society* 136 (2007).

Phelps, H. "The Asiatic Clam (*Corbicula Fluminea*) Invasion and System-Level Ecological Change in the Potomac River Estuary near Washington, D.C." *Estuaries* 17 (1994).

Rice, J. *Nature and History in the Potomac Country: From Hunter-Gatherers to the Age of Jefferson.* Baltimore, MD: Johns Hopkins University Press, 2009.

Puget Lowland Forests

Category: Forest Biomes.
Geographic Location: North America.
Summary: This ecoregion, dominated by temperate coniferous forest, is now a major population center.

The Puget Lowland Forest ecoregion (PLF) of the Pacific Northwest is home to awe-inspiring old-growth coniferous forests and to millions of people. The mild temperature and ample rainfall of this biome promote forest development. Primarily in Washington state, it is bordered by the Olympic Mountains to the west and the Cascade Mountains to the east. The PLF extends north to Canada, where it includes the Fraser Valley lowlands, the coastal lowlands, and several of the

Gulf Islands. To the south, it extends to the border between Washington and Oregon.

The PLF ecoregion covers an area of approximately 8,683 square miles (22,488 square kilometers). Today, it is home to 3.2 million people, and it is anticipated that an additional 1 million people will inhabit the area by 2025. The PLF ecoregion contains the Canadian cities of Vancouver and Victoria, British Columbia; and Bellingham, Seattle, Olympia, and Tacoma in Washington state. The San Juan Islands that divide Puget Lowland from the Strait of Georgia in British Columbia are also in this ecoregion.

The Puget lowland is a structural downwarp that was subsequently modified by glacial scour and fill from the Cordilleran ice sheet during the Fraser Glaciation. When the ice sheet retreated 14,000 years ago, meltwater flowed southward, depositing gravelly outwash plains. Portions of these outwash plains are prairies, some of which display mounded topography or Mima mounds, consisting of more or less evenly spaced and circular mounds 1 to 7 feet (0.3 to 2.1 meters) high and 8–40 feet (2.4–12.1 meters) in diameter.

Climate

The Pacific Ocean buffers the PLF climate from temperature extremes, and provides moisture carried by the prevailing winds. As a result, the Puget lowland has a mild maritime climate with a mean annual temperature of 48 degrees F (9 degrees C), ranging to 59 degrees F (15 degrees C) in the summer and 38 degrees F (3.5 degrees C) in the winter.

Annual precipitation averages 31–35 inches (800–900 millimeters). Areas in the Olympic Mountains rain shadow may receive only 18 inches (460 millimeters) of rain a year, while other areas may receive as much as 60 inches (1,530 millimeters) of precipitation annually. More than 75 percent of yearly precipitation falls between the beginning of October and the end of March.

Elevation in the region ranges from sea level to 1,509 feet (460 meters), yet typically is below 525 feet (160 meters). One of the most influential factors in the Puget Sound area is the limited solar radiation available to plants because the sky is overcast for approximately 229 days a year.

Historical Biome Changes

Although grass and oak (*Quercus garryana*) prairies and savannas were once extensive features of the southern portion of the PLF, the natural tendency of even this landscape is toward forest. Native prairie vegetation, formerly maintained by Native American burning, has been reduced significantly in size since European settlement. Major threats to prairies include urbanization, agriculture, and invasion of nonnative and woody vegetation resulting from fire suppression and livestock grazing.

The earliest artifacts of inhabitants date to approximately 8,000 years ago. There is evidence that indigenous cultures had a well-developed marine-based lifestyle by 1,000 B.C.E. Tribes in the Puget Sound include the Duwamish, Nisqually, Skagit, and Snoqualmie, which speak a variant of the Salish language family.

While marine waters traditionally provided the majority of the food for the tribes, the lowland forests and prairies complemented this diet with plant foods including salal (*Gaultheria shallon*), huckleberries (*Vaccinium* spp.), salmon berries (*Rubus spectabilis*), and bracken fern (*Polystichum munitum*). Open-canopied habitats, including the prairies that were anthropogenically managed through burning, provided habitat for a variety of game and shade-intolerant food and medicinal plants, including camas (*Cammasia quamash*).

In the 1900s, logging transformed the landscape in the Puget Lowlands. Although planting of clearcuts with trees has been practiced on federal and other lands since the first half of the 1900s, the Washington State Department of Natural Resources has required such planting within three years of harvest on all private holdings since 1975. Today, 95 percent of the PLF ecoregion has been culturally modified. Small, isolated islands of original habitat such as old-growth forest, bogs, and prairie-oak woodlands are surrounded by urbanization.

Before European settlement, old-growth forests of the lowlands were primarily covered with western hemlock (*Tsuga heterophylla*), western red cedar (*Thuja plicata*), and Douglas fir (*Pseudotsuga mensiesii*). Periodic flooding and infrequent fires were historically the predominant disturbance regimes. Centuries typically passed between large-scale fire events in the moist forest types. Drier forests (*Quercus* spp., *P. ponderosa*) and prairies, however, experienced more frequent fires. Today, forest remains the primary land-cover class, at 48.4 percent across the biome.

Vegetation

Three vegetation zones occur in the PLF and are named for their dominant trees at late-successional stages: the Western Hemlock Zone, from sea level to the lower montane slopes of the basin; the Pacific Silver Fir (*Abies amabilis*) Zone, in the midmontane altitudinal belt; and the Mountain Hemlock (*Tsuga mertensiana*) Zone, at the higher elevations. The most prevalent tree species in the lowland forests today is Douglas fir. Typical understory species include salal, western swordfern (*Polystichum munitum*), several huckleberry species, Oregon grape (*Mahonia* spp.), and Pacific rhododendron (*Rhododendron macrophyllum*).

The PLF has towering evergreen trees that led early European explorers and settlers to describe and sketch the landscape as a primeval landscape. The old-growth stage of lowland conifer forest succession is at its prime at 350 to 750 years, yet it begins to resemble old-growth starting at 175 to 250 years. In the old growth stage, trees vary in size from seedlings and saplings to large trees with massive trunks that are well spaced. The shaded understory retains enough light to support a substantial understory of tree seedlings, shrubs, and forbs.

Biodiversity

The old growth in the PLF is exemplary of an ecosystem in which all parts are connected. The living forest depends on dead logs, stumps, and other litter to retain water through the summers' drought and provide habitat for microorganisms and invertebrates, which in turn recycle nutrients for new plant life. The canopy of the forest is often covered with lichen, ferns, mosses, and algae that collect moisture and particulate matter from the air.

The riparian forests in the PLF continue to provide important spawning areas for salmonids (*Oncorhynchus* spp.), amphibians, bats, bald eagles (*Haliaeetus leucocephalus*), and much more wildlife. Many species of birds and mammals are characteristic indicators of old growth in the PLF. Seven bird species rely on this old-growth habitat, and the forest canopy supports six mammal species, including the northern flying squirrel (*Glaucomys sabrinus*) and tree vole (*Arborimus longicaudus*). The northern spotted owl (*Strix occidentalis caurina*) nests in the cavities of old trees.

Conservation Efforts

Major support for preserving old growth since the 1990s has come from the urgency to preserve habitat of the spotted owl and other old-growth-dependent species. The Northwest Forest Plan was developed in the 1990s and banned timber harvest in 10 million of the 17 million acres (4 million of the 6 million hectares) of national forests in the Pacific Northwest.

Climate change in the Puget Lowland Forest biome is projected to lessen snowfall and increase rainfall. The results from a habitat standpoint will be a stressful shift in watering regimes, as snowpacks will hold back less water during winter. This likely would lead to greater erosion and flooding events in spring, and will be combined with earlier sprouting and breeding events by resident and migratory species.

The outcomes of these relatively rapid changes will need to be assessed in real time, but it can be assumed that dominant tree canopy and understory species of the lowland biome will experience at least a moderate level of plant community alteration. Also affected will be fauna that must contend with greater stream and river turbidity, and changes in the population dynamic of insects, freshwater crustaceans, and other food-web components.

Daniela Shebitz

Further Reading

Kruckeberg, A. R. *The Natural History of Puget Sound Country.* Seattle: University of Washington Press, 1998.

Rickets, T. H., E. Dinerstein, D. M. Olson, and C. J. Loucks, eds. *Terrestrial Ecoregions of North America: A Conservation Assessment.* Washington, DC: Island Press, 1999.

Sorenson, Daniel G. "Summary of Land-Cover Trends—Puget Lowland Ecoregion." United States Geological Survey, May 2011. http://landcovertrends.usgs.gov/west/eco2Report.html.

Puget Sound

Category: Marine and Oceanic Biomes.
Geographic Location: North America.
Summary: A marine estuary linked to the Pacific Ocean, this biome is a waterway of exceptional beauty and ecological diversity, but is threatened by pollution, shoreline degradation, and population pressures.

Located in the northwest corner of Washington State, Puget Sound is at once a tidal basin, an estuary, and an inland sea. The sound is an arm of the Pacific Ocean linked to the ebb and flow of marine tides through the Strait of Juan de Fuca and Strait of Georgia. Freshwater enters the system through 19 river basins that drain the Cascade Mountains to the east and the rugged Olympics to the west. Puget Sound is the largest marine estuary on the west coast of the United States, and second only to Chesapeake Bay nationally. Its 2,500 miles of convoluted shoreline is nearly fractal in its complexity.

From sandy bluff to pebbled beach, from saltwater marsh to freshwater estuary, the ecology of Puget Sound is as diverse as the ways in which water meets shore. Much of that shore has been altered by human development, resulting in a significant loss of habitat. Toxic chemicals and other pollutants have tainted the once pristine waters, contributing to pressure on threatened salmon populations and other marine organisms. With over 4.4 million people living along this unique inland sea, and substantial growth expected in its future, the health of the Puget Sound ecosystem will remain dependent on coordinated conservation strategies.

Today, the shores of Puget Sound are punctuated by the drone of fog horns, the whistle of trains, and the rhythmic lapping of waves—but 11,000 years ago, there was only the groan of ice as the Vashon lobe of the Fraser Glacier retreated into Canada, leaving a deep north-to-south-running trough. The waterway that filled this trough was called *Whulj* or *Whulge,* meaning *the saltwater before us* in the dialect of indigenous tribes collectively called the Coastal Salish.

In 1792, Peter Puget, for whom the Sound was named, served as second lieutenant to Captain George Vancouver aboard the English ship *Discovery*. Puget led an expedition to explore the intricate system of bays, narrows, and inlets of the sound in search of a hypothetical inland passage. In 1990, a proposal to rename Puget Sound and adjoining Canadian waters the Salish Sea was met with a mixed response. However, many of the rivers that feed into the sound honor the region's rich Native American heritage including the Nisqually, Puyallup, Skagit, Stillaguamish, and Duwamish.

Puget Sound is a place of vistas, of broad bays, and narrow inlets, above which rise low forested hills, sprawling cityscapes, and on clear days, the snow-capped peaks of Mt. Baker, Mt. Rainier, and the Olympic Mountains. The dugout canoes of the Salish tribes have given way to kayaks, sailboats, fishing boats, tugboats, cargo ships, super tankers, submarines, and an extensive system of state-run ferries that serve local recreation, national defense, and international commerce. Beneath the surface, Puget Sound is a mixture of deep basins and higher sills—a topography as varied and complex as that of the terrestrial landscape. Depths average 450 feet (137 meters), and range from 900 feet (274 meters) near the mounded green jewels of the San Juan Islands in the north, to 300 feet (91 meters) in the muddy-bottomed inlets of the southern sound near Olympia. Twice daily, heavy tides circulate water through an intricate system of channels, passages, bays, and inlets including Budd Inlet, Nisqually Reach, The Narrows, Elliot Bay, and Deception Pass.

Biota

The ebb and flow of the tide flushes nutrients and pollutants alike over diverse habitats including subtidal, intertidal, and shoreline zones col-

A young seal rests on a Puget Sound beach strewn with pieces of kelp. Because of the high levels of pollutants in Puget Sound water and sediments, harbor seals on Puget Sound proper have been found to be seven times more contaminated with toxic chemicals than those found in Canada's nearby Strait of Georgia. (Thinkstock)

lectively called the marine near-shore. Limpets, mussels, oysters, crabs, anemones, sea stars, jellies, chitons, and tube worms are but a few of the thousands of invertebrates reliant on healthy near-shore environments. Shallow eelgrass beds, such as those of Padilla Bay, are crucial to juvenile salmon and other wildlife.

In areas with sufficient current, forests of bull-kelp, with medusa-like heads of flattened fronds, shelter a variety of Puget Sound denizens including shark, sculpin, and rockfish—which can live up to 100 years in these cold waters. The giant Pacific octopus is an elusive predator of rocky habitats here, and a favorite of recreational scuba-divers. Sandy sediments conceal flounder, English sole, and other flatfish.

Overall, the sound supports more than 200 species of fish, 100 seabird species, and 14 species of marine mammals including orcas or killer whales, endangered on the south sound. Canary and yellow-eyed rockfish were listed as threatened on Puget Sound in 2010—as are populations of steelhead, chum, and Chinook salmon; the latter is vital to tribal, recreational, and commercial fisheries, and an iconic symbol of the region.

Conservation and Threats

Puget Sound was designated an Estuary of National Significance by the U.S. Environmental Protection Agency (EPA) in 1998; however, the area is imperiled by pollutants, shoreline degradation, and a growing population. The Washington State Department of Ecology has identified 1,400 toxic clean-up sites within one-half mile (0.8 kilometer) of the sound. Bottom sediments, particularly those near urban centers like Seattle, are contaminated with oils, heavy metals, and other pollutants from decades of industrial activity. Toxic compounds have migrated through the sound's food chain to concentrate in the tissues of high-level predators.

The EPA routinely monitors toxins found in fatty tissues in harbor seals, and tracks liver lesions in English sole. Harbor seals on Puget Sound proper are seven times more contaminated with toxic chemicals than those found in Canada's nearby Strait of Georgia. Orca whales resident to the south sound carry more PCB (polychlorinated biphenyl) contaminants than any other mammal monitored world-wide.

The human impact on the Puget Sound ecosystem has been extensive, and the population is expected to climb. Roughly two-thirds of Washington State's residents live along this dynamic waterway, the majority in the dense urban corridor between the cities of Everett and Tacoma. Stormwater runoff from these urban areas is another significant source of pollutants. Sewage contamination has contributed to ongoing shellfish closures in a region noted for recreational clamming and commercial oyster beds.

Along the fjord-like inlet of Hood Canal, a portion of the sound treasured for its recreational and scenic attributes, runoff of fertilizers and a growing population has compounded problems of low dissolved oxygen. Following massive fish die-offs in the early 2000s, Hood Canal was identified as Washington's first Aquatic Rehabilitation Zone.

In addition to Hood Canal, more than 70 percent of estuaries, eel-grass beds, salt marshes, and other critical habitats have been impacted by development along Puget Sound's shores. Armoring of the shoreline with bulkheads and seawalls to protect residential and commercial properties has disrupted natural processes that contribute sediments to beaches. Species such as the sand lance or candle-fish, which makes up the bulk of the food base for juvenile Chinook salmon, rely on intertidal sand or sand-gravel beaches for spawning—and this habitat is now quite disrupted.

Conservation

Segments of natural shoreline, and access to state-owned submerged saltwater lands, have been preserved in over 80 state and regional parks around the biome. Federal refuges, including three national marine estuaries at Nisqually Delta, Skagit Bay, and Padilla Bay, have been the focus of concerted restoration efforts, much of which was influenced by the need to protect habitats vital to salmon species listed as threatened by the U.S. Fish and Wildlife Service in 1999.

Efforts to restore salmon populations on Puget Sound have had positive impacts on ecosystem protection in general, as native vegetation and water quality have been restored or improved with local and region-wide initiatives that began with the Shoreline Management Act of 1971. In 1996, federal, state, tribal, and local governments combined to form the Puget Sound Action Team, which set priorities for restoration and conservation efforts.

The Puget Sound Initiative, proposed by Washington's governor and passed by the legislature in 2005, sought to reinforce the ongoing commitment of public, business, environmental, and government sectors whose cooperation is key to protecting the unique natural resources of Puget Sound. The Initiative set in motion coordinated funding, research, monitoring, education, and conservation efforts with a goal to restore and protect the Puget Sound ecosystem by 2020. For the millions who will live along this complex waterway, a personal connection to the Sound, along with a sense of collective responsibility, will be essential to preserving the ecological diversity that has made this region unique.

Climate change is yet another peril to be faced. Lower snowfall on nearby mountains is projected; this will mean a diminishing seasonal flow of freshwater into Puget Sound. However, warmer temperatures mean more rainfall will occur, leading to inland and coastal washouts and over-sedimentation of many areas. From the ocean side, higher sea levels could punish many of the already-damaged shoreline environments, taxing species just on the verge of recovery.

CONNIE LEVESQUE

Further Reading

Kruckeberg, Arthur R. *The Natural History of Puget Sound Country*. Seattle: University of Washington Press, 1991.

Masson, D. and P. Cummins. "Observations and Modeling of Seasonal Variability in the Straits of Georgia and Juan de Fuca." *Journal of Marine Research* 62 (2004).

Puget Sound Partnership. "Leading Puget Sound Recovery." Washington State Government. http://www.psp.wa.gov.

Pulley Ridge Coral Reef

Category: Marine and Oceanic Biomes.
Geographic Location: North America.
Summary: An unusually deep coral reef ecosystem built atop a drowned barrier island chain, Pulley Ridge supports a mix of marine species drawn from both deepwater and shallow-reef communities.

The Pulley Ridge Coral Reef biome is a ridge of drowned barrier islands more than 60 miles (97 kilometers) long, located approximately 150 miles (241 kilometers) west of the southern coast of Florida in the Gulf of Mexico. The ridge system is important because of its abundance of reef-building corals (particularly lettuce corals of the genus *Agaricia*) and reef-associated fish. It is generally thought to be one of the deepest such coral reefs in the United States, with depths ranging to 180–250 feet (55–76 meters).

During past glacial periods, when the sea level was much lower, this area formed a north–south chain of barrier islands protecting the Gulf Coast of what is now Florida, from about Cape Coral to Cape Sable. This formation was not unlike the islands of the Outer Banks, which now protect the shoreline of the Carolinas. As glaciation retreated and the sea level rose, the island chain became submerged; it is now considered a drowned chain.

Coral Community

While drowned barrier island formations are not exceptionally rare, the Pulley Ridge system is somewhat unique because much of the southern region of this feature is covered with hermatypic stony corals. Hermatypic corals rely on a symbiotic relationship with photosynthetic cyanobacteria called zooxanthellae that live within the coral polyp and provide energy to the coral for growth. Thus, they are dependent on sunlight to survive.

While many slow-growing coral species, known as ahermatypic corals or octocorals, exist in deeper water, those types are not dependent on light and generally do not build reefs. Pulley Ridge, on the other hand, is the deepest known occurrence of reef-building corals in waters of the United States.

Lettuce corals (genera *Agaricia* and *Leptoseris*) are most abundant in the Pulley Ridge system, accounting for up to 60 percent of live-coral cover at some localities. Less common species include star corals (genus *Montastrea*) and finger corals (genera *Madracis* and *Porites*). Along with these key coral types, a variety of the slow-growing octo-corals are present here. The community also supports both calcareous and fleshy algae; a wide variety of green, red, and brown macroalgae species thrive here, despite receiving less than 1 percent of the sunlight available at the surface—or about 5 percent of the light observed on most shallow-water coral-reef systems. Sponges, too, have found a niche here to grow in substantial numbers.

Marine Life

Pulley Ridge is home to a wide array of fish, at least 60 species in all, with a mix of typical shallow-water reef fish and some deepwater species sharing this ecosystem. Among those more typically seen in shallower waters but represented here are bluehead, bicolor damselfish, rock beauty, coney, French angelfish, and hogfish. Deepwater denizens present here include wrasse bass, bank butterflyfish, spotfin hogfish, deepwater squirrelfish, and roughtongue bass. Red grouper (*Epinephelus morio*), scamp (*Mycteroperca phenax*), and snapper here are of particular importance to commercial fisheries in the Gulf of Mexico.

Several factors help account for the existence of this unique community. The underlying drowned barrier islands provided elevated hard substrate, which is ideal for the formation of reefs. Also, the region is located along the western edge of the

Loop Current that brings relatively warm water to the southern ridge, but located along a front that upwells nutrients from deeper water to fuel the relatively high productivity observed here.

Conservation Efforts

The Pulley Ridge system was classified by the Gulf of Mexico Fisheries Management Council as a Habitat Area of Particular Concern in 2005, granting it protection from certain damaging kinds of fishing, specifically bottom longlines. The biome is considered a deepwater refuge, a source of larval recruits, and a recruitment corridor, providing continuity for many shallower reef habitats.

The exposure of the corals of Pulley Ridge to climate change impacts will be mitigated in part by its depth. Rising surface temperatures will not be likely to penetrate here; indeed, there is no evidence of recent bleaching events on Pulley Ridge. Heavier storm action, too, will be unlikely to affect an ecosystem at this depth. However, changing acidity of the seawater—which can be a side effect of greater oceanic absorption of excess atmospheric greenhouse gasses—could potentially affect the reproductive and coral-building capacities of some species here. This factor will have to be monitored.

Jason Krumholz

Further Reading

Cowen, Robert. "New Research to Illuminate Connections Between Reefs in the Gulf of Mexico." University of Miami. http://www.rsmas.miami.edu/news-events/press-releases/2011/new-research-to-illuminate-connections-between-reefs-in-the-gulf-of-mexico.

Halley, R., V. Garrison, K. Ciembronowicz, R. Edwards, W. Jaap, G. Mead, et al. *Pulley Ridge—The United States' Deepest Coral Reef?* Washington, DC: United States Geological Survey, 2003.

Jarrett, B. D., A. C. Hine, R. B. Halley, D. F. Naar, S. D. Locker, A. C. Neumann et al. "Strange Bedfellows—A Deep-Water Hermatypic Coral Reef Superimposed on a Drowned Barrier Island; Southern Pulley Ridge, SW Florida Platform Margin." *Marine Geology* 214, no. 4 (2005).

Purus Varzea Forests

Category: Forest Biomes.
Geographic Location: South America.
Summary: Among the most extensive of the Amazon flooded forest biomes, the Purus Varzea offers intense species abundance and diversity.

The Purus Varzea Forests biome is a swath of flooded woodlands that line the western tributaries of the Amazon River Basin, including the Solimões River, much of the Juruá, central Purus, and Japura-Caquetá Rivers. The Purus varzea biome represents the flooded forests of the seasonally inundated river basins of the central Amazon. Bird and fish diversity are extraordinary, but terrestrial mammal diversity is low, due to the periods when the entire area is underwater.

This biome, because it encompasses aquatic super-highways where boats are the primary transportation method, is increasingly more affected by human activities than surrounding rainforests. Varzea biomes share a single common trait—that they are an interface between freshwater and terrestrial ecosystems, where seasonally a tree can be either on dry land or completely submerged in water. This unique and threatened area supports not only a plethora of endangered species such as manatee and giant otter, but also provides for many indigenous peoples who have become largely displaced from elsewhere.

Biodiversity

In the dry season, the Purus varzea is used by some of the giants of the Amazon: the giant armadillo and the giant anteater. However, when the region floods, all such terrestrial animals seek higher ground, as the river can rise as much as 20 feet (6 meters), making the varzea forest a submerged aquatic environment. During these flood stages, another set of giants replace the mammals—some of the largest fish in the Amazon call these forests home. These include the pacu (*Metynnis* and *Mylossoma*), tambaqui (*Colossoma macropomum*), pirarucu (*Arapaima gigas*), sardinha (*Triportheus angulatus*), and the piranha (*Serrasalmus* spp.).

Many aquarium fish that are familiar to hobbyists around the world originate from these flooded forests, including discus, many cichlids (*Cichlasoma* genus), tetras, and catfish. Fish greatly outnumber birds and mammals here, and often fill the ecological role as seed dispersers and grazers.

Two primates are almost entirely restricted to this biome: the white uakari monkeys (*Cacajao calvus calvus*) and blackish squirrel monkeys (*Saimiri vanzolinii*). Monkeys are able to take refuge in the trees even during the flood season, and thus remain permanent residents. Other primates include monk sakis (*Pithecia albicans*), endangered black-chinned emperor tamarins (*Saguinus imperator*),

tamarins (*S. mystax*), night monkeys (*Aotus nancymaae*), and titi monkeys (*Callicebus dubius*).

Over 600 species of birds have been reported from this region; many of the varzea residents are familiar aquatic species like egrets, herons, ducks, ibis, spoonbills, and herons. Other birds that are adapted for water are common residents here, including kingfishers.

The largest tree in the varzea biome, kapok (*Ceiba pentandra*), usually forms the emergent canopy, and is propped up by enormous buttress roots. Other trees characteristic of the biome are *Parkia inundabilis*, *Septotheca tessmannii*, *Coumarouna micrantha*, *Ceiba burchellii*, *Ochroma lagopus*, and *Manilkara inundata*.

Threats and Conservation

Many threats confront the forests of these and other varzea biomes. Their accessibility to humans—especially during flooding season—makes them susceptible to both deforestation and overfishing pressures. Overharvest of aquarium fish such as the discus is likely affecting local populations. Climate change is also a concern; ironically, this region is not immune from drought. With entire habitats and biota having evolved around dynamic seasonal flooding, a change to a drier regime—while not certain, but a future possibility—will wreak havoc on many species here and their interrelations.

Meanwhile, along many tributaries, people have set up gold mining operations; some of these use mercury in operations and cause contamination of the water, which impacts both human and wildlife populations. Increasingly, large-scale cattle operations are expanding into margin pasture areas, and often result in deforestation of varzea forests for seasonal grazing.

JAN SCHIPPER

The giant buttress roots of a kapok tree (Ceiba pentandra) growing in a South American forest. As the largest trees in the varzea biome, kapoks make up much of the forest's emergent canopy. (Thinkstock)

Further Reading

Daly, D. C., and J. D. Mitchell. "Lowland Vegetation of Tropical South America." In *Imperfect Balance: Landscape Transformations in the Precolumbian Americas*, edited by D. L. Lentz. New York: Columbia University Press, 2000.

Ducke, A. and G. A. Black. "Phytogeographical Notes on the Brazilian Amazon." *Anais da Academia Brasileira de Ciências* 25 (1953).

Prance, G. T. "Notes on the Vegetation of Amazonia III. The Terminology of Amazonian Forest Types Subject to Inundation." *Brittonia* 31 (1979).

Pyramid Lake

Category: Inland Aquatic Biomes.
Geographic Location: North America.
Summary: *Named for its striking tufa limestone formations, Pyramid Lake is home to fairly healthy waters and a limited but rich biota.*

The ice age that took place over 20,000 years ago in present-day Nevada was known as the Great Basin Ice Age. Unlike other ice ages, which are known for ice sheets and glaciers, this one was more marked by periods of increased rain and reduced evaporation. This large flow of water drained into the lower basin of the region, creating two vast lakes, Lake Lahontan and Lake Bonneville. During this time, the climate was much cooler and wetter than it is in present day Nevada, which is warm and arid.

Lake Lahontan had a depth of some 500 feet (152 meters) and stretched over 8,000 square miles (21,000 square kilometers), covering much of what is now Nevada. A warming and drying period followed, eventually shrinking and splitting the lake into smaller basins separated by mountain ranges. As one of the last remnants of ancient Lake Lahontan, today's Pyramid Lake first took shape by about 9,000 B.C.E. Today, as one of the oldest and largest natural lakes in North America, the lake is fed by the Truckee River in the Sierra Nevada Mountains. About 15 miles (24 kilometers) long north to south, and up to 11 miles (18 kilometers) wide, Pyramid Lake reaches a depth of 350 feet (107 meters).

Pyramid Lake takes its name from the limestone deposits known as tufa, which here have formed in some very large pyramid-like shapes; they are found everywhere from deposits on the lake bottom to freestanding structures within and around the lake. The most singular tufa conglomeration is Anaho Island, which itself has the appearance of a weathered pyramid.

Flora and Fauna

The cyanobacterium *Nodularia spumigena*, a blue-green algae, is a fundamental part of the food web of Pyramid Lake. Late summer here finds large blooms occurring, which help to spread nitrogen into lake waters, a valuable part of the nutrient load for higher species.

Marshes, mainly around the mouth of the Truckee River inflow, and semiarid shrubland elsewhere around the lake combine to provide habitat where a great variety of vegetation flourishes. Buckwheat, bunchgrass, wheatgrass, elderberry, larkspur, juniper, sumac, and a range of herbs and forbs thrive.

Among aquatic species in the lake, the cui-ui (*Chasmistes cujus*) is perhaps most emblematic. This member of family *Catostomidae*, or sucker fish, is now listed as endangered. As one of only three species in genus *Chasmistes*, the cui-ui is known to occur only in the lower Truckee River and in Pyramid Lake. Feeding on zooplankton, the cui-ui of Pyramid Lake is a long-lived species, with a lifespan estimated at up to 40–50 years.

The cui-ui, along with Lahontan cutthroat trout (*Oncorhynchus clarkii henshawi*), were the favored species of indigenous peoples here in ancient times. Both types of fish were part of sacred ceremonies. One tribe, the Kuyuidokado, even took its name, meaning *cui-ui eaters*, from this valuable fauna.

One of the largest colonies of American white pelican (*Pelecanus erythrorhynchos*)—which favors a diet of cui-ui, Lahontan cutthroat trout, and tui chub—is based on Anaho Island here. They are known also to range out to 100 miles (160 kilometers) to search wetland areas for other preferred fish. Other birds in the Pyramid Lake biome include snowy egrets, great blue herons, terns, gulls, and cormorants.

The native biota of the lake has been impacted by the spread of invasive freshwater mussels, the quagga (*Dreissena rostriformis bugensis*) and zebra (*D. polymorpha*). Their most pernicious effects include removing suspended particulate matter from the lake water, which permits deeper penetration of sunlight and resultant alteration of the natural balance of the lake's species; another damaging effect is the excretion of material that depletes oxygen in the water.

Threats and Conservation

Dams and water-diversion projects on the Truckee River have caused tremendous harm to many of the native species of Pyramid Lake. Several of the hallmark fish species, for example, were blocked from their natural spawning runs to smaller streams off the Truckee; whole years thus went by without a reproductive class of cui-ui or cutthroat trout. Some mitigation and recovery efforts, including hatchery, restocking activity, and construction of the Pyramid Lake Fishway and Marble Bluff Fish Handling Facility, have managed to preserve at least the cui-ui in Pyramid Lake, albeit in an endangered status for the present. Still, the permanent loss of access to various favorable habitats, the eradication of some riverine vegetation from the area, and the introduction of invasive species have combined to permanently alter the biota of the Pyramid Lake ecosystem.

Global warming is thought to already be exacting a toll on this biome. Cui-ui fish are particularly sensitive to water temperature, for instance. Research indicates that their larvae are most successful at surviving when water temperature is in the range of 48–59 degrees F (9-15 degrees C), and are at great risk with temperatures in the range of 64–75 degrees F (18–24 degrees C). Spawning by cui-ui is timed to coincide with peak snowmelt, when chilled waters rush downstream from the Sierra Nevada range.

Unfortunately, the incidence of lower snow levels and earlier onset of the melt season have combined to apply stress to the cui-ui spawning survival rate in the Pyramid Lake area. Scientific modeling projects more winter precipitation will come in the form of rain instead of snow in coming years, a trend already underway.

Water pollution from urban runoff, agricultural flows, and septic tanks have caused the most persistent recent problems with the waters of the Truckee River that flows down into Pyramid Lake.

A U.S. Fish and Wildlife Service biologist taking measurements of a Lahontan cutthroat trout as part of restoration efforts at the Marble Bluff Fish Handling Facility. (U.S. Fish and Wildlife Service/Jeannie Stafford)

Nevertheless, the water is considered in fairly strong condition. It is a priority for the people of the Paiute Pyramid Lake Reservation to help recovery of the fish species and the natural water quality of Pyramid Lake. Various agencies, such as the Resource Department of Pyramid Lake Fisheries, regularly sample water quality and monitor the health of the local fish in the area.

The Paiute Reservation covers over 700 square miles (1,813 square kilometers), including about 112,000 acres (45,400 hectares) of Pyramid Lake. Since the local governing body of the Paiute Reservation maintains official relations with both state and federal agencies, the Paiute are able to direct funds and support toward sustaining the health of the Pyramid Lake ecosystem. Much of the economy of the reservation relies on the fishing and leisure activities of tourists, such as hiking, camping, and kayaking. Some of the grants they receive have been used to help restore Pyramid Lake's natural habitat, which ties into various aspects of the tribal people's traditional ways.

In 2010, First Lady Michelle Obama selected Pyramid Lake as the first Preserve America Tribal Community in Nevada. This is part of a national initiative that supports community work aimed at preserving both cultural and natural heritage factors in a chosen area. Such an initiative adds momentum to the range of independent and government-sponsored scientific investigations and recovery initiatives that are generally thought to be improving the ecology, sustainability, and water quality of Pyramid Lake.

WILLIAM FORBES
KATHERINE SPARROW

Further Reading

Benson, Larry. "The Tufas of Pyramid Lake." United States Geological Survey. http://www.plpt.nsn.us/geology/index.html.

Knack, Martha C. "A Short Resource History of Pyramid Lake, Nevada." *Ethnohistory* 24, no. 1 (Winter 1977).

Lebo, M. E. and P. Wagner. "Managing the Resources of Pyramid Lake, Nevada, Amidst Competing Interests." *Journal of Soil and Water Conservation* 51, no. 2 (1996).

Nevada Department of Conservation & Natural Resources. "Nevada's Native Fishes." http://dcnr.nv.gov/documents/documents/nevadas-fishes-2/.